A PRACTICAL HANDBOOK FOR DRILLING FLUIDS PROCESSING

A PRACTICAL HANDBOOK FOR DRILLING FLUIDS PROCESSING

SAMUEL BRIDGES
Technical Service and Product Line Manager,
Houston, TX, United States

LEON ROBINSON[†]
Houston, TX, United States

Gulf Professional Publishing is an imprint of Elsevier
50 Hampshire Street, 5th Floor, Cambridge, MA 02139, United States
The Boulevard, Langford Lane, Kidlington, Oxford, OX5 1GB, United Kingdom

Copyright © 2020 Elsevier Inc. All rights reserved.

Published in cooperation with International Association of Drilling Contractors.

No part of this publication may be reproduced or transmitted in any form or by any means, electronic or mechanical, including photocopying, recording, or any information storage and retrieval system, without permission in writing from the publisher. Details on how to seek permission, further information about the Publisher's permissions policies and our arrangements with organizations such as the Copyright Clearance Center and the Copyright Licensing Agency, can be found at our website: www.elsevier.com/permissions.

This book and the individual contributions contained in it are protected under copyright by the Publisher (other than as may be noted herein).

Notices
Knowledge and best practice in this field are constantly changing. As new research and experience broaden our understanding, changes in research methods, professional practices, or medical treatment may become necessary.

Practitioners and researchers must always rely on their own experience and knowledge in evaluating and using any information, methods, compounds, or experiments described herein. In using such information or methods they should be mindful of their own safety and the safety of others, including parties for whom they have a professional responsibility.

To the fullest extent of the law, neither the Publisher nor the authors, contributors, or editors, assume any liability for any injury and/or damage to persons or property as a matter of products liability, negligence or otherwise, or from any use or operation of any methods, products, instructions, or ideas contained in the material herein.

Library of Congress Cataloging-in-Publication Data
A catalog record for this book is available from the Library of Congress

British Library Cataloguing-in-Publication Data
A catalogue record for this book is available from the British Library

ISBN: 978-0-12-821341-4

For information on all Gulf Professional Publishing publications visit our website at https://www.elsevier.com/books-and-journals

Publisher: Joe Hayton
Senior Acquisitions Editor: Katie Hammon
Editorial Project Manager: Joanna Collett
Production Project Manager: Selvaraj Raviraj
Cover Designer: Christian Bilbow

Typeset by TNQ Technologies

Contents

Disclaimer	*xv*
In Memoriam	*xvii*
Preface	*xxv*

Section I Drilling fluid treatment, properties, and functions   1

1. Rheology   3

Introduction	3
Drilling fluid rheology review	3
Drilling fluid functions	3
Viscosity dilemma	5
Measuring drilling fluid properties	9
Calculating viscosity	11
Using the bingham plastic model	12
Funnel viscosity	13
Other rheological models	14
Drilling fluid additives	17
Drilling fluid measurements	19
Mud weight	19
Special note	19
Gel strength	22
Filtration rates	22
Solids content	23
Sand content	23
Chemical measurements	24
Electric stability	25
Pilot testing	25
Problems	25

2. Mud tank arrangements needed for safety   27

Arrangement and equipment selection	27
Suction section	28
Appendix 2.A	44
Appendix 2.B	47
Problems	47

3. Addition section — 49

Mud hoppers	49
Mud hopper recommendations	53
Mud guns	54
Selecting nozzles and centrifugal pumps for mud guns	56
Situation	58
Proper application of mud guns	82
Appendix 3.A	84

4. Agitation — 87

Introduction	87
Mixing and blending drilling fluid	87
Mechanical agitators	87
Motors	88
Impellers	88
Proprietary blades	93
Compartments	98
Sizing agitators	98
Turnover rate	99
Mud guns	103

5. Drilled solids calculations — 105

Introduction	105
Calculation of low gravity solids from retort data for a fresh-water drilling fluid	107
Discussion	108
Retorts	111
Validation of the equation	121
Non-aqueous drilling fluids	123
Special note about potassium chloride drilling fluids	128
Calculation volume fraction of low gravity solids in non aqueous fluids	128
Procedure for determining the low gravity solids in non-aqueous fluids	130
Alternate method of calculating low gravity solids in an unweighted, water-based drilling fluid	131
Problems	136

6. Cuttings transport — 139

Hole cleaning	139
Empirical correlation for cleaning; vertical or near-vertical boreholes (up to 35 degrees)	139
Historical perspective	140
Empirical correlation	140
Using the correlation	143

	Diagnostics	144
	Practical suggestions	145
	Comments	146
	Hole cleaning for highly deviated wells	147
	Suggestion	150
	Appendix 6.A	150
	Appendix 6.B	151
	Appendix 6.C	155
	Cuttings transport problems	155
7.	**Dilution**	**163**
	Introduction	163
	True costs of drilling a well	163
	Dilution principles	164
	Application	168
	Problems	171

Section II Drilling fluid processing – mechanical separation of solids   175

8.	**Surface drilling fluid systems**	**177**
	Introduction	177
	Generic systems for unweighted drilling fluid	177
	Generic systems for weighted drilling fluid	178
	Alternate system	179
	By-pass troughs after the shale shakers	182
	Appendix 8.A	183
9.	**Removal section**	**185**
	Introduction	185
	Unweighted drilling fluid	185
	Weighted drilling fluid	187
	Distribution chamber	191
10.	**Centrifugal pumps**	**193**
	Introduction	193
	A centrifugal pump is a constant head device	196
	Pump description	197
	Pump curves	200

Application of pump curves		206
Sizing impellers		208
Operating point		208
Application for desilting drilling fluid		214
Net positive suction head (NPSH)		218
Atmospheric pressure		218
Flow into a centrifugal pump		219
Cavitation		221
Practical operating guidelines		224
Appendix 10.A		229
Appendix 10.B		230
Problems		243

11. Fraction of drilling fluid processed — 245

Introduction — 245
Calculating drilling fluid process efficiency — 246
Summary — 252
Problems — 253

12. Equipment solids removal efficiency — 267

Effect of equipment removal efficiency — 267
Reasons for drilled solids removal — 268
Diluting as a means for controlling drilled solids — 268
Chemical treatment — 269
Mechanical treatment — 269
Mechanical separation-basics — 270
Effect of solids removal system performance — 271
Relationship of solids removal efficiency to clean drilling fluid needed — 272
Five examples of the effect of equipment solids removal efficiency — 276
Solids removal efficiency for minimum volume of drilling fluid to dilute drilled solids — 284
Estimating equipment drilled solids removal efficiency for an unweighted drilling fluid from field data — 286
Estimating equipment drilled solids removal efficiency for a weighted drilling fluid — 289
Another method of calculating the dilution quantity — 292
API method — 292
Equipment solids removal efficiency problems — 293
General comments — 294

13. Cut points — 295

Introduction — 295
Cut points of shale shakers — 298
Cut points of hydrocyclones — 300
Cut points of centrifuges — 302
Comment on particle size presentations — 305

14. Operating guidelines for drilling fluid surface systems — 311

Objective — 311
Description of the surface drilling fluid system — 311
Design of the active surface processing system — 311

Section III Solids control equipment — 323

15. Drilled solids removal — 325

Introduction — 325
Roles of the drilling fluid processing system — 327
Drilled solids sizes — 328
Why control drilled solids? — 331
Stuck pipe — 331
Filter cakes in NADF — 334
Extra note — 335
Cement placement — 335
Surge and swab pressure — 336
Wear — 336
Disposal costs — 336
Lost circulation — 336
Log interpretation — 337
Formation damage — 338
Cementing — 338
Drilling performance — 339
Polycrystalline diamond compact bits — 341
Carrying capacity — 342
Summary — 343
Problems — 343

16. Shale shakers — 345

Introduction — 345
Panel screens — 347

Screen design	349
Screen weaves	351
Screen openings	352
Conductance	354
Screen labeling	357
Equivalent aperture opening size	358
How a shale shaker screens fluid	358
Discussion of API RP13C	361
Background	363
API screen number determination	363
Comparison with RP13E	365
Rig site performance	366
Labeling requirements	366
Putting the label to use	368
Conclusion	368
Frequently asked questions regarding RP13C	369
Screen types and tensioning systems	370
Gumbo and water-wet solids problems	372
Blinding and plugging of screens	373
Shale shaker design	376
Motion types	376
Scalping shakers	378
Fine screen shakers	381
Shaker capacity	383
Solids conveyance	384
"G" factor determination	384
Relationship of "G" factor to stroke and speed of rotation	389
Dryer shakers	389
Triple deck shakers	391
Gumbo removal	393
A shaker that doesn't shake	394
Maintenance	394
Power systems	396
Cascade systems	398
Summary	400
Shaker users guide	401
Summary	404

17. Sand traps — 407

Introduction	407
Settling rates	407
Bypassing shale shaker	411

18. Degassers and mud gas separators — 413

Introduction — 413
Effects of gas-cut drilling fluid — 417
For example — 417
Removing gas bubbles — 418
Installation — 419
Mud/gas separators — 420

19. Hydrocyclones — 423

Introduction — 423
Cut points — 431
Desanders — 432
Summary — 433
Desilters — 434
Maintenance — 435
Hydrocyclone tanks and arrangements — 436
Hydrocyclone operating tips — 436
Appendix 19.A — 438

20. Mud cleaners — 439

Introduction — 439
Uses of mud cleaners — 445
Situations where mud cleaners may not be economical — 447
Location of mud cleaners in a drilling fluid system — 447
Operating mud cleaners — 448
Estimating ratio of low gravity solid volume and barite volume in mud cleaner screen discard — 449
Mud cleaner performance — 451
Mud cleaner economics — 454
Accuracy required for specific gravity of solids — 456
Heavy drilling fluids — 457
Operation guidelines — 458
Installation — 459
Non-oilfield usage of mud cleaners — 461
Appendix 20.A — 462

21. Centrifuges — 475

Introduction — 475
Centrifuging NADF — 479
Operating tips — 482
Summary — 482

Rotary mud separator	483
Appendix 21.A	485
Problems	487

22. Solutions to chapter problems — 489

Preface quiz answers	489
Chapter 1 Rheology solutions	492
Problem 1.1 solution	492
Problem 1.2 solution	494
Chapter 2 Mud tank arrangements solutions	495
Problem 2.1 solution	495
Problem 2.2 solution	495
Problem 2.3 solution	495
Problem 2.4 solution	496
Chapter 5 Drilled solids calculations solutions	496
Problem 5.1 solution	496
Problem 5.2 solution	497
Problem 5.3 solution	498
Chapter 6 Cuttings transport solutions	499
Problem 6.1 solution	499
Problem 6.2 solution	500
Problem 6.3 solution	500
Problem 6.4 solution	501
Problem 6.5 solution	502
Problem 6.6 solution	503
Problem 6.7 solution	504
Problem 6.8 solution	504
Problem 6.9 solution	504
Problem 6.10 solution	505
Chapter 7 Dilution solutions	505
Problem 7.1 solution	505
Problem 7.2 solution	507
Problem 7.3 solution	509
Problem 7.4 solution	510
Chapter 10 Centrifugal pumps solutions	510
Problem 10.1 solution	510
Problem 10.2 solution	511
Problem 10.3 solution	513
Problem 10.4 solution	513
Chapter 11 Fraction of drilling fluid processed solutions	515
Problem 11.1 solution	515

Problem 11.2 solution	516
Problem 11.3 solution	516
Problem 11.4 solution	517
Problem 11.5 solution	520
Problem 11.6 solution	522
Problem 11.7 solution	522
Problem 11.8 solution	524
Problem 11.8 solutions A—D	527
Problem 11.9 solution	533
Problem 11.10 solution	534
Problem 11.11 solution	535
Problem 11.12 solution	537
Problem 11.13 solution	540
Problem 11.14 solution	543
Problem 11.15 solution	545
Problem 11.16 solution	546
Chapter 12 Equipment solids removal efficiency solutions	547
Problem 12.1 solution	547
Problem 12.2 solution	549
Problem 12.3 solution	549
Problem 12.4 solution	550
15 Control drilled solids solutions	550
Problem 15.1 solution	550
Problem 15.2 solution	552
Chapter 21 Centrifuge solutions	554
Problem 21.1 solution	554
Problem 21.2 solution	554
Appendices	*557*
Bibliography	*573*
Index	*575*

Disclaimer

This book was prepared under the auspices of the IADC Technical Publications Committee but has not been reviewed or endorsed by the IADC Board of Directors. While the committee strives to include the most accurate and correct information, IADC cannot and does not warranty the material contained herein. The mission of the IADC Technical Publications Committee is to publish a comprehensive, practical, and readily understandable series of peer-reviewed books on the petroleum drilling industry in order to educate and guide industry personnel at all levels.

In Memoriam

Dr. Leon H. Robinson, Jr. (1927–2019)
Former Chairman, IADC – Technical Publications Committee

On June 5, 1944, the day before D-Day, (75 years ago this past June), Leon joined the army about 10 days after graduating from high school in Greenville, South Carolina. Little did he know that about 6 weeks later he would be in France and later Germany helping liberate Europe from Nazi Germany. He served in the 288th Engineering Combat Battalion in Europe first as a Training Non-Com, then as a Demolition Sergeant, a Supply Sergeant, and finally as a Staff Sergeant in charge of Battalion Operations. He was discharged in December 1946 and entered Clemson University the next month–in January 1947.

In March of 1949, Leon met his future wife, Esther, when she came to Clemson to install a Sigma Pi Sigma Physics Honor Society. They married in May of 1950, had three children and remained married for 56 years before her passing.

Leon enjoyed camping and during the summers traveled all over the United States. One special place was Shady Lake in Arkansas's Ouachita mountains.

Leon was often found volunteering. A few of his lesser-known volunteer activities include
- YMCA Indian Guides for 10 years
- Girl Scout Leader Outdoor Trainer for 12 years
- Skipper of Boy Scout Sea Scout Ship 680 for 13 years
- Red Cross Sailing and Canoeing Instructor for 20 years
- Co-leader of Girl Scout Troop 472 for over 20 years

Leon was one of the last of the "Greatest Generation". After returning from WWII, he graduated with Bachelor's Degree in Physics in January, 1949 and started Graduate School. Since it was mid-way through the academic calendar, his big challenge was to take the second semester of graduate courses of Advanced Calculus and Theoretical Physics without having completed the prerequisite of the first semester of those courses!

With a stellar academic record, he was inducted into several honor societies, including
- Sigma Pi Sigma [Physics Honor Society]
- Sigma Tau Epsilon [Honorary Scholastic Fraternity]
- Sigma Xi "ksi" [Graduate Honor Society]
- Pi Epsilon Tau [Petroleum Engineering Honor Society].
- He was invited to join the Tau Beta Pi Engineering Honor Society at N. C. State but could not afford the fee.

Leon received a master's degree in Physics in 1950 and went on to North Carolina State to obtain a PhD in Engineering Physics. His thesis explored electrets—loosely speaking the electrostatic analog of magnetic poles.

Leon finished his academic work in 1953 and joined Humble Production Research. If you've been keeping track, that's three degrees, BS, MS, PhD in 6 years after returning from WWII. In 1954, he moved into the Research Center on Buffalo Speedway and stayed there for the rest of his 39-year career. However, he traveled extensively to implement research on the drilling rigs, to teach drilling courses and consult with Exxon Affiliates while visiting at least 34 countries before he retired in 1992. Being on the rigs and coming up with ways to help the roughnecks were some of the favorite parts of his work.

His diverse interests were manifest if not magnified by the various professional organizations he was active in during his career, including:
- American Physics Society
- American Society of Mechanical Engineers (ASME)
- Society for Professional Well Log Analyst
- International Measurement-While-Drilling Society
- Offshore Energy Society
- Rock Mechanics Society
- Society of Petroleum Engineers (SPE)
- American Petroleum Institute (API)
- International Association of Drilling Contractors (IADC)

- Society of Rheology
- Society of Explosive Engineers
- American Association of Drilling Engineers (AADE)

This is not an exhaustive list of all of Leon's awards, but below were the most significant to him:

1) SPE Drilling Legend, 2008 (only five professionals have this title)
2) Included in the first group to be inducted into the AADE Hall of Fame, 2006
3) Annual SPE Drilling Engineering Award, 1985 (this was the second one ever awarded)
4) Featured as Petroleum Engineering Magazine's Petroleum Person of the Month, May 1986 (only other Exxon Professional featured was a vice-president in the corporate office.)
5) PetroSkills Top of Class Award, 2014 (for the highest instructor evaluations out of 273 instructors during 2013) only time awarded
6) Exxon Distinguished Lecturer Award, 1990
7) Exxon Outstanding Instructor Award, 1975, 1976, 1980
8) Safety Awards: Exxon Belt Buckle and Jackets for working on offshore rigs long enough to participate in rig crews awards.
9) IADC Exceptional Service Award, 2013
10) Bureau of Alcohol, Tobacco, and Firearms Certificate of Appreciation, 1981
11) API Service Award for 40 years of service, 2006
12) American Association Drilling Engineering Meritorious Service Award, 1999
13) SPE Legion of Honor Award, 2003
14) IADC Presidential Award — 2019
15) Appreciation Award from SPE Section in Teraranganu, Malaysia, 1996

Those who know, do. Those who understand, teach.

Aristotle

Leon was quick to master complex technical areas and had a true gift for being able to teach the concepts of technology in very understandable ways. His major teaching gigs over his career included:
- Teaching infantry soldiers to be combat engineers
- Physics at Clemson, University of Chattanooga, and North Carolina State
- Sailing (for Boy Scout Sea Scouts, and the Red Cross)
- Camping (for Girl Scout leaders)
- Drilling on six continents (they don't yet drill in Antarctica)

One of the IADC technical committees that he organized and chaired in 1972 continued to produce technology for various groups until he actually retired in 2019 at age 92. This Committee authored the following:
- IADC Mud Equipment Manual, (12 handbooks)

- AADE Shale Shaker Handbook
- ASME Drilling Fluids Processing Handbook
- API Recommended Practice 13C (Solids Control)
- IADC Drilling Manual (Fluids processing chapter)
- Drillers Knowledge Book with Juan Garcia as co-author
- A Practical Handbook for Drilling Fluids Processing, IADC-TPC with Sam Bridges as co-author
- Drilling Encyclopedia (Committee chair, author)

While Doc had encyclopedic knowledge and experience, his last major committee undertaking—writing the "Drilling Encyclopedia"—was to capture many other "subject matter experts" knowledge. This last endeavor was started in October 2005, (when Doc was a youngster at about 78) and envisioned over 20 technical books about all aspects of drilling — a decade or two long project. He realized that he would probably not be able to complete this task personally, but the quest was challenging and worthwhile. These committees were characterized by civility and good, in-depth technical discussions without creating animosity. Because of the people involved in these committees, everyone enjoyed the work and most obtained a tremendous education from the discussions and arguments.

Leon leaves many, many, rich legacies and memories.

He leaves a legacy of leadership. At Exxon Research, several of us who came in early to beat the traffic would gather in his office for coffee and discussions of solving problems—whether drilling or the worlds. He had a way of getting people to volunteer to do more work on something and they might have even thought it was their own idea. One of his favorite operating rules was that, "There is no limit to the amount of good you can do if you don't care who gets the credit. This promotes team-effort and produces great results." A corollary to that was that the best way to get someone to work on something is to make them think it was their idea in the first place!

He leaves a legacy of beating the odds. At more than one of those early morning coffee discussions, he opined that he really didn't expect to live more than about 63 years of age—based on family history. He beat that by about 50%! He beat the odds in education—obtaining a PhD in a hard science, and in only 6 years! He beat the odds in employment—working for the same company from start to finish 39 years later. He beat the odds in shear accomplishments—by a very large margin.

He leaves an enormous legacy of teaching excellence. It has been said that whether teaching boy scouts to sail, girl scout leaders to camp, university student's physics, drilling engineers how to be better at their craft, or anything else, he was constantly seeking better ways to convey an understanding of the subject matter. After he retired from Exxon in 1992, he started teaching with Oil and Gas Consultants Inc or OGCI, the forerunner to today's Petroskills. In the 27 or so years he taught 2265 drilling students. To put that in perspective, there are only approximately 2087 working drilling rigs at work today.

He leaves a legacy with his family. Esther, his children, grandchildren, great grandchildren, and great-great grandchildren were a source of both smiles and pride he would talk of at the coffee roundtable meetings. Whether it was the pool and upcoming plans for a party for a grandchild, or building a tree house, or recounting a rope swing in the front yard, his face brightened noticeably when family came up.

He leaves a legacy of teamwork. He understood more than most then and almost everyone today that no matter how good a technology is, it has to be field friendly, embraced, and implemented by rig site personnel to be successful. This required enormous degrees of teamwork, with the team sometimes ranging from a rig hand who had no formal education at all and didn't speak English to someone like Leon with a PhD in physics and everyone in-between. He thought it was a particular complement when a rig-site field operation was going very smoothly, and the Rig Company Man on location came up to him and inquired who "the boss" was since he had not seen or heard anyone yelling orders or shouting at others — Leon told him that was called "teamwork".

He leaves a legacy of invention and innovation. 58 US and international patents—but that is just the tip of the iceberg. For every patent there were dozens filed-but-not-granted patents. For every patent filed there were dozens of solved technical issues—for every solved technical issue there were dozens of ideas—for every idea there were dozens of tests—for every test there were dozens of obstacles to overcome.

Below are just a few among the 24 international patents and 34 US patents:

US Patent Number	Date	Title
2,786,977	3/26/1957	Filtration and Electrical Resistivity Measuring Device
3,014,423	12/26/1961	Apparatus for Drilling Boreholes with Explosive Charges
3,022,824	2/27/1962	Method and Composition for Cementing Wells
3,070,010	12/25/1962	Drilling Boreholes with Explosive Charges
3,070,011	12/25/1962	Directional Drilling with Explosive Charges
3,078,934	2/26/1963	Drilling Earth Formations by Extrusion
3,083,778	4/2/1963	Rotary Drilling of Wells Using Explosives
3,104,716	9/24/1963	Treating Liquid Device for Gas Wells
3,104,716	9/24/1963	Method and Apparatus for Injecting Treating Liquids in Wells
3,112,233	11/26/1963	Drilling Fluid Containing Explosive Compositions
3,118,508	1/21/1964	Drilling of Off-Vertical Boreholes
3,123,143	3/3/1964	Freeing Pipe Stuck in a Borehole
3,138,206	6/23/1964	Perforating in Wells
3,163,242	12/29/1964	Drill Bits

Continued

US Patent Number	Date	Title
3,185,224	5/25/1965	Apparatus for Drilling Boreholes
3,125,074	11/2/1965	Apparatus for Well Drilling Operations with Explosives
3,274,933	9/27/1966	Apparatus for Explosive Charge Drilling
3,280,926	10/25/1966	Bit for Drilling Earth Formations
3,373,820	3/19/1968	Apparatus for Drilling with a Gaseous Drilling Fluid
3,461,965	8/19/1969	Fracturing of Earth Formations
3,491,841	1/27/1970	Method and Apparatus for the Explosive Drilling of Boreholes
3,533,471	10/13/1970	Method of Fracturing with Explosives using Reflective Fractures
3,766,997	10/23/1973	Method and Equipment for Treating Drilling Fluid - the Mud Cleaner
3,957,118	5/18/1976	Cable Gripper Slips
4,098,342	7/4/1978	Method and Apparatus for Maintaining Electric Cable inside Drill Pipe
4,250,974	2/17/1981	Apparatus and Method for Detecting Abnormal Drilling Conditions
4,271,908	1/29/1980	Tracked Weight Assembly to Store Wire Inside Drill Pipe
4,372,706	2/8/1983	Wireline Cable Head
4,375,310	3/1/1983	Self-Locking Cable Connector Member
4,416,494	11/22/1983	Technique of Storing a Variable Length Telemetry Wire Inside a Drill String
4,448,250	5/15/1984	Method of Freeing a Hollow Tubular
4,537,457	8/27/1985	Connector for Providing Electrical Continuity across a Threaded Connector

He leaves a legacy of gratefulness. As mentioned, Doc honestly thought—based on the data of lifespans of males in the Robinson family—that he would not live past 63. When he lived past that, he genuinely thought he was on additional time gifted by God and tried to make the best use of the time.

What was his secret? Many have observed and shared that it was simple: He stayed engaged in the things he loved to do, and he never slowed down. Just a couple of months ago before he passed away in November 2019, he was awarded the Presidential award from the IADC. After the ceremony, he stayed to the end of the committee meeting, he was continuing to write, to advise writers, and to collaborate with Juan Garcia and others on still-to-be-finished chapters. Then he drove himself home!

He leaves a legacy of inspiration and encouragement. Many truly great ideas were born in his office if not his own mind. Oftentimes for us younger generation, we might come into our morning meeting with what we thought was the best idea since sliced bread for drilling, only to find out that Leon had "been there done that one" twenty or more years earlier. However, he had a way of not dampening our enthusiasm.

After one of my ideas had been summarily shot down by his file cabinets of tests and reports, I was devastated. He pointed out, however, that just because someone beat me to the idea, it did not diminish the fact that I had independently thought of it—thus leaving the drive in us to find and explore new technologies completely intact.

As I was leaving his house many years ago after discussing a revision to a course we both taught at the time, I remarked about a beautiful large oak tree in his front yard—something like, "Wow, What a beautiful tree it is…I wonder how old it is? It must be 100 years or older." He answered and said, "I know exactly how old it is—I planted it!" When I asked what his green thumb secret was, he said, "It's simple. I just plant it, water it, feed it regularly, take care of it, and watch it grow—just like I do with people."

A special thanks to Ray and Candy for helping him so much over the years—both with his and your Mom's travel and organizing South Padre Island trips and excursion details and birthday parties on Clear Lake, to name a few. It clearly enabled him to keep going as long as he did. A special thanks to Marie, who selflessly moved back in to help her Dad live at home for the last year or so and took care of him every step of the way up to the very end.

Last, but certainly not least, thank you Leon for your time, your talent, and your treasures.

Mark Ramsey, PE
January 2020

Preface

Book description

This book concentrates on properly treating drilling fluid in the surface systems including a detailed discussion of the three required sections: suction, addition, and removal. The single, most important function of a drilling fluid is to prevent a blowout and to be able to control a kick if one does occur. For this reason, the first subject addressed in this book is the fluid processing required to handle a kick. For safety, in anticipation of a kick, the drilling fluid within the drill pipe should have the same density avoiding sag, settling, and uneven density in the annulus that could lead to a kick. To achieve these conditions, the surface system must be able to blend a sufficient quantity of homogeneous fluid. The agitation and blending of this fluid is discussed and proper calculations related to ensuring the proper properties.

In addition to safety, drilling fluid has many other functions to fulfill, including providing the optimal rheology to assist drilling operations. This book presents a basic understanding of how the drilling fluid processing operation affects the drilling process. Pertinent rheology facts are reviewed to avoid misunderstanding of the flow properties of the drilling fluid.

Agitation and blending of new additions to the surface system is explained. The mud hopper and mud gun plumbing and arrangements are discussed to assist in proper blending of the drilling fluid additives. A method of calculating slug volumes and densities for tripping is discussed in the suction section discussion.

The next section of the book involves the removal of drilled solids. The process of removing drilled solids actually starts at the bottom of the borehole. Cuttings need to be removed from the bottom of the hole before the next row of cutters arrive. This requires maximizing the hydraulic impact or hydraulic power of the fluid flowing through the nozzles. This will require adjusting the flow rate to the optimum value as will be presented in the IADC Drillers Knowledge Book on hydraulics and rheology. After removal from beneath the bit, cuttings should be transported out of the hole without tumbling and grinding into smaller pieces. Since the flow rate is established by drilling optimization conditions, the rheology of the drilling fluid must be adjusted to transport the cuttings. The rheology required is discussed in depth in Chapter 6.

Once the cuttings arrive at the surface, the removal section needs to be structured so that these drilled solids are removed before the fluid reaches the suction section. Each piece of drilled solids removal equipment is discussed and explained in detail.

The proper equation for calculation of drilled solids concentration depends upon the density of the drilled solids. A field method of determining the drilled solids density is presented, along with a method to evaluate the plumbing for proper fluid processing. A method of calculating the fraction of dirty drilling fluid processed is explained. Plumbing arrangements are frequently incorrect and, sometimes, can be corrected with relatively minor changes. Several tank arrangements that have been observed on drilling rigs are presented and calculations made for evaluation of these real-life examples.

The calculation of drilled solids concentration depends upon the density of the drilled solids. These densities change significantly from one geological formation to another. A field method to determine the drilled solids density is discussed in Chapter 5.

Summary of the role of solids control in good drilling practices

1. Provide sufficient mud tank volume and proper agitation to keep the fluid in the drill pipe homogeneous.
2. Arrange the tank plumbing so that all of the drilling fluid is processed by the surface equipment.
3. Strive to have a low plastic viscosity.
4. Provide a thin, slick, compressible filter cake and reduce incidents of stuck pipe.
5. Determine the flow rate and nozzle sizes which will cause the drilling fluid to strike the bottom of the hole with the most force or most power. (This must be done on the rig floor just prior to pulling a dull bit).
6. After dressing the new bit with the proper nozzles, find the founder point of the new bit after circulating bottoms-up. (Use MSE).
7. The annular velocity in the largest part of the hole is determined by the hydraulics optimization, and the rheological properties of the drilling fluid need to be adjusted to bring cuttings to the surface without tumbling. Cuttings should have sharp edges.
8. Screen the fluid through API 140, API 170, or API 200 screens.
9. Look for holes in the screens during connections.
10. Unplug hydrocyclones and determine how those large solids remained in the drilling fluid.
11. When using a centrifuge on weighted drilling fluids, do not return any of the light (or overflow) from the centrifuge back to the active pits, and replace the chemicals (filtration additives and rheology adjusters).
12. When using a centrifuge on weighted drilling fluids, replace the chemicals (filtration additives and rheology adjusters).

Economics of solids control

The actual money saved by using correct drilling fluids processing is difficult to document. Most operators regard cost of drilling wells as proprietary information. However, the economics of having the correct homogeneous fluid in the drill pipe in case of a kick is almost self-evident. The inability to determine the underbalanced pressure could be disastrous or could just require much extra rig time to handle. The benefit of having good solids control practice is more difficult to quantify. Very few drilling programs schedule stuck pipes or other problems caused by drilled solids in the drilling fluid. How can you prove something didn't happen because of some action taken?

Case histories are difficult to find when related to poor solids control. One event stands out in the past experience of one of the authors. A platform was scheduled to drill twelve wells. While drilling the sixth well, the operator was unhappy with the ability to control drilling fluid properties, especially the low gravity solids concentration. An evaluation of the plumbing revealed many problems. The contractor (who will remain anonymous) thought every pump should be able to pump from and to any compartment. The needed plumbing changes were so daunting that the recommendation from the rig was that it would cost too much. Operator management made the decision to modify the drilling fluid processing system. At the end of the sixth well, six welders were sent to the rig and spent several days modifying the system at a cost of 10% of the AFE of well number six. The savings on well number seven, because of the elimination of visible and invisible NPT, was much greater than the funds spent changing the system. That savings accrued on all of the rest of the wells.

Historical perspective

Drilled solids management has evolved over the years as drilling becomes more challenging and environmental concerns are paramount. Equipment changes and improvements have responded to the necessity of treating more and more expensive drilling fluids. Probably the largest impact was the recognition that polymers could make much better drilling fluids despite the cost. Polymer drilling fluids required lower drilled solids concentration so superior solids removal systems were developed to meet those demands. A historical perspective, including drilling fluid management, specifications, solids control, and auxiliary processes, provides a clear and complete picture of the evolution of current equipment and practices. The negative impact of drilled solids on drilling costs, performance, and non-productive time has been well documented in recent years. However, there have been many skeptics because the effects are not always immediately observed.

Drilling fluid was used in the mid 1800s in cable tool (percussion) drilling to suspend the cuttings until bailed from the drilled hole. (For a discussion of cable tool drilling, see History of Oil Well Drilling by J. E. Brantley.) With the advent of rotary drilling in the water well drilling industry, drilling fluid was well understood to cool the drill bit and to suspend drilled cuttings for removal from the well bore. Clays were being added to the drilling fluid by the 1890s.

At the time Spindle top was discovered in 1901, suspended solids (clay) in the drilling fluid were considered necessary to support the walls of the borehole. With the advent of rotary drilling at Spindle top, cuttings needed to be brought to the surface by the circulating fluid. Water was insufficient so mud from mud puddles, spiked with some hay, was circulated downhole to bring rock cuttings to the surface.

If the formations penetrated failed to yield sufficient clay in the drilling process, clay was mined on the surface from a nearby source and added to the drilling fluid. These were native muds, created either by "mud making formations", or as mentioned, by adding specific materials from a surface source.

Drilling fluid was recirculated and water was added to maintain the best fluid density and viscosity for the specific drilling conditions. Cuttings, or pieces of formation, "small rocks" that were not dispersed by water, required removal from the drilling fluid in order to continue the drilling operation. Based on the sole judgment of the driller or tool pusher, a system of pits and ditches were traditionally dug on-site to separate cuttings from the drilling fluid by gravity settling. This system included a ditch from the well, settling pits, and a suction pit from which the "clean" drilling fluid was picked up by the mud pump and recirculated (Figs. P.1 and P.2).

Drilling fluid was circulated through these pits, and sometimes a partition was used to accelerate settling of the unwanted sand and cuttings. Frequently, two or three pits would be dug and interconnected with a ditch or channel. Drilling fluid would slowly flow through these earthen pits. Larger drilled solids would settle and the cleaner fluid would

Fig. P.1 Settling compartment using mud pits dug into the ground

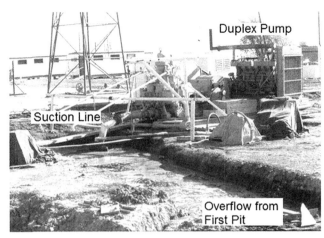

Fig. P.2 Settling compartment overflows into the suction tank after solids are removed

overflow into the next pit. Sometime later, steel pits were used with partitions between compartments. This partition extended to within a foot or two of the bottom of the pit; thereby, forcing all of the drilling fluid to move downward under the partition and up again to flow into a ditch to the suction pit. Much of the heavier material settled out, by gravity, in the bottom of the pit. With time the pits would fill with cuttings and the fluid became too thick to pump because of the finely ground cuttings being carried along in the drilling fluid. To remedy this problem, jets were placed in the settling pits to move the unusable drilling fluid to a reserve pit. Then water was added to thin the drilling fluid and drilling continued.

In the late 1920s drillers started looking at other industries to determine how similar problems were solved. Ore dressing plants and coal tipples were using:
1. Fixed Bar Screens Placed on an Incline
2. Revolving Drum Screens
3. Vibrating Screens

The fixed bar screens are currently used as "gumbo" busters, to remove very sticky clay from the drilling fluid. The latter two methods were selected for cleaning cuttings from drilling fluids.

The revolving drum, or barrel-type, screens were widely used with the early, low height substructures. These units could be placed in a ditch or incorporated into the flow line from the well bore. The drilling fluid flowing into the machine turned a paddle wheel which rotated the drum screen through which the drilling fluid flowed. These units were quite popular because no electricity was required and the settling pits did not fill so quickly. Currently, revolving drum units have just about disappeared from the scene. The screens were very coarse, API 4 to API 10, and sometimes with a fine screen (an API 12). The number after the "API" designation refers to the mesh size.

Mesh is the number of openings per inch. Common practice was to identify a "10-mesh" screen as one which has 10 openings in each direction—or a square mesh screen.

The vibrating screen, or shaker, became the first line of defense in the solids removal chain, and for a long time was the only machine used. Early shakers were generally used in dry sizing applications and went through several modifications to arrive at a basic type and size for drilling. The first modification was to reduce the size and weight of the unit for transporting between locations. The name "shale shaker" was adopted to distinguish the difference between shakers used in mining and the shale shakers used in oil well drilling. This nomenclature was necessary since both shakers were obtained from the same suppliers. The first publication about using a shale shaker in drilling operations described a "vibrating screen to clean mud" was in *The Oil Weekly*, October 17, 1930. The shaker screen was a 30-mesh, four by five feet, and supported by four coil springs.

1940s

These shale shakers had 4 ft. by 5 ft. hook strip screens mounted that were tensioned from the sides with tension bolts. The vibrators were usually mounted above the screens causing the screen to move with an elliptical motion. The axis of the ellipse pointed toward the vibrator. Since the axis of the ellipse at the feed end pointed toward the discharge end and the axis of the discharge end pointed toward the feed end, these shakers were called unbalanced elliptical motion shakers. The screens required a down-slope to move cuttings off the screen. Solids at the feed end, particularly with sticky clay discards, would frequently start rolling back up hill instead of falling off the shaker. Screen mesh was limited from about an API 20 to API 30 (838−541 μm). These units were the predominant shakers in the industry until the late 1950s. Even though superseded by circular motion in the 1960s and linear motion shale shakers in the 1980s, the unbalanced elliptical motion shale shakers are still in demand and are still manufactured today.

1950s

The smaller cuttings, or drilled solids, left in the drilling fluid were discovered to be detrimental to the drilling process. Another ore dressing machine was introduced from the mining industry—the cone classifier. This machine, combined with the concept of a centrifugal separator taken from the dairy industry, became the hydrocyclone "desander", and was introduced to the industry around the late 1950s. The basic principle of the separation of heavier (and coarser) materials from the drilling fluid is the centrifugal action of rotating the volume of solids-laden drilling fluid to the outer limit or periphery of the cone. Application of this centripetal acceleration caused heavier particles to move outward against the walls of the cones. These heavier particles exit the bottom of the cone and the cleaner drilling fluid exits the top of the cone. The desander, ranges in size from 6 to 12 inches in diameter, removes most solids larger than 30−60 μm. During the past few decades, desanders have been refined considerably through the use of more abrasion resistant materials and more accurately defined body geometry. Hydrocyclones are now an integral part of most solids separation systems today.

After the oilfield desander development, it became apparent that side wall sticking of the drill string on the bore hole wall was generally associated with soft, thick filter cakes. Using the already existing desander design, a 4-inch hydrocyclone was introduced in 1962. Results were better than anticipated. Unexpected beneficial results were longer bit life, reduced pump repair costs, increased penetration rates, less lost circulation problems, and lower drilling fluid costs. These smaller hydrocyclones became known as "desilters" since they remove solids called "silt" down to 15–30 μm.

George Ormsby, then with Pioneer Centrifuge Company, related a story about the first desilter that they installed on a drilling rig. The bank of 4-inch desilters was mounted on the berm of the duck's nest. (The duck's nest was an earthen pit used for storing excess drilling fluid and was usually an area of the reserve pit). The equipment was removing large quantities of drilled solids from an unweighted drilling fluid. After two days, however, the rig personnel called to have the equipment picked up "because it was no longer working". When George arrived at the location, the equipment was completely buried in drilled solids so there was no way more solids could be removed by the hydrocyclones.

During this period, research recognized the problems associated created with ultra fines (colloidal) in sizes less than 10 μm. These ultra-fines "tie up" or trap large amounts of liquid and create viscosity problems that was traditionally solved only by water additions (dilution). As large cuttings are ground into smaller particles, the surface area increases greatly even though the total volume of cuttings do not change. Centrifuges, which had been used in many industries for years, were adapted to drilling operations in the early 1950s to remove and discard colloidal solids from weighted drilling fluids. The heavy slurry containing drilled solids and barite larger than about 10 μm is returned to the drilling fluid system.

In recent years, centrifuges have been used in unweighted drilling fluids to remove drilled solids. In these fluids, the heavy slurry containing drilled solids down to around 7 to 10 μm is discarded and the light slurry with solids and chemicals (less than 7 to 10 μm) is returned to the drilling fluid. This application saves expensive liquid phases of drilling fluid. Dilution is minimized, thereby, reducing drilling fluid cost.

1960s

These hydrocyclones were usually loaded with solids because of the coarse screens on the shale shakers. Removing more of the intermediate-size particles led to the development of the circular motion shale shakers. These "tandem shakers" utilizing two screening surfaces were introduced in the mid-1960s. Development was slow for these "fine screen-high speed" shakers for two reasons:

1. Screen technology was not sufficiently developed for screen strength, so screen life was short. There was not sufficient mass in the screen wires to properly secure the screens without tearing.
2. The screen basket required greater development expertise than that required for earlier modifications in drilling fluid handling equipment.

Tandem shakers have a top screen with larger openings for removal of larger particles and a bottom screen with smaller openings (finer screen) for removal of the smaller particles. Various methods of screen openings were developed including oblong, or rectangular, openings. These screens removed fine particles and had a high fluid capacity. The screens can be made of larger wires so they had greater strength. Layered screens (fine mesh screen for good solids removal over a coarse mesh screen for strength) were developed in the 1960s. These layered screens were easier to build and had adequate strength for proper tensioning for increased screen life. This development made it possible for the shale shaker to remove particles greater in size than API 80 × API 80 (177 μm).

1970s

In the 1970s the mud cleaner was developed. At that time, no shale shaker could handle the full rig flow on an API 200 screen. Desanders and desilters were normally used after the shale shaker in unweighted drilling fluid; however, they discard large quantities of barite when used on a weighted drilling fluid. This meant that the drilled solids larger than an API 80 and API 200 (the upper limit of the barite size) could not be removed from the drilling fluid. API specifications currently allow three weight percent of barite larger than 74 μm, which is an API 200 screen. To solve this problem, the underflow from desanders and desilters reported to a pre-tensioned API 200 screen on a shaker. Much of the liquid from the underflow of the hydrocyclones and most of the barite passes through an API 200 screen. This was also the first successful oilfield application of a pre-tensioned fine screen bonded to a rigid frame. Some mud cleaners had screen cleaners, or sliders, beneath the screen to prevent screen blinding. Mud cleaners have also been used with API 250 screens in unweighted drilling fluids that have expensive liquid phases. The history of the development of the mud cleaner is discussed in the appendix of Chapter 20.

1980s

In the 1980s the linear motion shale shaker was developed. The first commercial unit, called a Shimmy Shaker, used hydraulic pistons to move screens in a straight line. This entry went out of business in a short time. This version had screens which sloped downward from the back tank instead of creating a pool of liquid and transporting solids up an incline. Two electric vibrators were introduced that caused the screen to vibrate with a linear motion. Linear motion is the best conveying motion to move solids off the screen.

Solids can be conveyed uphill out of a pool of liquid as it flows onto the screen from the flow line. The pool of liquid provided an additional head to help drilling fluid pass through the screen. Screens with smaller openings, such as API 200 (74 µm), can be used on linear motion shakers but they could not be used on any of the earlier types of shakers. Developments in screen technology have made it possible for pre-tensioned screens to be layered and, in some cases, have three dimensional surfaces.

When the linear motion shale shakers were introduced, several were frequently arranged in parallel to receive drilling fluid from scalping shakers. Since API 200 screens could be used on these primary shale shakers, mud cleaners were widely considered superfluous. Mud cleaner use diminished significantly. However, installation of mud cleaners, even with API 150 screens down-stream from these linear motion shale shakers, revealed that some removable drilled solids were still in the drilling fluid. In real situations sufficient drilling fluid bypasses linear motion shale shakers to make mud cleaner installation economical. In retrospect, since the lower apex discharge of desilters frequently plug down stream from linear motion shale shakers. This provides proof that all of the large solids are not removed by linear motion shakers.

These systems, or combinations of the various items discussed above, meet most environmental requirements and conserve expensive liquid phases. The desirable effect is to reduce the liquid content of the discarded drilled solids so that they can be removed from a location with a dump truck instead of a vacuum truck.

Prior to the current API RP13C (which is also an ISO standard), screens were identified by mesh size. The English unit of "inches" does not translate well into the ISO metric nomenclature. In addition to the change in units, a more compelling change was required because of the complexities of the new shaker screens. Screens ceased to be easily described with a simple measurement of openings in either direction. Screens are layered to form complex opening patterns. Screens are now described with the equivalent opening size in microns and an API number (which was formerly the mesh designation).

The next step was to cause the screen to move in a balanced elliptical motion. The motion is similar to an unbalanced elliptical motion shaker (from the 1940s) except that all axes of vibration are pointed toward the discharge end. The movement of the screen is similar to a linear motion shaker except the motion, however, makes an elliptical path. Solids are transported from a pool of liquid at the feed end of the shaker screen just as they are on a linear motion screen.

Unfortunately, many drillers continued to believe that drilled solids were not responsible for poor drilling performance. Consequently, drilling fluid systems were not properly arranged to take advantage of the benefits of good solids control. Mud tanks were frequently plumbed incorrectly because of indifference concerning the detrimental effects of drilled solids. Benefits were not really generally accepted until the mid 1980s. Inspection of drilling rig drilling fluid processing systems currently still reveal that proper plumbing was not well understood nor was it a priority.

2000s

With the advent of the concept of well-bore strengthening, triple deck shakers were developed. The drilled solids are removed from the top and lower decks and the solids (containing wellbore strengthening material and lost circulation additives) are recovered from the middle deck. These solids are sized to pack fractures created by lost circulation problems.

The latest commercial entry is a continuous belt of a fine mesh screen. A similar design was introduced about 30–40 years ago and did not become commercial. The new design uses a vacuum beneath the screen to cause the fluid to move through the screen. The solids are air-blasted from the screen after the liquid flows through the screen.

Emphasis on minimization of liquid discharges for environmental considerations has created techniques to remove liquid from the drilled solids discard. Since the decanting centrifuge is a very low shear-rate device for the drilling fluid (even though the drilling fluid is rotating over 15,000 RPM, it can be used to concentrate flocculated and coalesced solids). The light slurry, which is almost a clear stream of water, is returned to the drilling fluid. This has become an important part of the "closed mud system" system. Actually, the intent is to eliminate or reduce the quantity of liquid discarded.

A recent innovation for environmental purposes and minimization of liquid discharge is the dryer shaker. The discharge from linear motion shale shakers, desanders, and desilters flow onto another linear motion shaker which has even finer screens than the main shale shakers (screens as fine as API 450 or 32 μm) and usually has a larger screening surface. The dryer has a closed sump under the screen with a pump installed. Any liquid in the sump is returned to the active system through a centrifuge.

Another innovation introduced in the Gulf of Mexico in the 1990s is the gumbo conveyer. Before these were introduced, some drilling rigs would mount stainless steel rods about two to three inches apart on a downward slope. Gumbo would slide down these rods and be removed from the system. Drilling fluid would easily flow through the openings between the steel rods. At least two versions are currently marketed. One is a chain and the other is a continuous permeable belt. These special conveyors drag gumbo or large pliable sticky cuttings out of the drilling fluid before the drilling fluid encounters a shale shaker. This operation reduces the severe screen loading problems caused by gumbo.

Innovations of drilled solids removal equipment will probably continue. However, new, novel, spectacular equipment is useless if it is installed improperly and subjected to poor maintenance and operating procedures. This book concentrates on providing guidelines for practical operations of the surface drilling fluid system.

The authors would like to acknowledge the help, participation, and counsel provided by the professionals who reviewed the material and made many constructive suggestions about the text and graphics. They are, in alphabetical order:

Mary Dimataris
Juan Garcia, PhD
Fred Growcock, PhD
Bob Line
Nace Peard
Mark Ramsey, PE
Les Skinner, PE
Larry Waters

The authors would also like to thank Leah Hubenak for her expertise in editing, organizing, and formatting the book. She also provided excellent assistance in creating the excellent graphics which met the publisher's standards and elucidated the text material.

Drilling fluid processing quiz

Preamble: The questions below represent information that should be known before reading this book.

1. Define viscosity:
2. With Viscometer readings R600 = 60 R300 = 45; calculate PV and YP.
3. Calculate the viscosity at each reading in Question 2:
4. Why would you want Plastic Viscosity to have a high value?
5. Plastic Viscosity depends upon four things. They are:
6. A centrifugal pump is connected directly to a joint of 7 inch casing standing next to the derrick. The casing is open at the top. When the pump takes suction from a tank filled with water, the water stands to a height of 20 ft. above the liquid level in the tank. The water is drained from the system and the centrifugal pump suction is connected to a tank filled with 16.6 ppg drilling fluid (twice as heavy as water). When the pump is turned on, how high will the drilling fluid go in the casing (circle the correct answer)
   a. Not as high as the water did
   b. Same height as the water
   c. Higher than the water did
7. In problem 6, would the 16.6 ppg drilling fluid go any higher if the centrifugal pump motor horsepower was doubled? **Yes No**
8. In problem 6, if another identical pump was connected to bottom of the casing (install a tee), would the fluid in the casing go almost twice as high? **Yes No**
9. What is NPSH [Net Positive Suction Head]?
10. What is a flounder point?

11. What drilling fluid parameters control the flounder point?
12. What are the Bingham Plastic model and the Power Law model for drilling fluid?
13. How do you select the flow rate to use while drilling a well?
14. How should you select a flow rate to use while drilling a well?
15. What is "energy" (definition)?
16. What is energy per unit volume?
17. How do you make the drilling fluid have a low viscosity when it strikes the bottom of the hole to remove the cuttings and a high viscosity in the annulus to bring cuttings to the surface?
18. A surface casing is set at 3000 ft in a 9.0 ppg drilling fluid. What is the head at the casing seat?
19. A barite recovery centrifuge will separate barite from low gravity solids. **T or F**
20. A 3000 RPM centrifuge exerts a high-shear rate on the drilling fluid. **T or F**
21. Mud guns can be used to stir all mud tanks if the fluid is clean drilling fluid in the suction compartment. **T or F**
22. The shale shaker back tank should be dumped into the sand trap before each trip to prevent fluid from drying on the shaker screens. **T or F**
23. The desilter overflow should be returned to a compartment upstream from its suction so that the desilter can "look at the mud twice". **T or F**
24. Fluid loss gives a good indicator for the quality of filter cake. **T or F**
25. An API 200 (75 μm) screen has a cut-point of 75 μm. **T or F**
26. Drill strings will not experience differentially stuck pipe in an oil-base drilling fluid. **T or F**
27. Centrifuges and desilters separate drilled solids in the same size ranges in a weighted drilling fluid. **T or F**
28. Since drilled solids have much less effect on Yield Point in an oil-based drilling fluid than they do in a water-based drilling fluid, good solids control is not necessary in an oil-based drilling fluid system. **T or F**
29. A centrifugal pump produces a constant pressure. **T or F**
30. The flow rate from a centrifugal pump should be controlled with a valve on its suction. **T or F**
31. Closing the valve on the discharge pipe of a centrifugal pump for a few minutes will damage the pump. **T or F**
32. The head at a surface casing seat at 3000 ft. in 9.0 ppg drilling fluid is 1404 psi. **T or F**

SECTION I

Drilling fluid treatment, properties, and functions

CHAPTER 1

Rheology

Introduction

This book is designed for individuals who have knowledge about the various components of the drilling fluid system on a drilling rig. In order to understand the material, some understanding and working knowledge of the drilling fluid system is beneficial.

This book concentrates on the mechanical treatment of drilling fluid in the surface system to provide safety, to bring cuttings to the surface expediently, and to process the fluid correctly to remove the maximum amount of drilled solids. A review of pertinent rheology concepts is presented here to avoid misunderstanding of the importance of the flow properties of the drilling fluid. Mud tank arrangement will be discussed to promote proper surface treatment of the drilling fluid. All of the solids removal equipment will be discussed in detail. Since centrifugal pumps are used in the surface system and are frequently misunderstood, a discussion of sizing pump impellers will be presented. The agitation of the various compartments will be reviewed. Detail calculation procedures will be presented to evaluate the processing efficiency of the equipment and the importance of good solids removal efficiency will be highlighted. The target value for low gravity solids concentrations in drilling fluids is a complex subject and is discussed in Chapter 5.

Drilling fluid rheology review

Processing drilling fluid properly requires an understanding of controlling the flow properties of the fluid. This is described as the rheology of the fluid. Most drilling fluids (except for air and clear water) are Non-Newtonian fluids. That means that the viscosity changes with shear rate. A Newtonian fluid, such as water, salt water, or oil, has the same viscosity no matter how fast it is flowing or what the shear rate is.

Drilling fluid functions

Drilling fluids perform many functions in a borehole. Drilling fluid was initially used as a means of removing cuttings from a borehole. (This was in the Spindletop well near Beaumont, Texas in 1902). From this beginning, the number of required functions or duties has increased.

Drilling fluid should:
- Remove cuttings from the bottom of the hole as the drill bit creates them before the next bit tooth regrinds them. The fluid must be able to impact the cuttings as they are

created with sufficient force to move them from the bottom of the hole (*To do this, the drilling fluid needs to have a very low viscosity*).

- Transport the cuttings to the surface without regrinding. Cuttings that are tumbled in the annulus form rounded balls of rock. The small particles that are created by the grinding process are very difficult to remove from the drilling fluid after it reaches the surface. Drilling fluids should have the minimum possible drilled solids concentration. Cuttings and small formation particles are drilled solids. Drilling fluid must move fast enough in the annulus to bring cuttings and well bore sloughings to the surface without too much grinding. This also requires careful attention to the viscosity of the fluid in the annulus. Guidelines for cutting transport will be discussed (*To do this, the drilling fluid needs to have a high viscosity*).
- Prevent a blow-out. Drilling fluid density is increased so that the bottom hole pressure matches the formation fluid pressure. High density solids, such as barite or hematite, are added to increase the drilling fluid density (a.k.a. mud weight). This is also part of the desire to create a fluid with the minimum quantity of solids. As wells were drilled into pressured reservoirs, the density of the drilling fluid was increased to prevent formation fluid influx into the wellbore. Wells drilled with higher pressures in the wellbore than in the formation started experiencing problems associated with thick filter cakes (stuck pipe, surge and swab pressure problems, poor cementing jobs, etc.). This required careful attention to the deposition of a thin, slick, low permeability filter cake.
- Economically deliver a useable well bore. A well is drilled for specific reasons. Frequently, or hopefully most of the time, the well bore is drilled for production. Some wells are drilled for exploration and, offshore, may be expendable. Whatever the reason, the fluid should be selected to meet that objective.
- Prevent change in hole size (or hole instability). This helps with evaluation of the formations, with cementing casing in place, and decreases waste discard because of excess rock arriving at the surface.
- Create a "slick" hole to prevent stuck pipe. The filter cake created by the drilling fluid should be compressible, thin, and slick. These filter cakes cannot be created with "dirty" drilling fluids (or drilling fluids with a large quantity of drilled solids). Thick filter cakes form a good pressure seal around drill collars and anchor them the wall of the bore hole. Stuck pipe requires rig time that should be used for drilling.
- Control corrosion. If the pH of a water-based drilling fluid decreases below values around 8, the potential exists for corrosion of iron drill strings, casing, and tubing. If aluminum drill pipe is used, pH above 10 will cause degradation from reaction of the aluminum with caustic.
- Provide for good cement placement. If the wellbore has a thick and incompressible filter cake, the ability to have a good cement job is severely reduced. A thin,

compressible filter cake will allow cement to fully surround the casing and will prevent unintended lost circulation behind the casing.
- Prevent lost circulation. If the gel strength of the drilling fluid increases with time, the fluid at the bottom of a borehole during a trip could become very viscous. Attempts to break circulation with a new bit will result in pressures that break formations. If the fluid in the annulus is too viscous, the pressure losses from circulation may create pressures that exceed formation break down. These pressures are called the equivalent circulating pressures (ECD).
- Prevent near-well bore flushing of hydrocarbons. If too much filtrate escapes through the filter cake, the fluid in the formation will be displaced away from the wall of the well bore. Logging measurements will be required to "read" through the filtrate zone to determine the "true" resistivity of the formation. This becomes a severe problem if shale streaks exist in the production horizon.
- Prevent extensive "skin" damage. If the well is to be used for production, obviously care must be taken to prevent formation permeability damage. Frequently now, drilling fluids are changed to "drill-in" fluids before penetrating the production horizon. The value of the increased production usually more than compensates for changing the fluid.
- Control torque on drill strings; particularly in deviated wells. In long extended reach wells, if rotary steerable tools are not used, the drill string must be able to slide when changing well direction. This means that the filter cake must be thin and slick. Again, this cannot be achieved with drilled solids in the drilling fluid.
- Cool drill bit.
- Buoy the weight of the drill string. The last two effects are the result of using a fluid and no additives or treatment are required to achieve this.

Viscosity dilemma

How can a fluid have a very low viscosity when it hits the bottom of the hole to move cuttings from beneath the drill bit, and then have a very high viscosity as it brings those cutting to the surface?

The word "viscosity" requires a definition to understand how a fluid can act in such a manner. Viscosity is frequently described as "a resistance to flow". This is a description not a definition. Viscosity has "units"—usually centipoise. For example, water at room temperature has a viscosity close to one centipoise. Generally, if an entity has a "unit", the definition will include a way to calculate it. Love, hate, happiness and other feelings have no "units" and cannot be calculated from a formula. Speed, however, has "units". Traveling 60 miles in 1 h would require an average speed of 60 miles/hour. The "miles per hour" is a "unit". This is calculated from the definition of speed (which is the distance traveled per unit time). Viscosity has "units" (like centipoise), so there must be a definition that permits calculation of a number to associate with that unit.

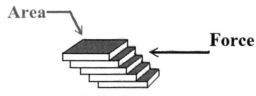

Fig. 1.1 Sheer stress.

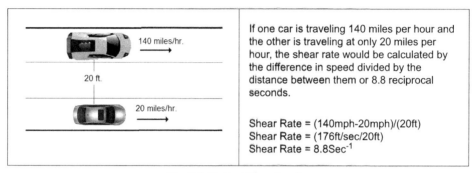

Fig. 1.2 Shear rate example.

Definition: Viscosity is the ratio of shear stress to shear rate. This definition, then, requires definitions of the two terms: shear stress and shear rate.

Shear Stress is force divided by an area parallel to the force. It has the same units as pressure (pounds per square inch or Newtons per square meter or dynes per square centimeter). This can be visualized by displacing a stack of books sideways with a force so that the stack is sheared (Fig. 1.1).

Shear Rate is the change in velocity divided by the distance over which that change takes place.

In Fig. 1.2, the shear rate can be visualized by two cars traveling down a freeway about 20 ft apart.

The units of shear rate are somewhat unusual: reciprocal seconds. To appreciate the value, consider the situation where the cars are only one inch apart (Fig. 1.3).

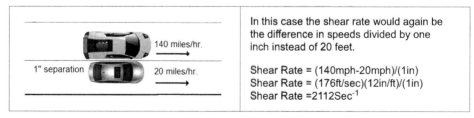

Fig. 1.3 Shear rate Example 2.

The shear rate increases from about 9 to over 2000 reciprocal seconds. The uneasy feeling you have on a freeway when someone passes you at a high rate of speed is caused by the high shear rate; not just from the difference in speed. When one car passes another with a very narrow distance of separation, the high shear rate can cause consternation.

The units of centipoise viscosity are actually metric units. If the shear stress is expressed in dynes per square centimeter and the shear rate in reciprocal seconds, the viscosity will be the unit "poise".

$$\text{Viscosity in Poise} = \frac{\text{Shear Stress in dynes/cm}^2}{\text{Shear Rate in sec}^{-1}}$$

The most common concept of viscosity is that the viscosity is constant regardless of the shear rate. This is called "Newtonian Viscosity". Water and oil are two fluids that have the relationship such that the shear stress is proportional to shear rate. If shear stress is plotted as a function of shear rate, the curve will be a straight line. The slope would be constant and, with the correct units for shear stress and shear rate, the viscosity.

Rheologists discuss the fluid characteristics in terms of what happens on a shear stress/shear rate curve.

In the graph shown in Fig. 1.4, the shear stress (SS) is measured in dynes/cm^2 and the shear rate (SR) is measured in reciprocal seconds. The ratio of SS to SR is constant all along the curve at a ratio of 2.0. This means the viscosity of this fluid is 2.0 P. This would be the viscosity of a 10 W 30 motor oil at about 60 °F. Normally viscosities are reported in centipoise (cp) instead of poise.

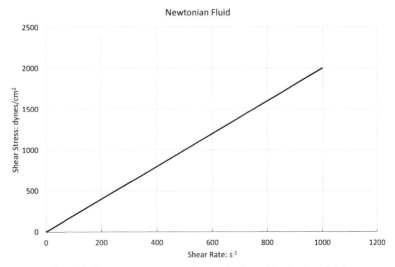

Fig. 1.4 Shear stress versus shear rate for a Newtonian fluid.

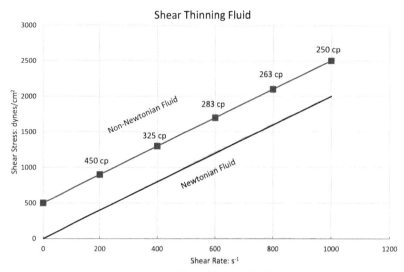

Fig. 1.5 Shear thinning fluid.

In the measurements shown in Fig. 1.5, the viscosity decreases as the shear rate increases. This is called a shear thinning fluid.

The change in viscosity is more vividly displayed in Fig. 1.6 which shows the viscosity of a water-based fluid containing a biopolymer. The viscosity in the low shear-rate range is quite large.

This is clearly a shear-thinning fluid. The viscosity increases when the shear rate decreases.

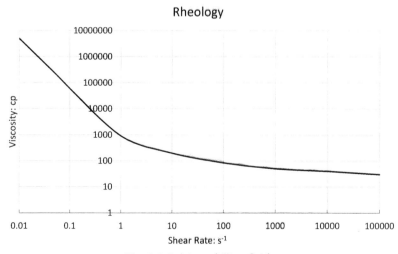

Fig. 1.6 Polymer drilling fluid.

How are these viscosities measured?

On drilling rigs, the relationship between shear stress and shear rate is determined with a concentric cylinder viscometer. The concentric cylinder viscometer consists of a bob retained by a spring and a cylinder which rotates around the bob. As the fluid between the cylinder and bob rotates, a shear stress is applied to the bob and indicated by the twist of the bob. The shear rate is the difference in velocity between the moving cylinder and the stationary bob. The bob ceases to rotate at equilibrium.

The outer cylinder rotates at various speeds to create different shear rates. The two most common speeds are 600 RPM and 300 RPM. The degree of rotation of the bob indicated by the dial reading is proportional to the shear stress. The outer layer of fluid between the cylinder and the bob rotates with the speed of the cylinder. The inner layer of fluid next to bob does not move relative to the bob. The change in velocity between the cylinder and bob divided by the distance between the cylinder and bob is a measure of the shear rate. The dial reading reflects the shear stress. This viscometer can be made with different bob diameters and different spring constants to cover a great variety of viscosities.

With the proper spring constant and bob diameter, the viscometer is used to determine the fluid plastic viscosity (PV) and yield point (YP). The concentric cylinder viscometer is calibrated to read the shear stress in pounds per one hundred square feet. The rotation speed of the outer cylinder is normally set at specific revolutions per minute: 600, 300, 200, 100, 6, or 3 RPM.

A "Fann" brand viscometer is shown in Fig. 1.7. The heat cup is shown on the left and is used to adjust the temperature of the drilling fluid before making measurements of shear stresses for various shear rates. Measurements are made at the same temperature so that trends in the values may be used for diagnosis of drilling fluid problems. Fann was the first company to develop this viscometer for use in the field. Consequently the name "Fann" is used frequently to describe this viscometer even though there may be other brands, such as the OFITE model shown in Fig. 1.8.

Measuring drilling fluid properties

The simplest rheological model can be represented by a straight line. This is called the Bingham Plastic Model. On the concentric cylinder viscometer, the shear stresses are measured at the 600 RPM and 300 RPM speeds. The viscometer readings for a drilling fluid are plotted in Fig. 1.9.

The viscometer is designed so that the difference between the 600 RPM reading and the 300 RPM reading is the plastic viscosity (PV, in centipoise)—or slope between those two readings. Subtracting PV from the 300 RPM reading calculates the yield point (YP, in pounds per 100 square feet).

Fig. 1.7 Viscometer A.

Fig. 1.8 Viscometer B.

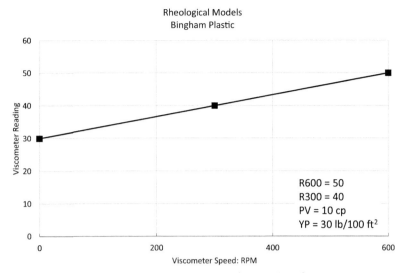

Fig. 1.9 Viscometer readings for a drilling fluid.

The equation of a straight line which describes this relationship has the form:

$$Y = mx + b$$

or

$$\text{Shear Stress} = (PV)\text{Shear Rate} + YP$$

where PV is called the plastic viscosity and YP is called the yield point.

If YP is zero, the fluid would be called a Newtonian fluid because then the viscosity would be constant for all shear rates.

Calculating viscosity

A standard concentric cylinder viscometer has six rotational speeds for the outer cylinder; 600 RPM, 300 RPM, 200 RPM, 100 RPM, 6 RPM, and 3 RPM. This provides a series of different shear rates so that the rheology profile can be determined. Drilling fluid does not have "a viscosity". The viscosity changes with each shear rate. Generally, the viscosity decreases as the shear rate increases. This is called a "shear-thinning" fluid.

The viscometer readings can be converted into shear stresses in dynes per square centimeter by multiplying the reading by 5.11. The shear rate in reciprocal seconds is 1.7 times the RPM of the outer cylinder. The ratio of shear stress in dynes per square

Table 1.1 Calculating viscosity for concentric cylinder viscometer.

Cylinder RPM	Viscometer reading	Viscosity: cp
600	50	25
300	35	35
200	30	45
100	27	81
6	10	500
3	6	600

centimeter to shear rate in reciprocal seconds gives the viscosity in poise. Generally, viscosities are reported in centipoise instead of poise; consequently, the conversion factor is 511 (Table 1.1).

$$\text{Viscosity} = \frac{\text{Reading} \times 511}{\text{RPM} \times 1.7} = \frac{\text{Reading}}{\text{RPM}} \times (300)$$

$$\text{Viscosity} = \frac{30 \times 511}{200 \times 1.7} = \frac{30 \text{ lb}/100 \text{ ft}^2}{200 \text{ RPM}} \times (300) = 45 \text{cp}$$

Using the bingham plastic model

The Bingham Plastic Model is useful for treating drilling fluids. It is not good for calculating pressure losses or trying to match viscosities of the fluid over a large range of shear rates. The Bingham Plastic Model examines the effect of the very high shear rate range (PV) and indicates something about the lower shear rate range (YP). Note YP is not a viscosity but the shear stress extrapolated to the zero shear rate. Because of this extrapolation problems with the drilling fluid can be more accurately diagnosed.

Plastic viscosity depends upon 4 things:
1. Liquid Phase Viscosity
2. Size of Particles
3. Shape of Particles
4. Number of Particles

Yield point is a function of electrochemical behavior or effect of long chain polymers. An increase in plastic viscosity usually means that drilled solids are increasing in the drilling fluid. An increase in yield point could indicate some chemical contamination or degradation of chemicals used to maintain the yield point. PV indicates solids problems. YP indicates chemical problems.

What does PV indicate?

The meaning of PV can be better understood by substituting the relationship for shear stress in the Bingham Plastic model into the definition of viscosity.

$$\text{Viscosity} = \frac{\text{Shear Stress}}{\text{Shear Rate}}$$

Shear stress is defined in the Bingham Plastic Model as $(PV)SR + YP$.

$$\text{Viscosity} = \frac{(PV)SR + YP}{SR}$$

$$\text{Viscosity} = PV + \frac{YP}{SR}$$

If shear rate goes to infinity, viscosity becomes equal to PV. So PV is the viscosity a fluid would have at a very high shear rate. Again, it is an important measurement which indicates information about the solids in the drilling fluid. If the PV is increasing but the volume percent of solids remains constant, the solids are becoming smaller and requiring more liquid phase to wet the surfaces that are being created by the solids grinding into smaller particles.

Funnel viscosity

On drilling rigs, the derrick man measures viscosity with a Marsh funnel. Drilling fluid poured into the funnel through a coarse mesh screen until it reaches the screen. The time, in seconds, for one quart to drain from the funnel is recorded as the "viscosity". In "metric" units, one liter is drained into the cup. Within 2 or 3 min a simple measurement can be made to determine something about the flow of the fluid. API RP13B indicates that water should require about 26 s to flow through the orifice at the bottom of the funnel.

As the fluid drains from the funnel, the head (or pressure) causing the flow decreases. The head provides the shear stress to cause flow. As the head decreases, the flow rate through the orifice decreases. That means that the shear rate is decreasing. Since neither the shear stress nor shear rate is constant or can be determined during the test, the funnel viscosity cannot be translated into centipoise viscosity. Rheologists might think that this is a useless tool. It indicates viscosity over a wide range of shear rates. However, if the PV and YP stay constant and the drilling fluid temperature does not change greatly, the funnel viscosity will remain constant. This means that if the Derrick man measures 46 s when he/she came on tour, and 46 s 6 h later, the PV and YP probably have not changed. However, if the funnel "viscosity" increased to 70 s, something happened to the drilling fluid. What happened cannot be ascertained from the funnel viscosity. The rheology needs to be examined with the concentric cylinder viscometer to see whether it is a solids problem or a chemical problem. It could indicate that the shale shaker screens have holes in them; it could indicate that a salt stringer was drilled. Drilling fluid should not be treated on the basis of the funnel viscometer. But, it is a wonderful monitoring tool. In a very short time, the person on the mud tanks can determine if the drilling fluid properties have changed with a funnel viscometer.

Other rheological models

The Bingham Plastic Model is the simplest possible rheological model that can be used to describe the relationship between shear stress and shear rate. It is wonderful for diagnosing problems with drilling fluids; but the lower end of the straight line does not match the actual readings. The lower end or low shear rate range of the viscometer readings match the shear rates achieved in the drill string, in the annulus and for particles settling through the drilling fluid. This would indicate that calculations of pressure losses or settling rates might not be very accurate because the viscosity of the fluid is not well represented with a simple Bingham Plastic Model. The second simplest model, a Power Law Model, would be to draw a curve through the 600 RPM and 300 RPM data points and the origin (Fig. 1.10).

These are the data from a previous problem where:
R600 = 60
R300 = 41

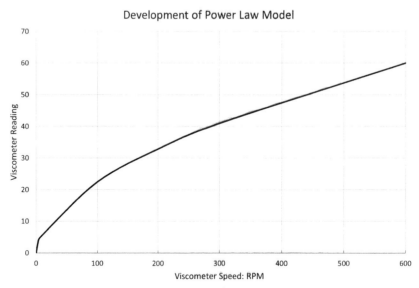

Fig. 1.10 Power law model.

PV = 19p; YP = 22 l b/100 sq ft
The equation for this curve would be:

$$\text{Shear Stress} = K(\text{Shear Rate})^n$$

This is a two parameter ("K" and "n") model just like the Bingham Plastic Model (PV and YP).

If n = 1, the "K" becomes a viscosity and the fluid is described as a Newtonian fluid.

$$K = \frac{SS}{(SR)^n}$$

If "K" is the viscosity of a Newtonian fluid in centipoise, the value of shear stress must be expressed in dynes/square centimeter and shear rate in reciprocal seconds. The viscometer reading must be multiplied by 511 for this conversion. This allows the value of "K" to be expressed in "effective viscosity". The units of "K" appear in all sorts of combinations. In this text, it will be such that, if n = 1, the value of "K" will be in centipoise. Frequently, the viscometer dial reading is used for the shear stress and the RPM used for the shear rate. If n = 1, this would give the viscosity in units of lb-RPM per 100 square feet. Few people would know how to relate this to something simple like water or oil. Many publications who treatise on calculating the value of "K", fail to convert the dial readings to dynes per square centimeter. Caution must be used when evaluating other documents and other calculations because of the failure to use units which will reduce to poise or centipoise if the exponent "n" is equal to one (Newtonian viscosity).

For drilling fluids the exponent "n" is usually less than one and is an indicator of how far removed the fluid is from being Newtonian.

At 300 RPM, the shear rate is (300) (1.7) reciprocal seconds. The 300 RPM viscometer reading is PV + YP and this must be multiplied by the conversion factor 511.

$$K = \frac{511 R_{300}}{1.7(300 RPM)^n}$$

The values for "n" and "K" can be calculated from the equations below:

$$n = 3.322 \log\left(\frac{2PV + YP}{PV + YP}\right)$$

$$K = (511)^{1-n}(PV + YP)$$

The shear stress in dynes per square centimeter is used to compute the value of "K". This converts the units to "effective viscosity". In the example above, "n" has a value of 0.55 and "K" has a value of 679 eff cp.

A comparison of the Bingham Plastic model and the Power Law model with the data from a six speed viscometer, illustrates that neither of these two simple models actually describe the lower shear rate ranges (Fig. 1.11).

The simplest rheological models have two constants to describe the relationship between shear stress and shear rate. Complex fluids sometimes require four, five, six, or more constants to describe flow. For example, injecting a molten plastic into a mold may require six constants to design the injection process.

Another commonly used rheological model is the three parameter Hershel Buckley model.

$$\text{Shear Stress} = K(\text{Shear Rate})n + \tau_0$$

Frequently, the 3 RPM viscometer reading is used for τ_0. This follows the actual readings more closely than either two parameter model and is becoming used more and more to calculate pressure losses – but does not account for variations in viscosity with temperature.

Oil and water are Newtonian Fluids but their viscosity is also a function of temperature. Oil viscosity is also a function of pressure as well as temperature. These functions, however, are well-known and predictable (Fig. 1.12).

Drilling fluids, however, may show a decrease, no change, or an increase in viscosity as temperature increases. This makes down hole hydraulic behavior very difficult to predict, no matter what rheological model is used. Examine the behavior of only the

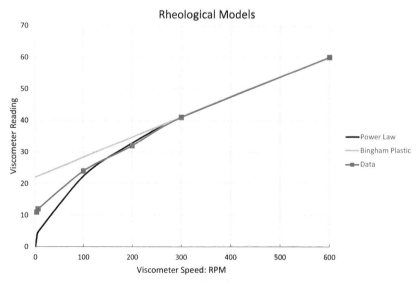

Fig. 1.11 Data compared with rheological models.

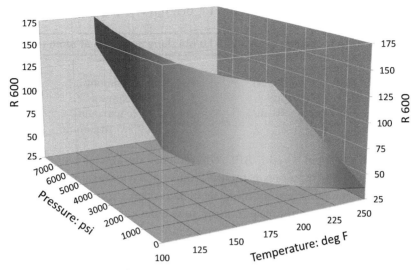

Fig. 1.12 Effect of pressure and temperature on the viscosity of an oil.

600RPM readings as a function of temperature and pressure. At a pressure of 7000 psi and a temperature of 100 °F the reading was 175 °F. Increasing the temperature to 250 °F results in a reading decrease to about 70 °F. This corresponds to a decrease in viscosity from about 83 cp to 35 cp.

Drilling fluid additives

The density increase is normally developed by adding a very dense material like barite (specific gravity of 4.2) or hematite (specific gravity of 5.02).

To create the low-shear-rate viscosities (LSRV), various additives are blended into the liquid phase. In a water-based drilling fluid, LSRV can be developed by using clay (bentonite), a clay-extender (partially hydrolyzed polyacrylamide), or a long chain polymer (like XC polymer, or HP007). Fluid loss control can be developed by using medium length polymers (like polyacrylates), or branched polymers (like starch), or well-deflocculated clay (like bentonite and lignosulfonate).

To create LSRV in a non aqueous fluid (NAF), medium length oil soluble polymers are added to the fluid. They work with the water droplets emulsified in the continuous phase. These droplets also help create the filter cake during the filtration process. Water droplets are emulsified into the NAF with long chain polymers that have water-soluble ions at one end and oil-soluble ions at the other end. Usually calcium is added to form a "hard" soap to increase the film strength of water droplet coating. An oil wetting additive is included in the formulation so that all solids additives are hydrophobic.

API has established some specifications for bentonite, barite, and hematite.

API specification 13A for bentonite for a 22.5 l b/bbl slurry aged 16 h at room temperature	
Suspension properties	
Viscometer dial reading at 600 RPM	30 minimum
Yield point/plastic viscosity ratio	3 maximum
Filtration volume (100 psi for 30 min)	15 cc maximum
Residue greater than 75 μm	4.0%wt maximum

The proposed changes for the API requirements are also contained in ISO/DIS 13500. Drilling grade bentonite is naturally occurring clay containing the clay mineral, smectite and may contain accessory minerals such as quartz, mica, feldspar, and calcite.

Barite is nominally barium sulfate. Barium sulfate has a specific gravity of 4.5 but the specifications acknowledge that some contamination can occur in the mined material. However, API Specifications 13A states that "Drilling grade barite is produced from barium sulfate-containing ores". Nowhere in the specifications does API require that the product solids as "barite" must be all barium sulfate. Many times, an inferior weight product is mined and brought up to specifications by the addition of a small amount of hematite.

API specification 13A for barite	
Density	4.2 g/cc minimum
Water soluble alkaline earth metals as calcium	250 mg/kg maximum
Residue greater than 75 μm	3.0%wt maximum
Residue less than 6 μm in equivalent spherical diameter	30.0%wt maximum

Hematite (Fe_2O_3) is becoming more prevalent in field operations as a weighting agent. Some hematite has a very hard and abrasive characteristic. Because of the rapid wear of expendables, this product has caused some drilling contractors to require a surcharge for their rigs if the drilling fluid has hematite. Since hematite has a higher density than barite, less product is required to raise drilling fluid density. From a drilling fluid point of view this is desirable because the plastic viscosity will be lower with fewer solids.

API specification 13A for API hematite	
Density	5.05 g/cc minimum
Water soluble alkaline earth metals as calcium	100 mg/kg maximum
Residue greater than 75 μm	1.5%wt maximum
Residue greater than 45 μm	15%wt maximum
Residue less than 6 μm in equivalent spherical diameter	15%wt maximum

The changes for the API requirements are contained in ISO/DIS 13500 and will contain the information above.

The ISO/DIS 13500 will also have standard tests for non-treated bentonite, OCMA-grade bentonite, attapulgite, sepiolite, technical grade low-viscosity CMC (carboxymethyl cellulose), technical grade high-viscosity CMC, and starch.

Drilling fluid measurements

To control drilling operations, certain measurements are made regularly to determine the drilling fluid properties. One source for routines to measure these properties may be found in API RP 13B-1 (Water-Based Fluids) and API RP 13B-2 (Oil-Based Fluids). This discussion will only identify the measurements and not discuss the methods in great detail.

Mud weight

Drilling fluid density is usually called mud weight. Normal pressure formations generally have a pressure gradient similar to a water gradient. For various reasons formation fluid pressures are frequently higher. Barite or hematite is used to increase the drilling fluid density. This increases the hydrostatic gradient in the well bore so that the pressure in the well bore is higher than formation pressure.

The density of a drilling fluid is determined with a simple beam balance (Fig. 1.13). A cup on one end of the beam holds a known—or specific—volume of liquid. A counterweight slider moves down a calibrated beam to balance the weight of the fluid in the cup.

To prepare the balance for field use, water fills the cup on the right. The lid has a weep hole to permit excess liquid to escape. The slider on the beam is set at the mud weight for water (8.34 ppg at 68 °F). Small BBs are added to or removed from the long cylinder at the right end of the beam until the beam balances.

Special note

The mud balance measures the mass of the content. If calibrated on earth, it will still read correctly on the moon. The mud cup and drilling fluid on one side of the fulcrum will have the same torque down (τ_{mud}) as the other side of the fulcrum (τ_{beam}) (Fig. 1.14).

Fig. 1.13 Mud balance.

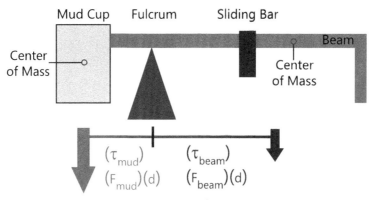

Fig. 1.14 Force diagram for mud balance.

Force is equal to mass times acceleration. The weight (or force) on the right side of the fulcrum would be mass of the right side times the acceleration of gravity. The weight (or force) on the left side of the fulcrum would be the mass of all parts of the beam times the acceleration of gravity. If the acceleration of gravity is smaller (as on the moon), the mud balance would still read the same mud weight as it does on earth.

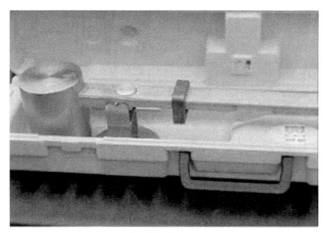

Mud balance.

More accurate mud weights can be determined with a pressurized mud balance (Fig. 1.15).

These balances are designed just like the unpressurized mud balance except fluid can be added to the cup with the plunger to increase the pressure in the cup to about 300–400 psi. This reduces the gas or air bubbles to about one-20th their size and gives a much better indicator of the bottom hole pressure.

Generally, the difference between pressurized and unpressurized mud weights will be at least two to three-tenths of a pound per gallon. Unfortunately, rig crews do not like

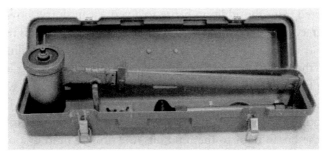

Fig. 1.15 Pressurized mud balance.

the extra work involved with using the pressurized mud balance, so they are frequently "lost". A mud guru (she was a laboratory technician at Exxon Production Research) acquired the name Bomo—for witch doctor when she found that the regular mud balance could measure mud weight within 0.05 ppg of the value obtained with the pressurized balance. A small amount of defoamer is added to a mud cup full of drilling fluid. Mix by flowing through the Marsh Funnel two or three times—this lets the air or gas break-out. Weigh in the regular balance.

The chart below (Fig. 1.16) indicates the maximum plastic viscosity (PV) value recommended for a water-based drilling fluid. This graph shows that the PV should increase as the mud weight increases. Usually, barite is added to increase mud weight. This addition of solids would naturally be reflected in an increase in plastic viscosity. Many recommended PV charts also include a lower line on the chart. The line might represent the

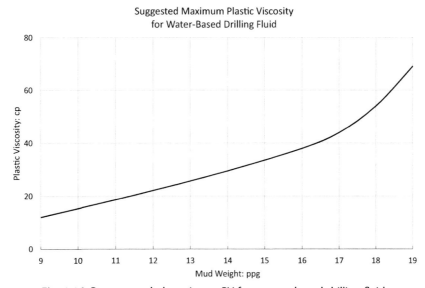

Fig. 1.16 Recommended maximum PV for a water-based drilling fluid.

minimum expected from the addition barite to a very clean slurry; however, the lowest possible value is recommended so no minimum line is included in this recommendation.

A similar chart to Fig. 1.16 is not available for non-aqueous drilling fluid (NADF) because most of the liquid phases have different viscosities depending upon the brand.

In very deep hot holes, 150 °F is selected for hot NADF. In deep water, the drilling fluid arriving at surface of the ocean is usually very cold in comparison. Deep ocean floor temperatures worldwide (including the tropics) are around 35 °F. As drilling fluid travels up a riser from the ocean floor to the mud pits, the drilling fluid is cooled significantly. Various companies use different temperature for measurements but the majority seems to continue to use 120 °F because changes are easier to interpret from experience.

Gel strength

Drilling fluid must suspend solids. Drilled solids moving up an annulus must be suspended when the pumps are turned off to make a connection. Barite or hematite weighting agents must remain suspended to maintain the bottom hole pressure. Drilling fluid needs to gel (like jelly) but the gels must be relatively easily broken. Very high gels will create a high resistance to flow. Excessive pressures may fracture formations after a trip when the pumps are turned on.

After the viscometer readings are measured, the drilling fluid is allowed to remain quiescent for 10 s. The outer cylinder is then rotated very slowly. The bob will turn with the cylinder until the gel strength is broken. The maximum reading is called the "10 s gel". The fluid is again stirred. Ten minutes after stopping, the outer cylinder is again rotated very slowly to determine the 10 min gel strength. Sometimes a "30 min gel" is also measured to make certain that the gels are not progressive. Some increase in gel strength can be tolerated; but if the gel strength becomes too high, lost circulation will result when the pumps try to move this fluid up the hole after a bit trip.

Filtration rates

When the pressure in a well bore is higher than the formation pore pressure, fluid will tend to enter the formation in the permeable zones. The solids in the drilling fluid will be deposited on the wall of the well bore. These solids are called a filter cake. In a well, filter cakes should be thin, slick, and compressible. The filter cake should be examined regularly to examine the filter cake that could be deposited on the wall of the well bore.

The room temperature water loss is measured by applying 100 psi to a fluid above a Whatman No. 50 filter paper which is supported by an API 200 screen. The amount of fluid collected in 30 min, the filter cake thickness, and a description of the filter cake are to be reported. This is called the API water loss test (APIWL).

The high temperature fluid loss is measured by applying a 100 psi differential across a Whatman No. 50 filter paper supported by an API 200 screen as well. The temperature is not specified in the API 13B documents. Generally, either a temperature about 50 °F higher than the bottom hole temperature or 300 °F is used. A back pressure is applied so that the liquid will not boil and vaporize. The API RP 13B-1 and 13B-2 give procedures for using temperatures to 300 °F but does not indicate what temperature should be used. (Since the 300 °F temperature is prominently mentioned, many think it is specified as the correct value for the test—it is not.)

Filter cake compressibility can be determined by comparing the 100 psi fluid loss measurement with a 500 psi pressure fluid loss from the high temperature cell but make both measurements at room temperature. Fluid loss through an incompressible filter cake is proportional to the square root of pressure. If the filter cake is incompressible, the ratio of fluid loss between the 500 psi and 100 psi pressure tests will be the square root of the ratio of 500 to 100. If the filter cake is compressible, the fluid loss in the 500 psi test can be the same as the fluid loss in the 100 psi test.

Another comment is appropriate here concerning compressible filter cakes. Frequently the filtration concept presented recommends that an assortment of particle sizes is most appropriate to decrease filtration rate. The theory is that if the holes between large diameter particles are filled with smaller diameter particles the fluid loss will be lower—and it will be. However, these cakes will be incompressible. If a filter cake is created from thin, platelets of clay that tend to mutually repel each other, additional pressure will press the clay platelets closer together and decrease the permeability. This can be achieved if the drilled solids are removed from the drilling fluid.

Solids content

Solids content is determined by removing all of the liquid from the drilling fluid with a retort. A known quantity of drilling fluid is heated to around 950 °F and the liquid condensed in a measuring tube. The results are reported as a percent volume. The quantity of both water and oil (or NADF) is reported.

Sand content

API sand is defined as any particle larger than 74 μm or particles that will remain on an API 200 screen. A measured quantity of drilling fluid is diluted with the drilling fluid continuous phase in a calibrated cylinder with a tapered bottom. The contents are poured through an API 200 screen and washed to remove all smaller particles. The screen is then inverted over a small funnel and the solids captured on the screen washed into the tapered bottom cylinder. The lower end is calibrated to read volume percent solids in the original drilling fluid sample.

Anything larger than 74 μm is called sand whether it is quartz, barite, gold, diamonds, or brass. Particles this size and larger make filter cakes thick and incompressible. With current solids control equipment, all solids this large should be removed from a correctly plumbed mud tank system.

Chemical measurements

One of the more important chemical measurements is the pH of the drilling fluid. By definition, pH is the negative logarithm of the hydrogen ion concentration.

$$pH = -\log_{10}[H^+]$$

The hydrogen concentration is measured in moles (A mole is the molecular weight of the atom in grams.) (Table 1.2).

If the pH is low, fluid is acidic; if the pH is high, the fluid is alkaline or basic. For drilling fluids, the pH needs to be high enough to prevent corrosion of the iron exposed to the fluid. At 75 °F the solution is neutral if the pH is 7. This means that there are as many hydrogen ions as hydroxyl ions. The neutral point, or concentration of hydrogen ions equal to the concentration of hydroxyl ions, varies somewhat with temperature. Close to the boiling point of water, the neutral point is around a pH of 5. Measuring devices can adjust for temperature effects of the probes but do not adjust for a change in the neutral point. If a 9 pH is measured in a hot drilling fluid, the pH might be as high as 11 when the fluid reaches room temperature. Many strings of aluminum pipe have been corroded in situations like this. Aluminum reacts with caustic to produce hydrogen gas and aluminum hydroxide (i.e. it eats it up).

Alkalinity is measured in water based drilling fluids. Both the whole drilling fluid and the filtrate are titrated with 0.02 normal sulfuric or nitric acid to determine the carbonate and bicarbonate contaminations in the drilling fluid.

P_f is the number of cubic centimeters of acid required to reduce the pH of the filtrate to the phenolphthalein end point (pH = 8.3) and M_f is the number of cubic centimeters of acid required to reduce the pH of the filtrate to the methyl orange end point (pH = 4.3).

The salinity is also measured as is the concentration of calcium and magnesium.

Table 1.2 The pH scale.

pH	Hydrogen concentration: mole	Hydroxyl concentration: mole	Acidic or basic
7	0.0000001	0.0000001	Neutral
5	0.00001	0.000000001	Acidic
3	0.001	0.000000000001	Acidic
9	0.000000001	0.00001	Basic
11	0.00000000001	0.001	Basic

The quantity of bentonite is determined by measuring the cation exchange capacity (CEC) with a methylene blue test (MBT). Methylene blue is a positively charged molecule that seeks the negative charges on the clay platelets. When it is adsorbed on a clay platelet, the liquid phase loses its blue color. Titration or adding small quantities of methylene blue to a treated sample of water-based drilling fluid indicates the quantity of clay in a drilling fluid.

In a NAF the salinity of the water phase is measured; the activity of the aqueous phase of the NAF. The aniline point determines the propensity of an oil to destroy rubber products and is determined by the temperature aniline becomes miscible with the drilling fluid. Aniline points should be higher than 140 °F.

Electric stability

Emulsion stability is determined by measuring the electric stability—or the voltage required to cause current to flow through the fluid.

Pilot testing

After measurements are made and treatment is needed, additions should be pilot tested. Each one gram of chemical or additive added to a 350 cc sample of drilling fluid is equivalent to adding one pound of the additive per barrel of drilling fluid. After the treatment, fluid properties of the 350 cc can be measured to insure that the proper amount of chemicals will be added to the drilling fluid in the pits. Frequently, if a massive or very large treatment is indicated, only half of the treatment will be added during one circulation. Drilling fluid can then be pilot tested again to validate the addition of the remainder of the treatment.

Problems

Problem 1.1

Calculate the PV and YP from the following readings:

FLUID #1:
R600 = 20
R300 = 10
PV =
YP =
FLUID #2:
R600 = 20
R300 = 15
PV =
YP =

FLUID #3:
R600 = 40
R300 = 35
PV =
YP =
FLUID #4:
R600 = 40
R300 = 15
PV =
YP =

Problem 1.2

Speed rpm	Reading lb/100 ft²	Viscosity cp
600	60	
300	41	
200	32	
100	24	
6	12	
3	11	

Calculate the conversion factor to convert the ratio of Readings to Speed to viscosity in centipoise:

$$\text{Viscosity} = \frac{\text{Reading} \times 511}{\text{RPM} \times 1.7} = \frac{\text{Reading}}{\text{RPM}} \times (300)$$

If the ratio is multiplied by 300, the viscosity at each shear rate can be determined.
1. Calculate the viscosity in centipoise for the fluid.
2. Calculate the PV and YP for this drilling fluid:

CHAPTER 2

Mud tank arrangements needed for safety

This chapter will concentrate one of the more important aspects of surface drilling fluid processing: SAFETY. Well control procedures require reading the stand pipe pressure to determine the amount of under balance at the bottom of the hole. The fluid in the drill pipe must be homogeneous if the well kicks.

Arrangement and equipment selection

Every drilling rig, no matter what size, should have a drilling fluid processing plant that has three clearly defined regions: removal, addition, and suction section (Fig. 2.1). These regions may be small compartments for very small drilling rigs or several tanks for the larger drilling rigs.

The size of each section depends upon the size of the hole being drilled and the depths of these holes. For example, a truck-mounted drilling rig drilling 3000 ft (1000 m), small diameter wells will need only a very small volume system. Larger drilling rigs drilling deeper, larger holes may need 3000 bbl or more drilling fluid processing plants. Each section will be discussed in depth. The addition and suction sections will be discussed first. Each component of the removal section will be discussed in individual chapters; then, the plumbing arrangement of each component will be discussed.

This chapter will concentrate one of the more important aspects of surface drilling fluid processing: SAFETY. One of the major functions of a drilling fluid is to prevent a blow-out. The density of the drilling fluid controls the pressure in the well bore.

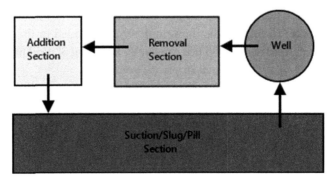

Fig. 2.1 Surface drilling fluid processing plant.

Well control procedures require reading the stand pipe pressure to determine the amount of under balance at the bottom of the hole. The fluid in the drill pipe must be homogeneous if the well kicks. In addition to this requirement, this chapter also discusses slug/pill tanks and trip tanks. A simple calculation process (read a graph) will insure an effective slug is placed in the drill string before tripping. A simple-to-use trip tank arrangement is discussed using a flow distribution tank.

Suction section

The suction section consists of the pill tank, the slug pit, the trip tank, and the suction compartment for fluid ready to go down hole.

Mud tank volume

The volume of the suction compartment needs to be sufficient for creating a homogeneous drilling fluid. The fluid in the drill pipe must have the same density (or mud weight) from the surface to the drill bit. This is necessary so that the drill pipe pressure measurement during a kick will read the underbalanced pressure.

Many rules-of-thumb have been proposed for creating the correct volume of drilling fluid needed on the surface when drilling a well. One suggestion, and possibly a regulation in some places, requires that $1.5 \times$ Hole Volume be available on the surface. However, no matter what guideline is used, the fluid in the drill string should have a homogeneous mud weight so that bottom-hole pressures may be calculated whenever a kick occurs. A suction tank volume should be several times the internal volume of the drill pipe to guarantee homogeneity of the fluid inside the drill string. Three methods of selecting suction tank volumes are presented below.

Plugged bit method

The plugged bit method determines the minimum size drilling fluid system based on the volume required to fill the hole when pulling a plugged bit and assumes all the fluid inside the drill string is lost. This means that the volume of the entire drill string and contents should be calculated to determine the minimum volume needed in the suction section.

For example, a rig rated to 20,000 ft is capable of handling 5″ drill pipe and 80,000 lb of drill collars to that depth. The total displaced volume is:

$$\text{Drill pipe:} \quad 20,000 \text{ ft of } 5'' \times 0.0243 \text{ bbl/ft} = 486 \text{ bbl}$$

$$\text{Drill collars:} \quad \frac{80,000 \text{ lb}}{\frac{2718 \text{ lb}}{\text{bbl}}} = 29 \text{ bbl}$$

$$\text{Total volume required:} = 515 \text{ bbl}$$

This method gives a close approximation of the maximum volume required to fill the hole when tripping a plugged string. Usually the volume is increased by about 20%, or 100 bbl, as a safety factor.

This method indicates that the minimum size suction section should be 615 bbl, plus a reserve to allow for lost circulation. Usually, the volume of the reserve system should be similar to that of the active system. Total system volume using the plugged bit method in this case is approximately 1230 bbl.

Cased hole method

The cased hole method simply doubles the volume contained in the final string of casing as a guideline for sizing a suction system. For example, consider a rig rated to 15,000 ft with 7 in casing as the final string. The total cased hole volume is:

$$15,000 \text{ ft of } 7'' \text{ casing} \times \frac{0.0390 \text{ bbl}}{\text{ft}} = 585 \text{ bbl}$$

Doubling this volume gives a total suction volume of approximately 1200 bbl.

Kick contingency method

The fluid in the suction section should be blended to achieve a homogeneous state so that, when pumped down hole, the mud weight is constant throughout the mud column. When the well is shut-in because of a kick, the stand pipe pressure is read to determine the under-balance pressure at the bottom of the hole. The drilling fluid in the drill string must have the same mud weight from top to bottom for these readings to have any meaning. The capacity of some common drill pipe sizes is presented in Table 2.1.

A 4″ diameter cylinder has a volume of 16 bbl/1029 ft. For estimation purposes, the 1029 ft can be considered to be 1000 ft and a quick estimate of the volume obtained. From Table 2.1, a $4\frac{1}{2}''$ drill pipe has an internal volume of 15.22 bbl/1000 ft. A 15,000 ft length of the $4\frac{1}{2}''$ drill pipe would have a volume of 228 bbl. The internal diameter of $4\frac{1}{2}''$ drill pipe could be approximated as 4″ and a volume estimated as 16 times 15[1] would indicate a volume of 240 bbl instead of the actual volume of 228 bbl. But, the estimated volume is convenient to provide an easy value for determining the

---

[1] The volume (in bbl) of a cylinder 1029 ft long can be calculated by squaring the diameter expressed in inches (See Appendix 2.B for derivation of equation.). For example, a four inch diameter cylinder would have a volume of: $(4'')^2/1029$ bbl/ft or 0.0155 bbl/ft For estimates, the diameter squared could be divided by 1000 ft instead of 1029 ft. This could be an easy esay mental exercise to estimate the volume of the 4″ cylinder as 0.016 bbl/ft. The calculation is useful for example in calculating the volume of drilling fluid in an annules between a $9\frac{7}{8}''$ hole and a 5″ drill pipe. Estimate the hole diameter as 10″, the approximate annulas volume would be: $(10'')^2-(5'')^2/1000$ bbl/ft = 0.075 bbl/ft. For a 10,000 drill pipe, the volume in the annulas would be 75 bbl.

Table 2.1 Capacity of internal upset drill pipe.

Drill pipe size (in)	Weight (lb/ft)	Capacity (L/m)	Capacity (gal/ft)	Capacity (bbl/1000 ft)
4	11.85	6.15	0.4930	11.75
4	14.0	5.65	0.4551	10.84
$4^1/_2$	13.75	7.94	0.6390	15.22
$4^1/_2$	16.6	7.42	0.5972	14.22
$4^1/_2$	20.0	6.72	0.5406	12.87
5	16.25	9.85	0.7928	18.88
5	19.5	9.27	0.7560	17.76
5	25.6	8.11	0.7245	17.25
$5^1/_2$	21.90	11.57	0.9314	22.18
$5^1/_2$	24.70	11.05	0.8898	21.19
$6^5/_8$	22.20	18.64	1.5008	35.73
$6^5/_8$	25.20	19.03	1.4517	34.56
$6^5/_8$	31.90	16.82	1.3541	32.24

volume required. Between two and three times this volume should be available in the suction section to ensure a constant mud weight in the drill string while drilling.

Plumbing

The generalized weighted drilling fluid processing plant shown in Fig. 2.2, shows the various components of the surface tank arrangement. A weighted drilling fluid is defined

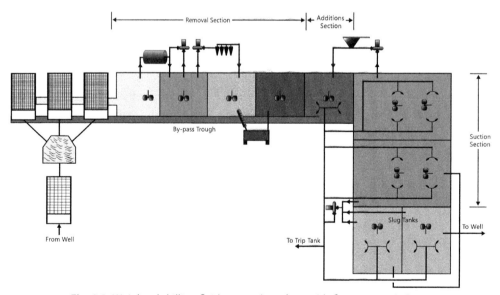

Fig. 2.2 Weighted drilling fluid processing plant with fine screen shakers.

as one in which commercial products have been added to increase the drilling fluid density. There are many variations to the processing plant. In the arrangement shown in Fig. 2.2, the fluid flows from the well to a scalping shale shaker to remove the very large particles that might break a fine wire screen on the main shakers. In this system, three main shakers are required to handle all of the rig flow. Consequently, a flow distribution chamber is positioned to receive the fluid flowing through the scalping shaker screen. On drilling rigs that do not have the main shakers which can process the fluid through very fine screens, a sand trap is beneficial. With three main shakers that can process the drilling fluid through very fine screens, the sand trap is not shown in Fig. 2.2. The fluid is degassed and then is processed through a mud cleaner. The fluid is then centrifuged before it reaches the addition compartment. All compartments are well agitated to provide uniform slurry to feed the equipment. The fluid is then treated with chemicals and necessary additives to meet the drilling fluid specifications as it flows into the suction section. Mud guns can be used in the addition and suction section to assist blending, but should not used in the removal section. All of these details will be discussed in this book.

The suction section can consist of several compartments depending upon the volume of fluid needed in the well, as shown in Fig. 2.3. The fluid in all compartments should be well blended and agitated to ensure a homogeneous slurry. Mud guns can be used for blending the fluid in the various compartments and to distribute the newly mixed fluid from the Additions compartment into the suction section. At least two slug (or pill) tanks should be available to mix relatively small volumes of a variety of fluids to be pumped down the well bore. The rig mud pumps should be plumbed to take suction from any of the compartments in the suction section.

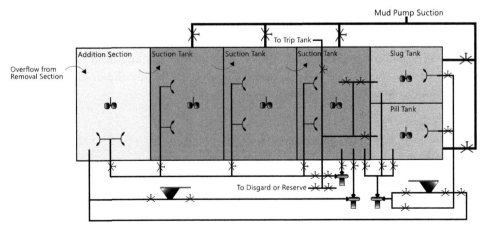

Fig. 2.3 Addition and suction sections of mud processing system.

Agitation

All of the fluid in the addition section and in the suction section needs to be well blended to provide a homogeneous slurry in the drill pipe. Additions to the drilling fluid surface system should be immediately blended with the fluid already in the tanks. Mud guns can be used effectively for this purpose. The next chapter discusses the use of mud guns as well as the operation of the mud hopper to make additions to the system.

Centrifugal pumps are used to circulate drilling fluid through a series of parallel nozzles. In Fig. 2.3, the mud gun suction is in the last suction compartment. Fluid from this compartment is distributed through the addition compartment and the rest of the suction section by the mud guns. This helps blend and homogenize the fluid in all of the compartments. Specifically, not only does the mud density need to be uniform in all of the suction section, the flow properties should also be uniform in all of the suction section. Uniformity of fluid properties in the suction section guarantees that the properties of a sample of the drilling fluid taken from this tank will represent the fluid actually pumped down the hole. Multiple mud guns split the drilling fluid from one compartment into many locations throughout the system. This provides a mechanism to properly blend the fluid.

The fluid exiting the nozzles of a mud gun entrains the adjacent fluid surrounding the nozzles into its flow stream. This is based on the Bernoulli principle that is discussed in Chapter 3 for mud guns and mud hoppers. The increase in velocity of the fluid exiting the nozzle creates a low pressure region around the flow stream. This pulls adjacent material into the flow stream. With a mud hopper, the low pressure pulls solids from the mud hopper into the flow stream which helps disperse the addition into the drilling fluid.

Most contractors are interested in supplying sufficient agitation to prevent settling of solids in these tanks. However, good blending means more than just keeping solids from settling to the bottom of the tank. Agitators must blend the fluid vertically as well as horizontally. Prevention of settling can occur if the solids are stirred radially but this fails to provide vertical homogeneity. Mud guns assist in moving drilling fluid from horizontal planes to vertical mixing. Chapter 4 discusses how agitators mix the drilling fluid blend the fluid and keep solids from settling.

Slug tanks

A slug and pill tank are typically small 20–50 barrel compartments within the suction section of the active system. These compartments are isolated from the active system and available for small volumes of specialized fluid. Most mud tank systems should have more than one of these small compartments.

Slug tanks can be used in many ways. Obviously, they can be used to mix pills for spotting on bottom of the hole, blending lost circulation ingredients to circulate as a pill, blending viscous pills to bring cuttings to the surface, and calibrating mud pump volumetric efficiencies.

Slug tanks are manifolded to a mixing hopper so that solids and chemicals may be added and are used to create a heavier slurry (a slug) to be pumped into the drill pipe before trips. This makes the fluid level in the drill pipe stand at a lower level than the fluid in the annulus. This prevents drilling fluid inside the pipe from splashing on the rig floor during trips. These compartments are also used to create various pills or viscous sweeps. The main pump suction is manifolded to the slug pit(s).

Hydraulic optimization procedures require accurate flow rate measurements be calculated from pump stroke measurements on the mud pump. The volumetric efficiency of mud pumps decreases when the drilling fluid contains air or gas. Usually, supercharged triplex pumps have a volumetric efficiency in the range of 97%. In one well, six percent air in the drilling fluid reduced the volumetric efficiency to 85%. The dimensions of the slug tank (minus any pipe volumes in the tank) can be used to calculate a volume of drilling fluid. While drilling, the mud pump suction can be switched from the suction tank to the slug tank. After timing a known volume removed from the slug tank, the pump suction can be switched back to the suction tank. Pumping down hole with the fluid provides back pressure on the pump discharge and an accurate volumetric measurement can be compared with the pump displacement. The volume of liquid in the slug tank needs to be calculated since the tank will contain gas and liquid. The volume fraction of gas in the drilling fluid can be calculated from the ratio of the difference between the pressurized mud weight and the unpressurized mud weight divided by the pressurized mud weight.

$$\%\text{vol Gas} = \frac{MW_{\text{Pressurized}} - MW_{\text{Unpressurized}}}{MW_{\text{Pressurized}}} \times 100$$

They are manifolded to a mixing hopper so that solids and chemicals may be added and often are used to create heavier slurries, or slugs, for use before trips. This makes the fluid level in the drill pipe stand at a lower level than the fluid in the annulus; drilling fluid inside the pipe is prevented from splashing on the rig floor during trips because the liquid level in the drill pipe is below the rig floor. Slug tanks are also used to create various pills or viscous sweeps. The main pump suction is manifolded to the slug pit(s).

The top of the fluid in the drill string while tripping should be about 100 ft below the surface. A slug of weighted drilling fluid is pumped into the drill pipe to keep the level in the drill pipe below the flow line. The density of the slug, or the increase in mud weight above the original density of fluid depends upon the inside diameter of the drill string and the initial mud weight.

Table 2.2 Height of slugs for various drill pipe diameters.

Drill pipe (in)	$4\frac{1}{2}$	5	$5\frac{1}{2}$	$6\frac{5}{8}$
Weight (lb/ft)	16.6	19.5	24.7	25.2
Capacity (bbl/1000 ft)	14.22	17.76	21.19	34.56
Height for 20 bbl (ft)	1406	1126	944	579
Height for 30 bbl (ft)	2110	1690	1415	868

Increase in mud weight needed for slug to lower the liquid level in $6\frac{5}{8}''$ drill pipe to 100 ft below the flow line.

The internal volumes of various drill pipes are available in many charts. A few are presented in Table 2.2 to use as illustrations of the calculation technique.

To create a liquid level inside of the drill pipe 100 ft below the flow line, the mud weight of the slug can be calculated from the equation below which assumes the height of the slug is given for a specific volume of the slug.

$$MW_{slug} = \frac{MW_{orig}(100 \text{ ft} + H_{slug})}{H_{slug}}$$

This equation is derived in Appendix 2-A in this chapter.

Sample calculations:

With a 10 ppg drilling fluid in a $4\frac{1}{2}''$ drill pipe, the mud weight of a 30 bbl slug should be:

$$MW_{slug} = \frac{10 \text{ ppg}(100 \text{ ft} + 2110 \text{ ft})}{2110 \text{ ft}} = 10.5 \text{ ppg}$$

With a 15 ppg drilling fluid in $6\frac{5}{8}''$, 25.2 lb/ft drill pipe, the mud weight of a 20 bbl slug would be:

$$MW_{slug} = \frac{15 \text{ ppg}(100 \text{ ft} + 579 \text{ ft})}{579 \text{ ft}} = 17.6 \text{ ppg}$$

With a 30 bbl slug, the mud weight of the slug should be:

$$MW_{slug} = \frac{15 \text{ ppg}(100 \text{ ft} + 868 \text{ ft})}{868 \text{ ft}} = 16.7 \text{ ppg}$$

The increase in mud weight required to lower the liquid level in the drill pipe can be calculated for different diameter drill pipes. The capacity of some internal upset drill pipes is presented in Table 2.1

Using the equations presented above, the mud weight increase for a variety of mud weights and slug volumes are plotted for four different pipe sizes in the graphs below (Figs. 2.4–2.7).

Mud tank arrangements needed for safety

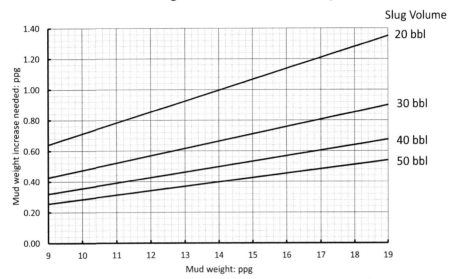

Fig. 2.4 Increase in mud weight needed for slug to lower the liquid level in 4½″ drill pipe to 100 ft below the flow line.

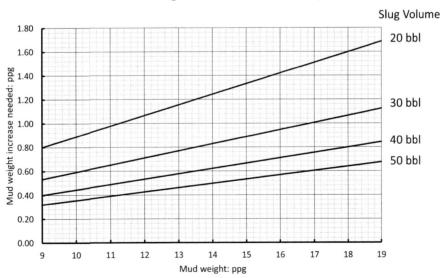

Fig. 2.5 Increase in mud weight needed for slug to lower the liquid level in 5″ drill pipe to 100 ft below the flow line.

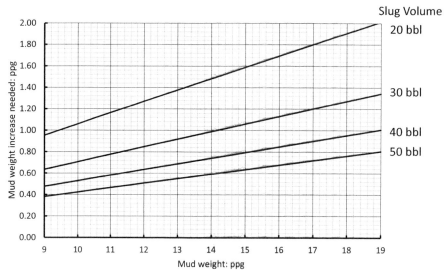

Fig. 2.6 Increase in mud weight needed for slug. to lower the liquid level in $5\frac{1}{2}''$ drill pipe to 100 ft below the flow line.

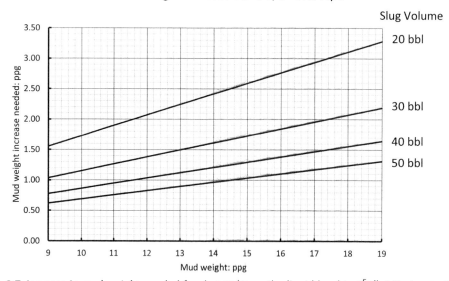

Fig. 2.7 Increase in mud weight needed for slug to lower the liquid level in $6\frac{5}{8}''$ drill pipe to 100 ft below the flow line.

Note: Non-aqueous drilling fluids (NADFs) have significantly higher compressibilities and coefficients of thermal expansion than water-base drilling fluids (WBDF), both of which affect the volume of a NADF down hole. However, near the surface, where the slug volume calculations above apply, the volumetric effects are minimal.

Mud weight increase guideline

The MW_{slug} equation can also be used to determine the location of the drilling fluid surface inside of the drill string for various increases in mud weight. Many drillers use an arbitrary guideline to increase mud weight by different amounts to create the slug. To do this, modify the MW equation, using X instead of 100 ft, and solve the equation for X:

$$MW_{slug} = \frac{MW_{orig}(100 \text{ ft} + H_{slug})}{H_{slug}}$$

$$MW_{slug} = \frac{MW_{orig}(X + H_{slug})}{H_{slug}}$$

$$X = \frac{(MW_{slug})(H_{slug}) - (MW_{orig})(H_{slug})}{MW_{orig}}$$

Calculate the depth of the top of a 20 bbl slug in a 15 ppg drilling fluid in $6^5/_8''$, 25.2 lb/ft drill pipe, using a slug mud weight of 16 ppg:

$$X = \frac{(MW_{slug})(H_{slug}) - (MW_{orig})(H_{slug})}{MW_{orig}}$$

From Table 2.2, a 20 bbl slug in a $6^5/_8''$, 25.2 lb/ft drill pipe would be 579 ft:

$$X = \frac{(16 \text{ ppg})(579 \text{ ft}) - (15 \text{ ppg})(579 \text{ ft})}{15 \text{ ppg}} = 53 \text{ ft}$$

In this case, the drilling fluid probably would not drain from the drill string, and the crew would say the slug did not work.

If the slug were 30 bbl in size, the top of the drilling fluid in the pipe would be below the flow line and still be ineffective.

$$X = \frac{(16 \text{ ppg})(868 \text{ ft}) - (15 \text{ ppg})(868 \text{ ft})}{15 \text{ ppg}} = 59 \text{ ft}$$

Frequently, a slug is ineffective because not all of the weighted fluid is placed in the drill string. The volume of the pump discharge line plus the stand pipe plus the rotary hose volumes should be pumped with a "chaser fluid" whose density is the same as the fluid in the annulus. The slug can be a hundred feet below the chaser fluid and still be effective. Enough chaser fluid should be pumped to ensure that the entire slug is in the drill string.

Trip tanks

A trip tank is used to measure the volume of drilling fluid entering or leaving the well bore during a trip. The volume of fluid that replaces the volume of the drill string (steel volume) is normally monitored on trips to make certain that formation fluids are not entering the wellbore. When one barrel of steel (drill string) is removed from the borehole, one barrel of drilling fluid should replace it to maintain a constant liquid level in the wellbore. If the drill string volume is not replaced, the liquid level may drop low enough to permit formation fluid to enter the wellbore due to the drop in hydrostatic pressure. This is known as a "kick". Usually, someone is assigned the responsibility of recording the volume required to fill the hole after each row of drill pipe is racked in the derrick (or alternately specified number of stands). Fluid may be returned to the trip tank during the trip into the well. The excess fluid from the trip tank should be returned to the active system across the shale shakers. Large solids can come out of the well and plug the hydrocyclones if this drilling fluid bypasses the shakers.

The addition of trip tanks to drilling rigs significantly reduced the number of induced well kicks. Trip tanks have replaced the obsolete or older system of drillers filling the hole with drilling fluid with the rig pumps by counting the mud pump strokes (the volume was calculated for the displacement of the drill pipe pulled). The problem here is that a certain pump efficiency is estimated in these calculations. If the mud pump is not as efficient as estimated, slowly but surely the height of the column of drilling fluid filling the hole decreases. This decreases hydrostatic head and if formation pressures are greater than the hydrostatic head of the drilling fluid a "kick" will occur. Another common cause of inducing a kick was to continue filling the hole with the same number of strokes used for the drill pipe even when reaching the heavy weight drill pipe, or drill collars were pulled. Both the heavy weight drill pipe and drill collars have more displacement per stand than the drill pipe. Therefore, a reduction in the height of the column of drilling fluid in the wellbore would occur and problems would result.

The next modification of the procedure to keep the hole full while pulling pipe was to have a tank that gravity fed fluid in to the well bore as the pipe was removed. The tank level was monitored and the driller kept track of the volume or fluid added to the well as the volume of metal was pulled from the well. Sometimes, a float would be installed on top of the liquid in the trip tank. A rod attached to the float would be placed next to a board marked with the volume. The driller then could visually see the tank level decrease as each stand was removed from the well. This system had some problems in very viscous, gelled drilling fluids. The drilling fluid would gel in the drain line and the well would not be filled with fluid from the tank.

Pulling out of hole

To continuously supply fluid to the well bore while pulling out of the hole, a small centrifugal pump circulates drilling fluid continuously into the well and overflow back into the trip tank. The volume of fluid required to keep the well bore full would be the volume of pipe removed from the well bore. If the volume required to keep the well bore full is less than the volume of pipe, fluid from the down hole formations are probably supplying the extra fluid. Stop pipe movement and observe the trip tank level. If it is increasing, return to bottom immediately and start well control procedures.

If more fluid is used to fill the well bore than the volume of drill pipe being removed, lost circulation is probably the cause. Again, stop and observe the liquid level in the trip tanks. If the level is decreasing, go back to bottom and cure the lost circulation problem. If left untreated, the liquid level in the well bore may fall so much that the hydrostatic pressure next to some formation is insufficient to prevent a kick. A kick in this situation usually means that there will be an underground blowout. Formation fluid will enter the well bore and exit into the lost circulation zone. This is a difficult well control problem. Fluid with sufficient density must be pumped faster that the fluid flowing up the annulus. Neither the flow rate required nor the mud weight will be known.

History

Most well bores do not take exactly the "calculated" value of fluid to fill the well bore for a row of drill pipe standing in the derrick. Usually, the calculated value is close to the actual value. However, a drilling fluid can contain 8% volume of air and not appear to be "foamy". The drill string stretches in the well bore because of the temperature and the load on the string below that point. Trip tank monitors should keep a record of each trip and use it as a guide for effectively keeping the well bore full.

Going in hole

When returning a new bit back to bottom, the drilling fluid displaced by the drill string can be measured with the trip tanks.

When going in the hole with large diameter drill collars, the plumbing should be sized so that the fluid displaced from the hole does not overflow the bell nipple. Drillers run the pipe in the hole at different speeds and the speed determines the displacement volume flow rate for different size pipe. A simple estimation of the volume rate of drill collars entering the hole (which is also the displacement velocity of the fluid), can be calculated from the equation which calculates the volume of a 1029 ft cylinder in barrels:

$$\text{Cylinder volume}/1029 \text{ ft} = \{\text{diameter of cylinder, in inches}\}^2$$

The derivation and discussion of the equation is presented in Appendix 2–B of this chapter. This equation makes it relatively easy to estimate the volume of a 1000 ft long cylinder. A $9^{7}/_{8}''$ hole is almost $10''$ in diameter and would have close to a volume of 100 bbl per thousand feet.

Table 2.3 Drill pipe displacement.

Drill pipe size		Displacement	
Inches	Wt/ft lb/ft	Gal/100 ft	Bbl/100 ft
3½	11.20	15.6671	0.3730
3½	13.30	18.8097	0.4479
4½	12.75	17.3403	0.4129
4½	13.75	18.7038	0.4453
4½	16.60	22.8959	0.5451
5½	19.00	25.2813	0.6019
5½	22.20	29.9347	0.7127
5½	22.25	34.8663	0.8301
6⅝	22.20	28.9939	0.6903
6⅝	25.20	33.9029	0.8072
6⅝	31.90	45.3926	1.0808
7⅝	29.25	39.9840	0.9520
8⅝	40.00	53.6942	1.2784

As the pipe is lowered into the hole, the expected volume of displaced liquid is shown in Table 2.3. A 90 ft stand of 5½″, 22.2 lb/ft drill pipe displaces 0.641 bbl of drilling fluid:

$$\left[\left(\frac{0.7127 \text{ bbl}}{100 \text{ ft}}\right)(90 \text{ ft}) = 0.641 \text{ bbl}\right]$$

A row of 10 stands should result in a displacement of about 6.4 bbl. This value would be the volume of fluid which overflows as the pipe is run back into the hole and the volume of fluid which should be added for a row of 10 stands as the pipe is tripped out of the hole.

Plumbing required to properly use trip tanks

Two schemes are presented here; but, many other arrangements can also provide good information as long as the basic guidelines are followed.

Basic Guidelines for pulling out of hole:
1. Fill the well-calibrated trip tanks with fluid from the suction section.
2. Connect one of the trip tanks to a small centrifugal pump.
3. Start circulating drilling fluid into the well bore and overflow back to the same trip tank.
4. Measure the volume of fluid required to keep the well bore full by measuring the decrease in volume of the trip tank. [This can be by rows of stands of drill pipe as each row is completed. However, for large diameter drill pipe and/or when large drill collars reach the surface, the volume can be reported by a specific number of stands.]

5. Record the volume measured in a tally book for future reference.
   Basic Guidelines for going in hole:
1. As the drill bit is lowered to bottom, drilling fluid should be displaced to a well-calibrated trip tank.
2. Reporting time for the volume displaced depends upon the size of the pipe entering the well bore.
3. When the trip tank is full, or when a sufficient number of stands of pipe have been lowered into the hole, the trip tank should be emptied into the active drilling fluid system through the main shaker. [Frequently, large solids, left in the well bore before the trip, are displaced into the trip tank and should be removed.]
4. When one trip tank is full, change the flow to the empty trip tank.
5. Record the volume measured in a tally book for future reference.

Plumbing scenario #1

This system is suitable for smaller drilling rigs. When drilling, all of the valves in the system are closed. After a slug has been pumped into the drill string, valves #1 and #2 are opened. Drilling fluid from the suction section is pumped into the trip tank from a pipe connected to the mud gun line. The pit levels are noted in each tank. The valve from compartment A is connected to a small centrifugal pump suction (Fig. 2.8). The valve below compartment B is closed. The centrifugal pump circulates fluid from compartment A, through valve #2, into the well bore. The fluid overflows through valve #1, back into compartment (A) As the pipe is tripped from the hole, the fluid level in compartment A will decrease according to the volume of metal pulled from the hole. After the fluid has been removed from compartment A, the valve below compartment A is closed and the valve below compartment B is opened. The overflow from the well, through valve #1 is directed into compartment (B) At this time, the drilling fluid in compartment A can be replenished.

On the trip back into the hole, the volume of fluid displaced by the drill string can be monitored (Fig. 2.9). In this situation, valve #1 is opened and valve #2 is closed. In Fig. 2.9, the fluid will overflow the well bore into compartment (B) While this is happening, the centrifugal pump can be used to empty compartment (A) The fluid displaced from the well will pass through the shaker screen to remove cuttings which may exit the well bore. After compartment B is filled, the overflow fluid is directed into compartment A and compartment B is emptied with the centrifugal pump.

Plumbing scenario #2

The second system uses the flow line as the overflow line from the well bore. This eliminates one valve on the BOP stack (Fig. 2.10). When drilling, all of the valves in the system are closed. After a slug has been pumped into the drill string, valves #1 and #2 are

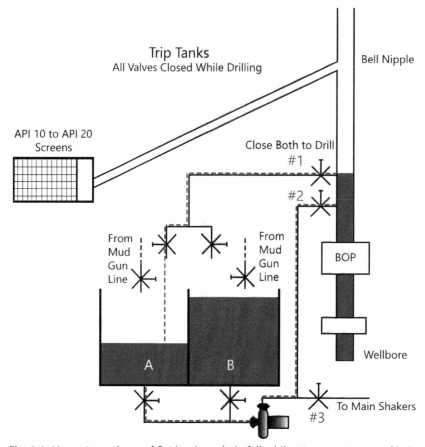

Fig. 2.8 Measuring volume of fluid to keep hole full while tripping pipe out of hole.

opened. Drilling fluid from the suction section is pumped into the trip tank from a pipe connected to the mud gun line. The pit levels are noted in each tank. The valve from compartment A is connected to a small centrifugal pump suction (Fig. 2.10). The valve below compartment B is closed. The centrifugal pump circulates fluid from compartment A, through valve #2, into the well bore. The fluid overflows down the flow line into the back tank of the scalping shaker. After passing through the scalping shaker, the fluid returns back to compartment (A). As the pipe is tripped from the hole, the fluid level in compartment A will decrease according to the volume of metal pulled from the hole. After the fluid has been removed from compartment A, the valve below compartment A is closed and the valve below compartment B is opened. The overflow from the well, through valve #1 is directed into compartment (B) At this time, the drilling fluid in compartment A can be replenished.

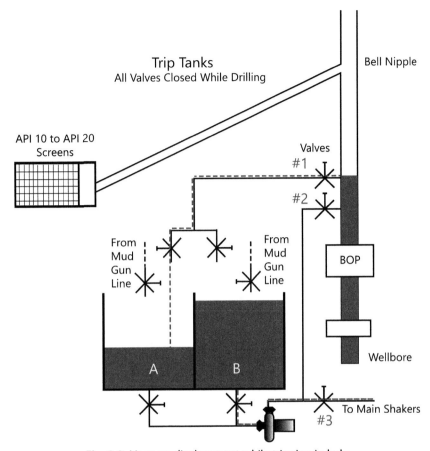

Fig. 2.9 Measure displacement while tripping in hole.

If a driller lowers a stand of 10 inch drill collars into the well (90′ instead of 100′) in about 1 min, the 10 bbl of fluid would flow up the annulus at a rate close to 10 bbl/min (or 400 gpm). The plumbing collecting this fluid must be capable of handling this type of flow rate without overflowing the bell nipple. If the driller lowers the drill collars in around 30 s instead of 60 s, the flow rate will be close to 20 bbl/min (or 800 gpm).

On the trip back into the hole, the volume of fluid displaced by the drill string can be monitored (Fig. 2.11). In this situation, valve #1 and valve #2 are opened. In Fig. 2.9, the fluid will overflow the well bore into compartment (A). While this is happening, the centrifugal pump can be used to empty compartment (B). The fluid displaced from the well will pass through a shaker screen to remove cuttings which may exit the well bore. After compartment A is filled, the overflow fluid is directed into compartment B and compartment A is emptied with the centrifugal pump.

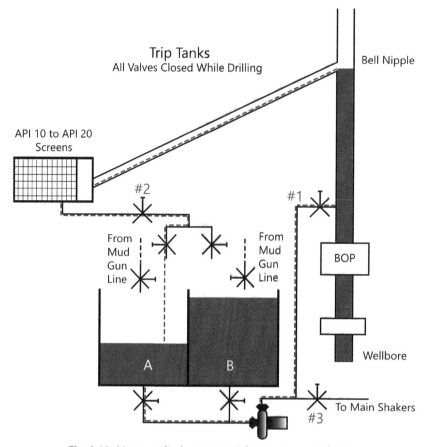

Fig. 2.10 Measure displacement while tripping out of hole.

Appendix 2.A
Derivation of slug effectiveness

Refer to Fig. 2.A.1 for the following derivations. The appropriate equations can be derived from considering the fact that P_1 is equal to P_2.

$$P_1 = P_2$$

$$0.052(MW_{slug})(H_{slug}) + 0.052(MW_{orig})(\text{Depth} - L_a - H_{slug})$$
$$= 0.052(MW_{orig})(\text{Depth})$$

$$\text{Let } L_a = 100 \text{ ft}$$

$$(MW_{slug})(H_{slug}) - (100 \text{ ft})(MW_{orig}) - (MW_{orig})(H_{slug}) = 0$$

$$MW_{slug} = \frac{MW_{orig}(100 \text{ ft} + H_{slug})}{H_{slug}}$$

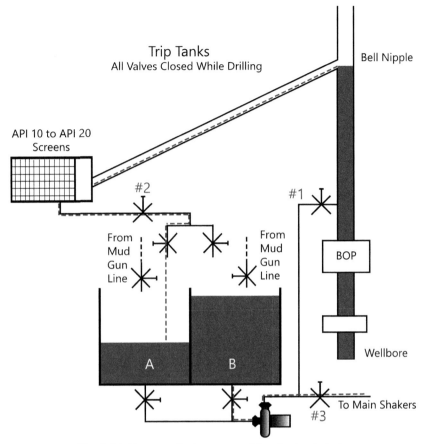

Fig. 2.11 Measure displacement while tripping in hole.

where:
MW_{slug} is the mud weight of the slug, ppg;
H_{slug} is the height or length of slug inside the drill string, ft;
Depth is the total vertical depth of the well, ft;
MW_{orig} is the mud weight of the drilling fluid in the hole, ppg; and
L_a is the length of the air gap, ft.

If an arbitrary or "standard" slug volume and density is used, the length of the air gap can be calculated from the equation:

$$(MW_{slug})(H_{slug}) \quad (L_a)(MW_{orig}) - (MW_{orig})(H_{slug}) = 0$$

$$(L_a) = \frac{[(MW_{orig}) - (MW_{slug})](H_{slug})}{MW_{orig}}$$

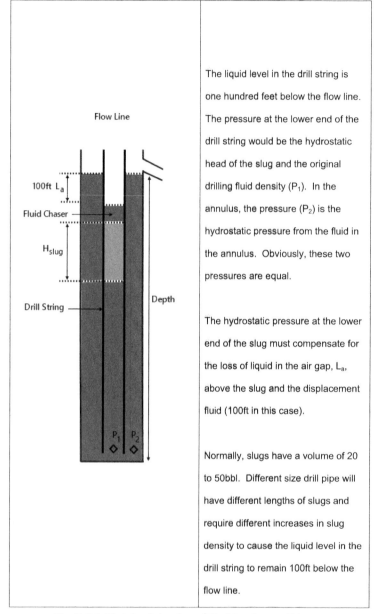

The liquid level in the drill string is one hundred feet below the flow line. The pressure at the lower end of the drill string would be the hydrostatic head of the slug and the original drilling fluid density (P_1). In the annulus, the pressure (P_2) is the hydrostatic pressure from the fluid in the annulus. Obviously, these two pressures are equal.

The hydrostatic pressure at the lower end of the slug must compensate for the loss of liquid in the air gap, L_a, above the slug and the displacement fluid (100ft in this case).

Normally, slugs have a volume of 20 to 50bbl. Different size drill pipe will have different lengths of slugs and require different increases in slug density to cause the liquid level in the drill string to remain 100ft below the flow line.

Fig. 2.A.1 Sketch of slug in a drill string in a well bore.

Appendix 2.B

Derivation of equation:

The volume, in bbl, of a cylinder 1029 ft long can be calculated quickly by squaring the cylinder diameter, in inches. For example, a 10 inch cylinder (or drill collar) 1029 ft long has a volume of 100 bbl.

Volume, inches^3 = [π/4] [(diameter, inches)^2] [length, inches]
Volume, gal^3 = [π/4] [(diameter, inches)^2] [length, inches] {1 gal/231 in^3}
Volume, bbl = [π/4] [(diameter, inches)^2] [length, inches] {1 gal/231 in^3}(1 bbl/42 gal)
Volume, bbl = [π/(4)(231)(42)] [(diameter, inches)^2] [length, inches]
Volume, bbl = [1/12,352.98 [(diameter, inches)^2] [length, inches]

To change the measured value of length from inches to feet, the volume, in bbl, could be calculated from: Volume, bbl = [1/12,352.98 [(diameter, inches)^2] [length, inches] {1 ft/12 inches}

Or.

Volume, bbl = [1/12,352.98 [(diameter, inches)^2] [length, inches] {1 ft/12 inches}
Volume, bbl = [1/1029.4] [(diameter, inches)^2] [length, ft]

This could also be written:

Volume, bbl/ft = [1/1029.4] [(diameter, inches)^2]

Or more simply stated, the square of a cylinder's diameter, measured in inches, is the volume of 1029 ft of the cylinder. For "rule of thumb" estimation, or back-of-the-envelope calculations, 1000 ft of 10 inch drill collars displace 100 bbl. A one hundred foot length would displace 10 bbl of fluid.

Problems

Problem 2.1

The driller is ready to trip pipe and needs to pump a 30 bbl slug down the 5″ drill pipe. The mud weight is 14.0 ppg. What should be the mud weight of the slug?

Problem 2.2

A driller is ready to trip 6⅝″ drill pipe and needs to pump a 30 bbl slug. The mud weight is 16 ppg. What mud weight should the slug have?

Problem 2.3

The driller is ready to trip pipe and pumped a 30 bbl slug of 15 ppg drilling fluid down the 4½″ drill pipe. The mud weight is 14.0 ppg. Will this provide a dry pipe when tripping pipe?

Problem 2.4

The driller in Problem 3 was next assigned a deep water rig and was ready to trip pipe. The mud weight was 14 ppg but the drill pipe was $6^5/_8''$. The 30 bbl slug of 15 ppg drilling fluid was pumped down the larger diameter drill pipe. Will this provide a dry pipe when tripping pipe?

CHAPTER 3

Addition section

Chemicals and fluids required to keep the drilling fluid properties properly adjusted to specifications are added to the surface drilling fluid system in the compartment upstream from the suction section described in Chapter 2. The fluid in this compartment should be thoroughly blended with the fluid in the suction section. Although not required, the fluid from the removal section upstream from the addition section should overflow into the addition section. The pipe connecting the removal section to the addition section could be connected at the bottom of the tank to an adjustable pipe which can be adjusted upward. This will keep the fluid in the removal section at the same level while drilling. All changes in fluid volumes (for detection of lost circulation or kicks) will occur in the suction section.

Solids required to adjust the drilling fluid properties are added through a mud hopper. The operation of the mud hopper is described in this chapter.

Mud guns should be used in the addition and suction sections to help blend the system so the fluid properties are the same in all compartments. The design of the mud guns and flow rates are discussed in this chapter.

Mud hoppers

Solids are added to the active system through a mud hopper (Fig. 3.1). The fluid velocity in a low-pressure mud hopper will be around 10 ft/s on the pressure side of the jet nozzle. The pressure line is reduced in size usually from 6 to 2 inches (152—51 mm) and exits the jet nozzle at a much higher velocity but lower pressure.

The high-velocity jet stream crosses the gap between the educator nozzle and the downstream venturi and creates a partial vacuum within the mixing chamber (or tee). This low-pressure area within the tee, along with gravity, actually draws mud materials from the hopper into the tee and fluid stream. The high-velocity fluid wets and disperses the mud additives into the fluid stream. This reduces lumping of material and is an initial shear of the additives.

The mud hopper works on the Bernoulli Principle. When the velocity of a fluid increases, the pressure next to the surface decreases. This is the principle that allows planes to fly and sailboats to sail close to the wind (Fig. 3.2).

Airplane wings generate lift by creating high and low pressure zones. For horizontal flow, ignoring losses for friction, the total energy at any point along the wing is equal to the sum of the energy from the pressure (P)/specific gravity and energy from the velocity ($V^2/2g$).

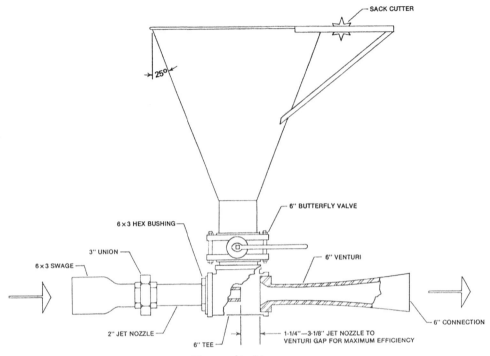

Fig. 3.1 Mud hopper.

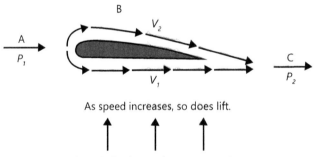

Fig. 3.2 Airplane wing cross section.

Because aircraft wings are curved on top, air travels farther and thus moves faster above the wing than underneath it. Therefore, velocity at point B is greater above the wing than below it. The law of conservation of energy indicates that pressure is affected inversely: If V increases, then P decreases. This creates a differential pressure, or ΔP: higher pressure beneath the wing adds lift. As speed increases, so does lift.

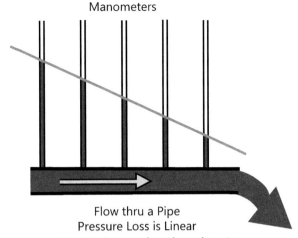

Fig. 3.3 Pressure loss through a pipe.

The Bernoulli concept can be demonstrated for flow through a pipe by looking at the pressure drop along a pipe. A fluid flowing through a straight pipe with a uniform inside diameter (Fig. 3.3), would have a uniform pressure loss through the pipe.

If a smaller section is introduced to the uniform pipe (Fig. 3.4), the manometers would show a slight decrease in the height of the fluid. If fluid flows through a constricted region, the velocity of the fluid increases and the pressure in that region would decrease (Fig. 3.4). The pressure at three locations shown with the manometers, would show the low pressure in the pipe where the fluid velocity increased.

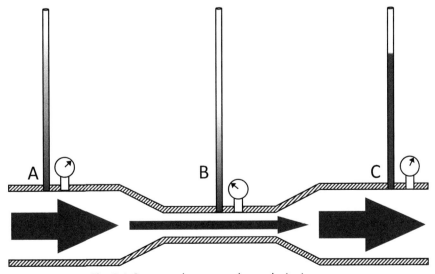

Fig. 3.4 Pressure decreases when velocity increases.

When the fluid carrying the drilling fluid additives exits the mud hopper, a venture straightens the flow profile to help regain some of the pressure head. Without the venturi, the mud hopper could not be placed on the ground outside of the mud tanks because too much pressure loss in the eductor or jet would leave insufficient pressure to make the drilling fluid flow over the mud tank wall.

At point A in Fig. 3.4, the pressure head is high, while the velocity head would be low. As the fluid moves downstream through the venturi, the total head at point C would theoretically remain the same, with the velocity and pressure head being equivalent to that at point A.

The main problem with this equation is that it ignores friction head losses that in practical applications can be about 50%. In actual practice, friction head must be accounted for, and if the venture were not present, there would be tremendous turbulence as the flow expanded into the larger-diameter pipe. The venturi simply helps to streamline the flow back to the large pipe with less turbulence. This results in minimum friction loss and will provide the maximum available head to push the fluid and newly added commercial additives from the hopper into the piping system. Mud hoppers come with valves to isolate the hopper from the mixing chamber. When closed, the space between the venturi chamber and the valve is less than atmospheric, i.e., a partial vacuum is formed. The amount of vacuum at point B influences the resultant addition rates of the device and is determined by several factors, including feed pressure, nozzle diameter and length, venturi, design, fluid properties, and downstream piping restrictions.

A better, more manageable system is to add whole drilling fluid into this compartment. An auxiliary tank can contain drilling fluid with the desired properties and added to the additions compartment. Occasional additions of solids may be necessary to maintain desired fluid properties. The hopper needs to be plumbed so it may continue to add material to the active system.

The fluid from the additions tank should be well-blended with the fluid in the next section—the suction section. Mud guns should be used to blend and mix all of the compartments in the addition and suction section (Fig. 3.5).

Mud hoppers use this principle to draw solids into the moving stream of fluid where the jet or educator is shown in Fig. 3.1. The increase in fluid velocity through the jet, reduces the pressure below the hopper and causes the content of the hopper to flow into the moving drilling fluid stream.

Mud hoppers also tend to pull air into the solids stream. A simple addition to the end of the hopper discharge line will assist in removing the air from the drilling fluid (Fig. 3.6).

This air separator can be easily constructed by a rig welder. A couple of feet of $13^{5}/_{8}''-16''$ casing is used for the separator. The top plate has a hole of about one-half the diameter of the casing and is welded to the top end of the separator. The flow line from the mud hopper is welded so that the fluid swirls inside of the separator. The centrifugal force cases the fluid to press against the casing and the air exits through the opening.

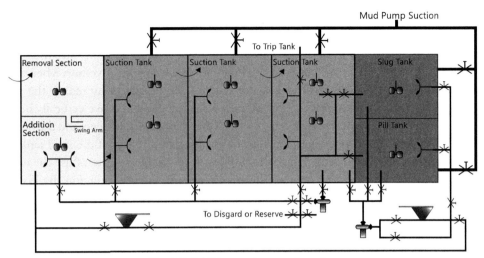

Fig. 3.5 Addition and removal section.

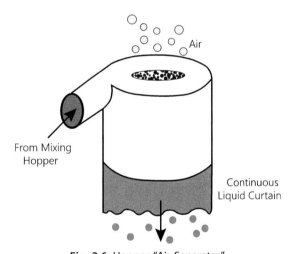

Fig. 3.6 Hopper "Air Separator".

The bottom is open and should be close to the drilling fluid surface. This device requires very little pressure drop and usually can be fitted onto most mud hopper lines.

Mud hopper recommendations

The following recommendations will promote efficient mud hopper installation and use:
- Select a mud hopper that is properly sized for the mud system. Generally, a single hopper is sufficient for most rigs. If the mud circulation rate is greater than 1200 gpm, then consider using a hopper with 1200 gpm capacity. Generally, there

is no need to add chemicals faster than this. For many operations, 600–800 gpm is adequate.
- Keep the lines to and from the hopper as short and straight as possible. Size the pump and motor based on the system head and flow rate requirements. A venturi is beneficial in all operations, but especially when the system back pressure may reduce the mud hopper efficiency and operation. The venturi will allow fluid to move vertically higher than the hopper height. On many rigs, hoppers are place at ground level and the downstream pipe is raised to a height equal to or greater than the top of the mud tanks.
- Use new or clean fittings to reduce friction loss. After each operation, flush the entire system with clean fluid to prevent the mud from drying and plugging the system. Clean the throat of the hopper to prevent material from bridging over that may cause poor performance the next time the hopper is used.
- A table should be attached to or located near the hopper to hold sacks of material. The table should be at a convenient height (36–42 inches) so that personnel can add material easily with minimal strain. Power-assisted pallet and sack handlers will enhance addition rates and minimize personnel fatigue.
- As with all equipment on the rig, develop a regular maintenance and inspection program for the mud hopper. The mud hopper is normally simple and easy to operate, but worn jets and valves hinder the operation. Inspect the entire system ever 30–60 days. Maintain an inventory of spare nozzles, valves, and bushings.

Muddy gun.

Mud guns

Both high-pressure and low-pressure mud guns agitate mud by means of rapid fluid movement through a nozzle. Pressure and flow are delivered via either high-pressure main mud pumps or, most often, centrifugal pumps.

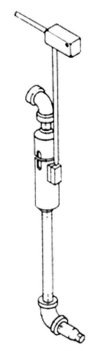

Fig. 3.7 Typical 2′ gun.

High-pressure guns typically come in 3000 and 6000 psi ratings (Fig. 3.7), and require heavy-walled piping. The gun nozzle sizes range from $1/4$ to $3/4$ inch. Mud supply is from a positive displacement rig pump. This type of mud gun is not used as frequently as the low-pressure mud gun. Low-pressure mud guns usually require around 75 ft of head for effective operation (Fig. 3.8). Nozzle sizes will range from $5/8$ to 1 inch. Mud is supplied by a centrifugal pump through low-pressure lines. The centrifugal pump can be run by either an electric motor or diesel engine.

Mud guns are usually placed about six inches from the tank bottom. Most guns come with a 360° swivel joint that allows directional positioning of the jet stream to permit stirring of dead areas within the mud tanks.

The operation of mud guns, whether high-pressure or low-pressure, depends on the movement of fluid through the jet nozzle. The greater the velocity, the more effective the agitation.

Fig. 3.7 illustrates a typical 2″ gun having full 2″ unrestricted opening and 2″ line pipe thread at the bottom connection. It is entirely spin proof, due to the exact and permanent alignment of the inlet, and discharge tube openings. The two swivel joints permit easy handling and a lock screw provides means for locking in set position. Maximum working pressure is 3000 psi.

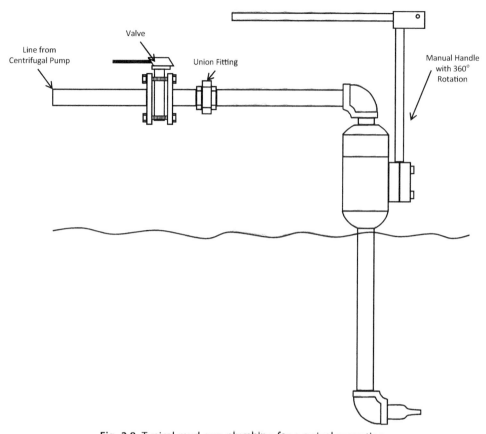

Fig. 3.8 Typical mud gun plumbing for a swivel mounting.

Some mud guns are arranged so that the nozzles rotate as the fluid flows through the nozzle (Fig. 3.9).

This augments the agitators and provides more movement of the drilling fluid across the bottom of the compartment.

Some mud guns are equipped with eductors to provide better blending (Fig. 3.10). The fluid exiting the nozzle passes through a venturi section. This utilizes the Bernoulli process to create a low pressure zone as the fluid leaves the mud gun. Drilling fluid will be entrained in the stream and provide better blending. The eductor can increase the flow volume by a factor of four times more than the mud gun nozzle flow rate (Fig. 3.11).

Selecting nozzles and centrifugal pumps for mud guns

Mud guns are very useful in the addition and suction sections of a surface drilling fluid system. They provide a method of blending all of the fluid in the tanks so the properties are homogeneous. Mud weight must be maintained at a constant value while drilling so that pressures read during a kick can be used to calculate bottom hole pressures. One way

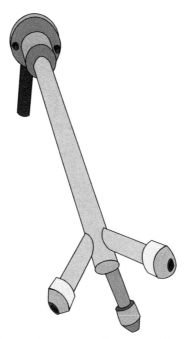

Fig. 3.9 Rotating nozzles on a mud gun.

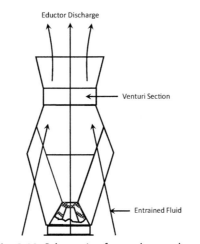

Fig. 3.10 Schematic of a mud gun eductor.

to do this is to use mud guns to split the drilling fluid into many streams in this section of the surface system and blend it with drilling fluid arriving from the removal section.

A centrifugal pump normally supplies the fluid for low pressure mud guns. The centrifugal pump used for the desilters or mud cleaner is sized to processes more fluid than is

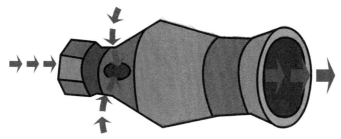

Fig. 3.11 Mud gun eductors.

being pumped downhole. Maintenance and inventory can be simplified if the same pump, motor, and impeller size are used for the mud guns.

Situation

If 400 gpm is circulated down a well bore and the flow through the desilters is 550 gpm. All of the drilling fluid can be processed completely if the plumbing is correctly arranged. The centrifugal pump supplying fluid to the desilters is a 6 × 8 × 14, 175 0RPM pump with an 11 inch impeller driven by a 100 HP motor. The same type of pump can be used to supply drilling fluid to the mud guns. The addition compartment contains 100 bbl of drilling fluid and the suction compartment consists of two 200 bbl tanks. Two agitators and two mud guns are in each 200 bbl tank to provide mixing and blending for every 100 bbl of drilling fluid (Fig. 3.12).

With this surface mud tank system, the drilling rig is pumping 400 gpm down hole. The addition and suction sections contain 500 bbl (or 21,000 gal) of fluid. Every

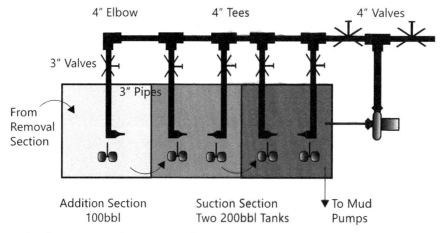

20 feet from pump to first 4" tee; 10 feet between 4" tees; 8 feet of 3" pipe for each mud gun.

Fig. 3.12 Addition and suction section plumbing.

53 min [21,000 gal/400 gpm], the tanks are filled with new drilling fluid from the well. To blend this system to be homogeneous, the initial flow selection for the 5 mud guns is 600 gpm of fluid (or 120 gpm each). This has a turn-over rate time of 35 min [4200 gal/120 gpm]. All of the fluid in each compartment will be blended with the fluid coming from the well.

To ensure good blending, each mud gun should circulate around 100 gpm. For the initial calculation, the pump output is assumed to be 600 gpm for the mud guns. The pump curve shows that this pump can up to 600 gpm at 110 ft of head. The centrifugal pump produces a constant head and the flow rate depends upon what is connected to the pump. The pressure, or head, losses in the pipe can be calculated by assuming the 600 gpm output and calculate the nozzle diameters to create the utilization of all of the head produced by the pump.

Each mud gun should supply around 100 gpm to each 100 bbl region that it will blend. In each 200 bbl tank, two mud guns and two agitators thoroughly blend the system. The operation point can be located on the centrifugal pump curve (Fig. 3.13). The curve is relatively flat at this point, meaning that if the flow rate is 600 gpm more or less than the targeted 500 gpm, the pump out put head will be the same.

Head losses, for water flowing through the pipe, are well documented in the literature and depends upon the flow rate and the length of pipe. The mud gun nozzle sizes will be

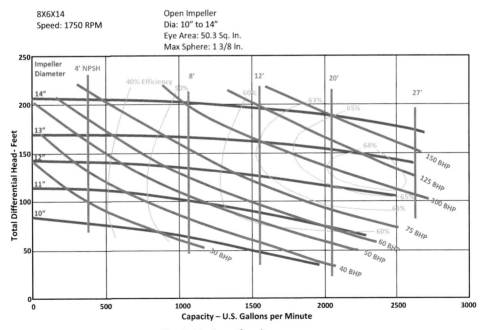

Fig. 3.13 Centrifugal pump curve.

based on the flow of water—a Newtonian fluid. Fortunately, great accuracy in the flow rate through the nozzles is not critical in this situation.

The next step is to size the nozzles so that the head at each nozzle will be sufficient to supply the required 120 gpm.

Find the head at each mud gun nozzle

The head generated by the centrifugal pump is 110 ft. As the fluid flows through the pipe to the first mud gun, there will be a head loss. The fittings and piping diagram for the tanks (Fig. 3.12), indicate that the fluid will flow through a 4″ schedule 40 steel pipe from the centrifugal pump to the first mud gun. The first mud gun will be supplied with fluid through a 4″ tee into a 3″ pipe. Each mud gun is separated by 10 ft of 4″ pipe.

Head loss for water flowing through pipe has been well published and is described in Table 10.2 in Chapter 10. These tables do not present data for small changes in flow rate. To properly solve the equations to determine nozzle sizes, these tables were converted to equations. A plot of the head loss for a 4″ diameter pipe (Fig. 3.14), shows a curve that

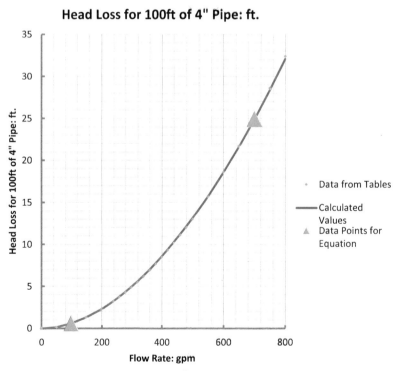

Fig. 3.14 Friction loss in 4″ pipe as a function of flow rate.

resembles a parabola; that is the head loss is a function of the square of the flow rate. However, plotting this on a log-log plot (Fig. 3.15), indicates that the slope of the curve is not a "2" but somewhat lower.

The equation for flow through the pipe is:

$$HL_{100 \text{ ft}} = \frac{Q^a}{b}$$

where HL is the head loss in 100 ft of pipe, in feet of head.

Q is the flow rate, in gpm; and

a and b are constants.

The data from Table 3.2 for head loss in 100ft of 4″ pipe is shown as blue dots in Fig. 3.14. The data looks like the equation should be a parabola; that is the head is a function of the square of the flow rate. Plotting the data on a log-log graph indicates that the equation shows that the head loss is proportional to the flow rate raised to some exponent.

To convert the numbers in Table 3.2 to equations, two data points were selected from the tables, shown in Fig. 3.14 as green triangles. The constants in the equation were calculated from those two points and the red curve was drawn from the equation not the data. Obviously the curve overlies the data points which validate the equation.

Table 3.2 was reduced to equations (Table 3.1), for several different pipe sizes. This permits calculation of head losses for any flow rate.

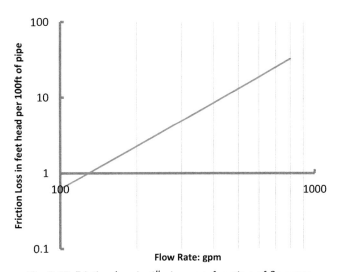

Fig. 3.15 Friction loss in 4″ pipe as a function of flow rate.

Table 3.1 Head loss through 100 ft of pipe.

Pipe diameter, inches	Equation
3	$HL_{100\,ft} = \dfrac{Q^{1.9}}{2605}$
4	$HL_{100\,ft} = \dfrac{Q^{1.9}}{10,200}$
5	$HL_{100\,ft} = \dfrac{Q^{1.9}}{30,019}$
6	$HL_{100\,ft} = \dfrac{Q^{1.9}}{68,350}$
8	$HL_{100\,ft} = \dfrac{Q^{1.9}}{320,390}$

Find the mud gun nozzle sizes

The piping diagram for the first three mud guns (Fig. 3.16), can be used to identify each component of the pressure (or head) loss in the pipes feeding the mud guns. For the purpose of this calculation, only one nozzle will be fitted into each mud gun.

Calculating the head available at each nozzle
First segment
The head loss through the pipe from the pump to the first 4″ tee can be calculated by determining the equivalent pipe length (Table 3.2).

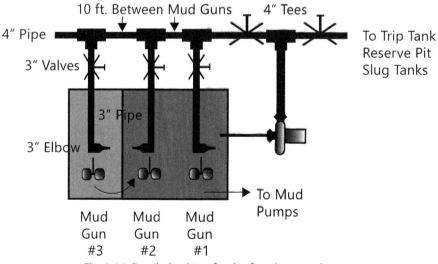

Fig. 3.16 Detail plumbing for the first three mud guns.

Table 3.2 Equivalent 4″ pipe length of the first segment.

	Equivalent length, ft
20′ of 4″ pipe	20.0
One 4″ butterfly valve	15.1
One branch 4″ tee	20.1
One thru flow 4″ tee	6.71
Total	**61.8**

From Table 3.1, the head loss for 100 ft of the 4″ pipe section could be calculated from the equation:

$$HL_{100} = \frac{Q^{1.90}}{10,200} = \frac{600\text{gpm}^{1.9}}{10,200} = 18.62 \text{ ft head per 100 ft pipe}$$

The equivalent length of 4″ pipe is 61.8 ft. The head loss from the pump to the first mud gun 3″ pipe is:

$$\text{Head Loss} = \frac{18.62 \text{ ft head}}{100 \text{ ft pipe}}[61.8 \text{ ft pipe}] = 11.5 \text{ ft head}$$

The head at the first 4″ tee would be the pump output head minus the head loss to the first mud gun line, or 110 ft − 11.5 ft = 98.5 ft

$$HL_{4''} = \frac{600\text{gpm}^{1.9}}{10,200}\left(\frac{61.8 \text{ ft head}}{100 \text{ ft pipe}}\right) = \frac{Q^{1.9}}{16,505} = 11.5 \text{ ft head}$$

The head loss in the 3″ mud gun feed line can be calculated by determining the equivalent 3″ pipe length (Table 3.3).

The equivalent length of 3″ pipe is 27.2 ft. The head loss from the 4″tee to the mud gun nozzle is:

$$HL_{100} = \frac{Q^{1.9}}{2605} = \frac{120 \text{ gpm}^{1.9}}{2605} = 3.42 \text{ ft head for 100 ft pipe}$$

Table 3.3 Equivalent 3″ pipe length for the first mud gun.

	Equivalent length, ft
8 ft of 3″ pipe	8.0
One 3″ butterfly valve	11.5
One 3″ elbow	7.67
Total	**27.2**

$$\text{Head Loss} = \frac{3.42 \text{ ft head}}{100 \text{ ft pipe}} [27.2 \text{ ft pipe}] = 0.93 \text{ ft head}$$

The head at the mud gun nozzle would be 98.57 ft−0.93 ft = 97.64 ft. The nozzle diameter can be calculated from the equation:

$$\text{Flow Rate, gpm} = 19.6 \left(\sqrt{\text{head, ft}}\right)(\text{nozzle diameter, inches})^2$$

$$120 \text{ gpm} = 19.6 \left(\sqrt{97.64 \text{ ft}}\right)(\text{nozzle diameter, inches})^2$$

$$(\text{nozzle diameter, inches})^2 = 0.620$$

$$(\text{nozzle diameter, inches}) = 0.46$$

Second mud gun

The calculation process will be the same. The head loss through the pipe from the first 4″ tee to the second 4″ tee can be calculated by determining the equivalent pipe length (Table 3.4).

The flow rate in this section of 4″ pipe is 600 gpm−120 gpm or 480 gpm.

From Table 3.1, the head loss for 10 ft of the 4″ pipe section could be calculated from the equation:

$$HL_{100} = \frac{Q^{1.90}}{10,200} = \frac{(600 \text{ gpm} - 120 \text{ gpm})^{1.9}}{10,200} = 12.18 \text{ ft}$$

The equivalent length of 4″ pipe is 16.7 ft. The head loss between the 4″ tees is:

$$\text{Head Loss} = \frac{12.18 \text{ ft head}}{100 \text{ ft pipe}} [16.7 \text{ ft pipe}] = 2.03 \text{ ft head}$$

The head at the second 4″ tee would be the head at the first 4″ tee minus the head loss to the second mud gun line, or 98.57 ft−2.03 ft or 96.54 ft.

The head loss in the 3″ mud gun feed line can be calculated by determining the equivalent 3″ pipe length (Table 3.5), [identical to Table 3.3 and first mud gun].

The equivalent length of 3″ pipe is 27.2 ft. The head loss from the 4″ tee to the mud gun nozzle is:

Table 3.4 Equivalent 4″ pipe length between the first and second mud gun.

	Equivalent length, ft
10′ of 4″ pipe	10.0
One thru flow 4″ tee	6.71
Total	**16.7**

Table 3.5 Equivalent 3″ pipe length for the second mud gun.

	Equivalent length, ft
8 ft of 3″ pipe	8.0
One 3″ butterfly valve	11.5
One 3″ elbow	7.67
Total	**27.2**

$$HL_{100} = \frac{Q^{1.9}}{2605} = \frac{120 \text{ gpm}^{1.9}}{2605} = 3.42 \text{ ft} \frac{\text{head}}{100 \text{ ft pipe}}$$

$$\text{Head Loss} = \frac{3.42 \text{ ft head}}{100 \text{ ft pipe}} [27.2 \text{ ft pipe}] = 0.93 \text{ ft head}$$

Each mud gun line is exactly the same. Consequently, all head losses in the 3″ pipes will be the same for all other nozzle calculations.

The head at the mud gun nozzle would be 96.47 ft −0.93 ft = 95.54 ft.

The nozzle diameter can be calculated from the equation:

$$\text{Flow Rate, gpm} = 19.6 \left(\sqrt{\text{head, ft}}\right) (\text{nozzle diameter, inches})^2$$

$$120 \text{gpm} = 19.6 \left(\sqrt{95.54 \text{ ft}}\right) (\text{nozzle diameter, inches})^2$$

$$(\text{nozzle diameter, inches})^2 = 0.626$$

$$(\text{nozzle diameter, inches}) = 0.791$$

The second mud gun nozzle diameter (0.791″) is larger than the first mud gun nozzle diameter (0.782″) because the head is lower at the second mud gun. Each mud gun will be slightly larger in diameter to accommodate the required 120 gpm needed through each much gun.

Third mud gun

The calculation process will be the same as for the second nozzle. After each interval of 4″ pipe, the flow rate decreases by 120 gpm. The flow rate in this section is 480 gpm−120 gpm or 360 gpm.

From Table 3.1, the head loss for 10 ft of the 4″ pipe section could be calculated from the equation:

$$HL_{100} = \frac{Q^{1.90}}{10,200} = \frac{(360 \text{ gpm})^{1.9}}{10,200} = 7.05 \text{ ft}$$

The equivalent length of 4″ pipe is 16.7 ft. The head loss between the 4″ tees is

Table 3.6 Equivalent 3″ pipe length for the third mud gun.

	Equivalent length, ft
8 ft of 3″ pipe	8.0
One 3″ butterfly valve	11.5
One 3″ elbow	7.67
Total	**27.2**

$$\text{Head Loss} = \frac{7.05 \text{ ft head}}{100 \text{ ft pipe}} [16.7 \text{ ft pipe}] = 1.18 \text{ ft head}$$

The head at the third 4″ tee would be the head at the second 4″ tee minus the head loss to the third mud gun line, or 96.47 ft−1.18 ft or 95.29 ft.

The head loss in the 3″ mud gun feed line can be calculated by determining the equivalent 3″ pipe length (Table 3.6), [identical to Tables 3.3 and 3.5, first mud gun and second mud gun].

Each mud gun line is exactly the same. Consequently, all head losses in the 3″ pipes will be the same for all other nozzle calculations for the same flow rate. The equivalent length of 3″ pipe is 27.2 ft.

The head loss from the 4″ tee to the mud gun nozzle is:

$$HL_{100} = \frac{Q^{1.9}}{2605} = \frac{120 \text{ gpm}^{1.9}}{2605} = 3.42 \text{ ft} \frac{\text{head}}{100 \text{ ft pipe}}$$

$$\text{Head Loss} = \frac{3.42 \text{ ft head}}{100 \text{ ft pipe}} [27.2 \text{ ft pipe}] = 0.93 \text{ ft head}$$

The head at the mud gun nozzle would be 95.30 ft − 0.93 = 94.37 ft.
The nozzle diameter can be calculated from the equation:

$$\text{Flow Rate, gpm} = 19.6 \left(\sqrt{\text{head, ft}}\right)(\text{nozzle diameter, inches})^2$$

$$120 \text{ gpm} = 19.6 \left(\sqrt{94.36 \text{ ft}}\right)(\text{nozzle diameter, inches})^2$$

$$(\text{nozzle diameter, inches})^2 = 0.630$$

$$(\text{nozzle diameter, inches}) = 0.794$$

Fourth mud gun

The calculation process will be the same as for the third nozzle. After each interval of 4″ pipe, the flow rate decreases by 83 gpm (Fig. 3.17).

The flow rate in this section is 360 gpm−120 gpm or 240 gpm.

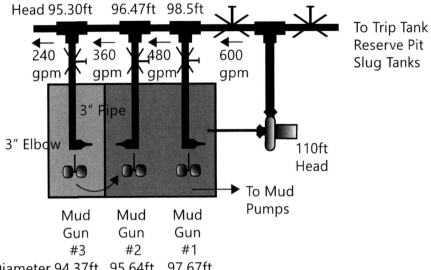

Fig. 3.17 Summary of calculations for the first three mud guns.

From Table 3.1, the head loss for 10 ft of the 4″ pipe section could be calculated from the equation:

$$\mathrm{HL}_{100} = \frac{Q^{1.90}}{10,200} = \frac{(240 \text{ gpm})^{1.9}}{10,200} = 3.26 \text{ ft}$$

The equivalent length of 4″ pipe is 30.1 ft. The head loss between the 4″ tees is:

$$\text{Head Loss} = \frac{3.26 \text{ ft head}}{100 \text{ ft pipe}} [16.7 \text{ ft pipe}] = 0.545 \text{ ft head}$$

The head at the fourth 4″ tee would be the head at the third 4″ tee minus the head loss to the fourth mud gun line, or 95.30 ft−0.545 ft or 94.76 ft.

Each mud gun line is exactly the same. Consequently, all head losses in the 3″ pipes will be the same for all other nozzle calculations or 0.93 ft.

The head at the mud gun nozzle would be 94.76 ft − 0.93 ft = 93.83 ft.

The nozzle diameter can be calculated from the equation:

$$\text{Flow Rate, gpm} = 19.6 \left(\sqrt{\text{head, ft}}\right)(\text{nozzle diameter, inches})^2$$

$$120 \text{gpm} = 19.6 \left(\sqrt{93.83 \text{ ft}}\right)(\text{nozzle diameter, inches})^2$$

$$(\text{nozzle diameter, inches})^2 = 0.632$$

$$(\text{nozzle diameter, inches}) = 0.795$$

Fifth mud gun

The calculation process will be the same as for the third nozzle except the fluid now flows through a 4" elbow instead of a 4" tee. After each interval of 4" pipe, the flow rate decreases by 120 gpm. The flow rate in this section is 240 gpm−120 gpm or 120 gpm.

From Table 3.1, the head loss for 10 ft of the 4" pipe section could be calculated from the equation:

$$\text{HL}_{100} = \frac{Q^{1.90}}{10,200} = \frac{(120 \text{ gpm})^{1.9}}{10,200} = 0.875 \text{ ft}$$

The equivalent length of 4" pipe is the 10 ft of pipe and the 4" elbow (10.1 ft) or 20.1 ft. The head loss between the 4" tees is:

$$\text{Head Loss} = \frac{0.875 \text{ ft head}}{100 \text{ ft pipe}} [20.1 \text{ ft pipe}] = 0.176 \text{ ft head}$$

The head at the fifth 4" tee would be the head at the fourth 4" tee minus the head loss to the fifth mud gun line, or 94.85 ft −0.176 ft = 94.67 ft.

Each mud gun line is exactly the same. Consequently, all head losses in the 3" pipes will be the same for all other nozzle calculations or 0.93 ft.

The head at the mud gun nozzle would be 94.67 ft −0.93 ft = 93.74 ft.

The nozzle diameter can be calculated from the equation:

$$\text{Flow Rate, gpm} = 19.6 \left(\sqrt{\text{head, ft}}\right) (\text{nozzle diameter, inches})^2$$

$$120 \text{ gpm} = 19.6 \left(\sqrt{93.74 \text{ ft}}\right) (\text{nozzle diameter, inches})^2$$

$$(\text{nozzle diameter, inches})^2 = 0.632$$

$$(\text{nozzle diameter, inches}) = 0.795$$

Summary

The nozzle diameters increase to provide the same flow rate as the pressure (head) in the pipe decreases because of the pressure (or head) losses in the 4" pipe (Fig. 3.18).

If two nozzles are planned for each mud gun, the above example can be modified by changing the head loss in the 3" mud gun line. The equivalent length of pipe would be calculated from the 8 ft of pipe, the butterfly valve, and a 3" tee instead of a 3" elbow. The equivalent length of a branch flow 3" tee is 15.3 ft and the equivalent length of a 3" elbow is 7.67 ft. This will make a very small change in the head loss through the mud gun line. The diameter of the nozzles could be estimated by using the areas presented in

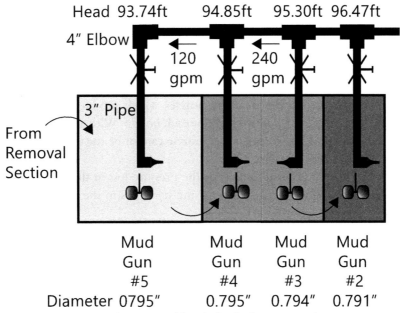

Fig. 3.18 Flow rate and heads for the last two mud guns.

Table 3.7. The area for the two nozzles will be the same as for the one nozzle. The diameters of the new nozzles could be closely estimated by dividing the diameters in the table by the square root of two.

Nozzles are not normally machined with the precision of the previous calculation. The exact flow rate is not that important. All of these nozzles are close to 0.75″ (or $^3/_4$th of an inch). Nozzles are readily available in this size range. The calculation of the flow through the nozzles is much more difficult because the pump output flow rate depends upon the piping connected to it. A centrifugal pump supplies fluid at a constant head which is independent of the flow rate up to the point at which the velocity head becomes significant. With the centrifugal pump used in this example, the velocity head becomes significant around 700 gpm, (Fig. 3.14).

Table 3.7 Nozzles to provide 120 gpm flow in each compartment.

Mud gun #	Diameter, in	Area, in²
1	0.782	0.4800
2	0.791	0.4912
3	0.794	0.4949
4	0.795	0.7961
5	0.795	0.7961

Calculation of fluid flow through $3/4''$ nozzles

The centrifugal pump produces 110 ft of head up to about 600 to 700 gpm. The flow rate depends upon the head (or pressure) drop within the piping system. The nozzle diameters will be selected and now the flow rate needs to be calculated. As the flow rate changes, the head losses changes and this changes the head available at the nozzles.

The $3/4''$ nozzles are smaller than the nozzles which produced 120 gpm for each nozzle. In that case, 120 gpm was specified for each nozzle. With a fixed nozzle diameter in all mud guns, the flow rate through each nozzle cannot be the same. Each nozzle will have a successively lower flow rate.

With the mathematical relationships for the pressure loss in the $4''$ and $3''$ pipes, an expression can be developed to account for the head loss in the piping system for the applied 110 ft of head.

First segment

The head loss consists of the head loss in the $4''$ pipe from the pump to the first tee, the head loss in the first $3''$ mud gun line, and the nozzle head loss. This can be expressed:

$$110 \text{ ft} = [4'' \text{ pipe head loss} + 3'' \text{ pipe head loss} + \text{nozzle head loss}]$$

$$110 \text{ ft} = HL_{4''} + HL_{3''} + HL_n$$

Each of the values on the right side of the above equation will be calculated below.
Calculation for $4''$ pipe head loss (HL_4);

The head loss in the $4''$ pipe can be calculated by determining the equivalent $4''$ pipe length (Table 3.8).

The equivalent length of $4''$ pipe is 61.8 ft.

From Table 3.1, the head loss for length of $4''$ pipe section could be calculated from the equation:

$$HL_{4''} = \frac{Q^{1.90}}{10,200} \frac{61.8 \text{ ft}}{100 \text{ ft}} = \frac{Q^{1.90}}{16,505}$$

Table 3.8 Equivalent $4''$ pipe length.

	Equivalent length, ft
20' of $4''$ pipe	20.0
One $4''$ butterfly valve	15.1
One branch $4''$ tee	20.1
One thru flow $4''$ tee	6.71
Total	**61.8**

The flow through most of the nozzles will be slightly less than 1/5th of the flow rate (Q). As the fluid moves through the 4″ manifold, the flow rate through each successive nozzle will be slightly smaller because the head at each nozzle will be slightly lower. However, a reasonable estimate can be made assuming, initially, that the flow rate from the pump will be five times the flow rate through the first nozzle.

$$HL_{4″} = \frac{(5Q_n)^{1.90}}{16,505} = \frac{Q_n^{1.9}}{775.5}$$

Calculation for 3″ pipe head loss (HL_3);

The head loss in the 3″ mud gun feed line can be calculated by determining the equivalent 3″ pipe length (Table 3.9).

The equivalent length of 3″ pipe is 27.2 ft.

$$HL_{3″} = \frac{Q_n^{1.9}}{2605} \frac{27.2 \text{ ft}}{100 \text{ ft}} = \frac{Q_n^{1.9}}{9577}$$

The flow rate through the 3″ line and the nozzle is not known. An assumption will be made that the flow rate through each nozzle will be 1/5th of the total flow.

Calculation for nozzle head loss (HL_n);

The equation for flow rate through the nozzle can be solved for HL_n.

$$\text{Flow Rate, gpm} = 19.6 \left(\sqrt{HL_n, \text{ ft}}\right)(\text{nozzle diameter, inches})^2$$

$$HL_n = \frac{Q_n^2}{[(19.6)(d^2)]^2} = \frac{Q_n^2}{121.6}$$

For the ¾″ nozzle diameters.

The equation for the head loss through the system is:

$$110 \text{ ft} = HL_{4″} + HL_{3″} + HL_n$$

$$110 \text{ ft} = \frac{Q_n^{1.9}}{775.5} + \frac{Q_n^{1.9}}{9577} + \frac{Q_n^2}{121.6}$$

Factoring the first two terms:

Table 3.9 Equivalent 3″ pipe length.

	Equivalent length, ft
8 ft of 3″ pipe	8.0
One 3″ butterfly valve	11.5
One 3″ elbow	7.67
Total	**27.2**

Table 3.10 Comparing calculated head losses with known pump head.

Pump head, ft	Nozzle flow rate, gpm	4" & 3" head loss, ft	Nozzle head loss, ft	Calculated head loss, ft	Compare calculated with 110 ft
110	120	12.43607	118.4211	130.86	20.85712
110	115	11.47003	108.7582	120.23	10.22825
110	110	10.54107	99.50658	110.05	0.04765
110	105	9.649355	90.66612	100.32	−9.684526
110	109	10.35974	97.70559	108.07	−1.934665
110	109.9	10.52287	99.32574	109.85	−0.151389
110	109.95	10.53197	99.41614	109.95	−0.051892
110	109.98	10.53743	99.4704	110.01	0.007828

$$110 \text{ ft} = \left[\frac{1}{775.5} + \frac{1}{9577}\right] Q_n^{1.9} + \frac{Q_n^2}{121.6}$$

$$110 \text{ ft} = [0.0012894 + 0.0001044] Q_n^{1.9} + \frac{Q_n^2}{121.6}$$

$$110 \text{ ft} = [0.001394] Q_n^{1.9} + \frac{Q_n^2}{121.6}$$

$$110 \text{ ft} = \frac{Q_n^{1.9}}{717.4} + \frac{Q_n^2}{121.6}$$

One way to solve this rather complicated algebraic equation is by iteration. A value of the nozzle flow rate is selected and the right side of the equation is compared with the known value of 110 ft. This operation is expedited if an Excel spread sheet is used. In the calculation below (Table 3.10), the initial flow rate was selected as 120 gpm. The calculated head loss was 130 ft instead of 110 ft. The trial value of flow rate was reduced by 5 gpm and the error decreased at a proposed flow rate of 110 gpm. Reduction to 105 gpm resulted in an error of −9.7 ft. The flow rate produced by the pump was slightly below 110 gpm.

The flow rates through the other four nozzles will be slightly smaller. The total flow produced by the pump should be close to 550 gpm. The flow through the 4" pipe from the first 4" tee to the second mud gun should be 440 gpm.

Second mud gun

The calculation process will be the same as the first segment.

Calculation for 4" pipe head loss (HL_4):

The head loss through the pipe from the first 4" tee to the second 4" tee can be calculated by determining the equivalent pipe length (Table 3.11).

Table 3.11 Equivalent 4″ pipe length.

	Equivalent length, ft
10′ of 4″ pipe	10.0
One thru flow 4″ tee	6.71
Total	**16.7**

The head loss from the pump to the 4″ tee for the first mud gun is:

$$HL_{4''} = \frac{Q^{1.90}}{10,200} \frac{61.8 \text{ ft}}{100 \text{ ft}} = \frac{550^{1.90}}{16,505} = 9.75 \text{ ft}$$

The head at the first tee is 110 ft−9.75 ft = 100.25 ft.

$$100.25 \text{ ft} = [4'' \text{ pipe head loss} + 3'' \text{ pipe head loss} + \text{nozzle head loss}]$$

$$100.25 \text{ ft} = HL_{4''} + HL_{3''} + HL_n$$

The flow rate in this section of 4″ pipe is 550 gpm−110 gpm or 440 gpm.

From Table 3.1, the head loss for 10 ft of the 4″ pipe section could be calculated from the equation:

$$HL_{100} = \frac{Q^{1.90}}{10,200} = \frac{(550 \text{ gpm} - 110 \text{ gpm})^{1.9}}{10,200} = 10.33 \text{ ft}$$

The equivalent length of 4″ pipe is 16.7 ft. The head loss between the 4″ tees is:

$$\text{Head Loss} = \frac{10.33 \text{ ft head}}{100 \text{ ft pipe}} [16.7 \text{ ft pipe}] = 1.72 \text{ ft head}$$

The head at the second 4″ tee would be the head at the first 4″ tee minus the head loss to the second mud gun line, or 100.25 ft−1.72 ft or 98.53 ft.

The flow rate through the 3″ pipe to the nozzle is not known.

Calculation for 3″ pipe head loss (HL_3):

The head loss in the 3″ mud gun feed line can be calculated by determining the equivalent 3″ pipe length (Table 3.12).

The equivalent length of 3″ pipe is 27.2 ft. The head loss from the 4″ tee to the mud gun nozzle is:

$$HL_{100} = \frac{Q_n^{1.9}}{2605}$$

$$HL_{3''} = \frac{Q_n^{1.9}}{2605} \frac{27.2 \text{ ft of pipe}}{100 \text{ ft of pipe}} = \frac{Q_n^{1.9}}{9577}$$

Table 3.12 Equivalent 3″ pipe length.

	Equivalent length, ft
8 ft of 3″ pipe	8.0
One 3″ butterfly valve	11.5
One 3″ elbow	7.67
Total	**27.2**

Calculation for nozzle head loss (HL_n):
The equation for flow rate through the nozzle can be solved for HL_n.

$$\text{Flow Rate, gpm} = 19.6 \left(\sqrt{HL_n, \text{ ft}}\right)(\text{nozzle diameter, inches})^2$$

$$HL_n = \frac{Q_n^2}{[(19.6)(d^2)]^2} = \frac{Q_n^2}{121.6}$$

The head at the second 4″ tee is applied to the 3″ pipe and the nozzle.

$$98.53 \text{ ft} = \frac{Q_n^{1.9}}{9577} + \frac{Q_n^2}{121.6}$$

Again, the head losses can be calculated from the equation and compared to the known value of 98.53 ft at the 4″ tee for the second mud gun (Table 3.13).

The fluid flow through the #2 mud gun will be about 109 gpm.

Third mud gun

The fluid flowing between the #2 mud gun and the #3 mud gun in the 4″ line is 440−109 gpm or 331 gpm.

Calculation for 4″ pipe head loss (HL_4):
The head loss through the pipe from the second 4″ tee to the third 4″ tee can be calculated by determining the equivalent pipe length (Table 3.14).

Table 3.13 Comparing calculated head losses with known head.

Head at #2 tee, ft	Nozzle flow rate, gpm	3″ head loss, ft	Nozzle head loss, ft	Calculated head loss, ft	Compare calculated with know, ft
98.53	115	0.85	108.76	109.61	11.08
98.53	110	0.78	99.51	100.29	1.76
98.53	105	0.72	90.67	91.38	−7.15
98.53	109.5	0.77	98.60	99.38	0.85
98.53	109.3	0.77	98.24	99.02	0.49
98.53	109.1	0.77	97.88	98.65	0.12
98.53	109.02	0.77	97.74	98.51	−0.02

Table 3.14 Equivalent 4″ pipe length.

	Equivalent length, ft
10′ of 4″ pipe	10.0
One thru flow 4″ tee	6.71
Total	**16.7**

From Table 3.1, the head loss for 10 ft of the 4″ pipe section could be calculated from the equation:

$$HL_{100} = \frac{Q^{1.90}}{10,200} = \frac{(331)^{1.9}}{10,200} = 6.91 \text{ ft}$$

The equivalent length of 4″ pipe is 16.7 ft. The head loss between the 4″ tees is

$$\text{Head Loss} = \frac{6.01 \text{ ft head}}{100 \text{ ft pipe}}[16.7 \text{ ft pipe}] = 1.00 \text{ ft head}$$

The head at the second 4″ tee would be the head at the first 4″ tee minus the head loss to the second mud gun line, or 98.53 ft−1.09 ft or 97.44 ft.

Calculation for 3″ pipe head loss (HL₃):

The flow rate through the 3″ pipe to the nozzle is not known.

The head loss in the 3″ mud gun feed line can be calculated by determining the equivalent 3″ pipe length (Table 3.15).

The equivalent length of 3″ pipe is 27.2 ft. The head loss from the 4″ tee to the mud gun nozzle is:

$$HL_{100} = \frac{Q_n^{1.9}}{2605}$$

$$HL_{3''} = \frac{Q_n^{1.9}}{2605} \frac{27.2 \text{ ft of pipe}}{100 \text{ ft of pipe}} = \frac{Q_n^{1.9}}{9577}$$

Calculation for nozzle head loss (HL$_n$):

The equation for flow rate through the nozzle can be solved for HL$_n$.

Table 3.15 Equivalent 3″ pipe length.

	Equivalent length, ft
8 ft of 3″ pipe	8.0
One 3″ butterfly valve	11.5
One 3″ elbow	7.67
Total	**27.2**

Table 3.16 Comparing calculated head losses with known head.

Head at #2 tee, ft	Nozzle flow rate, gpm	3" head loss, ft	Nozzle head loss, ft	Calculated head loss, ft	Compare calculated with known ft
97.5	110	0.80	99.51	100.30	2.80
97.5	105	0.73	90.67	91.40	−6.10
97.5	109	0.78	97.71	98.49	0.99
97.5	108	0.77	95.92	96.69	−0.81
97.5	108.5	0.78	96.81	97.59	0.09
97.5	108.3	0.78	96.45	97.23	−0.27
97.5	108.4	0.78	96.63	97.41	−0.09
97.5	108.45	0.78	96.72	97.50	0.00

$$\text{Flow Rate, gpm} = 19.6 \left(\sqrt{HL_n, \text{ ft}}\right)(\text{nozzle diameter, inches})^2$$

$$HL_n = \frac{Q_n^2}{[(19.6)(d^2)]^2} = \frac{Q_n^2}{121.6}$$

The head at the second 4" tee is applied to the 3" pipe and the nozzle (Table 3.16).

$$97.44 \text{ ft} = \frac{Q_n^{1.9}}{9577} + \frac{Q_n^2}{121.6}$$

The flow rate through the #3 nozzle is about 108.4 gpm.

Fourth mud gun

The fluid flowing between the #3 mud gun and the #4 mud gun in the 4" line is 331 gpm−108.4 gpm or 222.6 gpm.

Calculation for 4" pipe head loss (HL₄):

The head loss through the pipe from the third 4" tee to the fourth 4" tee can be calculated by determining the equivalent pipe length (Table 3.17).

From Table 3.1, the head loss for 10 ft of the 4" pipe section could be calculated from the equation:

$$HL_{100} = \frac{Q^{1.90}}{10,200} = \frac{(222.6)^{1.9}}{10,200} = 2.82 \text{ ft}$$

The equivalent length of 4" pipe is 16.7 ft. The head loss between the 4" tees is:

$$\text{Head Loss} = \frac{2.82 \text{ ft head}}{100 \text{ ft pipe}}[16.7 \text{ ft pipe}] = 0.4725 \text{ ft head}$$

Table 3.17 Equivalent 4″ pipe length.

	Equivalent length, ft
10′ of 4″ pipe	10.0
One thru flow 4″ tee	6.71
Total	**16.7**

The head at the fourth 4″ tee would be the head at the third 4″ tee minus the head loss to the fourth mud gun line, or 97.50−0.473 ft or 97.03 ft.

Calculation 3″ pipe head loss (HL₃):

The flow rate through the 3″ pipe to the nozzle is not known.

The head loss in the 3″ mud gun feed line can be calculated by determining the equivalent 3″ pipe length (Table 3.18).

The equivalent length of 3″ pipe is 27.2 ft. The head loss from the 4″ tee to the mud gun nozzle is:

$$HL_{100} = \frac{Q_n^{1.9}}{2605}$$

$$HL_{3″} = \frac{Q_n^{1.9}}{2605} \frac{27.2 \text{ ft of pipe}}{100 \text{ ft of pipe}} = \frac{Q_n^{1.9}}{9775}$$

Calculation for nozzle head loss (HL$_n$):

The equation for flow rate through the nozzle can be solved for HL$_n$.

$$\text{Flow Rate, gpm} = 19.6 \left(\sqrt{HL_n, \text{ ft}}\right)(\text{nozzle diameter, inches})^2$$

$$HL_n = \frac{Q_n^2}{[(19.6)(d^2)]^2} = \frac{Q_n^2}{121.6}$$

The head at the second 4″ tee is applied to the 3″ pipe and the nozzle (Table 3.19).

$$97.03 \text{ ft} = \frac{Q_n^{1.9}}{9775} + \frac{Q_n^2}{121.6}$$

Table 3.18 Equivalent 3″ pipe length.

	Equivalent length, ft
8 ft of 3″ pipe	8.0
One 3″ butterfly valve	11.5
One 3″ elbow	7.67
Total	**27.2**

Table 3.19 Comparing calculated head losses with known pump head.

Head at #4 mud gun, ft	Nozzle flow rate, gpm	3" head loss, ft	Nozzle head loss, ft	Calculate head loss, ft	Compare calculated with known, ft difference
97.03	107	0.73	94.15	94.89	−2.14
97.03	108	0.75	95.92	96.67	−0.36
97.03	109	0.76	97.71	98.47	1.44
97.03	108.5	0.75	96.81	97.56	0.53
97.03	108.1	0.75	96.10	96.85	−0.18
97.03	108.2	0.75	96.28	97.03	0.00
97.03	108.15	0.75	96.19	96.94	−0.09
97.03	108.195	0.75	96.27	97.02	−0.01

Table 3.20 Equivalent 4″ pipe length.

	Equivalent length, ft
10′ of 4″ pipe	10.0
One thru flow 4″ tee	6.71
Total	**16.7**

The flow rate through nozzle number four is 108.2 gpm.

Fifth mud gun

The fluid flowing between the #4 mud gun and the #5 mud gun in the 4″ line is 222.6−108.2 gpm or 114.4 gpm.

Calculation for 4″ pipe head loss (HL₄):

The head loss through the pipe from the fourth 4″ tee to the fifth 4″ tee can be calculated by determining the equivalent pipe length (Table 3.20).

From Table 3.1, the head loss for 10 ft of the 4″ pipe section could be calculated from the equation:

$$HL_{100} = \frac{Q^{1.90}}{10,200} = \frac{(114.4)^{1.9}}{10,200} = 0.7987 \text{ ft}$$

The equivalent length of 4″ pipe is 16.7 ft. The head loss between the 4″ tees is:

$$\text{Head Loss} = \frac{0.7987 \text{ ft head}}{100 \text{ ft pipe}}[16.7 \text{ ft pipe}] = 0.133 \text{ ft head}$$

The head at the fourth 4″ tee would be the head at the third 4″ tee minus the head loss to the fourth mud gun line is 97.03−0.13 ft or 96.9 ft.

Calculation for 3″ pipe head loss (HL₃):

The flow rate through the 3″ pipe to the nozzle is 124.34 gpm because this is the last nozzle in the system. This will also be the flow rate (Table 3.21).

The equivalent length of 3″ pipe is 27.2 ft. The head loss from the 4″ tee to the mud gun nozzle is:

Table 3.21 Equivalent 3″ pipe length.

	Equivalent length, ft
8 ft of 3″ pipe	8.0
One 3″ butterfly valve	11.5
One 3″ elbow	7.67
Total	**27.2**

$$HL_{100} = \frac{Q_n^{1.9}}{2605}$$

$$HL_{3''} = \frac{Q_n^{1.9}}{2605} \frac{27.2 \text{ ft of pipe}}{100 \text{ ft of pipe}} = \frac{Q_n^{1.9}}{9775}$$

Calculation for nozzle head loss (HL_n):
The equation for flow rate through the nozzle can be solved for HL_n.

$$\text{Flow Rate, gpm} = 19.6 \left(\sqrt{HL_n, \text{ ft}}\right)(\text{nozzle diameter, inches})^2$$

$$HL_n = \frac{Q_n^2}{[(19.6)(d^2)]^2} = \frac{Q_n^2}{121.6}$$

The head at the fifth 4″ tee is applied to the 3″ pipe and the nozzle (Table 3.22).

$$96.9 \text{ ft} = \frac{Q_n^{1.9}}{9775} + \frac{Q_n^2}{121.6}$$

The flow rate through the number 5 nozzle would be 108.1 gpm.

The initial estimated flow rate was 550 gpm from the pump. Iteration of the various nozzle flow rates calculated 543.8 gpm for the actual flow rate (Table 3.23).

Since the objective was to have each nozzle contribute between 100 gpm and 110 gpm to each compartment, the $3/4''$ nozzles seem to satisfy that need.

However, the calculated sum of the nozzle flow rates (544 gpm) does not match the presumed 550 gpm which was used for this calculation. The difference of 6 gpm means that each nozzle flow rate was somewhat higher than the calculated flow rate in the first iteration. The presumed flow rate of 550 gpm could be increased very slightly to make the head loss calculations agree with the head generated by the pump. However, the head reading from the pump chart is not very accurate. The head of 110 ft reading from the pump could have been in error by two or three feet. The purpose of the calculation was ensuring blending of the fluid in these compartments. No criteria have been established to guide the minimum flow needed to do that. A value of 100 gpm will quite sufficiently blend the compartments when used with the agitators in each 100 bbl volume of these tanks.

Note: The calculations are quite sensitive to small changes in either the head applied by the pump or the flow rate assumed for the iteration process. If enough time is spent with the iteration process, the equations can be balanced although this is not needed from a practical point of view. The variations of flow rate and heads are such that only about two significant figures represent the accuracy of the input data. The variation of 10 gpm is beyond the significant figure accuracy of the input data. However, for the sake of completeness, the following observations can be made to validate the procedure.

Table 3.22 Comparing calculated head losses with known pump head.

Head at #4 mud gun, ft	Nozzle flow rate; gpm	3″ head loss, ft	Nozzle head loss, ft	Calculate head loss, ft	Compare calculated with known, ft difference
96.9	107	0.76	94.15	94.91	−1.99
96.9	108	0.77	95.92	96.69	−0.21
96.9	109	0.78	97.71	98.49	1.59
96.9	108.5	0.78	96.81	97.59	0.69
96.9	108.1	0.77	96.10	96.87	−0.03
96.9	108.2	0.77	96.28	97.05	0.15
96.9	108.15	0.77	96.19	96.96	0.06
96.9	108.12	0.77	96.13	94.91	−1.99

Table 3.23 Summary of flow rates calculated for the five nozzles.

Nozzle number	Flow rate estimated, gpm
1	110
2	109
3	108.4
4	108.2
5	108.1
All nozzles	**543.7**

If the flow rate is assumed to be 540 gpm, the flow at the first nozzle will be 110.2 gpm. Five times this flow is 551 gpm or a difference of +11 gpm.

If the flow rate is assumed to be 560 gpm, the flow at the first nozzle will be 109.6 gpm. Five times this flow is 548 gpm or a difference of −12 gpm.

If the flow rate is assumed to be 550.5 gpm, the flow at the first nozzle will be 109.97 gpm. Five times this flow is 549.9 gpm or a difference of 0.1 gpm.

Proper application of mud guns
Removal section

Mud guns are not recommended in the removal section except for the last compartment. Agitation occurs because of impellers mounted on agitators. If agitators are not available on a particular rig, each centrifugal pump can be sized to stir its own suction tank. Fluid should not be pumped from compartments downstream through mud guns. If this is done, the solids removal equipment capacities must be greatly increased. This is discussed more thoroughly in the tank arrangement chapter.

If mud guns are supplied with fluid from the suction tank and are used to stir the desilter suction tank, the flow rate of the mud guns must be added to the desilter's processing rate. Flow rates through mud guns can be approximated with the equation {derived in the appendix to this chapter}.

$$\text{Flow Rate, gpm} = 19.6 \left(\sqrt{\text{head, ft}}\right)(\text{nozzle diameter, inches})^2$$

If the head supplied by the centrifugal pump is 81 ft, and a single nozzle diameter is one inch, each mud gun nozzle would supply about 176 gpm to the compartment. If mud guns are used in the removal section an increase in equipment will be needed to handle all of the flow. Agitation is absolutely needed, however, in the desilter's suction tank. Some rigs label their removal tanks as "settling pits" and do not provide agitation. THIS IS WRONG. The fluid supplied to the desilter cones should be uniform and homogeneous. This is not possible without agitation. Slugs of fluid with too many solids will plug a desilter apex and the cone will cease to remove drilled solids.

The section after the removal section should be the place where additions are made to the system. As solids and accompanying liquids are removed from the system with the solids removal equipment, additional drilling fluid must be added to maintain a constant pit level. A constant pit level in the removal section makes it easier to detect a kick by observing the change in volume to the addition and suctions sections.

Suction section

One of the major functions of the suction section is to contain enough uniform, blended, homogeneous drilling fluid so that well control measurements are always possible. After a kick is detected and the BOP closed, the drill pipe pressure reveals the amount of under-balanced at the bottom of the hole—BUT ONLY IF THE FLUID IN THE DRILL-PIPE HAS THE SAME DENSITY FROM TOP TO BOTTOM. When performing a pressure integrity test (PIT) or a leak-off test (LOT), the fluid in the drill string must have the same density from top to bottom. Otherwise, it is not possible to calculate the pressure at the end of the drill string.

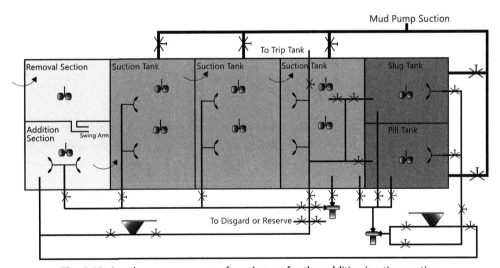

Fig. 3.19 Another arrangement of mud guns for the addition/suction section.

Appendix 3.A
Derivation of mud gun flow rate equation

The equation for flow through a mud gun in response to the head applied to the mud gun starts with the fundamental equation of the equality of energy.

$$\text{Potential Energy} = \text{Kinetic Energy}$$

Change to pressure because the correct definition of pressure is energy per unit volume:

$$\frac{\text{Potential Energy}}{\text{Volume}} = \frac{\text{Kinetic Energy}}{\text{Volume}} = \text{pressure}$$

$$\frac{mgh}{\text{Volume}} = \frac{\frac{1}{2}mv^2}{\text{Volume}}$$

Measurement of mass will be in terms of density (lb/gal) or weight per unit volume. To change mass (m) to weight (W), use Newton's Second Law of Motion:

$$W = ma = mg$$

$$\text{Pressure} = \frac{\frac{W}{g}gh}{\text{Volume}} = \frac{\frac{1}{2}\frac{W}{g}v^2}{\text{Volume}}$$

Weight per unit volume is called density or for drilling fluids it is called mud weight (MW).

$$\text{Pressure} = \frac{1}{2}\left(\frac{MW}{g}\right)(v^2)$$

To calculate the pressure in normal units of pounds per square inch, some unit conversion must be applied. For example, the left side of the equation has the units of (lb.ft./gal) instead of lb/in^2.

To change the units:

$$(MW, \text{lb/gal})(1\text{ gal}/231\text{ in}^3)(h, \text{ft})(12\text{ in/ft}) = .051948052(MW, \text{lb/gal})(h, \text{ft})$$

This should be a very familiar equation to people involved with well control and is usually written:

$$\text{Pressure, psi} = 0.052(MW, \text{lb/gal})(h, \text{ft})$$

Since this equation is to be used to calculate the flow rate (Q) through a mud gun nozzle, the velocity (v) must be changed to flow rate divided by area (A). The area will be calculated from the diameter (d) of the mud gun and will be

$$A = \frac{\pi}{4}d^2 = 0.7854d^2$$

Conversion of the units on the right side of the equation:

$$\frac{1}{2}\left(\frac{MW}{g}\right)\left(\frac{Q^2}{0.785\ d^2}\right)$$

The units in the equation must be modified to provide the pressure in pounds per square inch.

$$\frac{1}{2}\left(\frac{MW,\ \dfrac{lb}{gal}\left(\dfrac{gal}{231\ in^3}\right)}{g,\ \dfrac{ft}{s^2}\left(\dfrac{12\ in}{ft}\right)}\right)\left(\frac{\left(Q,\ \dfrac{gal}{min}\right)^2\left(\dfrac{231\ in}{gal}\right)^2\left(\dfrac{min}{60\ s}\right)^2}{(0.785\ d^2,\ in^2)^2}\right) = \frac{MWQ^2}{7429}$$

The equation:

$$MW\ (h) = \frac{1}{2}\left(\frac{MW}{g}\right)(v^2)$$

now becomes

$$0.052\left(MW,\ \frac{lb}{gal}\right)(h,\ ft) = \frac{\left(MW,\ \dfrac{lb}{gal}\right)\left(Q\dfrac{gal}{min}\right)^2}{7429}$$

Solving for the flow rate:

$$Q,\ gal/min = 19.6(d,\ in)^2\left(\sqrt{h,\ ft}\right)$$

CHAPTER 4

Agitation

Introduction

Drilling fluid has many functions to fulfill and requires careful attention to blending and homogenizing the fluid in the surface system. The suction tank must have a sufficient blended quantity of drilling fluid to maintain a uniform density of in the drill pipe in case of a kick. The pressure read at the surface when the blowout preventer is closed will not indicate the true pressure at the bottom of the hole unless the fluid in the drill pipe is homogeneous. When drilling fluid additives are being introduced to the system, the suction tank must be completely blended uniformly and well agitated. If additions are made of ready-to-use drilling fluid, these additions must also be blended uniformly. If non-aqueous drilling fluid (NADF) is being used, all ingredients must be made oil-wet to perform properly in the system. This requires rapid mixing of the additives into the system. The additions compartment must be agitated properly and quickly blended with the fluid in the suction section for both water-based and NADF. The removal section should be well agitated with mechanical stirrers to provide a homogeneous feed to the drilled solids removal equipment. Mud guns are not recommended in the removal section because they add additional fluid which must be treated by the drilled solids removal equipment.

There are several ways to determine if the mud in the tank is a homogenous mixture but this is typically not done on drilling rigs. One method of sampling liquid from different levels in a pit would be to mount a coffee can on a broom stick. The open end is pointed downward as the can is lowered into the pit. When it reaches the desired depth, it is capsized to fill with liquid. The simplest method would be to have a sample port (valve) between the charging pump and the mud pump and compare values of what is actually being pumped down the hole versus a sample from where it is typically taken. It is always good to know what is actually being pumped down the hole instead of making assumptions.

Many drilling rigs do not have their tanks agitated properly. The result is settled solids on the tank bottoms to some degree. Because of the high specific gravity of barite, a high percentage of the settled solids will be barite. This is money wasted. Additionally the actual working volume of the surface system is reduced so if a lost circulation event occurs for some reason there is less mud available to keep the hole filled.

Mixing and blending drilling fluid
Mechanical agitators

Drilling fluid agitators generally have similar components: a drive motor, a gear reducer, a gear output shaft, and an impeller. Some shafts may have two impellers mounted on them.

Motors

Usually, the drive motor is an explosion-proof electric motor (Fig. 4.1). These motors must meet specifications of the local codes and regulations.

Motors are available with horizontal or vertical mounts (Figs. 4.2 and 4.3). The motors generally have shafts that rotate much faster than the impellers need to rotate. A large gear box, or gear reducer, used to turn the shaft, is mounted at the end of the motor. The motor size depends upon the diameter of the impeller blades and the density of the slurry.

Impellers

Three different types of impellers are available for most mechanical agitators: flat blade, canted blade, and proprietary.

The flat blade impeller (Fig. 4.4), moves the fluid radially away from the blades. The blades are mounted vertically on the shaft and are in line with the shaft (Fig. 4.5). In radial flow, the fluid moves predominantly in a horizontal, circular pattern within the compartment. When the fluid meets the walls of the compartment, it moves upward and will maintain a homogenous slurry within the compartment. The blades are normally placed near the bottom of the compartment. If the blades are mounted higher in the tank, radial flow impellers create two zones of fluid movement: one above the blades and one below the blades. These two distinct zones of circulation create a boundary between the two zones and this decreases the ability of the agitator to blend the entire tank compartment.

Fig. 4.1 Horizontal drive for radial flow.

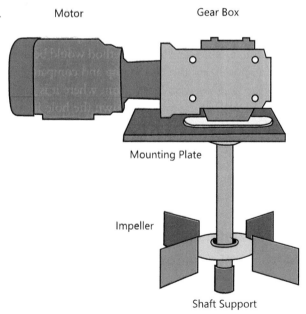

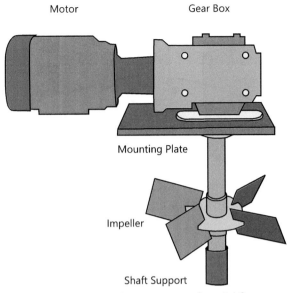

Fig. 4.2 Horizontal motor mount for axial flow.

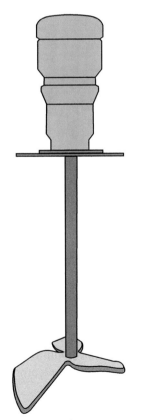

Fig. 4.3 Vertical motor mount.

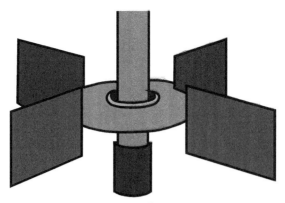

Fig. 4.4 Flat blade impellers create radial flow.

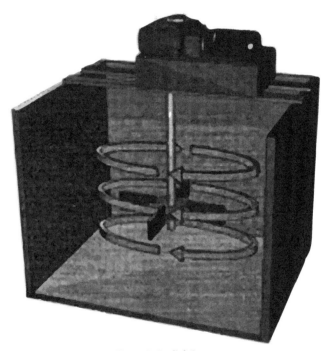

Fig. 4.5 Radial flow.

These impellers tend to produce vortexes in the slurry (Fig. 4.7). These vortex tend to blend air with the drilling fluid and may create problems with pumping. Centrifugal pumps may cease to pump if air collects in the center of the impeller. Mud pump efficiency may decrease significantly and not pump the expected flow rate of drilling fluid down hole. For this reason, baffles are highly recommended if these impellers are used.

The horizontal flow of fluid restricts the depth of the compartment which can be adequately stirred. Tank depths are normally limited to about 6 or 7 ft.

If drilling fluid overflows from the removal section into the downstream addition section, the liquid level in the removal section will remain constant. Usually the mud tanks are deep enough to prevent vortexing. The flow rate to the centrifugal pump determines the depth of fluid required. The required net positive suction head (NPSH$_R$) for a given pump can be determined from the manufacturer's pump curve illustrated below. Refer to Chapter 10 for determining the amount of net positive suction head is available (NPSH$_A$) based on the pump installation (Fig. 4.6).

The canted blade impeller (Fig. 4.8), tends to pump the drilling fluid downward toward the bottom of the compartment nook (Fig. 4.9). The blades are pitched at a 45—60 degree angle from the vertical shaft. The spinning motion causes some radial flow but the blades pull drilling fluid from the top of the compartment and force the fluid to strike the bottom of the compartment. The fluid then flows radially across the bottom of the tank and up the sides of the tank. These impellers should be placed about two-thirds to three-fourths of the blade diameter from the bottom of the compartment. Tanks that are deeper than 6 ft may require more than one blade on the shaft.

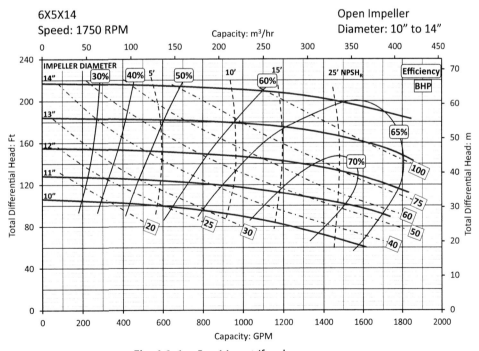

Fig. 4.6 6 × 5 × 14 centrifugal pump curve.

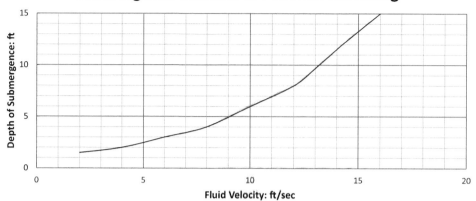

Submergence needed to prevent vortexing.

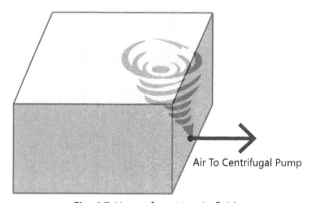

Fig. 4.7 Vortex formation in fluid.

Fig. 4.8 Canted blade impeller.

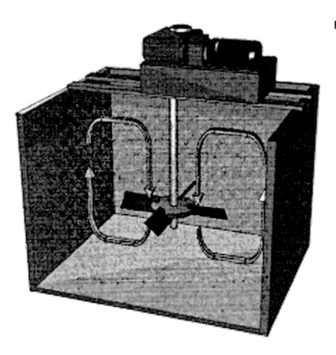

Fig. 4.9 Axial flow.

A simple way to improve the agitation on existing rigs with common oilfield type agitators would be to consider replacing a single large impeller with two smaller impellers (Fig. 4.10). For a minimal cost the impeller displacement could be improved by 39%. It could be higher with a better choice of canted blade design. For even better performance consider replace the existing impeller to one of the more efficient designs (or similar) discussed in this chapter and improve the performance by 70% or more.

Possible modifications to remember are:
1. If the shaft doesn't have a steady bearing the increased displacement may cause shaft deflection if the shaft is not stiff enough. To prevent deflection, a shaft stabilizer may be required.
2. Proper impeller placement is important. Improper placement of the upper impeller could cause vortexing which will induce shaft deflection and all the potential problems this could cause including bearing and gearbox failure.
3. This may not be viable option for shallow pits.

Proprietary blades

Some blades are now available which have variable pitch blades (Fig. 4.11). These blades promote both radial and axial flow patterns. The inclination and pitch of each blade determines whether more or less radial flow or axial flow will leave the blade. These impellers generally impart less shear force to the fluid than the single-plane blades.

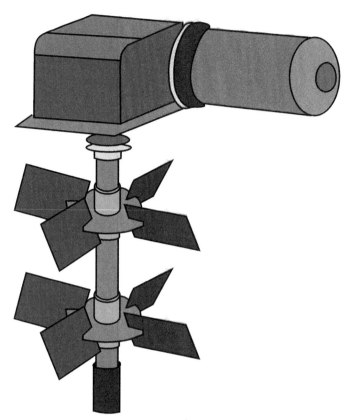

Fig. 4.10 One large impeller replaced with two smaller ones.

Fig. 4.11 Contour blades bolted to hub, mounted on hollow shaft.

Installation

The selection of motors and agitator blades depend upon the size of the tanks and the plumbing arrangement. The motors can be sized for specific mud weights but usually a contractor will plan on the drilling rig handling fluids up to very high densities so that the rig can meet unexpected demands. The goal of agitation is to keep all of the fluid blended, prevent settling, and eliminate dead spots in the compartment. The ability of the agitator to stir the entire compartment requires that the impeller be mounted correctly above the tank bottom. If agitator blades are located too far from the bottom of the compartment, the flow will not sweep the settling solids and return them back to the active system. Barite and other solids can settle if the agitators are not properly sized and positioned correctly. Settled solids in the mud tanks are difficult to clean out of the tanks when the rig is released. Barite settling also adds significantly to the drilling fluid cost. Settling in the suction tank also eliminates some of the volume available for properly treated drilling fluid.

In deep tanks where two or more impellers are mounted on one shaft, the shaft should be stabilized at the bottom of the compartment. A short piece of pipe can be welded onto the bottom of the tank. Drainage hole should be drilled into the short piece of pipe. The end of the shaft fits into this short piece of pipe and this prevents vibration of the shaft without interfering with the agitation. Vibration will cause excessive side loading of the bearings in the agitator and gear box. Stabilizing the bottom of the shaft is also beneficial if the agitators remain on the mud tanks when the rig is moved.

Axial flow impellers provide more flexibility in the choice of shaft lengths than with radial flow impellers. The axial flow impeller can be mounded about $^2/_3$ rds or $^3/_4$ ths of the impeller diameter above the bottom of the tank. In deep tanks, the shaft should also be stabilized as with the radial flow systems.

Since axial flow impellers are position closer to the drilling fluid surface in the tank, liquid levels must be maintained to prevent vortexes from forming. Excessive erosion of the tank bottom from bottom scouring can occur if the axial flow impellers are mounted too low in the tank. Mounting the axial flow impellers too low in the tank may also inhibit effective agitation.

The contour, or variable pitch, impellers require less horsepower than the axial or the more than one contact angle on the surface of the impeller. They can, therefore have larger impeller diameters and can be used in the very large compartments sometimes found in some drilling fluid systems. Because they impart more efficient movement to the fluid as it contacts the blade surface, less shear force is imparted to the fluid. Shear is desirable in the additions tank when new drilling fluid products are added to the fluid. In the removal section, however, high shear forces might deteriorate the drilled solids to sizes which cannot be removed with hydrocyclones. The contour impellers would satisfactorily keep the tanks stirred and all solids moving without shearing those solids. Slug tanks and pill tanks need to have good shearing forces to quickly blend ingredients into their fluid.

Placement of the impeller blades is very important. The lowest impeller should be close to the bottom of the tank to assist solids removal for those solids which settle in the tank. Each manufacturer has guidelines for proper sizing and installation of their agitators. Take advantage of their consultation.

Natural frequency determination

Consideration should be given when designing an agitation system with regards to its natural frequency. At times, the shaft strength can be overlooked causing structural damage to the shaft. Top entering agitators typically found in the oil and gas industry will oscillate at a natural frequency. If the rotating speed of the agitator is too close to its natural frequency, destructive oscillations may occur. A typical formula for calculating the first natural frequency, or critical speed of an agitator shaft takes into consideration the length, stiffness, rigidity, and weight of the entire shaft system:

$$N_c = \frac{37.8 d^2 \sqrt{E_y/\rho_m}}{L\sqrt{W_e}\sqrt{L+L_b}}$$

where
  N_c = critical speed, r/min.
  d = shaft diameter, in
  L = shaft extension, in
  W_e = equivalent weight (lb) of impellers and shaft at shaft extension
  L_b = spacing (in) of bearings that support shaft
  E_y = modulus of elasticity, lb/in^2
  ρ_m = density, lb/in^3

And for a two impeller situation (Fig. 4.12):

$$W_e = W_l + W_u \left(\frac{L_u}{L}\right)^3 + w_s \left(\frac{L}{4}\right)$$

where
  W_l = weight of impeller at the end of the shaft
  W_u = weight of upper impeller
  L_u = length from top of shaft to upper impeller
  w_s = weight of shaft
  L = overall length of shaft

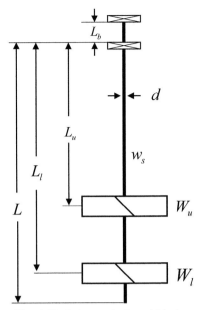

Fig. 4.12 Agitator shaft and blades.

Example

What is the natural frequency of a dual pitch blade impeller agitator constructed with the following:

Shaft Length: 125 in.
Shaft Diameter: 2.75 in.
Upper Impeller Location from top of shaft: 65 in.
Lower Impeller Location from top of shaft: 125 in.
Stainless Steel Shaft Weight: 1.724 lb/in = 215 lb
Shaft density: 0.29 lb/in^3.
Shaft weight: 0.228(d)2 lb/in.
Shaft modulus of elasticity: 28,000,000 lb/in^2.
Impeller/hub weight (ea): 62.5 lb
Spacing (in) of bearings that support shaft: 10.75 in.
Step 1: Determine the equivalent length of the impellers and shaft at shaft extension:

$$W_e = W_l + W_u \left(\frac{L_u}{L}\right)^3 + w_s \left(\frac{L}{4}\right)$$

$$= 62.5 + 62.5 \left(\frac{65}{125}\right)^3 + 0.228(2.75)^2 \left(\frac{125}{4}\right) = 125.17 \text{ lb}$$

Step 2: Determine the critical speed:

$$N_c = \frac{37.8\, d^2 \sqrt{E_y/\rho_m}}{L\sqrt{W_e}\sqrt{L+L_b}}$$

$$= \frac{37.8(2.75)^2 \sqrt{28{,}000{,}000/0.29}}{125\sqrt{125.17}\sqrt{125+10.75}} = 172\ \text{RPM}$$

For planning purposes, the actual rotational speed of the agitator shaft should be no more than 80% of the critical speed for stabilized shaft and no more than 65% of the critical speed for an un-stabilized shaft. In the above example, the maximum allowable rotational speed would be 138 RPM for a stabilized shaft and 112 RPM for an un-stabilized shaft.

Compartments

Agitators perform more efficient when they are placed in symmetrically sized round or square compartments. Round compartments are ideal because there is no place for the solids to settle. However, few contractors have round tanks — there are some available but, for some reason, most contractors have rectangular tanks.

Round tanks can be fitted with a center drain and the bottom tapered toward the drain. These tanks are easier to clean and require less wash fluid and time than rectangular or square tanks. Round tanks should be outfitted with 4 baffles located 90° from each other with a thickness of 1/12th of the tank diameter and offset from the tank wall as illustrated in Fig. 4.11.

Because the agitators impart a spinning motion to the drilling fluid, baffles should be installed around each agitator, in a square or rectangular compartment (Fig. 4.13). A typical baffle can be a plate of steel about $1/2$ to 1 in. thick and 12 in. wide and extends from the bottom of the tank to about a foot above the impeller blade. Four baffles should be installed around each agitator in a regular pattern at 90 degrees to each other with the same distance between each baffle. They can be about 6 in. from the tip of the blade oriented so they are in line with the agitator shaft. In a long rectangular tank, the tank should be divided into imaginary squares and the baffles pointed toward the agitator shaft. This prevents the drilling fluid from swirling in a manner that creates a vortex. A vortex will pull air into the drilling fluid. Air will create many problems, including corrosion of steel components and causing centrifugal pumps to cease pumping fluid. It will also greatly decrease the volumetric mud pump efficiency.

Sizing agitators

Regardless of what style of agitator or impeller is used, proper sizing of components is critical. Once compartment size has been determined, the impeller diameter and corresponding horsepower requirements must be calculated. If the maximum mud weight to

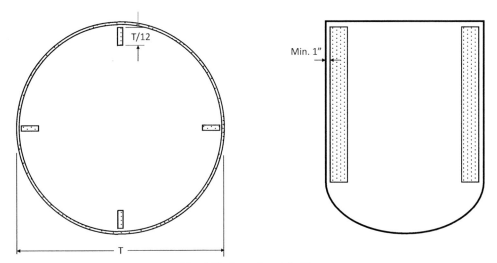

Fig. 4.13 Round tank baffles.

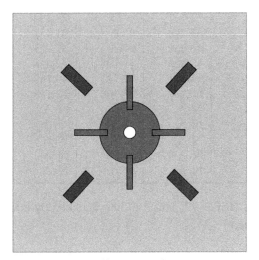

Fig. 4.14 Baffles: Pointed at corners.

be used with the rig is not know, it is best to base all calculations on 20 lb/gal fluid (2.4 specific gravity [SG]). This will give a sufficient safety factor to allow agitation of most any fluid without fear of overloading the motor. Most oilfield agitators range in shaft speed from 50 to 90 rpm.

Turnover rate

Impeller sizes are determined by calculating the TOR (sometimes called time of rollover) for each compartment. This is the time, in seconds, required to completely move the

Table 4.1 Typical turnover rate values in seconds.

Impeller type	Removal	Addition	Suction	Reserve	Pill
Canted/flat	50–75	50–75	65–85	50–80	40–65

Table 4.2 60-Hz impeller displacement values.

Diameter		Flat		Canted		Contour	
in	mm	gpm	lpm	gpm	lpm	gpm	lpm
20	508	1717	6500	1487	5629	N/A	N/A
24	610	2759	10,444	2389	9043	N/A	N/A
28	711	3407	12,897	2934	11,106	5861	22,185
30	762	4369	16,538	3670	13,892	N/A	N/A
32	813	5368	20,320	4636	17,549	N/A	N/A
34	864	6467	24,480	5600	21,198	8790	33,270
36	914	6839	25,888	6672	25,256	9180	34,746
38	965	7971	30,174	6903	26,131	10,604	40,136
40	1016	8376	31,707	7473	28,288	N/A	N/A
42	1067	9589	36,298	8305	31,438	13,940	52,762
44	1118	11,475	43,438	9966	37,725	N/A	N/A
46	1168	12,820	48,529	11,102	42,026	N/A	N/A
48	1219	14,691	55,611	12,723	48,162	20,020	76,776
50	1270	15,897	60,177	13,768	52,118	N/A	N/A
52	1321	N/A	N/A	N/A	N/A	24,852	94,063
54	1372	N/A	N/A	N/A	N/A	27,602	104,475
56	1422	N/A	N/A	N/A	N/A	30,353	114,887
60	1524	N/A	N/A	N/A	N/A	36,567	138,404
64	1626	N/A	N/A	N/A	N/A	43,533	164,771

fluid in a compartment (Table 4.1), and can be calculated by knowing the tank volume and impeller displacement:

$$\text{TOR} = \frac{V_t}{D} \times 60$$

where

V_t = tank volume, in gallons or liters

D = impeller displacement, in gpm or l pm (as displayed in Table 4.2).

For flat and canted impeller applications, TOR should range between 40 and 85 s. As the TOR approaches 40 s, the chance for vortex formation and possible air entrainment increases. At values greater than 85 s, proper suspension may be jeopardized and solids will begin to settle.

For contour impeller applications, values must be significantly faster (i.e., smaller numbers) to achieve the same results, but because of the impeller design, air entrainment is less probable. In symmetrical compartments, the fluid has a nearly equal distance to travel from the center of the impeller shaft or form the impeller blade tip before it contacts the vessel wall. Agitators should be placed where the shaft is centered in the tank or compartment.

When defining the area in which to mix, it is best to work with symmetrical shapes like squares or circles (as viewed in a plan drawing or overhead view of the tank layout). Rectangular tanks should be converted to nearly square compartments if possible. Maximum fluid working volumes in compartments should be higher than 1 foot (about 3/10 m) from the top of the tank. This will allow for a little extra capacity in emergencies, slightly out of level installations, and/or fluid movement on floating rigs.

Working volume for square or rectangular tanks is calculated by knowing dimensional values for length (L), width (W), and height (H; in feet for gallons, in meters for liters):

For gallons:
$$V_t = L \times W(H-1) \times 7.481$$

For liters:
$$V_t = L \times W(H-0.3.) \times 1000$$

The working volume for round tanks with flat bottoms is:
For gallons:
$$V_t = \pi r^2 (H-1) \times 7.481$$

For liters:
$$V_t = \pi r^2 (H-0.3) \times 1000$$

For round tanks with dish or cone bottoms, calculations for working fluid volume are based on straight wall height (i.e., this height is measured from the tank top to where the tank joins the cone or dish at the bottom). This leaves adequate free space above the maximum fluid operating level. In all cases, if $H < 5$ ft (approx. 1.5 m), a radial flow impeller should be specified.

Example

A compartment is 30 ft long, 10 ft wide, and 10 ft high (Fig. 4.14). Maximum mud weight is anticipated to be 16 lb/gal (1.92 SG). If the maximum mud weight is not known, use 20 lb/gal fluid density (2.4 SG).

All compartments will be solids-removal sections. Convert the compartment to symmetrical shapes. In this case, three compartments 10 × 10 ft (approx. 3 × 3 m) (Fig. 4.15). Determine the volume for one compartment.

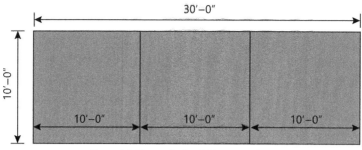

Fig. 4.15 Example for sizing exercise.

$$V_t = 10 \times 10 \times (10 - 1) \times 7.48$$

$$V_t = 900 \times 7.48$$

$$V_t = 6732 \text{ gal}$$

Since the tank is deeper than 6 ft, flat (turbine) impellers cannot be used; therefore, canted impellers are chosen. Locate the appropriate impeller diameter from the impeller displacement table D (Table 4.2) so that TOR is within the recommended range. As a rule, choose an initial impeller displacement value so that D is close to V_t increase the accuracy of the selection. In this case, the gpm value for 60 Hz service is close to the 38 in. impeller (6343 gal).

$$\text{TOR} = \left[\frac{V_t}{D}\right] \times 60$$

$$\text{TOR} = \left(\frac{6732}{6343}\right) \times 60$$

$$\text{TOR} = 1.06 \times 60, \text{ or about } 64 \text{ s}$$

Compare this TOR to values in Table 4.1 and determine suitability or compartment purpose. In this example, the TOR is sufficient for solids-removal compartments. If a lower numeric value to cause faster fluid movement is desired, then choose a larger impeller. If less movement is desired, choose a smaller impeller and recalculate until the appropriate values from Table 4.1 are achieved.

After determining which impeller will produce the effective TOR, locate the size in the Impeller Diameter columns in Table 4.2 In this case we anticipate using a 38 in. impeller and 16 ppg fluid. Since there is no 38 in. value in the 16 ppg column in Table 4.2, we must round up to the next highest value, that is, 40 in. This allows for a safety factor should the mud density increase slightly. Follow the 40 in. impeller value horizontally to the left in Table 4.2 and determine the horsepower needed for that application. In this case, a 10 hp motor is sufficient. Therefore, three 10 hp agitators with 38 in.canted impellers are suitable for the tank in this application.

It should be noted that this system is for one brand of agitator. And the values in the tables may not apply to all brands. Deep tank designs require that multiple blades be mounted on a single shaft. This involves more thorough calculations than those shown. Because of this, it is highly recommended that the manufacturer or supplier be consulted before final design or placing of an order for any application.

Mud guns

Mud guns were discussed extensively in Chapter 3. They are used in the addition and suction sections to assist blending these compartments to create a homogeneous fluid. A sketch of the addition and suction sections (Fig. 4.16), shows mud guns mounted in all of the compartments even though agitators are being used to prevent solids from settling and make each compartment well-mixed. The centrifugal pump serving the mud gun plumbing takes suction from the end compartment and divides it among all of the other compartments. This continuous splitting and blending does a good job of making the drilling fluid in each compartment have the same fluid properties.

Mud guns will also allow the addition and suction tanks to be well blended (Fig. 4.16).

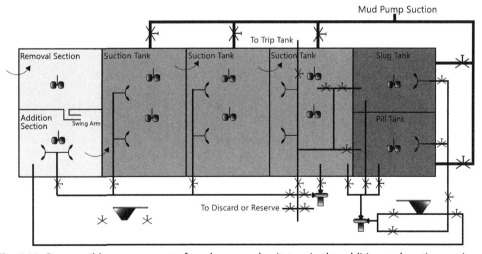

Fig. 4.16 One possible arrangement of mud guns and agitators in the addition and suction sections.

CHAPTER 5

Drilled solids calculations

The secret to drilling the cheapest wells safely is to remove drilled solids from the drilling fluid.

Introduction

Solids control actually starts at the bit. Drilled solids are easier to remove when they are large. Cuttings made by the bit should be removed before they are reground by the next set of bit teeth. This means that the nozzle selection and flow rate should be based on achieving the maximum hydraulic force or the maximum hydraulic power of the fluid striking the bottom of the hole.

Cuttings need to be transported up the annulus without tumbling and grinding into smaller solids. They should arrive at the shaker with sharp edges. This is discussed in the Chapter 6.

After the drill bit is equipped with the correct nozzles to provide the correct hydraulics, the founder point of the drill bit must be determined. This is discussed in depth in Chapter 14. The drilling fluid must have the lowest possible plastic viscosity to achieve the highest possible founder point (and drilling rate).

With the rig properly equipped with the correct drilling fluid processing equipment and the proper procedures implemented to achieve the highest founder point, a goal of 2%—4% volume drilled solids can be achieved economically. This should be the goal of the first well drilled in a new basin (or a wildcat well). Each basin has certain unique aspects that require a learning process to determine. After the first two or three wells, the economics of such a clean drilling fluid can be evaluated and the specifications for the maximum permissible drilled solids level determined with better precision. However, the economics should be based on both the visible and the invisible NPT.

The focus of evaluating solids control equipment frequently (and erroneously) turns to decreasing the cost of the drilling fluid. When a low value like 2%—4% volume low gravity solids is designated and the solids removal equipment is not functioning properly and/or the plumbing (or tank arrangement) is incorrect, these low values of drilled solids concentration can only be accomplished by dilution. That is a very expensive method to control drilled solids. When the low gravity solids, or drilled solids concentration increase because of the inability to remove drilled solids, the only method available is dilution.

This means removing dirty drilling fluid and replacing with clean drilling fluid. Dilution can be a large expense both from the new drilling fluid cost and the waste disposal cost.

Allowing the target concentration of low gravity solids to increase will decrease the cost of dilution and appear to be the cheapest solution. If the only fluid added to the mud pits is the volume necessary to keep the liquid levels constant in the pits, the drilled solids will build until it reaches an equilibrium. For example (as discussed in the following chapters), with a poor solids removal efficiency of about 60% and a solids concentration in the discard stream of 35% vol, the equilibrium drilled solids concentration will be 12% vol. This will decrease the cost of new drilling fluid additions to the active system. Unfortunately, this does not address the actual cost of the well. The consequences of too many drilled solids in the drilling fluid can be very expensive. Trouble costs will escalate.

For the lowest well cost, the question arises "What determines the target value of the low gravity solids to be recommended in a drilling program?" This question cannot be answered for all situations. The entire drilling process must be examined for the safe, lowest cost, trouble-free drilling. The drilling fluid properties play an important role and drilled solids removal is an important part of the process. The "perfect" solution would be to drill with only clean drilling fluid and discard the fluid coming from the hole. This would be too expensive. Economics dictate that as much drilling fluid be retained as possible but this should not be the primary economic evaluator. An economic balance must exist between the discarded solids and the recovered fluid. A low concentration of drilled solids has many advantages for good drilling practices. Retaining too many drilled solids can decrease the drilling fluid cost but the cost of the well may increase significantly.

The difficult question to answer is: "Exactly how many drilled solids is too many?" The answer is somewhat dependent upon the drilling location and the total cost of the well. Two types of problems can add to the cost of excess drilled solids (or trouble costs): visible and invisible non-productive time (NPT) as discussed in Chapter 15.

Now, the question is: "What level of drilled solids concentration should be specified?" A 2% to 4% volume drilled solids concentration is a great goal but achieving this goal requires many steps. Some drilling rigs are erroneously plumbed so that any centrifugal pump can pump from any tank to any other tank. This almost guarantees that the wrong valves will be opened (or leaking) and the low level of drilled solids cannot be achieved economically. Another common problem is irrational plumbing that fails to process the drilling fluid properly. These problems and solutions are well discussed in Chapter 15.

Another misconception is prevalent when rig designers erroneously think that each type of drilling fluid needs a special set of plumbing in the removal section. Water based drilling fluid (dispersed, non-dispersed, saline, polymer, or any other type) and non-aqueous drilling fluid (diesel oil, mineral oil, or synthetic) are processed the same way with the same equipment and with the same plumbing. Without solving these problems, dilution will be the only way to achieve a low level of drilled solids. This is a necessary condition but not a totally sufficient condition for creating a "clean" drilling fluid.

The importance of removal of drilled solids from the drilling fluid is emphasized throughout this book. The quantity of drilled solids remaining in the drilling fluid is determined with retorts. A known quantity of fluid is heated and the volume of liquid recovered is measured. Subtraction of the volume of liquid recovered from the known volume of drilling fluid heated provides a measure of the volume of soluble and insoluble solids in the drilling fluid.

In this chapter, retorts are discussed in depth. The information from the retorts allows the calculation of the low gravity solids concentration in the drilling fluid.

Calculation of low gravity solids from retort data for a fresh-water drilling fluid

The equation used to calculate the quantity of low gravity solids (LGS) can be derived by considering the masses of the constituents. The mass of the drilling fluid is equal to the mass of each of the components of the fluid (e.g. water, barite, and LGS).

$$\text{Mass mud} = \text{Mass water} + \text{Mass LGS} + \text{Mass barite}$$

$$\rho_{mud} \frac{\text{Volume}}{\text{Total vol}} = \frac{\rho_{water}\, \text{Vol water} + \rho_{LGS}\, \text{Vol}_{LGS} + \rho_B \text{Vol}_{barite}}{\text{Total vol}} \quad (5.1)$$

where:
ρ_{mud} is the density of the drilling fluid, in g/cc or specific gravity
ρ_w is the density of water, in g/cc or specific gravity
ρ_{LGS} is the density of the low gravity solids, in g/cc or specific gravity
ρ_B is the density of barite, in g/cc or specific gravity

$$\rho_{mud}(1.0) = \rho_{water}(\text{Vol fraction water})$$
$$+ \rho_{LGS}(\text{Vol fraction LGS}) \quad (5.2)$$
$$+ \rho_B(\text{Vol fraction barite})$$

The volume fraction of the water (v_w) plus the volume fraction of solids (v_s) is equal to 1.0.

$$v_w + v_s = 1 \quad (5.3)$$

The volume fraction of solids is equal to the sum of the volume fraction of each component.

$$v_s = v_{LGS} + v_B \quad (5.4)$$

where v_{LGS} is the volume fraction of low gravity solids and v_B is the volume fraction of barite.

Solve Eq. (5.3) for the volume fraction of water v_w and Eq. (5.4) for the volume fraction of barite, v_B. Substitute these values into Eq. (5.2).

$$\rho_{mud} = \rho_w(1 - v_s) + \rho_{LGS}v_{LGS} + \rho_B(v_s + v_{LGS}) \tag{5.5}$$

Combine terms:

$$\rho_{mud} = \rho_w(\rho_B - \rho_w)v_s + (\rho_{LGS} - \rho_B)v_{LGS} \tag{5.6}$$

Solve Eq. (5.6) for the volume fraction of low gravity solids.

$$(\rho_B - \rho_{LGS})v_{LGS} = \rho_w + (\rho_B - \rho_w)v_s - \rho_{mud} \tag{5.7}$$

To convert the density of drilling fluid, ρ_{mud}, from mud weight (MW) in pounds per gallon **to specific gravity (SG) or g/cc:**

$$\rho_{mud} = \frac{MW}{8.34} \tag{5.8}$$

To change the volume fraction (v) to volume percent (V), multiply all terms in Eq. (5.7) by 100.

$$(\rho_{LGS} - \rho_B)V_{LGS} = 100\rho_w = (\rho_B - \rho_w)V_s - \frac{100\,MW}{8.34} \tag{5.9}$$

$$(\rho_{LGS} - \rho_B)V_{LGS} = 100\rho_w + (\rho_B - \rho_w)V_s - 12\,MW \tag{5.10}$$

Discussion

For accurate determination of the low-gravity solids in a drilling fluid, drilled solids density must be known. As an example, calculate the low gravity solids in an 11 ppg water-based drilling fluid that has 15% volume total solids. The equation for calculating this value is:

$$(\rho_B - \rho_{LGS})V_{LGS} = 100\,\rho_w + (\rho_B - \rho_w)V_s - 12\,MW$$

This equation requires the density of the various components to be known. For a fresh-water liquid phase, the density of water is 1.0. The density of barite and other weighting agents commonly used is presented in Table 5.1.

Table 5.1 Weighting agents.

Barium sulfate	4.50
API barite	4.2 or 4.1
Galena	7.50
Hematite	5.26
API hematite	5.05

Table 5.2 Density of grains of common mineral/rocks.

Mineral	Density, g/cc
Anhydrite	2.96
Calcite	2.71
Chlorite	2.80
Dolomite	2.85
Gypsum	2.31
Halite	2.16
Kaolin	2.59
Limestone	2.1–2.7
Illite	2.66
Montmorillonite	2.1–2.6
Quartz (sand)	2.65
Smectite clay	2.1–2.6

The LGS densities have a much wider range, Table 5.2.

The weighting agent is normally barite, which has a density of 4.2, and the liquid phase is water with a density of 1.0. The equations in Table 5.3 are developed by inserting various values of the low gravity solids.

For the 11.0 ppg water-based drilling fluid with 15% volume total solids, the low gravity solids would range from 8% volume to 13.4% volume for low gravity densities ranging from 2.2 to 3.0 (Table 5.4).

With this large range of error possible, the low gravity solids in a drilling fluid should be determined to accurately calculate the drilled solids concentration in the drilling fluid. A method is presented in this chapter to permit this determination at a well site.

Table 5.3 Low-gravity solids (LGS) equations for LGS densities ranging from 2.2 to 3.0 g/cc.

$$(\rho_B - \rho_{LGS}) V_{LGS} = 100\, \rho_w + (\rho_B - \rho_w) V_s - 12\, MW$$

Density of LGS, g/cc	Equation
2.2	$V_{LGS} = 50.0 + 1.60\, V_s - 6.00\, MW$
2.3	$V_{LGS} = 52.6 + 1.68\, V_s - 6.34\, MW$
2.4	$V_{LGS} = 55.6 + 1.78\, V_s - 6.67\, MW$
2.5	$V_{LGS} = 58.8 + 1.88\, V_s - 7.06\, MW$
2.6	$V_{LGS} = 62.5 + 2.00\, V_s - 7.50\, MW$
2.7	$V_{LGS} = 66.7 + 2.13\, V_s - 8.00\, MW$
2.8	$V_{LGS} = 71.4 + 2.29\, V_s - 8.57\, MW$
2.9	$V_{LGS} = 76.9 + 2.46\, V_s - 9.23\, MW$
3.0	$V_{LGS} = 83.3 + 2.67\, V_s - 10.00\, MW$

Table 5.4 Volume percent LGS.

LGS density, g/cc	Equation	V_{LGS}, % volume
2.2	$V_{LGS} = 50.0 + 1.60\ V_s - 6.00\ MW$	8.0
2.3	$V_{LGS} = 52.6 + 1.68\ V_s - 6.30\ MW$	8.5
2.4	$V_{LGS} = 55.6 + 1.78\ V_s - 6.67\ MW$	8.9
2.5	$V_{LGS} = 58.8 + 1.88\ V_s - 7.06\ MW$	9.3
2.6	$V_{LGS} = 62.5 + 2.00\ V_s - 7.50\ MW$	10.0
2.7	$V_{LGS} = 66.7 + 2.13\ V_s - 8.00\ MW$	10.7
2.8	$V_{LGS} = 71.4 + 2.29\ V_s - 8.57\ MW$	11.5
2.9	$V_{LGS} = 78.9 + 2.46\ V_s - 9.23\ MW$	12.3
3.0	$V_{LGS} = 83.3 + 2.67\ V_s - 10.00\ MW$	13.4

If a low value of drilled solids is the goal of the solids removal system, the 15% volume of low gravity solids is too high regardless of the density of LGS selected. The problem may be caused by:

- Fluid by-passing the shale shaker screen—(are solids plugging the hydrocyclone underflow?).
- Holes in the shaker screen.
- Poor plumbing prevents the solids removal equipment from processing 100% of the fluid coming from the well.
- Equipment not operating properly.
- Inadequate hydraulics to remove cuttings from the bottom of the hole and/or transport them to the surface.

If the 11 ppg water-based drilling fluid contained only 12% volume of solids, the values for all LGS concentrations would be much lower (Table 5.5).

Generally, the low gravity solids concentration is calculated with the same LGS density no matter where the rig is drilling. Trends may be observed with this procedure but

Table 5.5 Low-gravity solids (LGS) equations for 11 ppg drilling fluid with 12% volume solids containing water, barite, and low-gravity solids.

$$(\rho_B - \rho_{LGS})V_{LGS} = 100\ \rho_w + (\rho_B - \rho_w)V_s - 12\ MW$$

LGS density, g/cc	Equation	V_{LGS}, % Volume
2.2	$V_{LGS} = 50.0 + 1.60\ V_s - 6.00\ MW$	3.2
2.3	$V_{LGS} = 52.6 + 1.68\ V_s - 6.30\ MW$	3.3
2.4	$V_{LGS} = 55.6 + 1.78\ V_s - 6.67\ MW$	3.6
2.5	$V_{LGS} = 58.8 + 1.88\ V_s - 7.06\ MW$	3.8
2.6	$V_{LGS} = 62.5 + 2.00\ V_s - 7.50\ MW$	4.1
2.7	$V_{LGS} = 66.7 + 2.13\ V_s - 8.00\ MW$	4.3
2.8	$V_{LGS} = 71.4 + 2.29\ V_s - 8.57\ MW$	4.5
2.9	$V_{LGS} = 76.9 + 2.46\ V_s - 9.23\ MW$	4.9
3.0	$V_{LGS} = 83.3 + 2.67\ V_s - 10.00\ MW$	5.3

the actual value of drilled solids will be unknown. However, looking for trends of values which indicate an increase in drilled solids requires accurate retort readings. As shown in the following section about retorts, 10 cc retorts are not sensitive enough for evaluation of changes in solids content.

Retorts

Retorts are normally used to determine the volume percent of solids in a drilling fluid. The first retorts were designed to measure the amount of liquid in 10 cc drilling fluid. Retorts were originally used to determine the amount of oil in water-based drilling fluid. The liquid is captured in a graduated cylinder and all of the dissolved and un-dissolved solids remain in the retort. The volume of liquid in the graduated cylinder subtracted from the volume of drilling fluid in the retort is volume of dissolved and un-dissolved solids. In a weighted saline water based drilling fluid these solids would be barite, the viscosifier/fluid loss additive (frequently bentonite), and salt.

Solids content is determined by removing all of the liquid from the drilling fluid with a retort. A known quantity of drilling fluid is heated to around 950 °F and the liquid condensed in a measuring tube. The results are reported as a percent volume. The quantity of both water and oil (or non-aqueous fluid) is reported.

Currently, three retort volumes are commercially available: 10 cc, 20 cc, and 50 cc. Each has a small cup which holds that amount of drilling fluid. This cup is attached to a long cylinder containing steel wool and placed in a special oven for heating, (Fig. 5.1). The upper end of the cylinder has long tube which connects it to a condenser

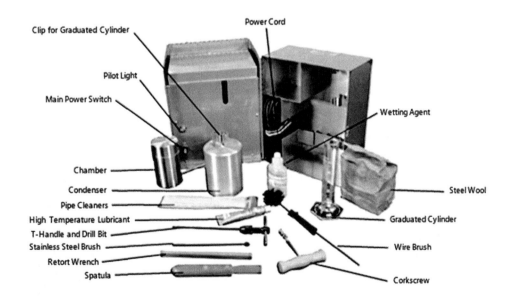

Fig. 5.1 A 50 cc retort.

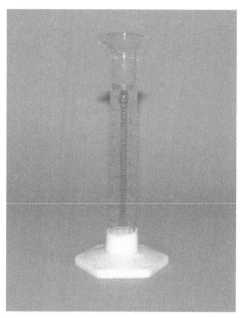

Fig. 5.2 10 cc graduated cylinder.

outside of the oven. Fluid that boils off of the sample is condensed and the volume measured with a graduated cylinder (Fig. 5.2). The steel wool helps prevent the rapid heating from splashing liquid from the small cup.

If no barite has been added to a water-based drilling fluid, the mud weight correlates exactly with the percent solids in the fluid.

Table 5.6 was developed from the equation:

$$V_{LGS} = 62.5 + 2.0\, V_s - 7.5(\text{MW})$$

Table 5.6 Volume percent of low gravity solids required to provide a specific mud weight.

Percent volume barite	Mud weight: ppg	%Volume low gravity
0	9.0	5.00
0	9.5	8.75
0	10.0	12.50
0	10.5	16.50
0	11.0	20.00
0	11.5	23.75
0	12.0	27.50

Since all of the solids are low gravity solids (no barite), $V_s = V_{LGS}$.

$$62.5 + V_{LGS} = 7.5(\text{MW})$$

$$V_{LGS} = 62.5 + 2.0\ V_s - 7.5(\text{MW})$$

An 11.0 ppg drilling fluid can be created by allowing drilled solids to increase to 20% vol.

The money saved on barite will be quickly consumed by the visible and invisible non-productive time (NPT). The NPT could include stuck pipe, lost circulation, very poor cementing jobs, and very low founder points.

Retort size required for accurate solids calculation

The 10 cc retort was adequate to determine how much oil was in the drilling fluid. Solids concentrations were not regarded as significant. Some drillers would allow the drilled solids to increase the mud weight because it was much cheaper than using barite. This is an easy number to calculate and it is apparently cheaper. However, the visible and invisible NPT was very costly. Stuck pipe was accepted as a normal event because it was common even though it was labeled as "trouble cost". The total solids concentration would not indicate a very large change and the NPT was not associated with drilled solids concentration. Thoughtful evaluations suggested that removal of these evil drilled solids might be beneficial. The 20 cc retort became of the greater accuracy. Finally, at the request of those interested in better determination of drilled solids in the drilling fluid, a 50 cc retort was developed.

The 50 cc retort was introduced to help provide better solids analysis of the drilling fluid. The amount of fluid removed from the sample changes very little with the small volume retorts and makes it very difficult to see an increase in drilled solids concentration. Furthermore, with non-aqueous invert emulsion drilling fluids, the relative amounts of water and non-aqueous base fluid must be determined with greater precision and accuracy than can be obtained with a 10 or 20 cc retort. 50 cc retorts should always be used with non-aqueous drilling fluids.

As an example of the sensitivity of the measurements in observing small changes in drilled solids, a calculation can be made for each type of retort (10 cc, 20 cc, and 50 cc) to determine the change in liquid collected when the drilled solids change by 2% in a weighted drilling fluid. For the purposes of illustrating these calculations, the drilling fluid will be 14.0 ppg fresh water with 4.2 specific gravity barite and 2.6 specific gravity low-gravity solids. The fluid will contain 18.2 ppb (or 2% volume) bentonite.

Fluid #1 Retort solids are 23% volume, retort liquid is 77% volume. The low gravity solids can be calculated for the equation:

$$V_{LGS} = 62.5 + 2.0\, V_s - 7.5(\text{MW})$$
$$= 62.5 + 2.0(23) - 7.5(14 \text{ ppg})$$
$$= 3.5\% \text{ volume}$$

Fluid #2 Will contain 2% more drilled solids.

$$V_{LGS} = 62.5 + 2.0(23) - 7.5(14 \text{ ppg})$$
$$3.5 + 2.0 = 62.5 + 2.0(23) - 7.5(14 \text{ ppg})$$
$$2\, V_s = 5.5 - 62.5 + 10.5$$
$$V_s = 24\% \text{ volume}$$

The retort liquid will be 76% volume.

Fluid #1 has 3.5% volume low gravity solids. Since 2% volume is bentonite, the drilled solids are 1.5% volume. The total solids are 23% volume and the liquid is 77%.

Fluid #2 has 5.5% volume low gravity solids. Since 2% volume is bentonite, the drilled solids are 3.5% volume (or more than twice as many as Fluid#1). The total solids are 24% volume and the liquid is 76% volume.

For the 10 cc retort, the liquid recovered for:

$$\text{Fluid \#1: } 0.77 \times 10 \text{ cc} = 7.7 \text{ cc}$$

$$\text{Fluid \#2: } 0.76 \times 10 \text{ cc} = 7.6 \text{ cc}$$

The drilled solids more than doubled, but the retort liquid graduated cylinder would only change by 0.1 cc. This would be very difficult to observe.

For the 20 cc retort, the liquid recovered for:

$$\text{Fluid \#1: } 0.77 \times 20 \text{ cc} = 15.4 \text{ cc}$$

$$\text{Fluid \#2: } 0.76 \times 20 \text{ cc} = 15.2 \text{ cc}$$

The drilled solids more than doubled, but only a 0.2 cc difference could be observed.
For the 50 cc retort, the liquid recovered for:

$$\text{Fluid \#1: } 0.77 \times 50 \text{ cc} = 38.5 \text{ cc}$$

$$\text{Fluid \#2: } 0.76 \times 50 \text{ cc} = 38.0 \text{ cc}$$

The drilled solids doubled and the retort recovered liquid changed by 0.5 cc — which is now readable, but requires careful work by the mud engineers. The below figure shows the difference is sizes between the 10 cc, 25 cc, 50 cc, and 100 cc graduated cylinders.

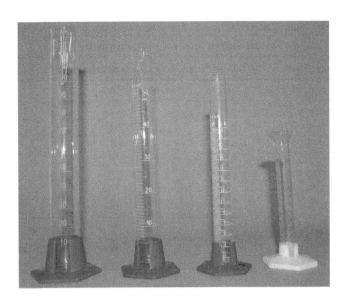

The tables below show calculations for a variety of mud weights with various amounts of total solids (V_s) in the drilling fluid while utilizing a 10 cc, 20 cc, and 50 cc retort (Tables 5.7–5.9). Five percent of the low gravity solids added to water will cause the drilling fluid to weigh 9.0 ppg. For a 10 ppg drilling fluid, 12.5% volume of the low gravity solids will be sufficient to provide the mud weight.

Table 5.7 Volume of liquid captured by a 10 cc retort for different mud weights.

Water-based drilling fluid with 2% volume bentonite				
V_s: %vol	V_b: %vol	V_{LGS}: %vol	Percent liquid	Liq vol. in retort
Retort volume: 10 cc mud weight: 9 ppg				
3.5	1.5	2	96.5	9.65
4.5	0.5	4	95.5	9.55
5.0	0.0	5	95.0	9.50
Retort volume: 10 cc mud weight: 10 ppg				
7.25	5.25	2	92.75	9.275
8.25	4.25	4	91.75	9.175
9.25	3.25	6	90.75	9.075
10.25	2.25	8	89.75	8.975
11.25	1.25	10	88.75	8.875
12.25	0.25	12	87.75	8.775
12.5	0.00	12.5	87.50	8.700

Continued

Table 5.7 Volume of liquid captured by a 10 cc retort for different mud weights.—cont'd

Water-based drilling fluid with 2% volume bentonite				
V_s: %vol	V_b: %vol	V_{LGS}: %vol	Percent liquid	Liq vol. in retort
Retort volume: 10 cc mud weight: 12 ppg				
14.75	12.75	2	85.25	8.525
15.75	11.75	4	84.25	8.425
16.75	10.75	6	83.25	8.325
17.75	9.75	8	82.25	8.225
18.75	8.75	10	81.25	8.125
19.75	7.75	12	80.25	8.025
20.75	6.75	14	79.25	7.925
Retort volume: 10 cc mud weight: 14 ppg				
22.25	20.25	2	77.75	7.775
23.25	19.25	4	76.75	7.675
24.25	18.25	6	75.75	7.575
25.25	17.25	8	74.75	7.475
26.25	16.25	10	73.75	7.375
27.25	15.25	12	72.75	7.275
28.25	14.25	14	71.75	7.175
Retort volume: 10 cc mud weight: 16 ppg				
29.75	27.75	2	70.25	7.025
30.75	26.75	4	69.25	6.925
31.75	25.75	6	68.25	6.825
32.75	24.75	8	67.25	6.725
33.75	23.75	10	66.25	6.625
34.75	22.75	12	65.25	6.525
35.75	21.75	14	64.25	6.425
Retort volume: 10 cc mud weight: 18 ppg				
37.25	35.25	2	62.75	6.275
38.25	34.25	4	61.75	6.175
39.25	33.25	6	60.75	6.075
40.25	32.25	8	59.75	5.975
41.25	31.25	10	58.75	5.875
42.25	30.25	12	57.75	5.775
43.25	29.25	14	56.75	5.675

Specific gravity of low gravity solids = 2.6; specific gravity of barite = 4.2.
V_b, percent volume barite; V_{lg}, percent volume low gravity solids; V_s, percent volume total solids.

Table 5.8 Volume of liquid captured by a 20 cc retort for different mud weights.

Water-based drilling fluid with 2% volume bentonite				
V_s: %vol	V_b: %vol	V_{LGS}: %vol	Percent liquid	Liq vol. in retort
Retort volume: 20 cc mud weight: 9 ppg				
3.5	1.5	2	96.5	19.3
4.5	0.5	4	95.5	19.1
5.0	0.0	5	95.0	19.0
Retort volume: 20 cc mud weight: 10 ppg				
7.25	5.25	2	92.75	18.55
8.25	4.25	4	91.75	18.35
9.25	3.25	6	90.75	18.15
10.25	2.25	8	89.75	17.95
11.25	1.25	10	88.75	17.75
12.25	0.25	12	87.75	17.55
12.50	0.00	12.5	87.50	17.50
Retort volume: 20 cc mud weight: 12 ppg				
14.75	12.75	2	85.25	17.05
15.75	11.75	4	84.25	16.85
16.75	10.75	6	83.25	16.65
17.75	9.75	8	82.25	16.45
18.75	8.75	10	81.25	16.25
19.75	7.75	12	80.25	16.05
20.75	6.75	14	79.25	15.85
Retort volume: 20 cc mud weight: 14 ppg				
22.25	20.25	2	77.75	15.55
23.25	19.25	4	76.75	15.35
24.25	18.25	6	75.75	15.15
25.25	17.25	8	74.75	14.95
26.25	16.25	10	73.75	14.75
27.25	15.25	12	72.75	14.55
28.25	14.25	14	71.75	14.35
Retort volume: 20 cc mud weight: 16 ppg				
29.75	27.75	2	70.25	14.05
30.75	26.75	4	69.25	13.85
31.75	25.75	6	68.25	13.65
32.75	24.75	8	67.25	13.45
33.75	23.75	10	66.25	13.25
34.75	22.75	12	65.25	13.05
35.75	21.75	14	64.25	12.85

Continued

Table 5.8 Volume of liquid captured by a 20 cc retort for different mud weights.—cont'd

Water-based drilling fluid with 2% volume bentonite				
V_s: %vol	V_b: %vol	V_{LGS}: %vol	Percent liquid	Liq vol. in retort
Retort volume: 20 cc mud weight: 18 ppg				
37.25	35.25	2	62.75	12.55
38.25	34.25	4	61.75	12.35
39.25	33.25	6	60.75	12.15
40.25	32.25	8	59.75	11.95
41.25	31.25	10	58.75	11.75
42.25	30.25	12	57.75	11.55
43.25	29.25	14	56.75	11.35

Specific gravity of low gravity solids = 2.6; specific gravity of barite = 4.2.
V_b, percent volume barite; V_{lg}, percent volume low gravity solids; V_s, percent volume total solids.

Table 5.9 Volume of liquid captured by a 50 cc retort for different mud weights.

Water-based drilling fluid with 2% volume bentonite				
V_s: %vol	V_b: %vol	V_{LGS}: %vol	Percent liquid	Liq. vol. in retort
Retort volume: 50 cc mud weight: 9 ppg				
3.5	1.5	2	96.5	48.25
4.5	0.5	4	95.5	47.75
5.0	0.0	5	95.0	47.50
Retort volume: 50 cc mud weight: 10 ppg				
7.25	5.25	2	92.75	46.375
8.25	4.25	4	91.75	45.875
9.25	3.25	6	90.75	45.375
10.25	2.25	8	89.75	44.875
11.25	1.25	10	88.75	44.375
12.25	0.25	12	87.75	43.875
12.5	0.00	12.5	87.50	43.750
Retort volume: 50 cc mud weight: 12 ppg				
14.75	12.75	2	85.25	42.625
15.75	11.75	4	84.25	42.125
16.75	10.75	6	83.25	41.625
17.75	9.75	8	82.25	41.125
18.75	8.75	10	81.25	40.625
19.75	7.75	12	80.25	40.125
20.75	6.75	14	79.25	39.625

Table 5.9 Volume of liquid captured by a 50 cc retort for different mud weights.—cont'd

Water-based drilling fluid with 2% volume bentonite				
V_s: %vol	V_b: %vol	V_{LGS}: %vol	Percent liquid	Liq. vol. in retort
Retort volume: 50 cc mud weight: 14 ppg				
22.25	20.25	2	77.75	38.875
23.25	19.25	4	76.75	38.375
24.25	18.25	6	75.75	37.875
25.25	17.25	8	74.75	37.375
26.25	16.25	10	73.75	36.875
27.25	15.25	12	72.75	36.375
28.25	14.25	14	71.75	35.875
Retort volume: 50 cc mud weight: 16 ppg				
29.75	27.75	2	70.25	35.125
30.75	26.75	4	69.25	34.625
31.75	25.75	6	68.25	34.125
32.75	24.75	8	67.25	33.625
33.75	23.75	10	66.25	33.125
34.75	22.75	12	65.25	32.625
35.75	21.75	14	64.25	32.125
Retort volume: 50 cc mud weight: 18 ppg				
37.25	35.25	2	62.75	31.375
38.25	34.25	4	61.75	30.875
39.25	33.25	6	60.75	30.375
40.25	32.25	8	59.75	29.875
41.25	31.25	10	58.75	29.375
42.25	30.25	12	57.75	28.875
43.25	29.25	14	56.75	28.375

Specific gravity of low gravity solids = 2.6; specific gravity of barite = 4.2.
V_b, percent volume barite; V_{lg}, percent volume low gravity solids; V_s, percent volume total solids.

Sensitivity to measurements

After examining the sensitivity of retorts to changes in low gravity solids, the effect of air or gas in the drilling fluid should be examined.

An unweighted water-base drilling fluid has a target value of 4% volume low gravity solids. With 2% volume bentonite and 2% volume drilled solids, the mud weight for a low gravity 2.6 specific gravity can be calculated from the equation:

$$V_{LGS} = 62.5 + 2 V_s - 7.5(\text{MW})$$

as discussed earlier in this chapter.

$$4 = 62.5 + 8 - 7.5(\text{MW})$$

$$\text{MW} = 8.9 \text{ ppg}$$

If the drilled solids doubled, the low gravity solids concentration would be 6% volume. The mud weight would be calculated from the equation:

$$V_{LGS} = 62.5 + 2V_s - 7.5(MW)$$
$$6 = 62.5 + 2(6) - 7.5(MW)$$
$$MW = 9.1 \text{ ppg}$$

A concentration of 2% volume gas or air is frequently not obvious from looking at the fluid. The volume percent of gas can be calculated from the equation:

$$V_{gas} = \frac{(MW_p - MW_{up})(100)}{MW_p}$$

where V_{gas} is the percent gas in the fluid.
MW_p is the pressurized (or true) mud weight, and
MW_{up} is the unpressurized mud weight.

After the drilled solids increased by 50% and air is entrained in the fluid, the mud weight as measured on the mud pits would be:

$$100\,MW_{up} = (9.1 \text{ ppg})(100) - (2)(9.1 \text{ ppg})$$
$$MW_{up} = 8.9 \text{ ppg}$$

The mud weight measured with an unpressurized mud balance on the mud pits would indicate no increase in drilled solids concentration even though the concentration had increased by 50%.

Field measurement of specific gravity of drilled solids

Good solids control procedures require calculation of the concentration of drilled solids in the drilling fluid or in the discharge streams from solids removal equipment. The specific gravity, or density, of the low gravity solids is necessary for accurate results. Many computer programs use a default number of 2.6 or 2.65 if a value is not available. If the opportunity to have the specific gravity is not available, an approximation can be made using the retort and the mud balance.

Collect representative samples of the drilled solids. One method is to capture some of the material being discharged from the shale shaker screens. This material can be washed with fresh water to remove drilling fluid and any foreign material (like barite, lost circulation material, etc.) Place the clean solids in a retort and dehydrate them using the same procedure which is used to determine the percent of solids in the drilling fluid.

1. Place dry drilled solids in the mud balance cup until the scale reads the density of water (8.34 ppg or 1.0 g/cc) with the lid on the mud cup. This step actually provides the mass of the solids. If the mud cup is full of water and has been properly calibrated, the volume of the fluid in the cup is also the weight of the fluid in the cup.

2. Carefully fill the mud cup of the mud balance with water and weigh it. Determine the "mud weight" of the slurry.
3. The Specific Gravity of the drilled solids can be calculated from the equation:

$$SG = \frac{1}{2 - MW}$$

where SG is the specific gravity of the drilled solids.
MW is the slurry mud weight in g/cc,
If the mud weight is in pounds per gallon:

$$SG = \frac{8.34 \text{ ppg}}{16.68 - MW(\text{ppg})}$$

Validation of the equation

Assume that clean quartz (sand) is added to the mud balance cup and the procedure performed according to the instructions above. Quartz has a density of 2.65 g/cc. If quartz is placed in the mud cup of the mud balance until the mud weight is indicated to be 8.34 ppg, the weight of the quartz will be the weight of the water required to fill the cup. For this example, the volume of the mud cup will be 140 cc. This means that the quartz can be calculated by dividing the weight by the density or 140 g/2.65 g/cc, which is 52.8 cc.

One mud cup volume is 140 cc. The amount of water added is the difference between the volume of the cup and the volume of quartz, or 140 cc−52.8 cc = 87.2 cc. The mud weight in the mud cup should be the mass of the quartz and the mass of the water divided by the volume of mud cup. The mass of water added is 87.2 g and the mass of the quartz is 140 g.

Mud weight = (140 g + 87.2 g)/140 cc = 1.622 g/cc or 13.5 ppg. This should be what the mud balance would read if the procedure had been followed.

Using the 1.62 g/cc mud weight and the equation just presented, calculate the specific gravity of the quartz:

$$SG = \frac{1}{2 - MW} = \frac{1}{2 - 1.622} = 2.65$$

This is the specific gravity of quartz.

Derivation of formula: In step 1, solids added to the mud balance cup have the same mass as the water which would fill the cup:

$$\text{Volume of Solids} = \frac{[\text{mass of solids}]}{(\text{SG of solids})}$$

$$\text{Volume of Solids} = \frac{[(\text{Volume of cup})(\text{SG of water})]}{(\text{SG of solids})}$$

where SG is specific gravity; g/cc.

After adding water to the dry solids:

$$\text{Volume of water added} = \text{vol of cup} - \text{vol of solids}$$

$$\text{Volume of water added} = \text{vol of cup} - [(\text{vol cup})(\text{SG water})/(\text{SG solids})]$$

$$\text{Slurry density} = \text{mass slurry volume cup}$$

$$\text{Mass slurry} = \text{mass solids} + \text{mass water}$$

$$\text{Mass solids} = (\text{Vol cup})(\text{SG water})$$

$$\text{Mass water} = \text{Vol water}(\text{SG water})$$
$$= [\text{vol cup} - \text{vol solids}]\text{SG water}$$

$$\text{Slurry density} = (\text{mass slurry})/\text{volume slurry}$$

$$\text{Slurry density} = (\text{Mass water} + \text{mass solids})/\text{vol cup}$$

$$\text{Slurry density} = \frac{[(\text{vol cup} - \text{vol solids})\text{SG water}] + [(\text{vol cup})(\text{SG water})]}{\text{vol cup}}$$

$$\text{Slurry density} = \frac{[(2\ \text{SG water})(\text{vol cup})] - [(\text{vol solids})(\text{SG water})]}{\text{vol cup}}$$

$$= [2\ \text{SG water}] - \left(\frac{\text{Vol solids}}{\text{Vol cup}}\right)\text{SG Water}$$

The solids in the cup have the same mass as the volume of water which will fill the cup.

$$= 2\ \text{SG water} - \frac{\text{SG water}}{\text{SG solids}}$$

For metric system:

$$\text{MW}\ (\text{g/cc}) = 2(1\ \text{g/cc}) - \frac{1\ \text{g/cc}}{\text{SG solids}}$$

Solve for SG solids:

$$\text{SG solids} = \frac{1}{2 - \text{MW}\ (\text{g/cc})}$$

In English units:

$$\text{SG solids} = \frac{8.34\ \text{ppg}}{16.68\ \text{ppg} - \text{MW(ppg)}}$$

Mud cups on the mud balances can have volumes from about 140 cc to 200 cc depending upon the brand. All balances should be calibrated with water. When the solids in the cup weigh the weight of water in the cup, the mud balance will read 8.34 ppg.

The drilled solids should be captured from the shale shaker discard. The sample can be placed on an API 40 to SPI 60 shaker screen and washed with the liquid phase of the drilling fluid. Cleaning of the cuttings is necessary to remove barite, salt, and other chemicals from the sample. A sufficient quantity of solids should be collected to almost fill the mud cup on the mud balance.

If the drilling fluid is water-based, the solids can be retorted or placed in a hot oven to evaporate the water. If the solids are collected from a non-aqueous drilling fluid (NADF), they should be washed with the clean base fluid, and then retorted to evaporate the remaining liquid. The solids could also be made water-wet by using a reverse wetting agent as is used in the spacer fluid while cementing. The excess oil, barite, salt, lime and other chemicals will be removed with the water and the solids baked in a hot oven to dry.

Non-aqueous drilling fluids

Calculating low gravity solids (LGS) in a non-aqueous drilling fluid
After removing all the liquid from a non-aqueous drilling fluid, the solids left are low-gravity solids, barite and salt. The liquid can contain water and oil.

Derivation of the equation for LGS In the drilling fluid, the total mass is a sum of the masses of all the components.

Mass of mud = mass of water + mass of LGS + mass of barite + mass of oil + mass of salt.

$$\tag{5.11}$$

or

$$M_m = M_w + M_{LGS} + M_B + M_O + M_{Salt} \tag{5.12}$$

Divide both sides by the volume of drilling fluid (V_m).

$$\frac{M_m}{V_m} = \frac{M_w}{V_m} + \frac{M_{LGS}}{V_m} + \frac{M_B}{V_m} + \frac{M_o}{V_m} + \frac{M_{Salt}}{V_M} \tag{5.13}$$

The mass of the barite is the product of the density of the barite (ρ_B) and the volume of barite (V_B).

The mass of the water plus the mass of the salt would give the mass of the fluid. The filtrate would have the same density. Calculating the mass of each component by multiplying the density of that component times the volume of that component. The densities are expressed g/cc in and the volumes in cc (or cm^3).

$$\frac{\rho_m V_m}{V_m} = \frac{(M_w + M_{salt})}{V_m} + \frac{\rho_{LGS} V_{LGS}}{V_m} + \frac{\rho_B V_B}{V_m} + \frac{\rho_O V_O}{V_m} \quad (5.14)$$

$$\frac{\rho_m V_m}{V_m} = \frac{\rho_f + V_{salt}}{V_m} + \frac{\rho_{LGS} V_{LGS}}{V_m} + \frac{\rho_B V_B}{V_m} + \frac{\rho_O V_O}{V_m} \quad (5.15)$$

To change the density of the drilling fluid (ρ_m) to the mud weight (MW) in pounds per gallon:

$$\text{MW (ppg)} = \frac{\text{Total mass (in g)}}{V_m \text{ (in cc)}} \left[\frac{8.34 \text{ lb/gal}}{1 \text{ g/cc}} \right] \quad (5.16)$$

The ratio of V_{LGS}/V_m is the fraction of the low gravity solids in the drilling fluid. This could be changed to percent low gravity solids in the drilling fluid by multiplying each ratio by 100.

$$\frac{V_{LGS}}{V_m} \times 100 = V_{LGS} \quad (5.17)$$

Eq. (5.15) then becomes—assuming no oil is added:

$$\frac{100}{8.34} \text{MW} = \rho_f V_W + \rho_{LGS} V_{LGS} + \rho_B V_B \quad (5.18)$$

The solids, V_s, consist of the low gravity solids V_{LGS} and barite V_B.

$$V_s = V_{LGS} + V_B \quad (5.19)$$

The volume fraction V_s of the solids is measure by the retort. The solids in the retort also contain the salts that may be in the drilling fluid. These soluble salts will remain in the filtrate from the drilling fluid and in Eq. (5.18), will be the reason the ρ_f is used instead of the density of water.

Solving Eq. (5.19) for the value of the percent volume of barite:

$$V_B = V_s - V_{LGS} \quad (5.20)$$

Substitute the V_B into Eq. (5.18).

$$12 \text{ MW} = \rho_f V_W + \rho_{LGS} V_{LGS} + \rho_B (V_s - V_{LG}) \quad (5.21)$$

The fluid in the retort initially consisted of water and solids, or one hundred percent was V_W & V_s:

$$V_W + V_s = 100 \quad (5.22)$$

Eq. (5.21) becomes:

$$12 \text{ MW} = \rho_f (100 - V_s) + \rho_{LGS} V_{LGS} + \rho_B (V_s - V_{LG}) \quad (5.23)$$

Solve Eq. (5.23) for V_{LGS}:

$$(\rho_B - \rho_{LGS})V_{LGS} = 100\,\rho_W + (\rho_B - \rho_F)V_s - 12\,\text{MW} \quad (5.24)$$

The density of the filtrate (or the liquid phase) of the water based drilling fluid can be calculated:

$$\rho_f = 1.0 + 6.45 \times 10^{-7}[\text{NaCl}] + 1.67 \times 10^{-3}[\text{KCl}] + 7.6^{10^{-7}}[\text{CaCl}_2] \quad (5.25)$$

where:
[NaCl] is the concentration of sodium chloride: mg/L
[KCl] is the concentration of potassium chloride: mg/L
[CaCl$_2$] is the concentration of calcium chloride: mg/L
The total suspended solids V_s, can be calculated from the equation:

$$V_s = \frac{100 - V_W}{\rho_f - ([\text{NaCl}] + [\text{CaCl}_2]) - 2.86 \times 10^{-3}[\text{KCl}]} \quad (5.26)$$

The mud weight used in these equations assumes no air or gas in the mud. Pressurized mud weights must be used. A mud weighing 9.8 ppg on the mud tank mud balance may weigh 10.2 ppg in a pressurized mud balance. This mud contains 4% volume air. The 9.8 ppg value for mud weight will not allow the low-gravity solids to be calculated accurately (Tables 5.10–5.12).

Table 5.10 Effect of temperature on properties of water.

Temperature (°F)	Density		Viscosity, cp	Vapor pressure, PSIA
	g/cc	ppg		
50	0.9997	8.343	1.308	0.18
60	0.9990	8.337	1.126	0.26
80	0.9966	8.317	0.861	0.51
100	0.9931	8.288	0.684	0.95
120	0.9886	8.250	0.560	1.69
140	0.9832	8.205	0.469	2.89
160	0.9773	8.156	0.400	4.74
180	0.9704	8.098	0.347	7.51
200	0.9629	8.036	0.304	11.53
220	0.9553	7.972	0.271	17.19
240	0.9469	7.902	0.242	24.97
260	0.9378	7.826	0.218	35.43
280	0.9282	7.746	0.199	49.20
300	0.9183	7.664	0.185	67.01
320	0.9077	7.575	0.174	89.66
340	0.8962	7.479	0.164	118.01
360	0.8843	7.380	0.155	153.04

Table 5.11 Sodium chloride solution densities.

NaCl solutions									
Mg per liter		Sp. Gr.	Parts per million		Volume %	Weight %	Fluid wt. per gallon		
Cl	NaCl		Cl	NaCl	NaCl	NaCl	NaCl	Water	Total
3,200	5,300	1.004	3,200	5,280	0.244	0.528	0.044	8.318	8.362
6,400	10,600	1.007	6,400	10,560	0.489	1.056	0.089	8.297	8.386
9,700	16,000	1.011	9,600	15,800	0.733	1.584	0.133	8.287	8.420
13,000	21,400	1.015	12,800	21,100	0.977	2.112	0.178	8.275	8.453
16,300	26,900	1.019	16,000	26,400	1.222	2.640	0.224	8.262	8.486
19,700	32,400	1.023	19,200	31,700	1.466	3.167	0.270	8.250	8.520
23,100	38,000	1.026	22,400	37,000	1.710	3.695	0.316	8.229	8.545
26,400	43,500	1.030	25,600	42,200	1.954	4.223	0.362	8.216	8.578
29,700	49,000	1.034	28,800	47,500	2.199	4.751	0.409	8.202	8.611
33,200	54,800	1.038	32,000	52,800	2.443	5.279	0.456	8.188	8.644
36,700	60,500	1.042	35,200	58,100	2.687	5.807	0.503	8.175	8.678
40,200	66,300	1.046	38,500	63,400	2.932	6.335	0.552	8.159	8.711
43,700	72,100	1.050	41,600	68,600	3.176	6.863	0.600	8.144	8.744
47,300	77,900	1.054	44,900	74,000	3.420	7.391	0.649	8.129	8.778
50,800	83,800	1.058	48,000	79,200	3.665	7.919	0.698	8.113	8.811
54,400	89,700	1.062	51,300	84,500	3.908	8.446	0.747	8.097	8.844
58,100	95,700	1.066	54,600	90,000	4.153	8.974	0.797	8.081	8.878
61,700	101,700	1.070	57,600	95,000	4.397	9.502	0.847	8.064	8.911
65,300	107,700	1.074	60,800	100,300	4.641	10.030	0.897	8.047	8.944
69,000	113,800	1.078	64,100	105,600	4.886	10.558	0.948	8.030	8.978
72,800	120,000	1.082	67,300	110,900	5.130	11.086	0.999	8.012	9.011
76,500	126,100	1.086	70,400	116,100	5.374	11.614	1.050	7.994	9.044
80,300	132,300	1.090	73,600	121,400	5.619	12.142	1.102	7.976	9.078
84,100	138,600	1.094	76,800	126,700	5.863	12.670	1.154	7.957	9.111
87,900	144,900	1.098	80,100	132,000	6.107	13.198	1.207	7.937	9.144
91,700	151,200	1.102	83,300	137,300	6.351	13.725	1.260	7.918	9.178
95,600	157,600	1.106	86,400	142,500	6.596	14.253	1.313	7.898	9.211
99,500	164,100	1.110	89,700	147,800	6.840	14.781	1.366	7.878	9.244
103,400	170,500	1.114	92,900	153,100	7.084	15.309	1.420	7.858	9.278
107,400	177,100	1.118	96,100	158,400	7.329	15.837	1.475	7.836	9.311
111,400	183,600	1.122	99,300	163,700	7.573	16.365	1.529	7.815	9.344
115,400	190,200	1.126	102,500	168,900	7.817	16.893	1.584	7.794	9.378
119,400	196,900	1.130	105,700	174,200	8.062	17.421	1.639	7.772	9.411
123,600	203,700	1.135	108,900	179,500	8.306	17.949	1.697	7.755	9.452
127,700	210,500	1.139	112,100	184,800	8.550	18.477	1.753	7.733	9.486
131,800	217,200	1.143	115,300	190,000	8.794	19.004	1.809	7.710	9.519
135,900	224,000	1.147	118,500	195,300	9.038	16.532	1.866	7.686	9.552
140,200	231,100	1.152	121,700	200,600	9.283	20.060	1.925	7.669	9.594
144,400	238,000	1.156	124,900	205,900	9.527	20.558	1.982	7.645	9.627
148,600	245,000	1.160	128,100	211,200	9.771	21.116	2.040	7.620	9.660
152,900	252,000	1.164	131,300	216,400	10.016	21.664	2.098	7.596	9.694

Table 5.11 Sodium chloride solution densities.—cont'd

NaCl solutions									
Mg per liter			Parts per million		Volume %	Weight %	Fluid wt. per gallon		
Cl	NaCl	Sp. Gr.	Cl	NaCl	NaCl	NaCl	NaCl	Water	Total
157,200	259,200	1.169	134,500	221,700	10.260	22.172	2.158	7.577	9.735
161,500	266,300	1.173	137,700	227,000	10.504	22.700	2.218	7.551	9.769
166,000	273,600	1.178	140,900	232,300	10.749	23.228	2.279	7.531	9.810
170,300	280,800	1.182	144,100	237,600	10.993	23.755	2.338	7.506	9.844
174,700	288,000	1.186	147,300	242,800	11.237	24.283	2.398	7.479	9.877
179,300	295,500	1.191	150,500	248,100	11.481	24.811	2.459	7.460	9.919
183,700	302,800	1.195	153,700	253,400	11.726	25.339	2.522	7.430	9.952
188,300	310,400	1.200	156,900	258,700	11.970	25.867	2.585	7.409	9.994
192,800	317,800	1.204	160,100	264,000	12.214	26.395	2.647	7.380	10.027

NaCl to Cl (ppm NaCl × 0.6066 or 0.61).
Cl to NaCl (ppm Cl × 1.6486 or 1.65).
Aqueous solutions at 77 °F.

Table 5.12 Potassium chloride solution densities.

Potassium chloride[a]									
Sp. Gr.	%KCl, wt.	%KCl, ppg	Density, ppg	Density, PCF	Lb. KCl/ Bbl. H$_2$O	KCl, ppm	K, ppm	Cl, ppm	
1.0046	1	0.50	8.37	62.60	3.52	10,050	5270.9	4779.1	
1.0110	2	1.01	8.42	62.99	7.09	20,220	10,604.8	9615.2	
1.0175	3	1.51	8.47	63.39	10.72	30,590	16,043.5	14,546.5	
1.0239	4	2.02	8.53	63.80	14.36	40,960	21,482.3	19,477.7	
1.0304	5	2.52	8.59	64.21	18.09	51,585	27,054.8	24,530.2	
1.0369	6	3.02	8.64	64.61	21.81	62,210	32,627.3	29,582.7	
1.0500	8	4.03	8.75	65.42	29.44	84,000	44,055.5	39,944.5	
1.0633	10	5.04	8.86	66.25	37.27	106,300	55,751.6	50,548.4	
1.0768	12	6.05	8.97	67.09	45.28	129,200	67,761.5	61,438.5	
1.0905	14	7.06	9.08	67.95	53.51	152,700	80,086.6	72,613.4	
1.1043	16	8.06	9.20	68.81	61.95	176,700	92,673.9	84,026.1	
1.1185	18	9.07	9.32	69.69	70.56	201,300	105,575.8	95,724.2	
1.1328	20	10.08	9.44	70.58	79.42	226,600	118,844.9	107,755.1	
1.1474	22	11.09	9.56	71.49	88.49	252,400	132,376.3	120,023.7	
1.1623	24	12.10	9.68	72.42	97.78	279,000	146,327.1	132,672.9	

[a] Aqueous solution at 68 °F.

Special note about potassium chloride drilling fluids

Potassium Chloride drilling fluids are frequently used for well bore stability. The potassium ion fits into the silicon tetrahedron matrix of clays and helps hold the structure together to buy more time before water absorption caused the shales to collapse into the well bore. When the potassium ion is absorbed on the clay surface another cation (mostly sodium) is replaced and enters the drilling fluid. This means that the *KCl* drilling fluid slowly becomes a *NaCl* (or salt) drilling fluid. If the chloride ion concentration is maintained constant, no additional *KCl* will be added. For proper salt calculations the sodium and potassium concentrations must be determined (Table 5.13).

Calculation volume fraction of low gravity solids in non aqueous fluids

Derivation of equation: Non Aqueous Drilling Fluid (NADF) generally has more components to consider than a water-based drilling fluid. NADF components are oil or synthetic fluid, water, barite, salt, lime and low gravity solids (LGS).

Mathematically this can be expressed by the fact the total mass of fluid is equal to the sum of the massed of each component:

$$\text{Total mass} = \text{Mass of oil} + \text{Mass of water} + \text{Mass of LGS} + \text{Mass of barite} + \text{Mass of salt} + \text{Mass of lime} \tag{5.27}$$

To convert this equation to an expression of the density (or mud weight), divide both sides of the equation by the volume of mud (V_m) in cubic centimeters (cc) being weighed. Measure the mass of each component in grams.

$$\frac{\text{Total mass(in g)}}{V_m(\text{in cc})} = \frac{\text{Mass water(in g)}}{V_m(\text{in cc})} + \frac{\text{Mass oil(in g)}}{V_m(\text{in cc})} + \frac{\text{Mass LGS(in g)}}{V_m(\text{in cc})} + \frac{\text{Mass barite(in g)}}{V_m(\text{in cc})} + \frac{\text{Mass salt(in g)}}{V_m(\text{in cc})} + \frac{\text{Mass lime(in g)}}{V_m(\text{in cc})} \tag{5.28}$$

The left side of the equation can be expressed as mud weight in pound per gallon (ppg).

$$\text{MW(ppg)} = \frac{\text{Total mass(in g)}}{V_m(\text{in cc})} \left[\frac{8.34 \text{ lb/gal}}{1 \text{ g/cc}} \right] \tag{5.29}$$

Each mass of the first four components on the right side of the equation can be converted into a density times volume. For example, the term for the low gravity solids will be:

$$\frac{\text{Mass LGS (in g)}}{V_m} = \frac{[\text{density LGS (in g/cc)}][\text{Volume of LGS (in cc)}]}{V_m(\text{in cc})} \tag{5.30}$$

Table 5.13 Weight and volume of sodium chloride (NaCl) solutions.

		Added to water, lb/bbl	In solution		NaCl Chloride (Cl), mg/L	Specific gravity	Solution weight, lb/gal	Volume increase factor	Freeze point, °F
Wt. %	Vol %		lb/bbl	mg/L					
1	0.49	3.53	3.52	10,050	6,100	1.0053	8.37	1.005	31
2	0.74	7.14	7.10	20,250	12,300	1.0125	8.43	1.008	30
4	1.43	14.58	14.40	41,100	24,900	1.0268	8.55	1.015	27.8
6	2.13	22.34	21.90	62,500	37,900	1.0413	8.67	1.022	25.5
8	2.87	30.43	29.61	84,500	51,300	1.0559	8.80	1.030	22.9
10	3.63	38.89	37.43	107,000	65,000	1.0707	8.92	1.038	20.2
12	4.36	47.73	45.65	130,300	79,100	1.0857	9.04	1.046	17.3
14	5.31	56.98	54.01	154,100	94,200	1.1009	9.17	1.057	14.1
16	6.28	66.67	62.58	178,600	109,000	1.1162	9.30	1.067	10.6
18	7.24	76.83	71.40	203,700	124,300	1.1314	9.43	1.078	6.7
20	8.17	87.50	80.47	229,600	140,000	1.1478	9.56	1.089	2.4
22	9.26	98.72	89.75	256,100	156,100	1.1640	9.70	1.102	−2.5
24	10.15	110.53	99.29	283,300	172,600	1.1804	9.83	1.113	−1.4
26	11.43	122.97	109.12	311,300	189,000	1.1972	9.97	1.129	27.9
26.3	11.74	126.00	110.45	315,500	192,100	1.1993	9.99	1.133	32

mg/L NaCl × 0.6066 = mg/L Cl$^-$
mg/L Cl × 1.65 = mg/L NaCl.
mg/L × Specific gravity of solution = PPM.
PPM × Specific gravity of solution = mg/L.

$$= \text{density LGS(in g/cc)} \left[\frac{V_{LGS}(\text{in cc})}{V_m(\text{in cc})} \right] \quad (5.31)$$

$$= [\rho_{LGS}(\text{in g/cc})][V_{LGS}] \quad (5.32)$$

where V_{LGS} is the volume fraction of low gravity solids in the drilling fluid.

Eq. (5.29) can now be written as a percentage of the volume of components by converting the fraction of each component into a percentage. For example, the term of the left side of Eq. (5.29) becomes (from Eq. 5.30):

$$\frac{\text{Total mass(in g)}}{V_m(\text{in cc})} = \frac{\text{MW(ppg)}}{8.34}[100] \quad (5.33)$$

$$\frac{\text{MW(ppg)}}{8.34}[100] = \rho_W V_W + \rho_{LGS} V_{LGS} + \rho_B V_B + \rho_O V_O + \left[\frac{\text{Mass lime}}{V_m(\text{in cc})}\right] + \left[\frac{\text{Mass salt}}{V_m(\text{in cc})}\right] \quad (5.34)$$

where:
ρ_W is the density of water
ρ_{LGS} is the density of LGS
ρ_B is the density of barite
ρ_O is the density of oil (or synthetic oil)
V_W is the volume percent of water
V_{LGS} is the volume percent of LGS
V_B is the volume percent of barite
V_O is the volume percent of oil (or synthetic oil)

The last two terms in Eq. (5.34) are measured by titration as described in API RP 13B-2. The salt concentration is determined by titrating for chlorides as described in Section 10 of the API RP 13B-2. The lime concentration [Ca (OH)$_2$] is determined by titration as described in Section 12 of the API RP 13B-2. Lime is often referred to as "whole mud alkalinity" or as "excess lime". The lime content is reported as the number of pounds of lime in one barrel drilling fluid. To convert this to g/cc, divide by 8.34. [8.34 lb/bbl = 1 g/cc]

Procedure for determining the low gravity solids in non-aqueous fluids

The solids left in a retort cup are low gravity solids, barite, salt and lime. The water and oil are collected in the graduated cylinder. The solids in the retort cup are weighed. The total volume of solids can be calculated by subtracting the volume of water and oil collected during the test from the volume of the retort cup. A 50 cc retort is highly recommended.

The weight of salt and lime, determined from the titration tests, is subtracted from the weight of the residue in the retort cup.

The densities of salt and lime are

$$\text{density of NaCl} = 2.165 \text{ g/cc}$$

$$\text{density of CaCl}_2 = 2.15 \text{ g/cc}$$

$$\text{density of lime} = 2.24 \text{ g/cc}$$

Eq. (5.24) becomes:

$$12 \text{ MW} = \rho_W V_W + \rho_{LGS} V_{LGS} + \rho_B V_B + \rho_O V_O \\ + \left[\frac{\text{Mass salt} + \text{lime}}{\text{Volume of mud}}\right] \text{Volume of retort} \quad (5.35)$$

$$V_s = V_{LGS} + V_B + C \quad (5.36)$$

Where C is the volume percent of salt and lime in the retort cup.

$$V_B = V_s - V_{LGS} - C_1 \quad (5.37)$$

$$12 \text{ MW} = \rho_W V_W + \rho_{LGS} V_{LGS} + \rho_B V_B + C_2 \quad (5.38)$$

$$12 \text{ MW} = \rho_W V_W + \rho_{LGS} V_{LGS} + \rho_B (V_s - V_{LGS} - C_1) + C_2 \quad (5.39)$$

where C_2 is the mass (in g) of salt and lime in the retort cup.

$$12 \text{ MW} = \rho_W V_W + \rho_{LGS} V_{LGS} + \rho_B V_s - \rho_B V_{LGS} - \rho_B C_1 + C_2 \quad (5.40)$$

$$(\rho_B - \rho_{LGS}) V_{LGS} = \rho_W V_W + \rho_B V_s - \rho_B C_1 + C_2 - 12 \text{ MW} \quad (5.41)$$

The density of barite should be 4.2 to meet API RP13A specifications unless it is certified to be 4.1. The density of hematite is 5.05 if it meets API RP13A specifications.

A field method of determining the density of the low gravity solids was discussed earlier in this chapter. Usually a value of ρ_{LGS} is selected as the 2.6 g/cc if no other data is available. Dry densities of formation material can range from 2.3 to 2.9. This will make a significant difference in the calculated concentration of low gravity solids.

Alternate method of calculating low gravity solids in an unweighted, water-based drilling fluid

If a drilling fluid consists of water and a solid, the density of the slurry provides a method of determining the concentration of solids in the fluid.

The weight of drilling fluid (Wt_m) is the weight of the water (Wt_w) plus the weight of the solids (Wt_{LGS}).

$$Wt_m = Wt_w + Wt_{LGS}$$

Density times volume is weight.

$$v_w SG_w + v_o SG_o + v_{lgs} SG_{LGS} = v_m SG_m$$

$$ASG_s = \frac{(v_m SG_m - v_w SG_w - v_o SG_o)}{v_{LGS}}$$

ASG = average specific gravity
v = volume, mL or %
SG = specific gravity
m = mud
w = water
o = oil
LGS = low gravity solids

With the average specific gravity of the solids, ASG_s, the percent of low-gravity solids in the fluid, V_{LGS} can be calculated with the equation:

$$v_{LGS} = \frac{(SG_B - ASG_s)v_s}{SG_B - SG_{LGS}}$$

where

SG_B is the specific gravity of the weighting agent (usually 4.2)
AGS_s is the average specific gravity of the solids,
v_s is the volume percent of undissolved solids, and
SG_{LGS} is the specific gravity of the low gravity solids (usually 2.6)

The pounds per barrel concentration of the low-gravity solids, PPB_{LGS} can be calculated with the equation:

$$PPB_{LGS} = v_{LGS}(SG_{LGS})(3.505)$$

Example A 10 cc retort is used to measure the fraction of un-dissolved solids in a 14.0 ppg drilling fluid. The retort indicated 6.5 cc of water and 0.5 cc of oil ($SG_o = 0.83$) were removed from the fluid. The undissolved solids in the retort would be:

$$10 \text{ cc} - 7 \text{ cc} = 3 \text{ cc or } 30\%$$

Determine the fraction of low-gravity solids:

$$ASG = [v_m SG_m - v_w SG_w - v_o SG_o]/ v_s$$

$$SG = 14.0 \text{ ppg}/ 8.345 \text{ ppg} = 1.7$$

$$ASG = \frac{(10)\left(\frac{14.0}{8.345}\right) - (1.0)(6.5) - (0.83)(0.5)}{3.0}$$

$$ASG = 3.3$$

$$v_{LGS} = \frac{(SG_B - ASG_s)v_s}{SG_B - SG_{LGS}}$$

$$v_{LGS} = \frac{(4.2 - 3.3)30}{4.2 - 2.6}$$

$$v_{LGS} = 16.9\% \text{ vol}$$

Pounds per barrel of low-gravity solids = 16.9(2.6) (3.505) = 154 ppb.

The amount of barite in the drilling fluid would be the difference between the total solids in the drilling fluid and the low-gravity solids:

$$\text{Percent volume of barite} = 30\% - 16.9\% = 13.1\% \text{ vol}$$

The concentration of barite in pounds per barrel of drilling fluid would be: (13.1) (4.2) (3.505) = 193 ppb.

Usually, the amount of bentonite is also measured with a methylene blue titration test. This is in an effort to determine the quantity of active commercial solids in the drilling fluid. The drilling fluid is boiled in acid to convert all of the cations to hydrogen for the titration test. This means that the measurement just measures the available clay platelets not the active clay platelets. In the case of this drilling fluid, the CEC (cation exchange concentration) was 5.5 cc of sulfuric acid. The quantity of available bentonite would be calculated from the equation:

$$\text{Bentonite equivalent} = 5 \, (\text{CEC})$$

where CEC is the number of cubic centimeters of methylene blue solution required to cover all of the surfaces on the clay. In this drilling fluid, it required 5.5 cc of methylene blue to attach to each negative site on the clay.

$$\text{Bentonite equivalent} = 5 \, (5.5) = 27.5 \text{ ppb}$$

or 27.5 ppb/9.1 ppb/% or 3% volume.

This means that the drilled solids concentration in this fluid is 16.9%–3% or 13.9% volume.

WARNING: With this many drilled solids in the drilling fluid, the well is going to cost much more to drill than it should. Visible and invisible NPT will raise the cost significantly.

Solids calculations The following equations allow calculation of all components of a water-based drilling fluid. Final accuracy is dependent upon accuracy of the various measurements including titrations, retort, and density (mud weight).

I. Mud weight, g/L = $\frac{\text{Mud weight, ppg} \times 1000}{8.34}$

II. Oil, g/L = $\frac{\text{Oil \% retort}}{10 \text{ mL}} \times \frac{6.7}{8.34} \times 10^3$

III. Water, g/L = $\frac{\text{Water \% retort}}{10 \text{ mL}} \times \frac{8.34}{8.34} \times 10^3$

IV. Salt (as NaCl) g/L = $\frac{(\text{Cl mg/L}) (1.65) (\text{Corrected Water Volume*})}{1000 \text{ mg/g}}$

*See Table 5.14.

"Corrected Water Volume" = Water, g/L/1000

(1 − Corrected Water Volume) = Corrected Solids Volume

V. Weight undissolved drilling fluid solids, grams/L = I−(II + III + IV)

VI. Average specific gravity undissolved solids = $\frac{\text{Weight undissolved mud solids, g/L}}{(1-\text{Corrected Water Volume}) \times 1000}$

Solids Fractions, barite and low gravity solids Assuming barite specific gravity 4.2 and low gravity solids specific gravity 2.6, then (4.2)B + 2.6(1−B) = 1 (Avg Specific Gravity Undissolved mud solids)

where B is the volume fraction barite.

(1−B) is the volume fraction low gravity solids and Avg Specific Gravity Undissolved mud solids is from the equation above.

VII. Barite and low gravity solids weight fraction, pounds per barrel

Barite, lbs bbl = (4.2) (350 lbs/bbl) (B) (Corrected solids volume).

Low gravity solids, lbs/bbl = (2.6) (350 lbs/bbl) (1−B) (Corrected solids volume).

Example Following properties are taken from the morning report:

Mud weight 13.4 ppg

Retort Oil 0%

Retort Solids (uncorrected) 22%

Cl, mg/L 8000

Assuming 4.2 for barite specific gravity and 2.6 for gravity of low density solids, what are:

a) Corrected solids volume
b) Barite content, lbs/bbl
c) Low gravity solids content, lbs/bbl?

I. Mud weight, g/L = $\frac{\text{Mud weight, ppg} \times 1000}{8.34} = \frac{13.4 \times 1000}{8.34} = 1607$ g/L

II. Oil, g/L = $\frac{\text{Oil \% retort}}{10 \text{ mL}} \times \frac{6.7}{8.33} \times 10^3 = \frac{0}{10 \text{ mL}} \times \frac{6.7}{8.33} \cdot 10^3 = 0.$

III. Water, g/L = $\frac{\text{Water \% retort}}{10 \text{ mL}} \times \frac{8.33}{8.33} \times 10^3$

$= \frac{7.8 \text{ mL}}{10 \text{ mL}} \times \frac{8.33}{8.33} \times 10^3 = 780$ g/L

Table 5.14 Data for common sodium chloride solutions.

Specific gravity	Weight, ppg	NaCl, ppm	Cl, ppm	NaCl, % wt	NaCl, % vol	Final volume factor
1.0000	8.34	0	0	0	0	1.0000
1.0053	8.37	10,050	6100	1	0.36	1.0036
1.0125	8.43	20,250	12,290	2	0.73	1.0074
1.0268	8.55	41,070	24,920	4	1.50	1.0152
1.0413	8.67	62,480	37,910	6	2.27	1.0232
1.0559	8.80	84,470	51,260	8	3.07	1.0317
1.0707	8.92	107,100	64,990	10	3.89	1.0405
1.0857	9.04	130,300	79,070	12	4.37	1.0497
1.1009	9.17	153,100	92,900	14	5.60	1.0593
1.1162	9.30	178,600	108,370	16	6.49	1.0694
1.1319	9.43	202,700	123,000	18	7.41	1.0800
1.1478	9.56	229,600	139,320	20	8.36	1.0912
1.1640	9.70	256,100	155,400	22	9.33	1.1029
1.1804	9.83	283,300	171,910	24	10.33	1.1152
1.1972	9.97	311,300	188,890	26	11.35	1.1282

IV. Salt (as NaCl), g/L $= \dfrac{(Cl,\ mg/L)\ (1.65)\ (\text{Corrected water volume*})}{1000\ mg/g}$.

$= \dfrac{(8000)(1.65)(0.78)(1.0048)}{1000\ mg/g} = 10.35\ g/L$

V. Weight undissolved mud solids, g/L = I − [II + III + IV] = 1607
− (0 + 780+10.35) = 816.65 g/L

VI. Average specific gravity undissolved solids $= \dfrac{\text{Weight undissolved mud solids, g/L}}{(1 - \text{Corrected water volume}) \times 1000}$

$= \dfrac{861.65\ g/L}{1 - (0.78\ \times\ 1.0048)\ (1000)} = 3.775$

VII. Solids Fractions, barite and low gravity solids.

Assuming barite specific gravity 4.2 and low gravity solids specific gravity 2.6, then:

$(4.2)B + 2.6(1 - B) = 1(\text{Avg Specific Gravity Undissolved Mud Solids})$

$4.2B + 2.6 - 2.6B = 1 \times 3.775$

$B = 0.73$

VIII. Barite and low gravity solids weight fraction, pounds per barrel.
Barite lbs/bbl = (4.2) (350 lbs/bbl) (B) (Corrected solids volume)
= (4.2) (350 lbs/bbl) (0.73) (00.216) = 232 lbs/bbl.
Low gravity solids, lbs/bbl = (2.6) (350 lbs/bbl) (1−B) (Corrected solids volume)
= (2.6) (350 lbs/bbl) (00.27) (00.216) = 53 lbs/bbl.
a. Corrected solids volume = 21.6%
b. Barite content = 232 lbs/bbl
c. Low gravity solids content = 53 lbs/bbl

Comment The mud engineer was using a 10 cc retort which means the solids analysis is not very accurate. It will be impossible to detect an increase in drilled solids. See discussion at the beginning of this chapter.

Problems

Problem 5.1 For a fresh water drilling fluid (no salt added), derive the equation to calculate the volume percent of low gravity solids from retort data for 4.2 SG Barite and a series of different densities of low gravity solids. Very young (geologically speaking) sediments can have specific gravities as low as 2.3. Very old sediments can have specific gravities as high as 2.9.

LGS specific gravity	Equation
2.3	$V_{LGS} =$
2.4	$V_{LGS} =$
2.5	$V_{LGS} =$
2.6	$V_{LGS} =$
2.7	$V_{LGS} =$
2.8	$V_{LGS} =$
2.9	$V_{LGS} =$

Problem 5.2 A 13.0 ppg fresh water drilling fluid with 20% volume total solids, calculate the V_{LGS} for a 2.3, a 2.6 and a 2.9 specific gravity drilled solid and a 4.2 SG Barite.

Problem 5.3 For the 13.0 ppg fresh water drilling fluid in problem #2, what would be the V_{LGS} if the barite specific gravity was 4.1 (which has been approved by API)?

CHAPTER 6

Cuttings transport

Drilled solids are difficult to remove from the drilling fluid unless they are brought to the surface with the minimum amount of deterioration. Two discussions are presented here. The first is a "field-useable" empirical correlation for almost vertical well bores. This correlation has been successful for bore holes up to 35 degrees from vertical. The second is a series of guidelines for holes from about 35 degrees to horizontal. These were developed in a laboratory model and have been field-tested.

Hole cleaning

Drilling fluid must perform many functions as described initially in almost any course or introduction to drilling fluid theory. One of the important functions is transporting cuttings and sloughings to the surface. Usually drilling fluids are non-Newtonian, and complex from a rheological perspective. This can enhance the carrying capability of a drilling fluid but it also makes it difficult to develop theoretical guidelines to adequately describe parameters necessary to clean a borehole. One approach that has been widely explored has been an evaluation of cutting slip velocities and a variety of correlation techniques.

Theoretical solutions of the carrying capacity can become very complex and is a fruitful area for graduate students to develop analytical skills. Most of the correlations and certainly the flow equations are much too complicated to be satisfactorily used on a drilling rig. The rules of thumb presented here are workable and in no way should diminish the excellent theoretical work by so many excellent scholars.

Empirical correlation for cleaning; vertical or near-vertical boreholes (up to 35 degrees)

One of the principle functions of a drilling fluid is to bring the cuttings and well-bore sloughing (or drilled solids) to the surface. Although this is a simple concept, the mathematical analysis of this function is very complex. The drilled solids have a variety of sizes, shapes, and densities. The fluid flow must be described with complex rheological mathematical models. The drill string is seldom concentric with the well-bore and the annulus dimensions are seldom known. Most of the pioneer work to understand the transport of drilled solids has been with models as well as mathematical analysis. Many PhD degrees have been awarded for analysis and mathematical solutions to this perplexing problem. The solution presented here is a simple empirical method that was evaluated for 10 years before initial publication.

Historical perspective

In 1951, a paper by Williams and Bruce called "Carrying Capacity of Drilling Muds" revealed that cuttings of different shapes move differently in the annulus. The fluid continuously moves upward but the cuttings ride upward on different velocity profiles, rotate sideways, and fall downward. Udo Zeidler demonstrated this validity in tests at the University of Tulsa as he completed his PhD thesis on the subject. The mathematical description of the process is obviously quite complex. These involved the calculation of slip velocity, models that described transport constants, and descriptions of non-Newtonian flow in annuli.

One significant paper by Sifferman, Myers, Haden, and Wahl, "Drill-Cutting Transport in Full-Scale Vertical Annuli", described some laboratory tests in which artificial cuttings were made with two different densities and three different sizes. They presented their results in terms of a cuttings transport ratio (the ratio of the velocity of the cuttings to the velocity of the fluid). This ratio is also equal to the ratio of concentration of cuttings in the feed to the cuttings in the annulus. In other words it represents the "storage" of cuttings. Fig. 6.1 from their paper is reproduced below. The low annular velocities required to transport cuttings with the "thick" drilling fluid evoked considerable controversy when it appeared.

Empirical correlation

In the field, many factors that affect carrying capacity predictions cannot be determined which decreases the practical application of precise results:
1. The size of the cuttings cannot be changed or known.

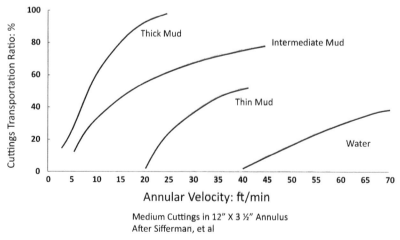

Fig. 6.1 Laboratory data for cuttings transport as a function of annular velocity.

2. The annulus eccentricity is unpredictable.
3. The size of the annulus varies unpredictably.
4. The downhole drilling fluid properties are not usually known.
5. The flow patterns in the annulus are not known.

At the rig only three "hole cleaning" variables can be controlled: the mud weight, the viscosity of the drilling fluid, and the annular velocity. Increasing any one of these variables increases hole cleaning. After watching shale shakers for years and the effects of various rheological characteristics of the drilling fluid, a simplistic equation was developed to help predict good hole cleaning. The product of the three variables (in US oil field dimensions) is equal to a value of around 400,000 when the cuttings were arriving properly at the surface. Good hole cleaning is indicated when the cuttings arrive at the surface with sharp edges. Rounded edges indicate tumbling action in the annulus because the cuttings are not transported expediently to the surface. Frequently the quantity of cuttings is misleading so the appearance of the cuttings is a better indicator of good hole cleaning.

A new term called "CCI" (carrying capacity index) was developed to describe hole cleaning:

$$\text{CCI} = \frac{(\text{MW})(K)(\text{AV})}{400,000}$$

Good hole cleaning is expected when CCI is equal to 1. The cuttings are sharp-edged and generally large. When the CCI has a value of 0.5, the cuttings are rounded and generally very small. When the CCI has a value of less than 0.3, the cuttings can be grain-sized.

Please note that the "400,000" has only one significant figure, so CCI values are only approximate. (This means the "real number" could be 380,000 or 430,000.)

The viscosity selected is the "K" value viscosity from the Power Law Rheological Model:

$$\text{SS} = K\,(\text{SR})^n$$

where
  SS is the shear stress.
  SR is the shear rate.
  K and n are power law model constants.

The values of K and n can be calculated from the PV and YP used in the Bingham Plastic Model numbers.

$$K = [511]^{1-n}[\text{PV}+\text{YP}], \text{ and } n = 3.322\,\text{Log}\frac{2\text{PV}+\text{YP}}{\text{PV}+\text{YP}}$$

The term [PV + YP] is the 300 RPM reading used in the Bingham Plastic Model. The term [2 PV + YP] is the 600 RPM reading used in the Bingham Plastic Model. The 300 RPM and the 600 RPM readings are not usually reported on a morning report form. Expressing K and n in terms of the numbers. which are reported on the morning report form, makes the calculation simpler.

(These equations are derived in Appendix 6.A.) In this form, the K has a value of effective centipoise viscosity. For convenience a simple chart is presented below to obtain the value of K for various combinations of PV and YP (Fig. 6.2).

For convenience of solving most problems in the field, the lower portion of the graph is expanded in Appendix 6.C. This makes it easier to determine values of yield points needed to correct problems in the field.

An interesting comparison of the CCI can be made with published data from Sifferman et al. Table 6.1 illustrates typical K-values from various types of drilling fluid with varying viscosity and gel strengths. The drilling fluid had the properties listed below:

CCI values are indicated on the chart and indicate an excellent agreement. The maximum value of the CCI is 1.5 for the highest value of the annular velocity for the

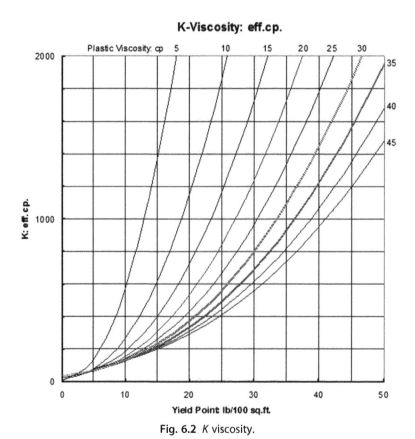

Fig. 6.2 K viscosity.

Table 6.1 Sifferman's data.

Description	PV	YP	Initial gels	10 min gels	K-values
Thick mud	16	37	13	29	2520
Inter mud	14	21	13	22	870
Thin mud	8	8	2	3	210
Water	1	0	0	0	1

thick mud. Frequently in the field, values as high as 2.0 are found. The annular velocity can frequently be decreased or the K-value can be decreased without creating problems in the well. Again, an examination of the cuttings will provide clues concerning the ability to clean the bore hole (Fig. 6.3). Figure 6.3 demonstrates the relationship between the increase in CCI values as a function of the increase in annular velocity with all other things being equal. Note that, for example, with a thick mud, a CCI value is obtained at an annular velocity of 18 ft/min.

Using the correlation

From a practical viewpoint the K-value chart is used in the field to determine what value of Yield Point is needed to clean the hole. To illustrate the use of the equation, consider a 13.6 ppg fluid (PV = 22, YP = 13) flowing at 62 ft/min in the casing/drill pipe annulus. The "K" value from the chart is 220 cp. The carrying capacity index (CCI) would be:

$$\text{CCI} = \frac{(13.6)(220)(62)}{400,000} = 0.46$$

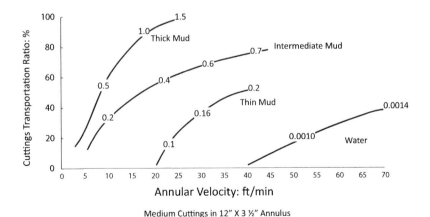

Medium Cuttings in 12" X 3 ½" Annulus
After Sifferman, et al

Fig. 6.3 Comparison of CCI with Sifferman's laboratory data.

The value less than one would indicate very poor hole cleaning. The CCI equation could be solved for the value of "K" needed to clean the hole:

$$K = \frac{400,000}{(13.6) \times (62)} = 474 \text{ cp}$$

If the plastic viscosity is unchanged, the yield point should be increased to 19—20 lb/100 sqft. With a PV of 22 cp, a yield point of 13 lb/100 sqft would provides a "K"-viscosity of only about half enough (Fig. 6.4).

Diagnostics

The correlation can also be used as a diagnostic tool. Consider a typical drilling situation (this was an actual case history—the names of the participants and the company are not revealed to protect those guilty of "knee-jerk" reaction):

Drilling along at 12,500 ft with a KCl-XC polymer drilling fluid, the annular velocity next to the $4\frac{1}{2}''$ drill pipe was 130 ft/min. Formation pressure is 9 ppg equivalent and the mud properties are: 9 ppg, PV15; YP10, gels 4/8.

The morning report reads: "drag on connections, tight hole, hole closing in, need mud weight to hold formation back".

With this diagnosis, the mud weight was increased to 12.5 ppg. (The morning report indicated the drilling fluid properties changed to 12.5 ppg, PV23, YP15, gels 6/10, with no drag, no tight hole further reported).

The assumption was that the hole was "closing in" and holding the drill collars. The formation was envisioned to be "Plastic" and "flow" into the well bore. The increase in mud weight would push the formation back. (Of course the only problem with this

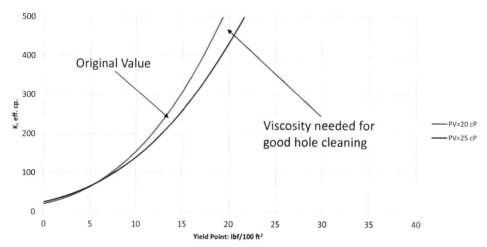

Fig. 6.4 Determining the yield point needed to clean a vertical hole.

concept is that no one has ever calipered 10 ft of shale under gauge. Usually shales fall into the well bore). The assumption that increasing mud weight "opened" the hole up, was clearly incorrect.

However, suppose at the shaker the cuttings were well-rounded in the 9 ppg drilling fluid case and had sharp edges after the mud weight was increased. (Unfortunately most morning report forms do not have a blank for cuttings description.) To test to see if this is a possible alternative, the CCI could be calculated for each mud weight. The K-value for the 9 ppg drilling fluid is 186 equiv cp and for the 12.5 ppg drilling fluid is 283 equiv cp. For the 9.0 ppg drilling fluid:

$$\text{CCI} = \frac{(9.0 \text{ ppg})(130 \text{ ft/min})(186 \text{ equiv cp})}{400,000} = 0.54$$

For the 12.5 ppg drilling fluid:

$$\text{CCI} = \frac{(12.5 \text{ ppg})(130 \text{ ft/min})(283 \text{ equiv cp})}{400,000} = 1.1$$

With a CCI around 0.5, a large number of cuttings were not reaching the shale shaker until they were ground smaller. The increase in mud weight however, also changes the viscosity of the fluid in the annulus and created a carrying capacity that would clean the hole. Now was the hole "closing up" or were the cuttings settling around the tool joints and creating a significant amount of drag? This question could probably be answered in many cases by looking at the caliper log in the shales. Strangely, very few shales, if any, are under gauge. The only three documented cases of under gauge shales have been found after an exhaustive 10 year search of caliper logs where shales were "closing the hole up". In all of the three cases, a four-arm caliper was used. Only two of the arms indicated a diameter less than the bit size. In all three cases, the small diameter was indicated where it appeared that the caliper was rotating so that the large diameter arms moved into the small diameter direction.

Practical suggestions

When applying the CCI correlation, several precautions should be taken to use the correct data.

The annular velocity should be selected for the lowest value in the bore hole. Frequently daily report forms calculate the annular velocity in the open hole next to the drill pipe and the drill collars. The lowest annular velocity may be near the top of the hole inside casing or riser.

The flow rate should be determined from pump calibration values, not from the assumption that the mud pump is 95% or 98% efficient Use the rate of drop in the level in the slug pit while drilling as a means of calibrating the pump. Be sure to account for the air or gas remaining in the drilling fluid. Measure the mud weight with a pressurized mud balance. This value can also be used determine the fraction of gas in the drilling fluid. The volume fraction of gas in a drilling fluid is the difference between the pressurized mud weight and the un-pressurized mud weight divided by the pressurized mud weight. When the pit level falls in the slug tank, some of the volume of the fluid is air or gas and not liquid.

Personal communications from a company man on penrod rig #77.

Comments

This hole cleaning correlation has been field-tested in several areas of Texas, Louisiana, Mississippi, California, Wyoming, Utah, Colorado, the Middle East, the Far East, Africa, and even a few foreign places for both water-based and oil-based drilling fluids. In one polymer system in Oklahoma, a mud engineer was trying a new system and had trouble controlling the yield point in the surface hole. The drilling fluid vacillated about every 6 h from very thin to too thick to circulate in the pit. This provided a rare opportunity to view a wide variety of K-values in a relatively short hole interval. The CCI predicted what the cuttings would look like as they left the shale shaker each time the properties changed.

The lowest annular velocity is in the riser or in the last long string of casing in a well. These values of annular velocity seem to work well for calculating CCI when starting to drill below casing seats. At times the value of CCI = 1, seems to no longer work because the cuttings cease to have sharp edges. At other times, when the CCI is calculated for the first time after drilling a considerable distance below a casing seat, the value of one does not result in sharp cuttings reporting to the surface. Increasing the required value of CCI to 1.2–1.5 has been needed to correctly transport cuttings. Usually, the reason has been that the bore hole has enlarged in diameter and the AV is much smaller than the value used in the calculation. If large cuttings are reporting to the surface (which is an indication of bore hole washout), the CCI may need to be increased to values larger than one.

This hole cleaning correlation is strictly empirical. There is no derivation except for the fact that it accounts for most of the variables which control hole cleaning in almost vertical holes. The criterion appears to do a reasonable job predicting hole cleaning in wells up to about 35 degree, and could also be used in the vertical part of directional wells. No matter what philosophy is used to clean bore holes with angles greater than 35 degree, solids must still be transported through the vertical or near-vertical part of the hole.

Finally, it would be also appropriate to discuss the surprising effect of drilled solids on the carrying capacity index. As drilled solids in a drilling fluid increase, plastic viscosity increases. For a constant YP, increasing the PV decreases the K-value (Refer to Fig. 6.2). This decreases the value of the CCI. Referring to the chart of K-values at a YP of 20 lbs/100 sqft, the K-value decreases from 1152 cp, to 722 cp, to 532 cp, to 432 cp, to 372 cp, as the PV increases from 10 cp, to 15 cp, to 20 cp, to 25 cp to 30 cp respectively. The message here is that removal of the evil drilled solids will assist cuttings transport. This is just another reason that good drilled solids management is crucial for trouble-free drilling.

Hole cleaning for highly deviated wells

In a high angle well solids need only fall a few inches to reach the bottom of the hole. In vertical wells the settling distance is thousands of feet. Two primary cleaning parameters seem important in any analysis: flow rate and pipe rotation.

With simpler drilling fluid systems of many years ago, most drilling contractors subscribed to the concept that the fluid velocity was the only parameter that would prevent settling in pipes. For example in the lower back-flow lines between mud tanks, if the velocity was less than 5 ft/s, barite would settle. Generally, the barite would settle and plug the lower part of the line until the velocity was greater than 5 ft/s (They also tried to prevent the velocity from exceeding 10 ft/s to decrease the likelihood of turbulent flow.) The wisdom of that era was that nothing could prevent settling except velocity. Fortunately, rheological means have been found that will allow several thousand feet of horizontal hole to be cleaned. These techniques are still evolving.

Some interesting work is underway regarding the viscoelastic behavior of drilling fluids. The difference between a liquid and a solid seems rather clear on the surface or from a superficial examination. Consider, however, the addition of Jello to hot water. The initial slurry is liquid. After a short time the mixture sets into a gelled structure. Is it a solid? At what point does the liquid become a solid? Many complex mixtures are described mathematically by examining the relationship between shear stress and shear rate. A simple liquid which has a shear rate directly proportional to the shear stress is called a Newtonian liquid. The constant of proportionality is called the viscosity. An elastic solid has a shear displacement directly proportional to the shear stress. Hooke's law describes the solid by stating that the strain is directly proportional to the stress.

Some materials exhibit characteristics of both liquids and solids. If such material are subjected to an oscillatory stress, the measured strain would not be either exactly in-phase with the applied stress (like an elastic solid) or exactly out-of-phase with the applied stress (like a liquid). The measured strain would be some intermediate angle between zero degrees and ninety degrees out of phase. Some energy would be stored in each cycle and some energy would be dissipated as in a liquid. So the material acts as a viscous material for part of the cycle and an elastic material for part of the cycle. This provides the term: "viscoelastic".

The rheological equation which describes this behavior involves relating the shear stress, γ, to a complex shear relaxation modulus, G. The stress on the material under oscillation at a frequency of $\omega/2\pi$ with a maximum amplitude, τ, could be represented with the equation:

$$\tau = \gamma(G' \sin \omega\tau + G'' \cos \omega\tau)$$

where G' is the shear or elastic modulus (the in-phase component), and G'' is the viscous modulus (the out-or-phase component).

The problem is that the equipment to make these field measurements has only recently been developed and has not yet been applied in the field. The exploration of the behavior of various polymers and clay blends indicate that the yield point, the gel strength, and the 3 or 6 RPM viscometer readings do not predict the responses from viscometers that measure the G' and G'' components. These measurements will probably be eventually used to aid in prediction of horizontal hole cleaning.

Horizontal holes can be cleaned effectively if the drilled solids are prevented from falling from the drilling fluid. If the fluid does not have a reasonably large elastic component of viscosity, G', solids will settle. For example, drilled solids suspended in molasses will eventually fall to bottom. Drilled solids suspended in grape jelly will not fall. Why? The elastic component of viscosity of the molasses is zero. It is a Newtonian fluid (the viscosity is constant no matter what the shear rate is). The grape jelly has a very high elastic component of viscosity and will suspend solids. The question has always been:

"how to produce a fluid that flows easily and has a very high gel structure (or high elastic component) when the flow stops?". Two ways are now available: high concentrations of XC-polymer or Dow Chemicals' mixed metal hydroxides (MMH).

Some of the most significant work to improve removal of drilled solids from high-angle wells was the recent development of the concept of critical polymer concentration (CPC). XC-polymers have been used in drilling fluid for many years; primarily to increase the low-shear rate viscosity, to clean vertical holes. Usually the concentration of XC was relatively low because the specified yield point was easily attained. Powell, Parks, and Seheult in 1991 reported that increasing the concentration of XC to a CPC increased the G', or the elastic component of viscosity. High concentrations of XC-polymer (above the 1.75 lb/bbl CPC) behave differently because of self interference. Some of the benefits of high concentrations of XC in fluids compared with conventional fluids are listed in the Powell et al. paper:

- Pump pressures will be lower for the same flow rates.
- Circulation lag time is reduced.
- Torque and drag is reduced due to improved hole cleaning.
- Fewer problems running logging tools, casing, or liners.

The second development that has been achieved recently is the use of the MMH fluid. MMH is a highly positively charged man-made additive that creates some unusual drilling fluid properties. An MMH drilling fluid in a well in East Texas had a funnel viscosity of 45 seconds yet it would support a 2″ diameter rock picked up on location. The turnkey contractor claimed that they were sinking record wells because of several benefits of the fluid - primarily better hole cleaning. MMH behaves differently from most fluids with a very high gel structure. Bentonite requires the application of high shear rates to lose the gel structure. MMH breaks with shear strain: therefore very low standpipe pressures required to initiate flow after a trip. The MMH fluid will also provide a very large elastic component of viscosity. The additive is very sensitive to treatment on the surface and competent mud engineers are an absolute necessity.

The best recent thoughts on cleaning horizontal holes is represented by the three papers:

- Powell, Parks, and Seheult have found that a critical (and much higher than normal) concentration of XC polymer can effectively clean horizontal holes—"Xanthan and Welan: The Effects of Critical Polymer Concentration on Rheology and Fluid Performance 1991" SPE Paper # 22,066
- Zamora, Jefferson, and Powell reported on their results comparing XC polymer with hydroxyethyl cellulose. They found rheological "fingerprints" that provide optimum hole-cleaning and suspension. "Hole-Cleaning Study of Polymer-Based Drilling Fluids" SPE Paper # 26,329

- Zamora and Hanson published some "Rules of Thumb to Improve High-Angle Hole Cleaning" that are based on their years of research work. Zamora, M. and Hanson, P. 1991. More Rules of Thumb to Improve High Angle Hole Cleaning. Pet. Eng. Intl. 22.

Suggestion

Try CCI and see if it works for you. All data should be available at the rig without requiring additional measurements. Remember the appearance of the cuttings, not the quantity, should be used to determine whether CCI correctly describes hole cleaning.

The symptoms of unstable shales or swelling shales are usually related to an increase in torque and drag and particularly "tight spots" while tripping pipe. It is easy to envision the shale creeping into the wellbore and causing problems; so the increase in mud weight will prevent the creep. It is not easy to find caliper logs that confirm this situation.

Appendix 6.A
Derivation of the effective viscosity term "K"

Viscosity by definition is the shear stress divided by shear rate. If shear stress has the units of dynes/sq cm. and shear rate has the units of reciprocal seconds, the viscosity will have the units of poise. With the viscometer, the dial reading may be converted into dynes/sq cm by multiplying by 5.11 and the viscometer RPM can be converted into reciprocal seconds by multiplying by 1.703. For the Power Law equation:

$$\tau = K(\gamma)^n$$

A similar conversion would yield:

$$K = \frac{R_{300}(5.11)(100 \text{ cp/poise})}{(300 \text{ RPM})(1.703 \text{ s}^{-1}/\text{RPM})^n} = (511)^{1-n} R_{300}$$

The K-value would therefore have units similar to viscosity except for the exponent on the shear rate term. The term "equivalent cp" described these units in the *Applied Drilling Engineering* SPE Textbook Series, Volume 1, by Bourgoyne, Millheim, Chenevert and Young, Pg. 476. Many publications use the K-value with units where the RPM has been converted to reciprocal seconds, but the shear stress is left with the units of lb/100 ft².

The value of "n" can be found without unit conversion since it is a dimensionless number. For two cases, the 600 RPM and the 300 RPM viscometer readings:

$$R_{600} = K(600)^n$$
$$R_{300} = K(300)^n$$

These two equations could be solved for "n" by dividing the second equation into the first and taking the logarithm of both sides. This produces

$$n = 3.322 \, \text{Log} \frac{R_{600}}{R_{300}}$$

or in terms of PV and YP:

$$n = 3.322 \, \text{Log} \frac{2\text{PV} + \text{YP}}{\text{PV} + \text{YP}}$$

Appendix 6.B

The annular velocity curves in this Appendix were calculated from the dimensions listed below in Figs. 6B.1–6B.8. Table 6B.1 shows the actual casing IDs for casing sizes ranging from 7 to 30 inches (nominal).

Table 6B.1 Annular spaces for calculating annular velocities.

Nominal diameter, in.	7	8⁵⁄₈	9⁵⁄₈	10³⁄₄	11³⁄₄	13³⁄₈	16	20	30
Casing ID, in	5.92	7.511	8.535	9.76	10.85	12.415	15.01	19	30

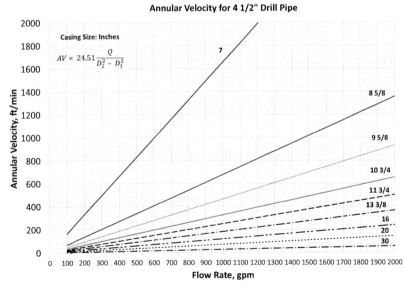

Fig. 6B.1 Annular velocity for $4\frac{1}{2}$" drill pipe #1.

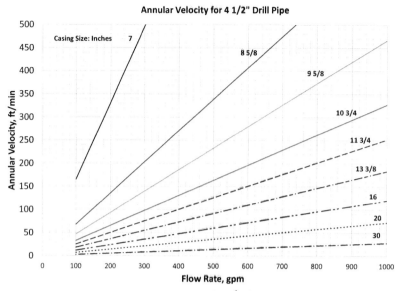

Fig. 6B.2 Annular velocity for $4\frac{1}{2}$ " drill pipe #2.

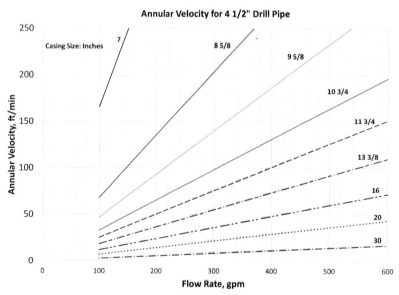

Fig. 6B.3 Annular velocity for $4\frac{1}{2}$ " drill pipe #3.

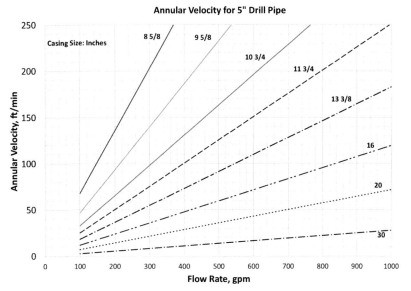

Fig. 6B.4 Annular velocity for 5″ drill pipe #1.

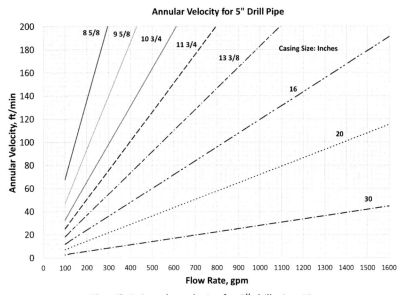

Fig. 6B.5 Annular velocity for 5″ drill pipe #2.

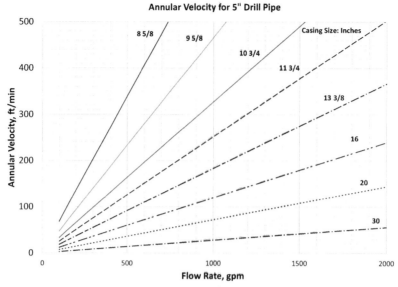

Fig. 6B.6 Annular velocity for 5″ drill pipe #3.

Fig. 6B.7 Annular velocity for $6^5/_8''$ drill pipe #1.

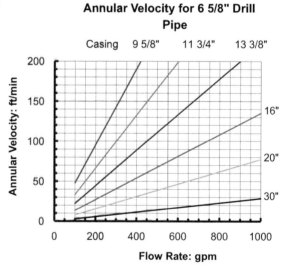

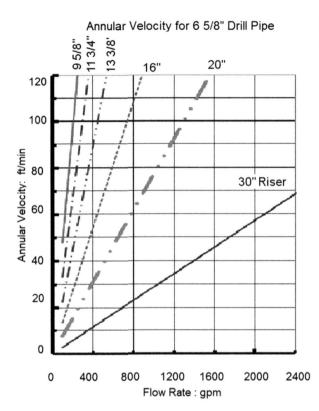

Fig. 6B.8 Annular velocity for $6^5/_8''$ drill pipe #2.

Appendix 6.C

Cuttings transport problems

Problem 6.1

Part 1

The lowest annular velocity for this well is 60 ft/min. For the drilling fluid properties listed below, determine which wells will have a problem with hole cleaning.

Fluid #	MW, ppg	PV, cp	YP, lb/100 sqft	K, eff cp	CCI
1	9.0	10	10		
2	9.0	5	5		
3	9.0	5	10		
4	15.0	30	5		
5	15.0	30	20		
6	10.0	10	15		

Part 2
The yield point of a drilling fluid can be changed without a significant change in the PV. What should the YP be to make certain that all drilling fluids are cleaning the borehole?

Fluid #	K needed, eff cp	YP, lb/100 sqft
1		
2		
3		
4		
5		
6		

Sample calculation:

$$K = \frac{400,000}{(\text{MW, ppg})(\text{AV, ft/min})}$$

From the K-value graph (Fig. 6C.1), read the YP needed.

Problem 6.2

An 11.0 ppg water-based PHPA drilling fluid was being used to drill a $9^7/_8''$ hole below an $11^3/_4''$ casing. The circulation rate was 550 gpm down the $4^1/_2''$ drill pipe (annular velocity 134 ft/min). The drilling fluid properties have been deteriorating slowly during the past week. The plastic viscosity has been slowly increasing from 11 cp to 22 cp; the yield point was held constant at 11 lb/100 sq ft. and the solids content has increased from 13% vol. to 19% vol. The filter cake quality was continually decreasing.

As this trend started, Joe Stumpem was diligently plotting the variable and had several conversations with the drilling foreman, Heizen Schwanz. They slowly increased the yield point of the drilling fluid from 11 to 15 lb/100 sqft. At the same time the gels increased from 4/10 to 9/20. To take full advantage of their linear motion shakers, the screens were changed from 120 mesh to 200 mesh. During the week, the change in drilling fluid properties required that the end of the shaker be elevated eventually to the maximum height possible to prevent loss of drilling fluid from the end of the shaker. The derrick man was instructed to thoroughly examine the shaker screens for holes or rips. The mounting of the screens on the decks were evaluated for leaks between shaker and the screens. The mud cleaner was inspected. No mechanical problems could be located which would explain the rise in drilled solids during this interval of the well. Other wells in the field, using different mud systems, had drilled this interval without the increase in drilled solids and generally only 100 mesh screens had been used by off-set operators.

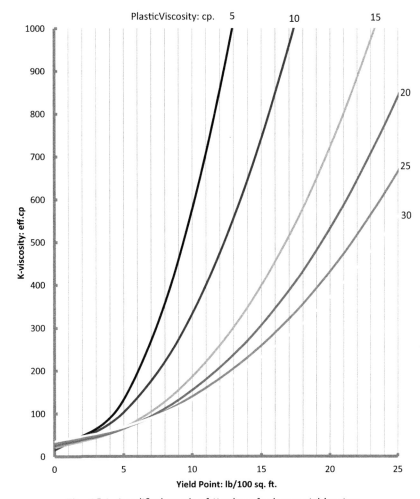

Fig. 6C.1 Amplified graph of *K*-values for lower yield points.

Ineffective hole cleaning could account for the symptoms plaguing this rig. The cuttings coming from the shale shaker were cubic pieces of shale about 1/16″ on each side. Small cuttings normally mean that the hole is not being cleaned; although in this case the edges of the cutting were sharp not rounded. Poor hole cleaning causes cutting to tumble in the hole and rounds the edges of the cuttings. The Carrying Capacity Index (described in PEI, Sept. 1993) in the largest annulus was initially above 1.0 (PV 11, YP 11 ($K = 293$)) (AV 134, MW 11). When Joe arrived at the rig, he increased properties (PV 11, YP 15 ($K = 284$)) to return the CCI to a value above 1.0.

What should be done? Maybe the CCI doesn't help in this situation. Or, do they need a centrifuge, another linear motion shaker, or some other equipment?

Problem 6.3

You arrive at a drilling rig and the tool pusher greets you with enthusiasm and asks that you to immediately go to the rig floor. He and the driller have had some serious discussions about the situation. They are observing the classic symptoms of a problem they have encountered many times. In this case, they have not been able to get the office drilling group to listen to their recommendations.

The rig is drilling at 11,200 ft with a $14^3/_4''$ bit, IADC Code 127. The 16" surface casing was set deep, 4000 ft, because this well will be exploring formations below current production. They have 5" drill pipe and the bottom five collars are 10" diameter. The drilling fluid properties are reported to be: Mud weight 11 ppg, PV 20 cp, YP 10 lb/100 sqft, gels 10/30, pH 10.0, funnel visc 52 s. The flow rate is 800 gpm. The tool pusher points out that he was told to increase the viscosity of the fluid to help bring the cuttings out, so he has allowed the PV to go to 20 cp. He also has some friends who are selling him "hole slickeners" at $200/drum and he is adding two of those per tour. He is keeping the yield point to a modest value of 10 so it will be easier to pump the fluid down the hole and blend it in the mud tanks.

They want you to watch a connection and see the drag on the pipe when the pump is off. The driller says it feels like the shale is closing in around the bit. The tool pusher and driller say that this is a common occurrence in this area and they have seen it many times. Turning on the pumps applies pressure to the bottom of the hole and opens the hole up. This, they report, only cures the problem for a short period of time and then they must increase the mud weight to continue to be able to make connections. When they reported this to the office, they claimed that the weevil operations people thought it might be a hole cleaning problem. From the rig floor, the cuttings could be observed so deep on the shaker that it was hard to see the shaker screen, so they did not believe this explanation.

In their experience, this situation will get worse unless several things are done immediately. They want to trip out of the hole and remove those large drill collars before the shale squeezes in around them so they can't move the pipe. They say this happens all the time by people not heeding their warning. They want to increase the mud weight to 13 to 14 ppg. They admit that the morning report shows an 11 ppg drilling fluid, but it actually weighs 12.0 ppg. They say they were trying to help the inexperienced people in the office by increasing the mud weight; but they can't go much higher before the barite use reveals what they are doing. In their experience, the shale will finally squeeze in around the bit so they can't even pump. Then, after a connection, when they turn on the pumps, they will lose circulation. The liquid level in the annulus will drop and the well will start flowing. Closing the well in will guarantee stuck pipe and this disaster will require about a month to correct. They also say that the shale that creeps into the well bore gets harder and harder. They already see this—as proof—because the drilling rate has dropped from 50 ft/h to 30 ft/h. After they increase the mud weight to hold the shale back, they expect to be drilling about 5–8 ft/h. All the symptoms are there and they encourage you to act now before it becomes too late.

Are you going to trip the bit and remove the big drill collars before increasing the mud weight? Do you increase the mud weight before the trip? Are you going to agree with these seasoned professionals who have seen this series of events many times in this drilling area? What are you going to do?

To help sort through these suggestions: First, what do the cuttings look like on the shaker screen?

Problem 6.4

Drilling a $17\frac{1}{2}''$ hole with $5''$ drill pipe through $20''$ surface casing with a 9.5 ppg drilling fluid. Pumping 600 gpm of NADF at a depth of 5300 ft, plastic viscosity is 15 cp, the yield point is 5 lb/100ft^2, and the gels are 2/6. Are the cuttings being transported correctly? If not, what changes should be made?")

Problem 6.5

Gel/Lignosulfonate Drilling Fluid.
   MW = 10.5 ppg
   VS = 14% volume
   AV = 100 ft/min
   MBT = 27.5 lb/bbl
   pH = 10.5

Viscometer readings		
RPM	Dial reading	Viscosity, cp
600	33	
300	20	
200	18	
100	13	
6	2	
3	1	

1. Describe cuttings.
2. Calculate change that needs to be made.

Problem 6.6

Drilling a 17½" hole at 6000 ft below 20" casing using 5" drill pipe circulating a 9.0 ppg drilling fluid at 1000 gpm.

1. Calculate the carrying capacity (CCI) for the following drilling fluid properties:

PV, cp	YP, lb/100 sqft.	K, eff cp	CCI
15	10		
15	15		
15	20		
10	5		
10	10		
10	15		

2. What K value is needed?

Problem 6.7

Drilling a 12¼" hole at 10,000 ft below 16" casing using 5" drill pipe circulating a 10.0 ppg drilling fluid at 1000 gpm.

1. Calculate the carrying capacity (CCI) for the following drilling fluid properties:
   AV=_____

PV, cp	YP, lb/100 sqft.	K, eff cp	CCI
20	20		
20	15		
20	10		
15	20		
15	15		
15	10		
10	15		
10	10		
10	5		

2. What K value is needed?

Problem 6.8

Drill fluid properties for 9800 vertical well.
   PV = 25 cp
   YP = 10 lb/100 ft^2
   MW = 10.0 ppg
   AV = 100 ft/min

Are the cuttings being transported properly? If not, what changes should be made?

Problem 6.9

PV = 10 cp
YP = 12 lb/100 ft^2
MW = 11.0 ppg
AV = 100 ft/min
Are the cuttings being transported properly? If not, what changes should be made?

Problem 6.10

PV = 20 cp
YP = 10 lb/100 ft^2
MW = 15.0 ppg
AV = 60 ft/min
Are the cuttings being transported properly? If not, what changes should be made?

CHAPTER 7

Dilution

Introduction

Drilling performance is directly affected by the quantity of drilled solids contained in the drilling fluid. Maintaining a low concentration of drilled solids in the drilling fluid is required to drill with the lowest cost per foot overall. It is crucial to understand where solids control begins—at the bit—and that if the rheological properties can be manipulated by maintaining a low concentration of drilled solids in the fluid, there are several key drilling factors that can be optimized:

1. Hydraulic Impact: To assist in the removal of cuttings from the bit, the hydraulic impact or power to assist in the removal of cuttings from beneath the bit must be maximized. The plastic viscosity should be as low as possible in able to assist in this effort.
2. Weight on Bit: Drill off tests should be conducted to determine the founder point and ensure that the weight on roller cone bits or torque for PDC bits do not exceed their limit.
3. Hole Cleaning: Both plastic viscosity and yield point affect the ability of the drilling fluid to efficiently carry the drilled cuttings to the surface in both vertical and deviated wells. To effectively clean the hole, the plastic viscosity must be kept to a minimum which means eliminating drilled solids. This will create a thin, slick, compressible filter cake and decrease the instances of stuck pipe, fracture-induced lost circulation, and provide for reduce surge and swab pressures.
4. Drilled Solids Removal: The first opportunity to remove the drilled solids from the fluid at the surface is at the shale shakers. Only API compliant screens should be used on the shakers to do this.
5. Solids Control Equipment Arrangement: The drilling fluid solids removal section should be arranged such that the maximum equipment solids removal efficiency is possible.
6. Waste Volumes: Good equipment solids removal efficiency will decrease the quantity of fluid required to maintain a targeted drilled solids concentration. This also reduces the drilling fluid waste volumes generated when dilution is used to maintain the targeted concentration.

True costs of drilling a well

Drillers today recognize that too many drilled solids in the drilling fluid can create expensive wells overall. The true cost of poor solids control performance is frequently overlooked and misunderstood due to the fact that it is incorporated in other costs such as drilling fluid, waste disposal, and non-productive time (NPT). The overall drilling performance suffers if too many drilled solids remain in the system.

Although the cost of poor drilling performance can far exceed trouble costs, rig down-time [NPT] is more obvious and is usually cited as the reason for improving the equipment solids control efficiency.

In General, the true cost of drilling increases as the concentration of drilled solids in the drilling fluid increases (Fig. 7.1).

There are 4 methods for controlling drilled solids in a mud.

1. Dilution
2. Chemical removal
3. Solids Settling
4. Mechanical Separation

Often a combination of all of these methods is performed in order to have the most efficient and cost effective solution. The remainder of this chapter is dedicated to discussing dilution and the basics involved. Dilution is often thought of as a negative, but in all wells it happens to some extent. The trick is to find the sweet spot, of dilution, solids removal, and solids concentration. The sweet spot would be maintaining the lowest possible LGS percentage, while only having to add mud to account for the hole volume drilled.

Dilution principles

Removing some of the "dirty" drilling fluid and replacing the volume with clean drilling fluid containing no drilled solid scan control drilled solids. This is an effective method of solids control however, it is an expensive method.

Fig. 7.1 Drilling costs increase with increases in drilled solids concentration.

The formula used for determining the amount of clean drilling fluid required to dilute a volume of drilling fluid containing a certain percentage of low gravity solids to a targeted percentage is defined by:

$$V_{wm} = \frac{V_m(F_{ct} - F_{cop})}{F_{cop} - F_{ca}}$$

where:

V_{wm} = barrels of dilution water or whole mud required

V_m = barrels of drilling fluid in circulating system

F_{ct} = %vol low gravity solids in system

F_{cop} = %vol targeted low gravity solids

F_{ca} = %vol commercial solids (bentonite, LCM, Barite, and/or chemicals added)

Similarly, to determine the amount of drilling fluid in a system that must be jetted to allow for dilution volume can be determined by the following equation:

$$V_{jet} = \frac{V_m(F_{ct} - F_{cop})}{F_{ct} - F_{ca}}$$

where:

V_{jet} = barrels of circulating system drilling fluid to be jetted

For example, envision 2000 barrels of drilling fluid in a well and in the mud tanks collected into a single tank. Assume the drilling fluids specifications require 5% volume targeted drilled solids (which would be 100 bbl). After drilling 1250 ft of a $9^7/_8''$ hole without removing any drilled solids, the volume of drilled solids would increase by approximately 100 bbl of solids if the formation had 15% porosity. This would double the volume of drilled solids in the system (Fig. 7.2).

To meet the required drilling fluid specification of 5% volume drilled solids, one-half of the drilling fluid must be discarded (Fig. 7.3):

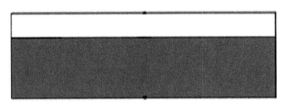

Fig. 7.2 2000 bbls drilling fluid containing 200 bbl of drilled solids or 10% volume.

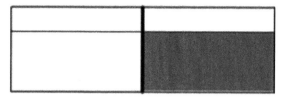

Fig. 7.3 1000 bbls drilling fluid discarded leaving 1000 bbls of drilling fluid containing 10% volume drilled solids.

$$V_{jet} = \frac{V_m(F_{ct} - F_{cop})}{F_{ct} - F_{ca}} = \frac{2000(10-5)}{10}$$

If clean drilling fluid is now added to the system, the 10% by volume of drilled solids in the 1000 bbls (or 100 bbl of drilled solids) will now be spread throughout the drilling fluid system of 2100 bbl. The new hole volume has increased by 100 bbl. This meets the specifications for the drilling fluid as required by the drilling program. If the drilling fluid costs only $20/bbl, the cost of decreasing solids in this manner is prohibitive (Fig. 7.4).

After drilling only 1250 ft of new hole, 1000 bbls of drilling fluid must be discarded to bring the drilled solids back into a reasonable value. A lower concentration of drilled solids would be better but far too expensive when dilution is used to control drilled solids. Two costs are associated with this process: the cost of the new drilling fluid (1000 bbl) and the cost of disposal of the dirty 1000 bbl discard. With drilling fluid costs ranging from $30 to $600 per barrel, the cost would be prohibitive to use this method of solids control except for the cheapest of the cheap drilling fluids. Because it is so expensive, a compromise is frequently made to allow the drilled solids to increase to levels above 10%–12% by volume.

Solids removal efficiency is defined as the volume of drilled solids removed divided by the volume of drilled solids reporting to the solids control equipment. If 100 bbl of drilled solids report to the solids control equipment and 60 bbl are removed, the equipment solids removal efficiency is 60%. Because of the fact that the drilled solids don't all arrive at the surface in the same order in which they were drilled, the calculation of solids removal efficiency requires that a long interval be drilled. If $12^1/_4''$ hole is drilled from

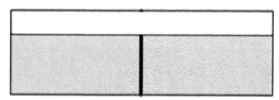

Fig. 7.4 After dilution, the drilling fluid once again contains only 100 bbl of drilled solids for the 2100 bbl.

4000 ft to 8400 ft, the total drilled solids reporting to the surface can be determined with a caliper log and a calculation of the volume of cuttings removed. If nothing is removed from the system, the pit levels will raise proportional to the volume of drill pipe inserted into the hole through the interval. In general, drilled cuttings have a negligible coefficient of expansion when they are brought to the surface, thus the volume of drilled solids and liquid exactly matches the volume of new hole drilled (Fig. 7.5).

Pit levels will only decrease when fluid or solids are removed from the system through losses in the solids control system, evaporation, or entering formation. Cuttings removed from the system through the solids control system will be wet with drilling fluid, thus both drilling fluid and solids will be removed. Generally, the best system is one in which the volume removed (or decrease in pit volume) exactly matches the volume of fluid required to dilute the remaining drilled solids to a target concentration (Fig. 7.6).

The amount of clean drilling fluid required depends upon the amount of solids removed along with the adsorbed fluid. Fig. 7.6 depicts the amount of clean drilling fluid required in barrels per barrel of hole drilled to maintain 5% drilled solids in the active system as a function of the solids removal efficiency and the discard makeup. In this case, the concentration of drilled solids in the discard is assumed to be 35% volume. This is discussed more completely in Chapter 12.

Frequently, when the solids control equipment is inadequate or, more often, plumbed incorrectly, the drilled solids will increase somewhat more slowly. If the target drilled solids concentration can be raised to a much higher concentration, less drilling fluid needs to be used to meet the specifications. The NPT, (visible and invisible),

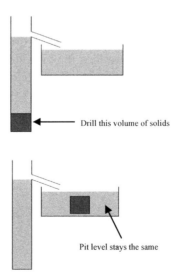

Fig. 7.5 Hole volume generated is exactly the volume of solids and liquid that reach the pits.

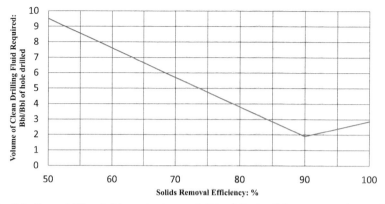

Fig. 7.6 Clean drilling fluid required to maintain drilled solids concentration at 5%.

however, will notice the relaxation of the stringent requirements. The out-of-pocket money for treating the drilling fluid will be lower but the total cost of the well (and long term effects) will be significantly higher.

While removing drilled solids with the solids control equipment, a portion of the drilling fluid system will be removed along with those drilled solids. For example, if he drilled solids removal efficiency is 60%, the volume of solids remaining in the system can be calculated. The actual volume of total discard depends upon the quantity of liquid and drilled solids removed or the drilled solids concentration in the discarded fluid. Assume 1000 bbl of drilled solids arrive at the surface during drilling of one interval. With the 60% removal efficiency, 400 bbl of drilled solids would remain in the system and 600 bbl of drilled solids would be discarded. If the discarded solids constitute 35% volume of the total discard, the volume removed from the system would be 1714 bbl [or 600 bbl/ 0.35]. This would be the volume available to add clean drilling fluid to dilute the remaining solids. If only this volume of clean drilling fluid is added to the system, the 400 bbl of remaining drilled solids would be dispersed in 1714 bbl of clean drilling fluid. The new drilling fluid volume would be 2114 bbl (400 bbl solids + 1714 bbl clean drilling fluid). The drilled solids concentration in this new addition would be 18.9% volume. Conversely, if the solids removal efficiency was increased to 80% in this same scenario, a total volume of 2286 bbl of discard replaced with that same amount of clean drilling fluid and blended with the 200 bbl of remaining drilled solids would yield an ending concentration of 8% volume in the new addition.

Application

The above calculations assume that the drilled solids concentration in the pits remain constant throughout the interval. In field operations, this is extremely difficult to do.

Pits are dumped regularly to allow addition of clean drilling fluid to either maintain low drilled solids concentration or modify the fluid rheology. Typically on a drilling rig, drilling fluid is dumped for a variety of reasons and the decisions are predicated upon many different concepts. Often times, the drilling fluid is dumped because it would be deemed as "worn-out" most likely due to poor solids control as derrick men or mud engineers have a difficult time maintaining a low drilled solids concentration. Fluid is dumped and diluted on a time schedule, on whims of the rig personnel, on a footage basis, on perceived increases in solids content, or for a variety of other reasons. Regardless, the fluid is dumped and diluted with some sort of periodicity.

For illustration purposes and comparative analysis, consider a case in which the targeted drilled solids concentration in a drilling fluid system is 5% volume. If a hole is drilled from 4000 ft to 8400 ft, the caliper log indicates a volume of 0.146 bbl/ft, and the solids removal efficiency is assumed to be 60%, a determination of the amount of drilled solids retained in the active system can be predicted per interval before diluting. For a 2000 bbl system, if the interval drilled before diluting is set at 1000 ft, the drilled concentration profile is illustrated in Fig. 7.7. Note that in this case, the concentration is allowed to climb as high as 8.7% in the first interval. Note that the average drilled solids concentration over the 4400 ft interval is 6%.

When the drilled solids concentration is calculated for 100 ft intervals, the drilled solids increase from 5% volume to approximately 5.2% volume resulting in an overall average of 5.1% volume throughout the 4400 ft interval as illustrated in Fig. 7.8.

The effect of postponing the adjustment can be observed at various interval depths in a well by comparing the maximum drilled solids concentration for different drilling intervals before adding dilution. (Fig. 7.9) depicts the effect of diluting at various interval

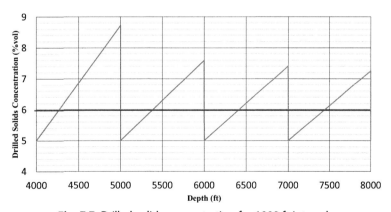

Fig. 7.7 Drilled solids concentration for 1000 ft intervals.

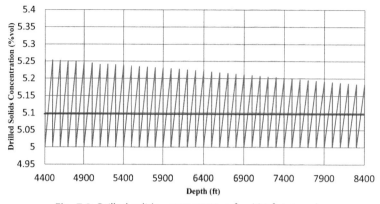

Fig. 7.8 Drilled solids concentration for 100 ft intervals.

depths on the ending drilled concentration for the interval before the actual occurrence of dilution. For example, if the starting concentration of drilled solids in a mud system is 5% vol and the decision is made to dilute every 600 ft of hole drilled, the concentration of drilled solids would be allowed to climb to 6.41% vol based on.

One item of consideration when determining the interval drilled before dilution should be the amount of clean drilling fluid required. Of course in each instance, the quantity of clean drilling fluid decreases as the interval between dilution additions is increased as in Fig. 7.10. However, that would be expected as less fluid is required for dilution when the drilling fluid is allowed to acquire more drilled solids.

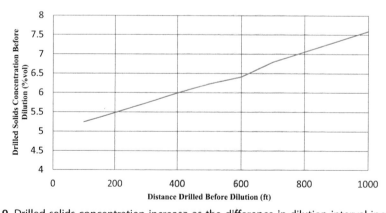

Fig. 7.9 Drilled solids concentration increase as the difference in dilution interval increases.

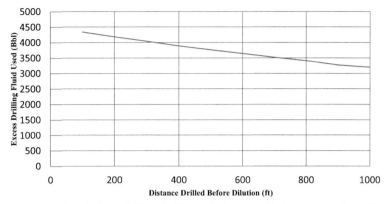

Fig. 7.10 Quantity of clean drilling fluid correlation to drilled interval before dilution.

Problems
Problem 7.1
Assume the Equipment Solids Removal Efficiency (SRE) is known.
  Situation:
  Drilling a $9^7/_8''$ hole at 44 ft/h for 24 h with a 15 ppg drilling fluid which has a target of 5% volume low gravity solids. The total volume of drilling fluid (active surface and hole) is 2000 bbl. Solids are discarded in a flow stream containing 35% volume drilled solids.
  Assume density of barite is 4.2 g/cc and of low gravity solids is 2.6 g/cc. Also, assume a 60% SRE.

1. Calculate the volume of drilled solids discarded and the volume of drilled solids remaining in the system.
2. Calculate the volume of fluid discarded from the system.
3. Calculate the volume of fluid needed to keep the pit levels constant while drilling this interval.
4. Calculate the quantity of barite needed in the clean drilling fluid added to keep the pit levels constant.
5. Calculate the increase in drilled solids concentration while keeping the pit levels constant.
6. Calculate the volume of excess drilling fluid required to keep the drilled solids concentration at the targeted value.
7. Calculate the quantity of barite which would have been added to the system to keep the drilled solids concentration at 5%vol.
8. Calculate the changes in the retort readings that a mud engineer would see after drilling this interval of hole.

Problem 7.2

Assume the Equipment Solids Removal Efficiency (SRE) is not known and needs to be determined. Using the same hole interval and using the information about additives to the drilling fluid, calculate the SRE.

Situation:

Drilling a $9^7/_8''$ hole at 44 ft/h for 24 h with a 15 ppg drilling fluid which has a target of 5% volume low gravity solids. The total volume of drilling fluid (active surface and hole) is 2000 bbl. Solids are discarded in a flow stream containing 35% volume drilled solids. Assume density of barite is 4.2 g/cc and of low gravity solids is 2.6 g/cc.

After drilling this interval, the mud engineer reports that 604 sacks of barite was required to keep the clean mud additions at 15.0 ppg.

1. How much clean drilling fluid volume would require this quantity of barite?
2. What volume of drilled solids was discarded while drilling this interval?
3. What is the solids removal efficiency?
4. The mud engineer reports that the drilling fluid in the tanks after drilling this interval of hole has a solids concentration of 29.1% volume. Find the volume percent of drilled solids in this fluid.
5. How much fluid needs to be discarded to make room for clean drilling fluid to reduce the drilled solids back to the target value of 5% volume?
6. The mud engineer reported that 374 sacks of barite were used after the pit levels were decreased by 406 bbl to allow for the clean dilution drilling fluid to be added. Was this correct? How much barite must be used in the clean drilling fluid after discarding the 406 bbl?

Problem 7.3

Never-wrong Drilling Company (the name has been changed to protect the guilty) is drilling a $12^1/_4''$ hole from a casing seat at 5000 ft to the next casing seat at 10,000 ft. Assume the formations have an average porosity of 10% volume. The 12.0 ppg water-base drilling fluid has a 18% volume solids content with an MBT of 18 lb/bbl. After drilling the well, a caliper indicates that the borehole washed-out to an average diameter of 14''. While drilling this 5000 ft interval, clean drilling fluid was added to the system to maintain a constant 18% volume solids concentration. The discard from all of the solids removal equipment was captured (to haul off location) and was 1840 bbl. The discard contained 35% volume drilled solids.

1. What would be the solids removal efficiency?
2. How much clean drilling fluid would be necessary to maintain a constant 18% volume solids while drilling this interval?
3. How much excess fluid must be discarded because the solids removal efficiency is so low?

4. If the solids removal efficiency was increased to 85%, how much clean drilling fluid would be required?

Comment: This is an "academic" type problem because it assumes that the average porosity of the formations is known. It also assumes that the solids content of the drilling fluid is constant during the entire time this 5000 ft was drilled. As explained in the text, this might happen if the removal efficiency is the optimum value for the target drilled solids concentration. However, generally, the drilled solids concentration will increase for some period of time and then measures will be implemented to eliminate the excess drilled solids. The purpose of the calculations is to provide some understanding of the high cost of failure to correctly remove drilled solids from the drilling fluid using properly plumbed equipment.

Problem 7.4

With this inefficient drilled solids removal system, the decision was made to drill with a polymer drilling fluid system that required a 4% volume drilled solids concentration. If disposal costs $40/bbl and drilling fluid costs $90/bbl, find the cost of reducing the drilled solids concentration only 2.5%.

SECTION II

Drilling fluid processing – mechanical separation of solids

CHAPTER 8

Surface drilling fluid systems

Introduction

The tank arrangement and equipment in the suction section is the same for all types of weighted or unweighted drilling fluids. However, the equipment arrangement for solids removal is different for weighted and unweighted drilling fluids. Some of the choices of solids removal also depends upon the type of shale shakers available on the drilling rig generic tank arrangements will be discussed in this chapter. In Chapter 9, the removal section will be discussed in much more detail.

Generic systems for unweighted drilling fluid

A basic surface processing system for unweighted drilling fluid, (Fig. 8.1), consists of a system that sequentially removes the larger solids first and smaller ones afterward. In some areas, gumbo is formed in the annulus and needs to be removed before the drilling fluid reports to shale shakers. Gumbo is usually a calcium montmorillonite that is a "sticky" clay. If these clay particles tumble in the annulus on the way out of the bore hole, they will stick together forming large clumps. After the gumbo is removed, the drilling fluid

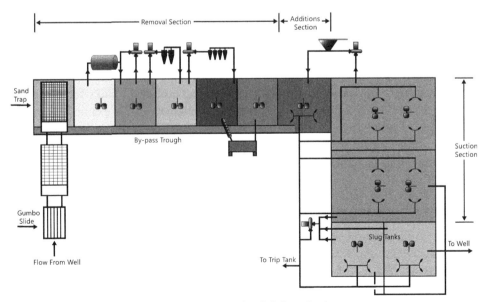

Fig. 8.1 Unweighted drilling fluid.

then reports to a scalping shaker. These shakers have very coarse screens and are designed to remove large chunks of rock that might damage a very fine shaker screen. The fluid is then processed through a main shaker which will remove the larger solids from the fluid. The scalping shaker and the gumbo removal equipment are not needed on all rigs. A fine screen shaker is very desirable and will greatly assist in creating a trouble-free drilling fluid.

A sand trap is useful beneath shakers that cannot process drilling fluid through fine screens. If linear motion or balanced elliptical motion shakers are mounted on the drilling rig, the sand trap does not effectively remove the small solids which pass through the fine screens. All of the compartments, except the sand trap, should be well-agitated and mud guns should not be used in the removal section.

The drilling fluid from the removal section can overflow into the additions/suction section. This helps keep the liquid level high so that the centrifugal pumps in the removal section will not be subjected to air vortexing into their suctions. This, also, assists observations of changes in pit levels to detect lost circulation or kicks. The additions section and the suction section should be well agitated and are blended together by using mud guns to homogenize the drilling fluid in these two compartments. Drilling fluid additives are added into the compartment downstream of the removal section.

Degassers are mounted on the mud tanks to remove gas from the drilling fluid so that the centrifugal pumps in the system can function properly.

If coarse screens are mounted on the main shakers, the solids load passing through the screens will generally be too large for the desilters to process. Desanders should process the drilling fluid before the fluid is pumped to the desilters. If fine screens are mounted on the main shale shakers, the desander will not be needed, (Fig. 8.2). These fine screens remove the solids that were normally removed by the desander.

If a centrifuge is required, it will process the fluid which has passed through the desilters. Generally, the centrifuge does not process all of the rig flow rate as all of the preceding equipment does. This completes the removal section.

Surface system suction section should have a couple of small tanks for mixing and pumping pills or small volumes of specially prepared drilling fluid. For example, when tripping pipe, a weighted slug can be prepared in the pill/slug tanks to prevent drilling fluid from splattering on the rig floor during the trip. A viscous slug of drilling fluid can be prepared to help clean the hole if cuttings are not being transported from the hole properly.

Generic systems for weighted drilling fluid

A properly arranged unweighted drilling fluid tank system can be converted relatively easily into a system which will process weighted drilling fluid. A weighted drilling fluid is any fluid which has had commercial ingredients added to increase the mud weight. Again, if the shale shakers are not capable of handling drilling fluid flow through screens finer than API 80, sand traps can be effectively used in either weighted or unweighted

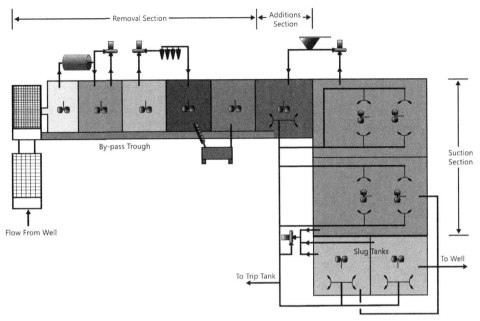

Fig. 8.2 Unweighted drilling fluid with Fine screen shakers.

drilling fluids, (Fig. 8.3). If fine screens are mounted on the shale shakers (such as API 140 screens), most of the settled solids will be the weighting agent—like barite.

If the shale shakers can process the fluid through fine screens, like API 140 or finer, the settling pit is not needed for either the unweighted or weighted drilling fluid, (Fig. 8.3). The desander is not used for weighted drilling fluids because it would discard too much of the weighting agent. The desilters will also concentrate the weighting agent along with the drilled solids. Most of the barite should be finer than 75 microns. An API 170 or API 200 screen mounted on the mud cleaner will return most of the barite and some drilled solids back to the drilling fluid. All of the larger particles will be removed by the mud cleaner shaker screen.

The rest of the system for a weighted drilling fluid will be exactly the same as used with the unweighted drilling fluid system.

Alternate system

Frequently, on large drilling rigs, several shakers are required to process the drilling fluid because the flow rates are so high. After the scalping shaker, a flow distributor can equally divide the flow rate from the well so that each shaker has an equal amount, (Fig. 8.4).

Drilling fluid from the well flows down the flow line to a scalping shaker with a very coarse screen and then into the flow distribution chamber, (Fig. 8.5). From this chamber, the flow to each shaker can be easily adjusted.

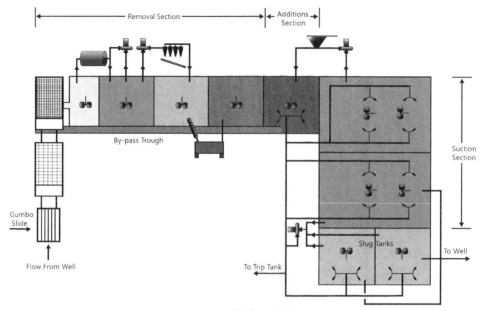

Fig. 8.3 Weighted drilling fluids with fine screen shakers.

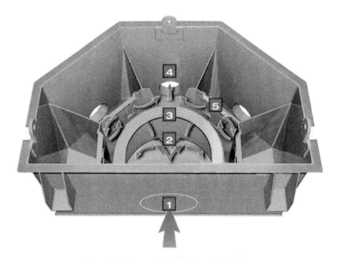

Fig. 8.4 Flow distribution chamber.

As the well gets deeper and usually into pressurized formations, one of the main shakers is not needed for processing the volume flow rate of the weighted drilling fluid circulated in the well bore. In this case, one of the main shakers can be converted into a mud cleaner, (Fig. 8.6). A bank of desilters can be mounted above the shale shaker. In the unweighted portion of the bore hole, the desilter underflow will be discarded. When weighting agent is

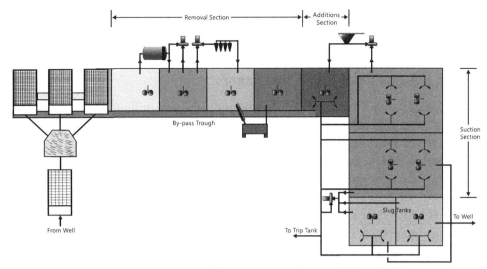

Fig. 8.5 Unweighted drilling fluid processing plant with flow distributor.

added to the drilling fluid, screens can be mounted on the shaker. Valves can be installed to change the desilters over flow from returning directly back to the proper tank and will allow the overflow from the desilters to flow under the screen to carry the screen throughput

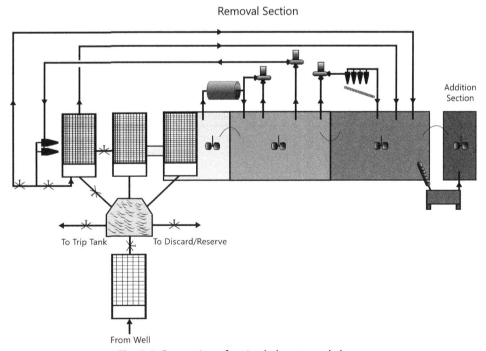

Fig. 8.6 Conversion of main shaker to mud cleaner.

back to the proper mud tank. The desilter underflow and the fluid passing through the shaker screen will not have sufficient low-shear-rate viscosity to transport solids.

By-pass troughs after the shale shakers

By-pass troughs (or ditches) are a common means of moving drilling fluid during drilling fluid swap-outs. When the drilling fluid is changed from the fluid in the tanks to another fluid, the removal section is not used. The water-based drilling fluid could be changed to a non-aqueous drilling fluid (NADF), or the NADF could be changed to a water-based drilling fluid, or a drill-in fluid might be needed, or a completions fluid could be needed

Table 8.1 Fluid flow sizing charts.

Fluid flow rate				Required slope of trough,
gpm	bbl/min	ft³/s	Avg. liquid velocity, ft/s	inches drop/running ft
Size: 8″ wide × 8″ high with 4″ liquid depth and clean bottom				
400	9.5	0.9	4	1/8″[a]
500	11.9	1.1	5	3/16″[a]
600	14.3	1.3	6	1/4″
750	17.8	1.7	7.5	3/8″
1000	23.8	2.2	10.0	1/2″
Size: 12″ wide × 12″ high with 6″ liquid depth and clean bottom				
900	21.4	2.0	4	1/8″[a]
1123	26.7	2.5	5	3/16″[a]
1347	32.1	3.0	6	1/4″
1684	40.1	3.75	7.5	3/8″
2245	53.5	5.0	10.0	1/2″
Size: 15″ wide × 15″ high with 9″ liquid depth and clean bottom				
1570	37.4	3.5	4	1/8″[a]
2110	50.2	4.7	5	3/16″[a]
2510	59.9	5.6	6	1/4″
3140	75.8	7.0	7.5	3/8″
4210	100.2	9.4	10.0	1/2″
Size: 18″ wide × 18″ high with 12″ liquid depth and clean bottom				
2690	64.1	6	4	1/8″[a]
3360	80.2	7.5	5	3/16″[a]
4040	96.2	9	6	1/4″
5050	120.3	11.25	7.5	3/8″
6730	160.4	15	10.0	1/2″

[a]Minimum slope of $1/4$ inch per foot is the recommended minimum to prevent frequent clogging.

in the hole. By-pass troughs are simple, effective, easy to follow and easy to clean. The only real problem with troughs is the tendency of barite and cuttings to settle and clog the troughs. Troughs should be sized so that the average velocity of the fluid is no less than 5 ft/s and no more than 10 ft/s. Frequent clogging will occur if the velocity is less than 5 ft/s. Excessive slopes and messy splashing will occur if the velocity exceeds 10 ft/s (Table 8.1).

Troughs should have at least $1/4''$ per foot of slope so they will tend to be self-cleaning. (If a trough is used between the bell nipple and the shakers, its slope may need to be 1 inch or more per foot.) The following charts maybe used as guidelines for sizing these troughs.

Appendix 8.A

Table 8.A.1 Recommended flow line diameters and trough sizes.

Recommended flow line size maximum circulation rate size (gpm)	Pipe diameter (inches)	Trough cross-section
750	10	10 inch W × 10 inch T
1250	12	12 inch W × 12 inch T
2000	14	14 inch W × 14 inch T
3000	16	16 inch W × 16 inch T

CHAPTER 9

Removal section

Introduction

Drilling fluid circulated from the well is processed through solids removal equipment to eliminate drilled solids. This section of the surface drilling fluid processing tanks is called the removal section. This is the crucial section for changing the economics of drilling operations. Many "theories" and false concepts prevail about how to treat the drilling fluid. The proper treatment is simple, but from a practical point of view, seldom done. The financial benefits for the operator are great. In a specific incidence, a major operator was planning to drill at least twelve wells on a platform. The first six wells were all much over budget and an evaluation of the removal system revealed many, many errors. The operator elected to correct most of the errors after drilling the sixth well. Six welders and two hundred fifty thousand dollars later, the seventh well was spudded. The savings accrued on the seventh well was more than the amount spent correcting the plumbing problems. This book explains how to achieve the lowest cost well by eliminating both the visible and invisible non productive time (NPT). Attention to the removal section is the primary focus for drilling trouble-free.

Unweighted drilling fluid

A complete generic surface drilling fluid processing system for an unweighted drilling fluid (Fig. 9.1), includes three sections: the removal section, the addition section, and the suction section. Mud guns are only used in the addition and suction section and none are used in the removal section. Valves are not shown in this sketch but will be discussed in the detailed analysis of the removal system. The by-pass trough is used primarily after drilling is complete. Completion fluid, cementing fluid, packer fluid and other fluids that do not need to have solids removed can be circulated in the well bore without passing through the removal section.

In the unweighted drilling fluid removal section (Fig. 9.1), all of the removal equipment is described in this chapter and may not be needed in all wells. An unweighted drilling fluid system is one in which no commercial products have been added to increase the mud weight. The tank arrangement of the equipment is independent of the drilling fluid type. The drilling fluid from the well flows into gumbo removal equipment as it exits the well. Gumbo is not a 'formation' but is formed in the well bore. When drilling very young (geologically speaking) formations containing a lot of clay, the clay tends to form large rings/balls or agglomerations as it moves up the well bore. Gumbo does

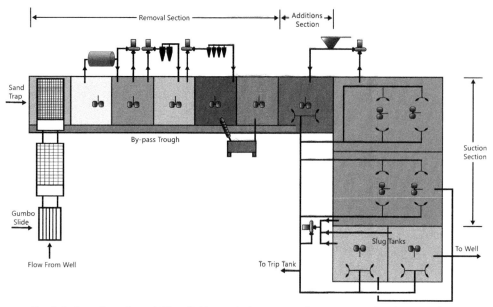

Fig. 9.1 Generic surface drilling fluid processing system for an unweighted drilling fluid.

not appear in all wells and it can be best removed by equipment specifically dedicated to addressing this problem. Inhibitive drilling fluids with good carrying capacity tend to mitigate this. However, a gumbo slide can remove these large masses of sticky clay before they reach the shaker. Gumbo will not easily be transported off the end of a linear motion or a balanced elliptical motion screen.

The first shaker is called a "scalping" shaker and usually has a very coarse screen mounted on it. The fluid then is processed through the main shaker which should have very fine screens to remove as many drilled solids as possible. Scalping shakers remove very large cuttings and solids which have sloughed into the hole before the fluid is screened by the main shaker. Measurements have shown that more solids are removed if the scalping shakers are dressed with large opening screens (such as API 10 or API 20 screens). Trying to screen the drilling fluid through finer screens seems to break solids apart so they cannot be removed with API 170 or API 200 screens.

Below the main shaker may be a sand trap. This is very effective only if coarse screens are mounted on the main shaker and no fine screen shakers are available. Solids settle in this compartment and are discarded frequently. Recently, however, the low-shear-rate viscosity of drilling fluid has been elevated to assist in transporting cuttings to the surface. The solids do not settle rapidly in transit in the well bore because of this viscosity. The residence time in the sand trap is short and very few solids will settle. If fine screens are mounted on the main shaker, the settling rate of solids smaller than 75 μm is very low.

See a discussion of settling rates in the Centrifuge, Chapter 21. Many rigs are now eliminating the sand trap when sufficiently fine screens can be mounted on the main shakers.

The fluid passing through the shaker screen may have gas in it. Centrifugal pumps cannot pump gas-cut drilling fluid very effectively. The gas tends to accumulate in the center of the impeller and eventually vapor locks the pump. If gas does enter and collect in a centrifugal pump, cavitation bubbles destroy the pump impeller. Since centrifugal pumps are needed to process the drilling fluid, a good degasser is necessary. A vacuum in the chamber causes atmospheric pressure to push fluid into the degasser. The fluid flows down some baffle plates and the gas does not have to travel a long distance through the fluid to enter the vacuum chamber. A jet pump is used to cause the fluid to leave the degasser and flow into the next compartment. The fluid driving the jet pump removing fluid from the degasser is from a centrifugal pump getting its fluid from a compartment of degassed drilling fluid downstream from the vacuum degasser.

A bank of desanders is used to decrease the solids loading of the desilters when fine screens are not used on the main shaker. These were necessary before the advent of the linear motion or the balanced elliptical motion shakers were available. If API 140 or finer, screens are mounted on the main shaker, they will remove the solids that were normally removed by the desanders. In this case, the desanders are no longer needed and can be deleted along with the related desander suction tank.

A centrifuge is used as if it is a super-desilter. The heavy (or underflow) slurry is discarded and the light (or overflow) slurry is retained. This eliminates solids which are larger than about 10 µm for the fluid processed. All solids removal equipment except the centrifuge should process about 110%–125% of the flow rate pumped down hole. The hydrocyclones and the degasser should process more fluid than is entering the suction tank of that equipment. This is discussed in depth in Chapter 11, Fraction of Drilling Fluid Processed.

Not every rig has a shaker which will process the fluid through the fine screens. Some still have the unbalanced elliptical or the circular motion shakers as their main shaker. In this case, API 70 to API 80 screens will be used on that shaker. With fine screen shakers, the sand trap and the desander are no longer needed, (Fig. 9.2).This system will be discussed in detail in the next section.

Weighted drilling fluid

A weighed drilling fluid is defined as a drilling fluid which contains commercial additives to increase the drilling fluid density. Usually, barite is added to increase the mud weight. There is currently no rig equipment available which will separate barite from drilled solids. Consequently, the drilled solids roughly in the same size range as the barite cannot be removed. This is discussed in Chapter 21, Centrifuges.

The tank arrangement for a weighted drilling fluid is almost identical to the tank arrangement used for unweighted drilling fluid and does not necessarily depend on the

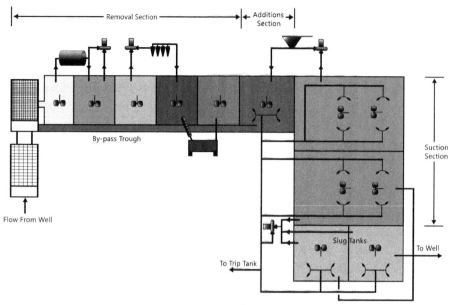

Fig. 9.2 Unweighted drilling fluid with fine screen shakers.

drilling fluid type (Figs. 9.3 and 9.4). The primary difference is the economic restriction of discarding all of the desilter underflow. To retain the barite in the desilter underflow, a screen is mounted on a shaker to allow all of the barite and some drilled solids to return to the active system. The screen discard will have some large barite particles (that are undesirable) and mostly drilled solids. This equipment is called a mud cleaner.

When the linear motion shale shakers were introduced and API 170 or API 200 screens could be used on the main shaker, mud cleaners were deemed as superseded by the new technology. After a period of time, mud cleaners became popular again because all of the drilling fluid is not always processed through the main shaker. This should have been obvious with the experience of finding desilters plugged with solids which are much larger than the screen openings. Solids plugging desilters underflow openings are much larger than the screen openings on the shale shaker. Yet, finding all cones unplugged on a rig is a rare occurrence.

The drilling fluid enters a distribution chamber, Fig. 9.5, which can be used to distribute the flow equally to all main shale shakers. This replaces the back tanks of shale shakers which have caused problems with the removal system. Derrick men frequently dump shale shaker back tanks into the "settling pits" before a trip. If drilling fluid dries on a screen during a trip, the screens usually have to be replaced because of plugging. Unfortunately, drilling fluids currently in use are designed to prevent settling. These cuttings progress downstream and plug the hydrocyclones when drilling is resumed.

When multiple fine screen shakers are used, the capacity requirements deeper downhole frequently will allow one of the shakers to be removed from the flow system.

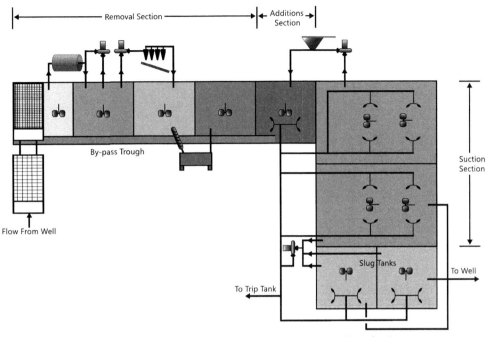

Fig. 9.3 Generic tank arrangement for weighted drilling fluids.

Usually, by this time the mud weight has increased. The surplus shaker can be converted into a mud cleaner (Fig. 9.6). The fluid that passes through the mud cleaner screen does not have much carrying capacity. To transport the solids back to the appropriate tank, the

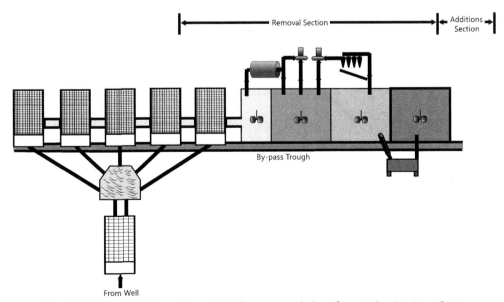

Fig. 9.4 Removal section with multiple fine screen shakers for weighted drilling fluids.

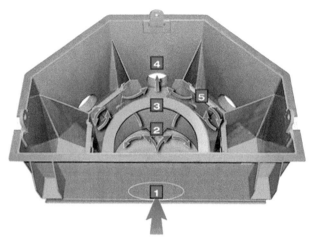

Fig. 9.5 Flow distribution chamber.

overflow from the desilters can be routed back under the shaker screen. This will transport the screen throughput back to the appropriate compartment.

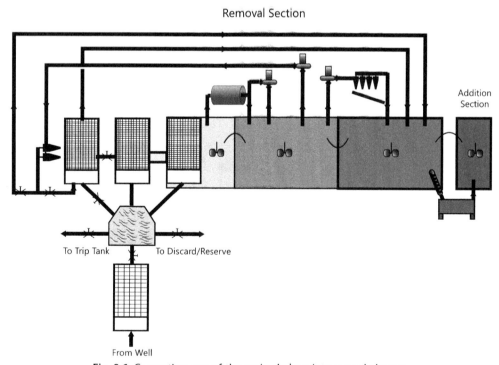

Fig. 9.6 Converting one of the main shakers into a mud cleaner.

Distribution chamber

The distribution chamber can also be used to provide a circulation chamber for trip tanks. While pulling pipe, drilling fluid can be circulated into the top of the well and overflow down the flow line. A valve near the bottom of the distribution chamber can be opened to allow excess drilling fluid to continuously flow back into the trip tank. See the arrangement in Chapter 2 (Fig. 2.8).

The removal section should be as small as possible. Most of the drilling fluid should be in the suction compartments, uniformly blended and ready to be pumped down hole.

CHAPTER 10

Centrifugal pumps

A Short Quiz to Determine Personal Understanding of Centrifugal Pumps:
A centrifugal pump is connected to a joint of $4\frac{1}{2}''$ casing standing next to a derrick. The pump is supplied with water from a tank as shown below (Fig. 10.1):

The water rises 20 ft up the casing and stops. The liquid level stands at 20 ft as long as the pump is running.

The water is drained from the tank and a 16.6 ppg drilling fluid is placed in the tank. The pump is started again. How high will this fluid, that is twice as dense as water, rise in the joint of casing (Fig. 10.2)?

Answer: See Appendix 10.A at the end of this chapter.

Introduction

This discussion will focus primarily on centrifugal pumps used in drilling fluid systems. Discussion of basic principles of operating and sizing impellers and motors can be used as a preamble for discussing centrifugal pumps used for other applications.

Centrifugal Pumps are constant head devices. The impellers impart kinetic energy to a fluid and create a constant head. This can perhaps be illustrated by taking a diversion from talking about pumps and talk about a simple, basic physics problem.

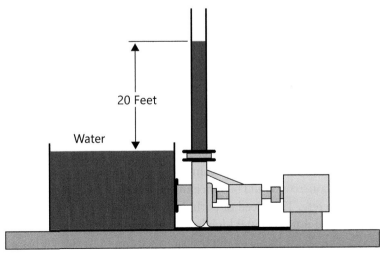

Fig. 10.1 Quiz diagram #1.

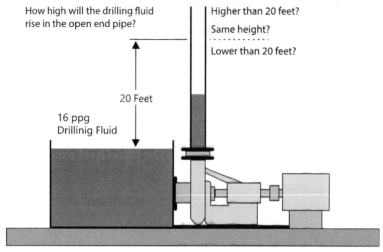

Fig. 10.2 Quiz diagram #2.

A 40 lb chunk of concrete is catapulted from the ground to the top of a 50 ft high building. Neglecting air friction, what is the velocity of the chunk when it leaves the catapult?

This is solved by considering a conservation of energy. Kinetic energy of the concrete released from the catapult is converted to Potential Energy when the chunk of concrete reaches the top of the 50 ft building.

$$\text{Kinetic Energy} = \text{Potential Energy}$$

$$\frac{1}{2} mv^2 = mgh$$

$$v^2 = 2gh$$

$$v = \left[2 \left(\frac{32.17 \text{ ft}}{s^2} \right) (50 \text{ ft}) \right]^{0.5}$$

$$v = \frac{56.7 \text{ ft}}{s}, \text{ or } 45.2 \text{ mph}$$

Observe that it doesn't matter whether it was a 20 lb or a 40 lb chunk of concrete. The mass cancels from the equation. The amount of energy (or work) required to throw the 40 lb chunk of concrete would be twice the amount of energy required to throw the 20 lb chunk of concrete. The speed of both, however, would be the same. (Note that if either chunk dropped from the 50 ft high building, they would both hit the ground with the same speed. The 40 lb chunk would probably hurt more because it has more energy than the 20 lb chunk).

The velocity is independent of mass. This is the same principle that can be used to describe the behavior of centrifugal pumps. A centrifugal pump impeller accelerates fluid along an impeller until it reaches the tip velocity of the impeller. The impeller is supplying the kinetic energy. The fluid will rise until it reaches the potential energy given by the kinetic energy.

Input: Kinetic energy

Output: Potential energy

The HEAD, or height the liquid will rise, is independent of mud weight!

What is HEAD?

At the surface casing seat 2500 ft deep, a rig is pumping with a 3000 psi standpipe pressure down 5 inch drill pipe while drilling a $17^1/_2$ inch diameter hole with a 12.0 ppg, oil-based drilling fluid that has a PV of 40 cp and a YP of 12 lb/100 sqft.

Answer: 2500 ft.

Head and Pressure are frequently confused. In the problem above, head is simply the height of fluid above a certain point and is independent of the mud weight. The pressure, however, depends upon the mud weight and the depth of the well (or more precisely, the height of fluid above the point of interest). In well control, drillers learn to calculate bottom hole pressure from the equation:

$$\text{Pressure, psi} = 0.052(\text{MW, ppg})(\text{Depth, ft})$$

the equation could also be written:

$$\text{Pressure, psi} = 0.052(\text{MW, ppg})(\text{Head, ft})$$
$$= (0.052)(12.0 \text{ ppg})(2500 \text{ ft}) = 1560 \text{ psi}$$

A little later, absolute pressure will be discussed. Atmospheric pressure is around 14.7 psi. This is applied to the top of the drilling fluid column, so the absolute pressure at 2500 ft is 1575 psia.[1]

Review the concrete problem above and apply to a rotating impeller. Calculate how high a tennis ball would rise if it leaves the impeller with the tip velocity (Fig. 10.3).

The velocity of the tip of the impeller determines the kinetic energy given to something moving along the impeller blade. If the tip velocity (and speed of the tennis ball when it leaves the impeller) is 60 miles per hour, how high would the ball rise? (60 miles per hour is 88 ft per second). The height would be:

$$\text{height} = \frac{v^2}{2g}$$

---

[1] Normally, atmospheric pressure is about 14.7 psia. ("psia" is used as an abbreviation to indicate absolute pressure, or the pressure in pounds per square inch above absolute zero).

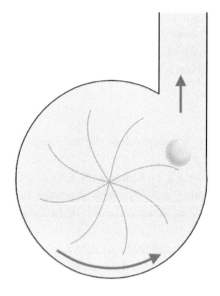

To calculate how high the ball would go in a vacuum, write the equations for the conservation of energy.

The kinetic energy would be changed to potential energy when the ball reaches the maximum height.

Kinetic Energy = $\frac{1}{2}mv^2$

Potential Energy = mgH

Where:

m = mass of ball

v = initial velocity

H = height of rise.

$gH = \frac{1}{2}v^2$

The height is determined by the velocity of the tip of the impeller and is independent of MASS.

Fig. 10.3 Impeller throwing a tennis ball.

$$\text{height} = \frac{[88 \text{ ft/s}]^2}{2(32.2 \text{ ft/s}^2)g} = 120 \text{ ft}$$

Fluid would rise to the same height if this was a centrifugal pump. Fluid will flow up to a certain height independent of the mass or density of the fluid. The mass of the ball or the mass of the fluid canceled in the equation relating the conservation of energy.

A centrifugal pump is a constant head device

The mass of the ball or the mass of the fluid would make a significant difference in the amount of energy, or work, required to cause the ball or fluid to rise to 120 ft. The motor

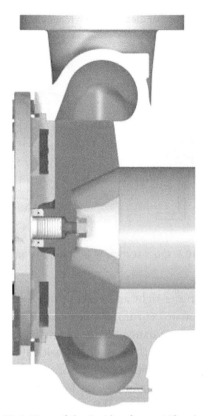

Fig. 10.4 View of the inside of a centrifugal pump.

size to rotate the impeller will control the amount of power or energy provided to move the fluid.

Pump description

In Fig. 10.4, fluid enters through the flange in the left and exits through the flange on the top. The shaft in the center of the right side rotates the impeller.

Centrifuges are described in terms of these flange sizes. A 8 × 6 × 14 centrifuge has a six inch pipe discharge (the flange on top), an eight inch pipe suction (the flange on the left), and can rotate a 14 inch impeller. The impeller size will depend upon the head needed for the pump to function correctly. Centrifugal pumps on drilling rigs seldom have a 14 inch impellers. The impellers do not touch the outer housing. They function only to increase the fluid velocity up to tip speed (Fig. 10.5).

Fig. 10.5 illustrates that the shaft moving the impeller extends to the left of the pump. A motor would be connected to the shaft on the left side of Fig. 10.5. Fluid enters the

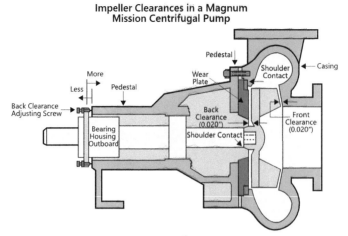

Fig. 10.5 Cross-section of a centrifugal pump.

pump through the flange on the right and exits through the flange on the top. These flange sizes are used to designate centrifugal pumps. A 3 × 4 × 14 pump would have a 4" suction flange, a 3" discharge flange, and a 14" diameter housing.

The fluid velocity depends upon impeller tip velocity—which depends upon the diameter of the impeller and the rotational speed of the impeller.

Since the impeller tip velocity controls the head produced by a pump, the same impeller in different pumps will produce the same head. This assumes that the pump housing construction remains the same. If the housing diameter is changed, the head produced by the pump will not be the same for the same size impellers. For initial consideration, the concept of a constant head at no flow conditions will make the theory easier to understand.

Since the head generated by the centrifugal pump is independent of the mud weight and depends only on the impeller tip velocity, the same size impeller in a 3 × 2 or a 8 × 6 pump should create the same head if the rotational speed is constant. Before most drilling rigs switched primarily to electric power, many pumps were mounted with their own diesel engines. Speed could be varied with the throttle of the motor. Now, however, most drilling rigs use either a 1750 RPM or an 1150 RPM electric motor to supply power the centrifugal pumps. Since the speed will be constant, the tip velocity will be constant. The head generated will depend only on the impeller diameter. As shown in Fig. 10.6, a 10 inch impeller generates a 120 ft head for all pumps with no flow. At about 400 gpm, the head curve for the 3 × 2 pump decreases to about 50 ft of head. This is caused by the internal head (or pressure) loss with in the pump. At 400 gpm the head

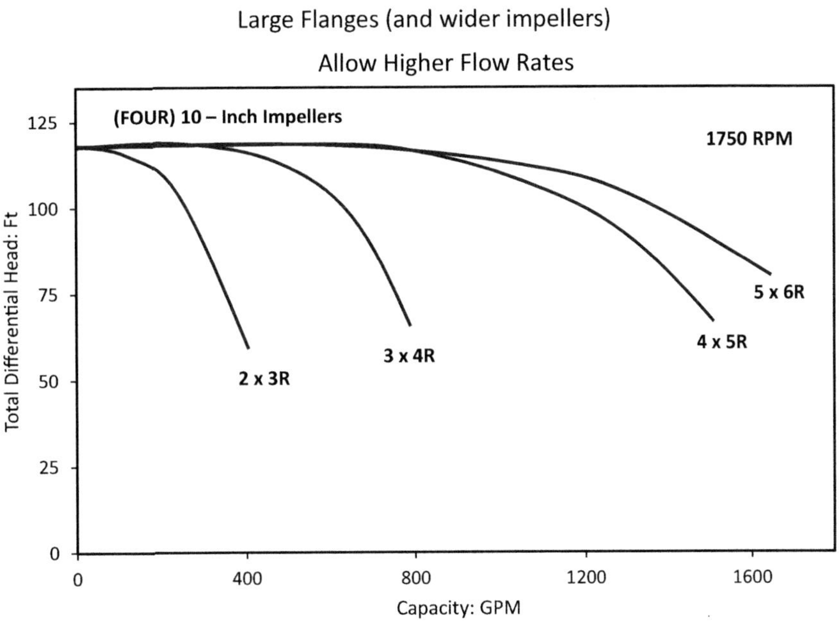

Fig. 10.6 A 10-inch impeller produces the same no-flow head regardless of the size of the pump flanges.

generated by the 5 × 4 pump is still 120 ft. If the pumps were turned off and simply connected to a pipe flowing 400 gpm, the pressure loss across the small 3 × 2 pump would be significantly higher than the pressure loss across the 5 × 4 pump.

A good model of a centrifugal pump is a standpipe that is continuously filled with fluid. The head supplied to a pipe connected to it would be constant (Fig. 10.7).

A model like this would very quickly answer the question about connecting two centrifugal pumps in parallel. Consider a standpipe connected to 2 miles of 1″ diameter pipe. Would the flow rate through the pipe increase if two standpipes were connected to the long pipe (Fig. 10.8)?

Would the flow rate through the pipe increase if two standpipes were connected to the long pipe?

The answer is obviously "NO". Usually one standpipe would have a lower head than the second standpipe. In this case, fluid would flow in the reverse direction. Or to state it in terms of centrifugal pumps: If one pump generates a slightly lower head, fluid will flow in the reverse direction through that pump when they are connected in parallel.

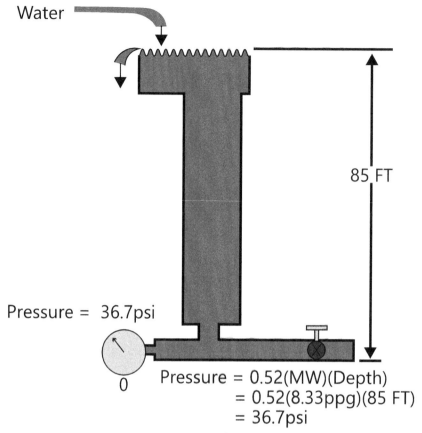

Fig. 10.7 A good model of a centrifugal pump.

Pump curves

Curves relating the head generated by a centrifugal pump to the flow rate achieved by the pump are created by the pumps. The curves look complicated. The flow rate is usually labeled as "capacity" which is confusing to drilling engineers. The curves will be developed one concept at a time. For example, the curve for a 13″ impeller in a 8 × 6 × 14 Mission Magnum, in Fig. 10.9, indicates that the head generated by this impeller with no flow is about 170 ft. This head remains relatively constant until the flow from the pump reaches about a 1000 gpm. Then the head starts to decrease because of the internal friction within the pump. The impeller is still producing the same head, but the pressure drop (or head loss) within the pump is becoming more significant. The flow from the pump will depend upon what piping and equipment is connected to the pump. At a flow rate of 2500 gpm, the head available has decreased to about 140 ft.

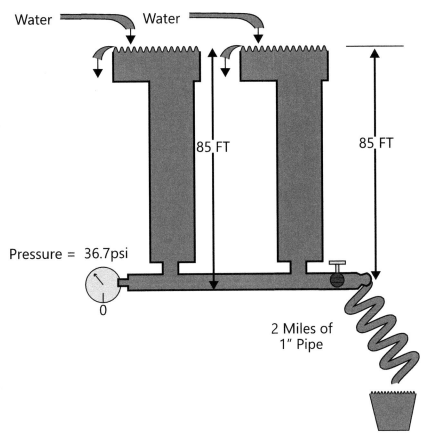

Fig. 10.8 Two centrifugal pumps connected in parallel.

At a zero flow rate, the 13″ impeller generates about a 170 ft head. This means that if a open-ended 200 ft vertical pipe was connected to the pump discharge, fluid would rise 170 ft up the pipe. As the pump starts moving fluid through piping and equipment, the head remains reasonably constant until the pressure loss through the pump becomes noticeable at a flow rate of about 1000 gpm The head loss (or pressure drop) through the pump increases as the flow rate through the pump increases. The pump curve no longer shows a constant 170 ft head (Fig. 10.10).

Imagine connecting the pump to a standpipe full of fluid and letting the fluid flow through the pump without starting the pump. The pressure loss through a very small pump (3 × 2) would be much larger than the pressure loss through a larger pump (8 × 6) at the same flow rates. Smaller pumps have a higher velocity head loss than bigger pumps.

In terms of the stand pipe model, the velocity head loss could be described by the illustration in Fig. 10.11. As fluid starts flowing from the standpipe, some pressure loss

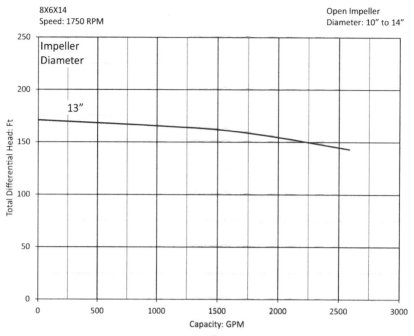

Fig. 10.9 The pump curve for a 13″ impeller rotating at 1750 RPM.

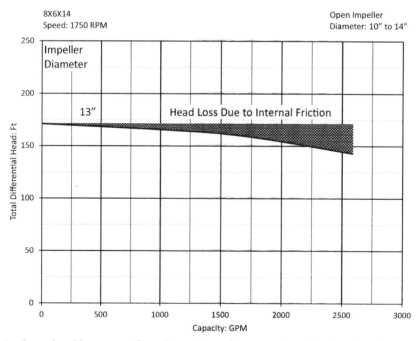

Fig. 10.10 Some head loss occurs from the pressure drop caused by flowing through a centrifugal pump.

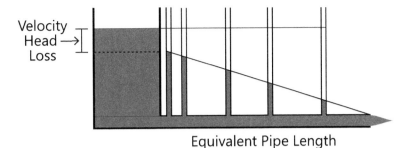

Fig. 10.11 Velocity head loss.

results as the fluid enters the pipe. This is called the velocity head loss. If the flow is stopped (close a valve at the discharge end of the pipe), the fluid would stand at the same height, in the sight tubes, as the fluid in the standpipe. This would be the "no-flow" head shown on centrifugal pump curves at zero flow rate.

Many different size impellers can be mounted in this centrifugal pump. Pump curves (Fig. 10.12), show values of head-vs-flow rate for impellers from 14 inches in diameter down to 10 inches. Impellers can be machined to diameters in between the standard sizes to match the head needed for a particular application. Usually, the diameters are changed in $1/4''$ increments.

The power required to provide the head and flow rate chosen is indicated on the pump curves (Fig. 10.13). These curves indicate the power required to pump water.

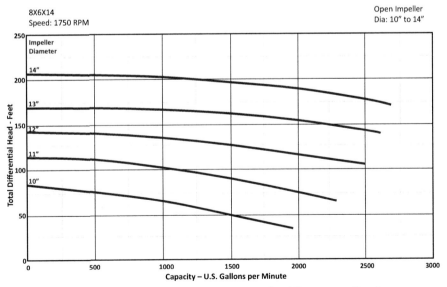

Fig. 10.12 Head curves for the centrifugal pump for different impeller diameters.

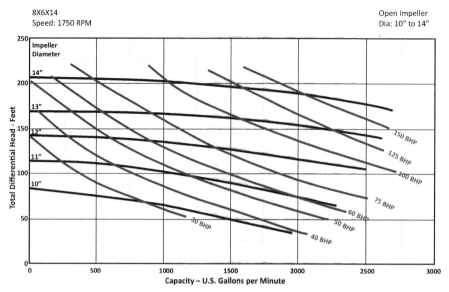

Fig. 10.13 Pump curves with the required input brake horsepower.

The power required increases as the mud weight increases. The power required increases by a ratio of the actual mud weight (in ppg) divided by the mud weight of water (8.34 ppg).

Power is work per unit time. As discussed in the beginning part of this discussion, the work required to move a heavy object (or a heavy drilling fluid) is greater than the work required to move a lighter object (or a lighter drilling fluid).

This increased requirement of power can be calculated by multiplying the required brake horsepower by the specific gravity of the fluid. For drilling fluid, a ratio of the mud weight, in ppg, to 8.33 ppg will calculate the specific gravity of the drilling fluid. On most drilling rigs, the motor size is selected to be large enough to handle a 20 ppg drilling fluid (Specific Gravity = 2.4). In many processing plants, the feed material does not change, so eliminating excess power becomes important to control the cost. On a drilling rig, the pump must be able to process whatever drilling fluid is used to drill any well.

The head curve for the 10″ diameter impeller indicates a decrease in head almost as soon as the flow starts through the pump. This smaller diameter impeller creates more head loss at low flow rates than the larger vane impellers. Perhaps this is a clue that the conversion of electric power input compared to hydraulic power output is not always efficient. Over laying most head curves will be another set of curves indicating an efficiency, as shown in Fig. 10.14. These values are very important in plants where the same fluid is processed daily. On drilling rigs, the proper head and flow rates are much more important than trying to optimize efficiency. The hydraulic power produced by

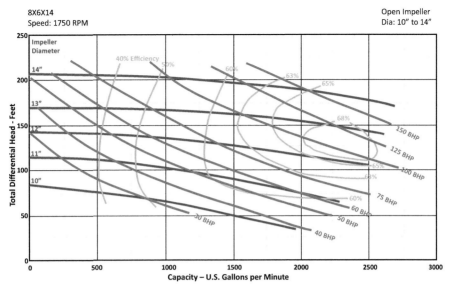

Fig. 10.14 Efficiency curves added to the pump curves.

the pump varies from about 40% to slightly over 68% of the motor brake horsepower needed to produce the needed head and flow rate. The hydraulic horsepower (HHP) can be calculated using the relationship:

$$\text{HHP} = \frac{PQ}{1714}$$

where P is the pressure in psi, and Q is the flow rate in gpm.

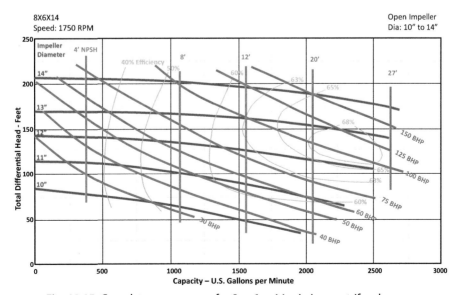

Fig. 10.15 Complete pump curve for $8 \times 6 \times 14$ mission centrifugal pump.

The final set of curves are almost vertical lines labeled with the number of feet of a head of water above absolute zero pressure which is required to prevent pump cavitation (Fig. 10.15). For pumps without many fittings in a very short suction line which is located below the liquid level of the drilling fluid, these lines will not be very important. For the initial discussion of centrifugal pump application to processing drilling fluid at the surface, the pressure inside of the suction flange will always be above atmospheric pressure. For centrifugal pump uses where the pump is located above the liquid level in the suction tank or where long suction lines are used, the net positive suction head (NPSH) must be calculated and is an important part of the calculation. The concept and a discussion will be presented later for many other applications of centrifugal pumps.

Application of pump curves

Pump curves are used to decide on the impeller size needed to produce the proper head and flow rate needed for various applications. In Fig. 10.16, a centrifugal pump is supplying fluid to a bank of hydrocyclones. Most 4″ desilters require 75 ft of head to properly separate solids from a drilling fluid. At this head, the cones can process between 50 and 80 gpm of drilling fluid each. For this bank of eight cones processing 50 gpm each, the centrifugal pump must supply 75 ft of head at the input header of the cones while pumping

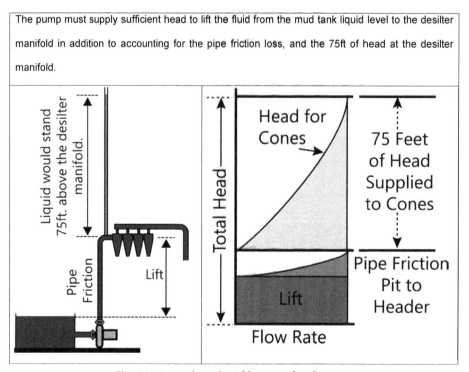

Fig. 10.16 Head produced by centrifugal pump.

Table 10.1 Pump pressure for different mud weights for situation described above.

Mud weight, ppg	Pressure, psi
9	35
10	39
11	43
12	47
13	51
14	55
15	59

400 gpm. Some head (or pressure) loss occurs as fluid flows through the suction and discharge pipes.

Head losses are calculated in feet of head; the pressure losses will depend upon the mud weight. For example, if the desilter manifold is 10 ft above the mud pit liquid level the lift head loss is 10 ft, if the suction and discharge pipe friction loss is 5 ft, and the head at the manifold of ten 4″ desilters is 75 ft, the pump must generate 90 ft of head while pumping 500 gpm to the desilter bank. Look at the pump curves and find the pump size and head curves which will indicate the impeller diameter.

The pressure generated by the pump will be governed by the mud weight from the equation:

$$\text{Pressure (psi)} = 0.052 \text{ (mud weight, ppg) (head, ft)}.$$

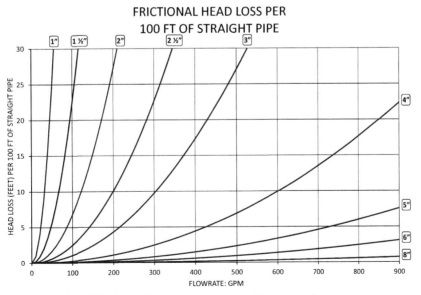

Fig. 10.17 Head loss in new steel pipe for various flow rates.

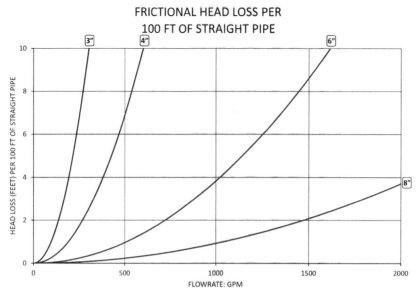

Fig. 10.18 Friction curves for 100 ft of schedule 40 steel pipe.

In the above problem, pump output pressure as a function of mud weight is listed in Table 10.1.

As a rule of thumb, the pressure (PSI) at the cone manifold should be 4 times the mud weight in ppg.

Sizing impellers

The fluid flow rate produced by a centrifugal pump depends upon the plumbing attached to the pump. The head loss, created by flowing water through Schedule 40 standard iron pipe, has been well documented. A series of curves (Fig. 10.17), published by the Hydraulics Institute shows the head loss for one hundred feet of schedule 40 steel pipe as a function of the flow rate of water through the pipe.

This information is replotted in Fig. 10.18 for common sizes of pipe used in connecting solids control equipment.

These curves are plotted from data supplied by the Hydraulics Institute for water. A more expanded version of these friction curves is presented in Tables 10.2–10.4.

Operating point

When a centrifugal pump supplies water to a piping system, the pump automatically creates a constant head. As flow is increased in a piping system, the head required to cause this flow also increases. When the flow produces the head created by the centrifugal

Centrifugal pumps 209

Table 10.2 Friction loss for water in steel pipe.

FRICTION LOSS FOR WATER IN FEET HEAD PER 100 FEET OF PIPE LENGTH

	TABLE CP-7				TABLE CP-8		
3 INCH NOMINAL	STEEL SCHEDULE 40 ID - 3.068 INCHES ε/D - 0.000587			3-1/2 INCH NOMINAL	STEEL SCHEDULE 40 ID - 3.548 INCHES ε/D - 0.000507		
FLOW RATE GPM	V ft/sec	V²/2g feet	Friction Loss in Feet Head Per 100 ft of Pipe	FLOW RATE GPM	V ft/sec	V²/2g feet	Friction Loss in Feet Head Per 100 ft of Pipe
25	1.09	0.02	0.19	35	1.14	0.02	0.17
30	1.30	0.03	0.26	40	1.30	0.03	0.22
35	1.52	0.04	0.35	45	1.46	0.03	0.27
40	1.74	0.05	0.44	50	1.62	0.04	0.33
45	1.95	0.06	0.55	60	1.95	0.06	0.46
50	2.17	0.07	0.66	70	2.27	0.08	0.60
55	2.39	0.09	0.79	80	2.60	0.11	0.77
60	2.60	0.11	0.92	90	2.92	0.13	0.96
65	2.82	0.12	1.07	100	3.25	0.16	1.17
70	3.04	0.14	1.22	110	3.57	0.20	1.39
75	3.25	0.17	1.39	120	3.89	0.24	1.64
80	3.47	0.19	1.57	130	4.22	0.28	1.90
85	3.69	0.21	1.76	140	4.54	0.32	2.18
90	3.91	0.24	1.96	150	4.87	0.37	2.48
95	4.12	0.26	2.17	160	5.19	0.42	2.80
100	4.34	0.29	2.39	170	5.52	0.47	3.15
110	4.77	0.35	2.86	180	5.84	0.53	3.50
120	5.21	0.42	3.37	190	6.17	0.59	3.87
130	5.64	0.50	3.92	200	6.49	0.66	4.27
140	6.08	0.57	4.51	220	7.14	0.79	5.12
150	6.51	0.66	5.14	240	7.79	0.94	6.04
160	6.94	0.75	5.81	260	8.44	1.11	7.04
170	7.38	0.85	6.53	280	9.09	1.28	8.11
180	7.81	0.95	7.28	300	9.74	1.47	9.26
190	8.25	1.06	8.07	320	10.4	1.68	10.48
200	8.68	1.17	8.90	340	11.0	1.89	11.8
220	9.55	1.42	10.7	360	11.7	2.12	13.2
240	10.4	1.69	12.6	380	12.3	2.36	14.6
260	11.3	1.98	14.7	400	13.0	2.62	16.2
280	12.2	2.29	16.9	420	13.6	2.89	17.8
300	13.0	2.63	19.2	440	14.3	3.17	19.4
320	13.9	3.00	22.0	460	14.9	3.46	21.2
340	14.8	3.38	24.8	480	15.6	3.77	23.0
360	15.6	3.79	27.7	500	16.2	4.09	25.0
380	16.5	4.23	30.7	550	17.8	4.95	30.1
400	17.4	4.68	33.9	600	19.5	5.89	35.6
420	18.2	5.16	37.3	650	21.1	6.91	41.6
440	19.1	5.67	40.9				
460	20.0	6.19	44.6				

NOTE: No allowance has been made for age, differences in diameter, or any abnormal condition of interior surface. Any factor of safety must be estimated from the local conditions and the requirements of each particular installation.

Recommended flow rates (shaded) for suction and discharge pipes are to avoid sanding up at lower flow rates and to avoid too much friction at higher flow rates.

Continued

Table 10.2 Friction loss for water in steel pipe.—cont'd

FRICTION LOSS FOR WATER IN FEET HEAD PER 100 FEET OF PIPE LENGTH

	TABLE CP-9				TABLE CP-10		
4 INCH NOMINAL	STEEL SCHEDULE 40 ID - 4.026 INCHES e/D - 0.000447			5 INCH NOMINAL	STEEL SCHEDULE 40 ID - 5.047 INCHES e/D - 0.000357		
FLOW RATE GPM	V ft/sec	$V^2/2g$ feet	Friction Loss in Feet Head Per 100 ft of Pipe	FLOW RATE GPM	V ft/sec	$V^2/2g$ feet	Friction Loss in Feet Head Per 100 ft of Pipe
40	1.01	0.02	0.12	70	1.12	0.02	0.11
50	1.26	0.02	0.18	80	1.28	0.03	0.14
60	1.51	0.04	0.25	90	1.44	0.03	0.17
70	1.76	0.05	0.33	100	1.60	0.04	0.20
80	2.02	0.06	0.42	120	1.92	0.06	0.29
90	2.27	0.08	0.52	140	2.25	0.08	0.38
100	2.52	0.10	0.62	160	2.57	0.10	0.49
110	2.77	0.12	0.74	180	2.89	0.13	0.61
120	3.02	0.14	0.88	200	3.21	0.16	0.74
130	3.28	0.17	1.02	220	3.53	0.19	0.88
140	3.53	0.19	1.17	240	3.85	0.23	1.04
150	3.78	0.22	1.32	260	4.17	0.27	1.20
160	4.03	0.25	1.49	280	4.49	0.31	1.38
170	4.28	0.29	1.67	300	4.81	0.36	1.58
180	4.54	0.32	1.86	320	5.13	0.41	1.78
190	4.79	0.36	2.06	340	5.45	0.46	2.00
200	5.04	0.40	2.27	360	5.77	0.52	2.22
220	5.54	0.48	2.72	380	6.09	0.58	2.46
240	6.05	0.57	3.21	400	6.41	0.64	2.72
260	6.55	0.67	3.74	420	6.74	0.71	2.98
280	7.06	0.77	4.30	440	7.06	0.77	3.26
300	7.56	0.89	4.89	460	7.38	0.85	3.55
320	8.06	1.01	5.51	480	7.70	0.92	3.85
340	8.57	1.14	6.19	500	8.02	1.00	4.16
360	9.07	1.28	6.92	550	8.82	1.21	4.98
380	9.58	1.43	7.68	600	9.62	1.44	5.88
400	10.10	1.58	8.47	650	10.4	1.69	6.87
420	10.6	1.74	9.30	700	11.2	1.96	7.93
440	11.1	1.91	10.2	750	12.0	2.25	9.05
460	11.6	2.09	11.1	800	12.8	2.56	10.22
480	12.1	2.27	12.0	850	13.6	2.89	11.5
500	12.6	2.47	13.0	900	14.4	3.24	12.9
550	13.9	2.99	15.7	950	15.2	3.61	14.3
600	15.1	3.55	18.6	1000	16.0	4.00	15.8
650	16.4	4.17	21.7	1100	17.6	4.84	19.0
700	17.6	4.84	25.0	1200	19.2	5.76	22.5
750	18.9	5.55	28.6	1300	20.8	6.75	26.3
800	20.2	6.32	32.4				

NOTE: No allowance has been made for age, differences in diameter, or any abnormal condition of interior surface. Any factor of safety must be estimated from the local conditions and the requirements of each particular installation.

Recommended flow rates (shaded) for suction and discharge pipes are to avoid sanding up at lower flow rates and to avoid too much friction at higher flow rates.

Table 10.2 Friction loss for water in steel pipe.—cont'd

FRICTION LOSS FOR WATER IN FEET HEAD PER 100 FEET OF PIPE LENGTH

TABLE CP-11				TABLE CP-12			
6 INCH NOMINAL	STEEL SCHEDULE 40 ID - 6.065 INCHES e/D - 0.000293			8 INCH NOMINAL	STEEL SCHEDULE 40 ID - 7.981 INCHES e/D - 0.000226		
FLOW RATE GPM	V ft/sec	$V^2/2g$ feet	Friction Loss in Feet Head Per 100 ft of Pipe	FLOW RATE GPM	V ft/sec	$V^2/2g$ feet	Friction Loss in Feet Head Per 100 ft of Pipe
100	1.11	0.02	0.08	60	1.03	0.02	0.05
120	1.33	0.03	0.12	180	1.15	0.02	0.06
140	1.55	0.04	0.16	200	1.28	0.03	0.08
160	1.78	0.05	0.20	220	1.41	0.03	0.09
180	2.00	0.06	0.25	240	1.54	0.04	0.11
200	2.22	0.08	0.30	260	1.67	0.04	0.13
220	2.44	0.09	0.36	280	1.80	0.05	0.14
240	2.66	0.11	0.42	300	1.92	0.06	0.16
260	2.89	0.13	0.49	320	2.05	0.07	0.18
280	3.11	0.15	0.56	340	2.18	0.07	0.21
300	3.33	0.17	0.64	360	2.31	0.08	0.23
320	3.55	0.20	0.72	380	2.44	0.09	0.25
340	3.78	0.22	0.81	400	2.57	0.10	0.28
360	4.00	0.24	0.90	450	2.89	0.13	0.35
380	4.22	0.28	1.00	500	3.21	0.16	0.42
400	4.44	0.31	1.10	550	3.53	0.19	0.51
420	4.66	0.34	1.20	600	3.85	0.23	0.60
440	4.89	0.37	1.31	650	4.17	0.27	0.70
460	5.11	0.41	1.42	700	4.49	0.31	0.80
480	5.33	0.44	1.54	750	4.81	0.36	0.91
500	5.55	0.48	1.66	800	5.13	0.41	1.02
550	6.11	0.58	1.99	850	5.45	0.46	1.15
600	6.66	0.69	2.34	900	5.77	0.52	1.27
650	7.22	0.81	2.73	950	6.09	0.58	1.41
700	7.77	0.94	3.13	1000	6.41	0.64	1.56
750	8.33	1.08	3.57	1100	7.05	0.77	1.87
800	8.88	1.23	4.03	1200	7.70	0.92	2.20
850	9.44	1.38	4.53	1300	8.34	1.08	2.56
900	9.99	1.55	5.05	1400	8.98	1.25	2.95
950	10.5	1.73	5.60	1500	9.62	1.44	3.37
1000	11.1	1.92	6.17	1600	10.3	1.64	3.82
1100	12.2	2.32	7.41	1700	10.9	1.85	4.29
1200	13.3	2.76	8.76	1800	11.5	2.07	4.79
1300	14.4	3.24	10.2	1900	12.2	2.31	5.31
1400	15.5	3.76	11.8	2000	12.8	2.56	5.86
1500	16.7	4.31	12.5	2200	14.1	3.09	7.02
1600	17.8	4.91	15.4	2400	15.4	3.68	8.31
1700	18.9	5.54	17.3	2600	16.7	4.32	9.70
1800	20.0	6.21	19.4	2800	18.0	5.01	11.20
1900	21.1	6.92	21.6	3000	19.2	5.75	12.8
2000	22.2	7.67	23.8	3200	20.5	6.55	14.5

NOTE: No allowance has been made for age, differences in diameter, or any abnormal condition of interior surface. Any factor of safety must be estimated from the local conditions and the requirements of each particular installation.

Table 10.3 Friction loss for pipe fittings.

Nominal pipe size	Actual inside diameter	Friction loss in pipe fittings in terms of equivalent feet of straight pipe									
		Gate valve full open	90 degrees elbow	Long radius 90 degrees or 45 degrees std. elbow	Std. tee thru flow	Std. tee branch flow	Close return bend	Swing check valve full open	Angle valve full open	Globe valve full open	Butterfly valve
1½	1.61	1.07	4.03	2.15	2.68	8.05	6.71	13.4	20.1		
2	2.067	1.38	5.17	2.76	3.45	10.3	8.61	17.2	25.8	7.75	7.75
2½	2.469	1.65	6.17	3.29	4.12	12.3	10.3	20.6	30.9	9.26	9.26
3	3.068	2.04	7.67	4.09	5.11	15.3	12.8	25.5	38.4	11.5	11.5
4	4.026	2.68	10.1	5.37	6.71	20.1	16.8	33.6	50.3	15.1	15.1
5	5.047	3.36	12.6	6.73	8.41	25.2	21	42.1	63.1	18.9	18.9
6	6.065	4.04	15.2	8.09	10.1	30.3	25.3	50.5	75.8	22.7	22.7
8	7.981	5.32	20	10.6	13.3	39.9	33.3	58	99.8	29.9	29.9
10	10.02	6.68	25.1	13.4	16.7	50.1	41.8	65	125	29.2	29.2
12	11.938	7.96	29.8	15.9	19.9	59.7	49.7	72	149	34.8	34.8
14	13.124	8.75	32.8	17.5	21.8	65.6	54.7	90	164	38.3	38.3
16	15	10	37.5	20	25	75	62.5	101	188	31.3	31.3
18	16.876	16.9	42.2	22.5	28.1	84.4	70.3	120	210	35.2	35.2
20	18.814	12.5	47	25.1	31.4	94.1	78.4	132	235	39.2	39.2

Calculated from data in Crane Co. – Technical Paper 410.

Table 10.4 Head loss for 1000 gpm in pipe.

Size, in	Velocity, ft/s	Head loss, ft
4	25.5	64.8
5	16	15.8
6	11.1	6.171
8	6.41	1.56
10	4.07	0.50

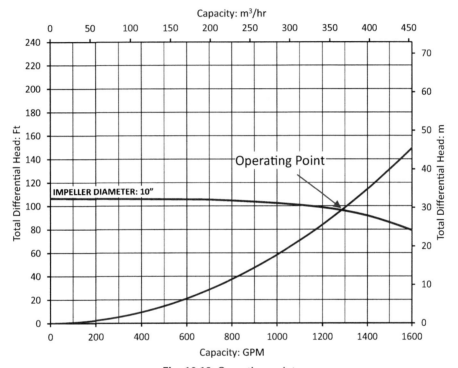

Fig. 10.19 Operating point.

pump at that flow rate, that flow and head is called the "operating point". In the illustration in Fig. 10.19, a friction curve crosses the pump head curve at about 1300 gpm. When this pump is connected to this piping system, the head produced by the pump will be about 100 ft and the flow rate will be 1300 gpm.

As an illustration, the 8 × 6 × 14 centrifugal pump with a 10″ impeller, described in Fig. 10.13, is connected to 375 ft of 3″ pipe. The pipe friction curve for this pipe is presented in Figs. 10.17 and 10.18. Plotting the pipe friction curve on the pump curve (Fig. 10.20), indicates that about 300 gpm will flow through the pipe.

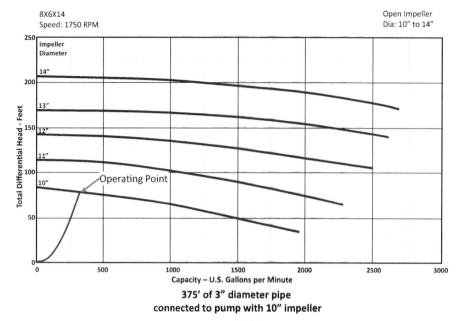

Fig. 10.20 Flow through 3″ diameter pipe.

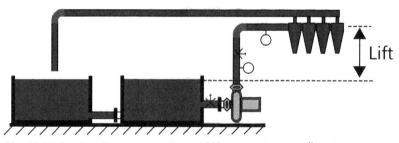

Fig. 10.21 6 × 5 × 13 pump supplying 1000 gpm to twenty 4″ hydrocyclones.

Application for desilting drilling fluid

A centrifugal pump is needed to process drilling fluid through a twenty cone bank of four inch desilters. The header is located about 10 ft above the drilling fluid level in the mud tanks, as shown in Fig. 10.21. The suction line is short with only one butterfly valve. The discharge line is very simple: 25 ft of pipe, one butterfly valve, and one elbow. A pressure gauge is mounted below the valve in the discharge piping. This is recommended for diagnostics of pump problems and will be discussed later.

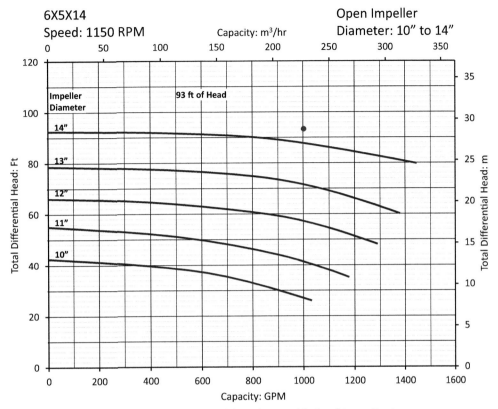

Fig. 10.22 This pump would not be suitable for this application.

From Tables 10.3–10.5:
Suction Line: 6″ Schedule 40 Pipe 6 ft Long
One 6″ Butterfly Valve, Equivalent Length = 22.7 ft
Discharge Line: 5″ Schedule 40 Pipe 25 ft Long
One 5″ Butterfly Valve, Equivalent Length = 18.9 ft
The total equivalent length of 6″ pipe is 28.79 ft.
The total equivalent length of 5″ pipe is 43.9 ft.
From Table 10.1:
With 1000 gpm, the head loss in the 6″ pipe is 6.17 ft/100 ft of pipe.
With 1000 gpm, the head loss in the 5″ pipe is 15.8 ft/100 ft of pipe.
Friction head loss in the suction piping will be:

$$= \left[\frac{6.17 \text{ ft of head}}{100 \text{ ft of pipe}}\right](24.9 \text{ ft of pipe}) = 153 \text{ ft of head}$$

Table 10.5 Equivalent pipe lengths for various fittings.

USE FOR ANY LIQUID

Nominal Pipe Size	Actual Inside Dia Inches	Gate Valve FULLY OPEN (Std Wt)	Long Ell Threaded or 45° Ell Threaded	Std Tee Threaded / Square Entrance / Cone Entrance	90° Ell Threaded	Square Ell / Reduced Tee Threaded / Std Tee Threaded / Extended Entrance	Butterfly Valve FULLY OPEN	Swing Check Valve FULLY OPEN	Globe Valve FULLY OPEN	Std Wt Welding Ell 90° Short (R/D = 1)	Std Wt Welding Ell 90° Long (R/D = 1.5)
½	.622	.4	.8	1.1	1.7	3.3	—	5	19	—	—
¾	.824	.4	1.0	1.4	2.1	4.2	—	7	23	—	.9
1	1.049	.6	1.4	1.8	2.6	5.3	—	8	29	—	—
1¼	1.380	.7	1.8	2.3	3.5	7.0	—	11	39	—	—
1½	1.610	.9	2.2	2.7	4.1	8.1	—	13	45	—	—
2	2.067	1.	3.	4.	5.	10.	11	15	58	3	2.
2½	2.467	1.	3.	4.	6.	12.	11	19	69	4	3.
3	3.068	2.	4.	5.	8.	15.	8	23	86	5	3.
4	4.026	2.	5.	7.	10.	20.	9	32	113	7	4.
5	5.047	3.	7.	8.	13.	25.	7	40	142	8	5
6	6.065	3.	8.	10.	15.	30.	11	48	170	10	6.
8	7.981	4.	11.	13.	20.	40.	13	33	224	13	8.
10	10.020	5.	13.	17.	25.	50.	16	42	281	17	10.
12	12.000	6.	16.	20.	30.	60.	22	50	336	20	12.

For 45° Ells Use 65% of Values Given Below

[4] Most of this data was taken by permission from page 49, Cameron Hydraulic Data, Ingersoll-Rand Company, Washington, N.J. 1962.

[5] The butterfly valve data was furnished by TRW Mission Mfg. Co., Houston, TX.

The friction loss for the discharge piping will be:

$$= \left[\frac{15.8 \text{ ft of head}}{100 \text{ ft of pipe}}\right](37.5 \text{ ft of pipe}) = 5.93 \text{ ft of head}$$

The total pipe friction loss will be 1.77 ft + 6.94 ft = 8.71 ft of head.

If the 20 cone 4″ desilter manifold is 10 ft above the liquid level in the mud tanks and the cones require a manifold head of 75 ft to perform properly, the total head requirement for the pump can now be calculated:

Total head requirement = Lift + Friction loss + Manifold head

Total head requirement = 10 ft + 7.46 ft + 75 ft = 93 ft

A 6 × 5 × 14—1150 RPM pump is available to supply drilling fluid to the desilters.

The 1150 RPM centrifugal pumps seem to have less maintenance problems but in this case, it could not produce the 94 ft of head required to supply the hydrocyclones with the proper head.

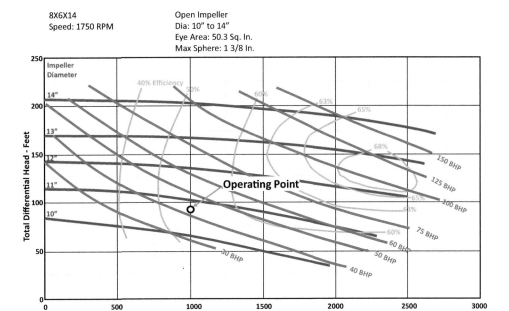

Fig. 10.23 This pump could be used.

The pump described in Fig. 10.23 is also available for this application. Locating the operating point of the pump curve (Fig. 10.20), indicates that this pump could supply the flow rate needed and the head required to properly operate the desilters.

This pump could supply 1000 gpm at 94 ft of head with a $10^3/_4''$ impeller. This is an $8 \times 6 \times 14$ pump that is probably too large for this specific application. To pump water, a brake horsepower of slightly over 40 HP would be required. To anticipate a mud weight over twice the density of water, a 100 HP motor would probably be installed. This pump rotates at 1750 RPM which is the reason that this pump will produce the required head; whereas, the 1150 RPM pump fails to do so.

The $6 \times 5 \times 14 - 1750$ RPM centrifugal pump (Fig. 10.24), could supply 1000 gpm at 94 ft of head with a $10''$ impeller. The pump curve indicates that water would require about 35 BHP and the pump would be operating at about 68% efficiency.

A short, large diameter suction line with only one valve produced only a 1.5 ft head loss as 1000 gpm was moved by the pump. This means the head (or pressure) at the impeller remained above atmospheric pressure because of the head of fluid in the mud tank. This is called a flooded suction. Because of this situation, the NPSH curves were not used to decide on the pump or impeller size. This will be discussed in the next section.

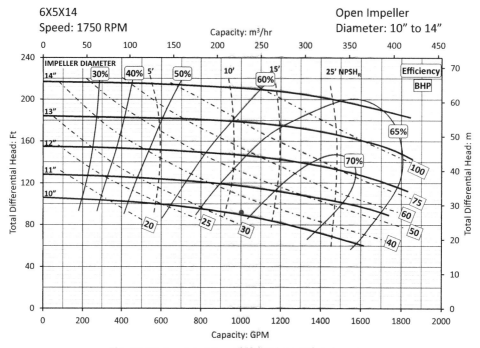

Fig. 10.24 6 × 5 × 14 – 1750 RPM centrifugal pump.

Net positive suction head (NPSH)

A centrifugal pump cannot "suck" fluids into the housing. Fluid must flow into the housing because of a positive pressure. A low pressure exists at the center of an impeller. Atmospheric pressure must push the fluid into the pump. If the pump suction requires that the fluid be lifted up into the pump housing, the head available many be insufficient. Pressures in the center of the impeller are measured from absolute zero pressure. Any pressure above absolute zero is called a net positive suction pressure.

Atmospheric pressure

To explain NPSH, a digression is appropriate here to discuss atmospheric pressure. When weather reports are made, the barometric pressure is usually reported in millimeters of mercury. This is the height of a column of mercury that would be supported by atmospheric pressure (Fig. 10.25).

If a three foot tube, closed at one end and open at the other, is filled with mercury, and inverted into a container with mercury exposed to the atmosphere, the mercury will drain until the height in the tube is supported by atmospheric pressure.

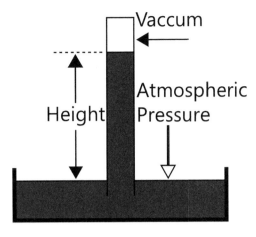

Fig. 10.25 Mercury barometer.

The pressure at the top of the column of mercury is zero and the pressure on the surface of the mercury is atmospheric pressure. The height, or head, of mercury can be calculated from the equation:

$$\text{Pressure} = 0.052[\text{mud weight, ppg}][\text{head, ft}]$$

The density of mercury is 13.59 gm/cc, or 113.37 ppg.

$$14.7 \text{ psi} = 0.052[113.37 \text{ ppg}][\text{head, ft}]$$

The head would be 2.49 ft or 760 mm.

If a long column of 12 ppg drilling fluid was suspended with the open end below the liquid level, how high would the drilling fluid stand in the column (Fig. 10.26)?

$$\text{Pressure} = 0.052[\text{mud weight, ppg}][\text{head, ft}]$$

$$14.7 \text{ psi} = 0.052[12.0 \text{ ppg}][\text{head, ft}]$$

Height, or head would be 23.6 ft. When the mud pumps are turned off, a 12.0 ppg drilling fluid will stand about 24 ft above the liquid level in the bell nipple. At the top of the column of drilling fluid would be a vacuum—or, more precisely, vapor pressure.

Flow into a centrifugal pump

The centrifugal pump impeller accelerates fluid as previously described. The pump, however, cannot move the fluid from the suction tank into the impeller. The fluid must be pushed into the impeller with the only pressure available: atmospheric pressure and available head from the height of fluid above the impeller.

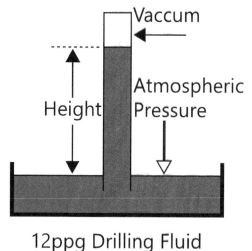

Fig. 10.26 Drilling fluid standing in a long tube.

Long suction lines produce a significant pressure loss from fluid flowing through them. In the sketch shown in Fig. 10.27, a mud tank supplies drilling fluid to a centrifugal pump. The suction line with all of the valves, elbows, swedges, tees, and length is represented by a long 'suction' line to the pump. In the example above, the equivalent lengths of the various fittings are converted into equivalent lengths of pipe. As the fluid flows through this line to the centrifugal pump, a pressure differential must exist to cause the fluid to flow. The only

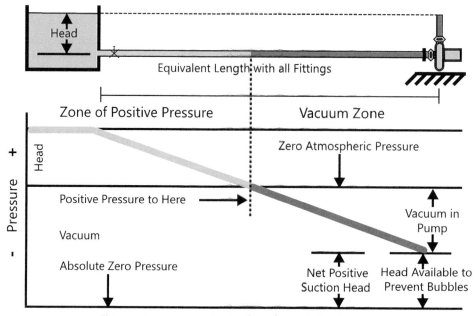

Fig. 10.27 Vacuum at the suction of a centrifugal pump.

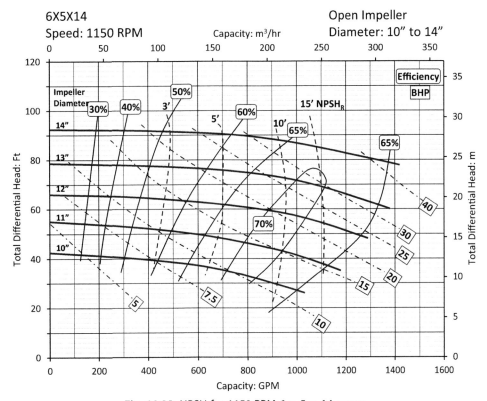

Fig. 10.28 NPSH for 1150 RPM 6 × 5 × 14 pump.

pressure that is available to make the fluid flow in the suction line is the head of the fluid in the mud tanks and atmospheric pressure. At some distance along this 'equivalent' suction line, the pressure in the line is less than atmospheric. By the time the fluid gets to the centrifugal pump, the pressure in the line is below atmospheric pressure.

The head available to prevent bubbles is the NPSH available. Each centrifugal pump has a minimum value of NPSH that is required to prevent bubbles from forming. This minimum value increases when the flow rate through the pump increases.

A flow rate of 500 gpm from a 6 × 5 × 14 pump at 1150 RPM (Fig. 10.28), indicates that the NPSH must be 3 ft of head to prevent a vacuum in the suction line at the pump suction. This is the head above absolute zero pressure that is required for the pump. The NPSH increases as the flow rate increases. At a flow rate of 1100 gpm from this pump, the NPSH is 15 ft of head.

Cavitation

If the pressure at the impeller suction drops below the NPSH, cavitation occurs (Fig. 10.29). The low pressure in the fluid at the eye of the impeller produces bubbles

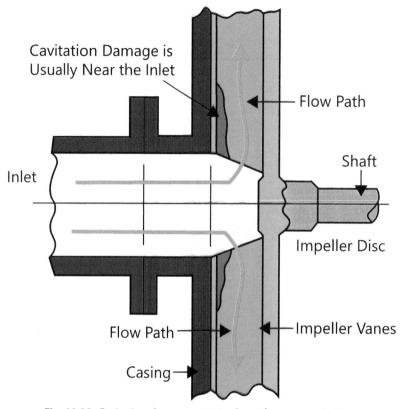

Fig. 10.29 Cavitation damage occurs where the vacuum is high.

of vapor. These bubbles travel along the impeller as the fluid is accelerated. As the fluid is accelerated, the pressure along the impeller blade increases. The vapor bubbles collapse. This implosion causes fluid to rush into the space formerly occupied by the vapor bubble. The pressure wave from the implosion is strong enough to apply a tensile load on the face of the impeller or other surfaces that actually will remove metal. The effects of cavitation are very costly because a centrifugal pump can be virtually destroyed in a short time because of these vapor bubble collapsing. Holes will develop in the pump housing, impeller, and even the stuffing box. From a practical standpoint, it also frequently sounds like gravel is being pumped instead of drilling fluid.

Cavitation can be caused by too much head loss between the suction tank in the mud system and the centrifugal pump. It can also be caused by pumps located above the surface of the fluid (Fig. 10.30). If the lift is too high, insufficient pressure (NPSH too low) will cause the pumps to cavitate.

The pumps normally used in the oil field are not self-priming. When they are started, no fluid will enter the impeller. This is the reason that centrifugal pumps used in the

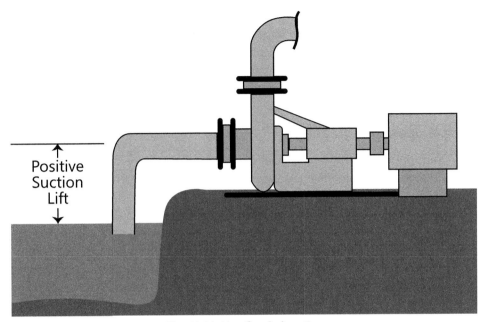

Fig. 10.30 Non-flooded suction.

drilled solids removal section of a drilling fluid processing system should have a flooded suction. The suction lines should be as short as possible with a straight section at least two to three pipe diameters in front of the centrifugal pump.

Standard centrifugal pumps are not self-priming and require the fluid end to be primed prior to activation. Install centrifugal pumps in a location that provides a flooded suction. Priming devices are available but flooded suctions are preferable. Once the pump casing is full of fluid, the pump can be started. Operating a pump dry or restricting the suction flow can severely damage the fluid end, mechanical seal or packing. Self-priming pumps have a turbulent flow pattern and experience a short life when handling drilling fluids.

Once a pump is primed for a non-flooded suction, and the motor started, the pressure at the eye of the impeller decreases below atmospheric pressure. If the suction is flooded, the pressure at the eye of the impeller will decrease but perhaps not below atmospheric pressure. The actual pressure at the eye of the impeller varies by pump size and flow and is shown on the pump curves as Net Positive Suction Head Required (NPSH$_R$). The inlet head to the pump will be atmospheric pressure or head (at sea level about 34 ft) plus or minus the head of liquid above or below the pump centerline minus the suction friction losses minus the velocity head. The sum of this calculation is known as Net Positive Suction Head Available ($NPSH_A$). If the $NPSH_A$ is greater than $NPSH_R$ the pump will function as designed. If the $NPSH_A$ is equal to or less than the $NPSH_R$, the pump will cavitate.

As the fluid enters the pump the impeller accelerates it. The diameter of the impeller and RPM the impeller is rotated directly affect the velocity of the fluid. The casing of the pump contains this velocity and converts it into pressure head. Casing size and impeller width control the volume that the pump is able to produce.

Practical operating guidelines

This section provides guidance to avoid common errors observed in using centrifugal pumps on drilling rigs, and practical suggestions about operating centrifugal pumps on drilling rigs.

- Eliminate manifolding of pumps in the mud tank solids removal section. This sounds simple—but too many rig designers want to be able to suck drilling fluid from anywhere and pump it to anywhere in the system. Each piece of removal equipment (degassers, desanders, desilters, mud cleaners, and centrifuges) has only one compartment that should be used for suction and only one compartment that should be used for discharge. Providing a means to do this incorrectly does not provide good operating procedures. A single switch should turn on each pump. When the desilting switch is turned on, the desilter should desilt correctly. Centrifugal pumps in the suction/check section and in the addition section can be used for multiple service: mud guns, mud hoppers, filling trip tanks, or slug tanks. Removal equipment centrifugal pumps should never be available as transfer pumps.
- A very common problem involves a labyrinth of pipes and valves emerging from the removal section, particularly on offshore drilling rigs. The probability that all of the valves are adjusted correctly is very small. Who changes the position of the valves? Usually the newest member of the drilling crew is sent below to open and close valves to allow the fluid to be processed correctly. Some of the manifolds are so complicated that more than an hour of examination is required to determine which valve should be open and which valve should be closed. After several months of use, the possibility of a valve leaking is great and can not be observed in most systems. Solution: one suction pipe, one pump, one switch, no manifolding or complicated flow systems. Eliminate the nonsense that someone needs to be able to pump any one tank to another tank.
- Never close, or 'pinch', or adjust a valve in the suction of a centrifugal pump to change the discharge pressure, or for any other reason. This could cause cavitation in the pump.
- Install centrifugal pumps with flooded suctions if at all possible.

- Never connect two centrifugal pumps in parallel; that is, to supply fluid to the same piece of equipment. Some experts will install two pumps to feed two identical banks of desilters or other equipment with a valve between the two pump discharges. Unfortunately all personnel on drilling rigs are not experts and will probably open that valve for a variety of incorrect reasons. Don't provide them with the opportunity to make a mistake.
- Select centrifugal pump piping by flow rate requirements, not by the size of flanges, or the pump.
- Never permit air or gas to enter the suction of a centrifugal pump in the drilling fluid system. As the impeller starts spinning the fluid, the centripetal acceleration causes air or gas to concentrate at the center of the impeller. Gradually, even a small amount of air bubbling into the suction will collect sufficient volume that it covers most of the pump suction inlet. The centrifugal pump will cease to move liquid.
- When drilling fluid falls from a line above the fluid level in the mud tanks, air is entrained in the fluid. If this line is above the suction to a centrifugal pump, air will gather in the center of the impeller (Fig. 10.31).
- Suction lines should be straight for two or three pipe diameters before entering a centrifugal pump (Fig. 10.32). If an elbow is positioned too close to the pump suction, the fluid does not impact the impeller uniformly all the way around the impeller. This unbalanced load eventually causes premature pump failure.
- Never replace an old impeller with a new one just because of a change in discharge pressure as shown on a pressure gauge.
- Install a pressure gauge connection, or a pressure gauge, on the discharge side of the centrifugal pump just below a butterfly valve. While the pump is running, close this valve briefly, and measure the pressure. This can be a great diagnostic tool during the life of the pump. When header pressures cease to indicate the value that they should, the no-flow pressure at the pump will indicate whether the trouble lies with the

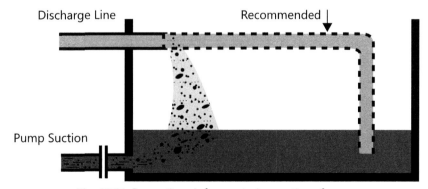

Fig. 10.31 Preventing air from entering suction of pump.

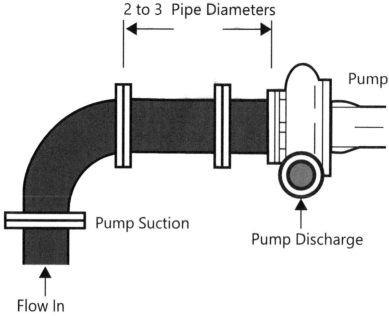

Fig. 10.32 Straightening flow pattern before fluid enters pump.

pump or with the piping system downstream. This frequently saves a lot of work caused by dismantling the pump to inspect the impeller.
- The centrifugal pump discharge pressure will change as the mud weight changes. This is a constant head device. As the mud weight goes up, the pressure will increase even though the head remains constant.
- Never walk away from a discharge valve that is closed to measure pump discharge pressure. A good rule to follow is the requirement that the hand can not leave the valve handle while the valve is closed. Centrifugal pumps use less power when the valve is closed, so it will not hurt the motor; however, the fluid in the pump will start to boil because of the internal friction and possibly destroy the seals.

Many things fall into a mud tank. Open suction lines can get clogged with a variety of things. One of the most common pieces of debris is a work glove. These clog hydrocyclones or other removal equipment and the results are sometimes difficult to diagnose. A large screen mounted around the suction line from a tank will filter out much of this type of debris (Fig. 10.33). The screens can be lined up with the suction lines by welding guides onto the side of the mud tank. After the pits are full of mud, positioning the screen would be difficult, if not impossible, without some type of guide. Weld a $1'$ to $1^1/_2\,''$ pipe onto the top of the screen. Align the pipe guides so that once the pipe is place into the slots of the pipe guides, the screen will automatically be located

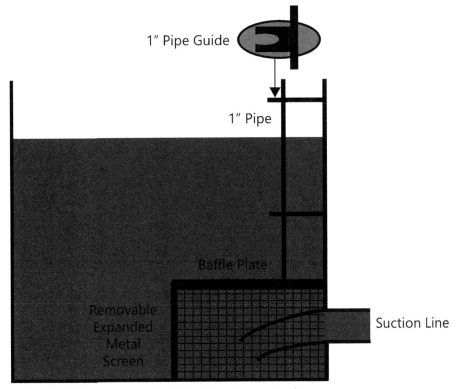

Fig. 10.33 Screening suction line.

in the correct position in the pit. A baffle plate can be built into the top of the screens to prevent vortexing.

Vortexing, or creating whirlpools in the liquid, can occur when the fluid velocity in the suction pipe is high (Fig. 10.34). The depth of fluid above the suction pipe must be increased significantly if the fluid velocity is above 10 ft/s in the pipe. This is another reason that the fluid level in the removal section of the drilling fluid processing system should be maintained at a constant level.

A flow rate of 600 gpm produces a velocity of 6.5 ft/s in a 6 inch suction pipe. From Fig. 10.34, the entrance to the suction line should be $3^1/_2$ ft below the surface of the liquid in the mud tank. When the liquid level drops below that value, a whirlpool will feed air into the suction line. The air will accumulate in the center of the impeller and vapor lock the pump. The pump will cease to pump drilling fluid.

Degassers remove gas from the drilling fluid. Many drilling personnel think that it is removed to keep the bottom hole pressure from decreasing too much because of the gas-cut drilling fluid. Actually, a significant decrease in mud weight because of gas fails to be reflected in a decrease in bottom hole pressure. After a clear water drilling fluid reaches a

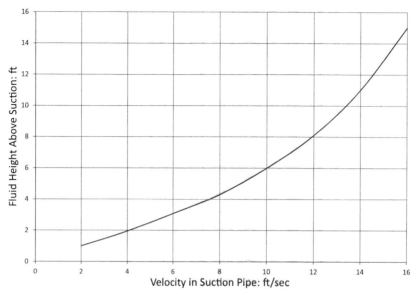

Fig. 10.34 Head of fluid needed to prevent vortex in mud tanks.

depth of about 3000 ft, the pressure had increased about 100 fold. Stated another way, the gas bubbles have decreased to about 1/100th of their original size. This has very little effect on the bottom hole pressure. Degassers remove gas so that the centrifugal pumps will not vapor lock.

When pumping into a long empty line, partially close the discharge valve until the line is full. A centrifugal-pump motor requires less power when the discharge valve is completely closed. If no fluid is in the line, the pump will attempt to create a constant head. Electric motors will continue to increase the current until the head is reached. Usually, a circuit breaker will blow before that happens. After the line is full, fully open the discharge line.

Do not try to reduce the centrifugal pump pressure by adding a by-pass line. As more fluid is moved by a centrifugal pump, the internal pressure losses increase (moving out on the pump curve). The head can be reduced by opening a by-pass valve, but the quantity of drilling fluid pumped will increase greatly. This will also require more power from the motor.

When the suction is not flooded (Fig. 10.35), a centrifugal pump must be primed—or filled with fluid before it will pump fluid. A foot valve is normally used to allow the suction line to be filled with fluid prior to starting the pump. A foot valve works like a check valve, fluid can flow up the suction line but not down the suction line. After the suction line is filled, the pump can be started. The pressure at the center of the pump impeller will be lower than atmospheric pressure. The head above absolute zero is the NPSH available for the pump. If this head is less than the NPSH required, cavitation will quickly follow.

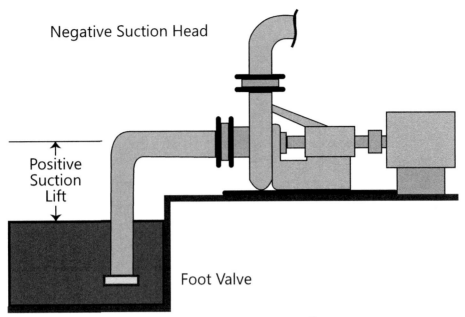

Fig. 10.35 Centrifugal pumps are not self-priming.

Appendix 10.A
Quiz solution

The centrifugal pump is a CONSTANT HEAD device. The height of the fluid will be the same whether it is blue smoke, water, oil, or heavy-weight drilling fluid. If this simple question was answered incorrectly, you do not understand centrifugal pumps.

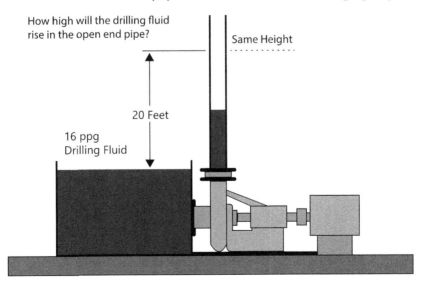

Appendix 10.B
Centrifugal pump head curves

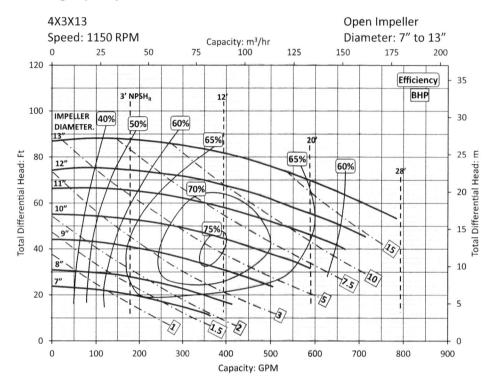

Centrifugal pumps 231

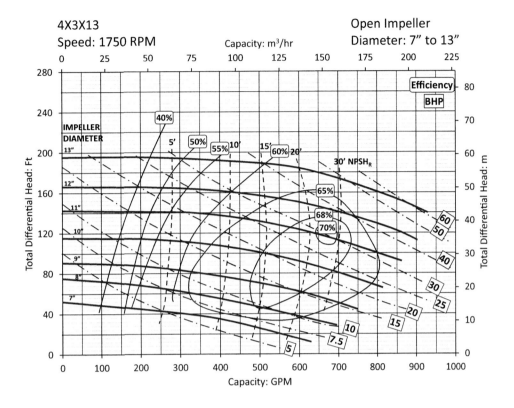

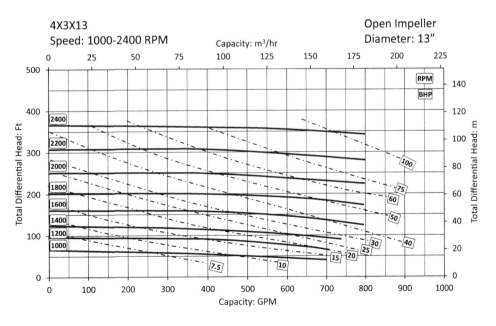

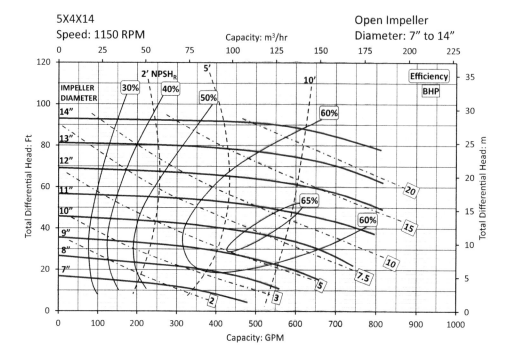

Centrifugal pumps 233

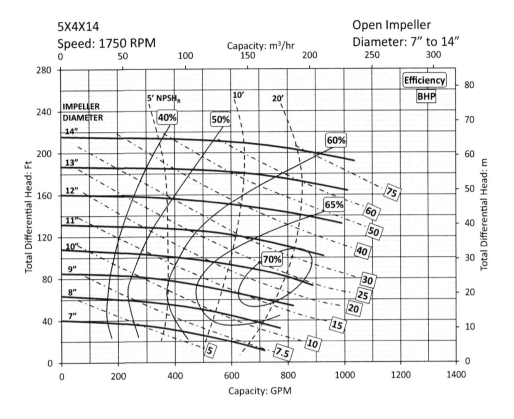

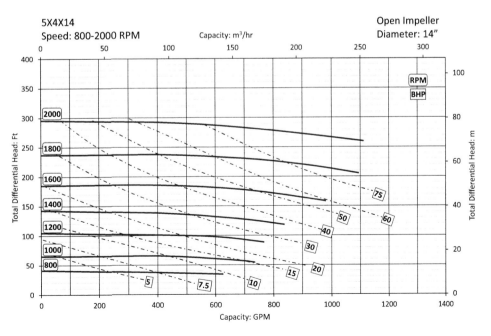

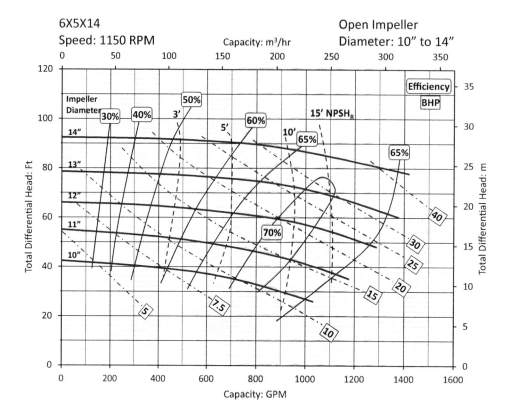

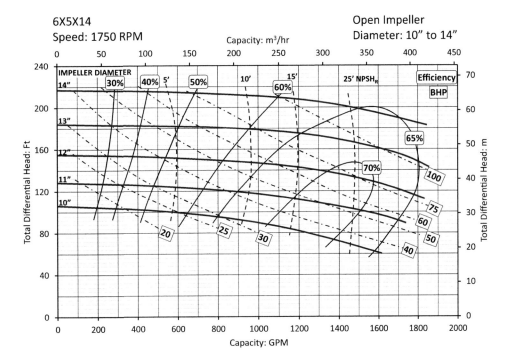

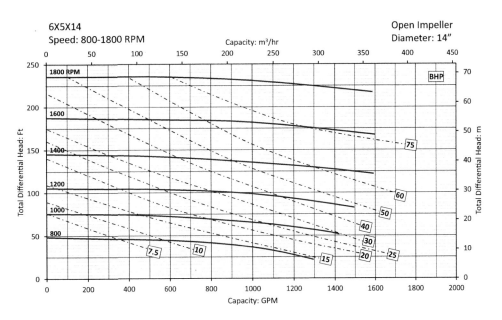

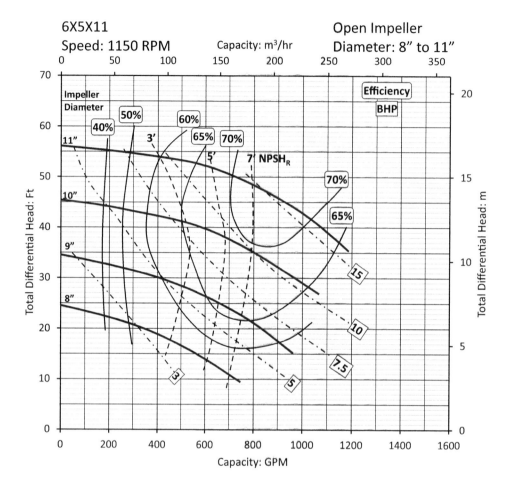

Centrifugal pumps

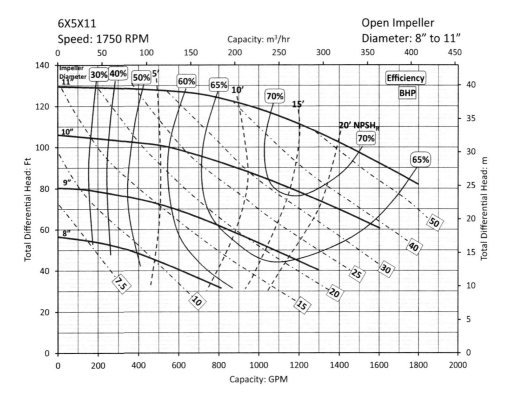

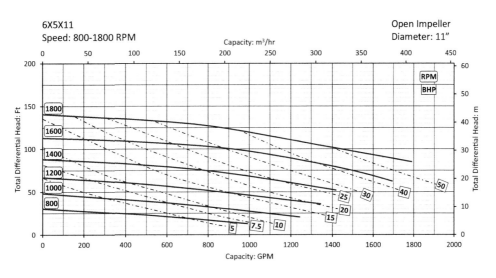

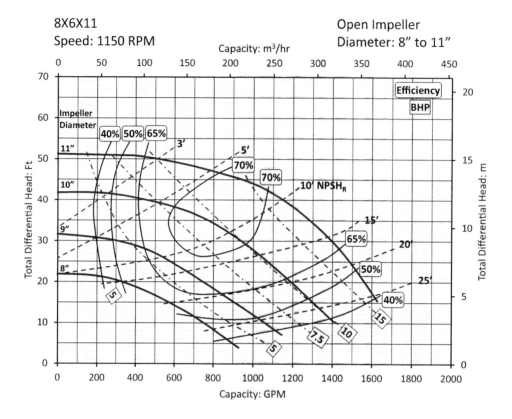

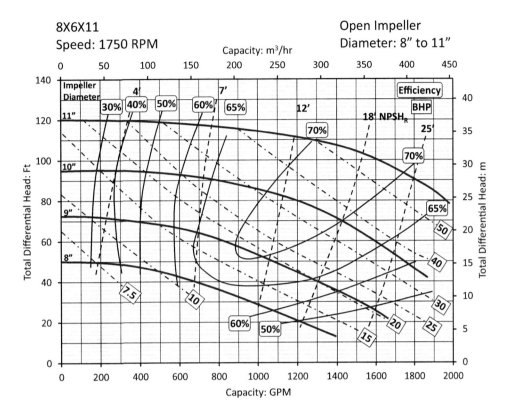

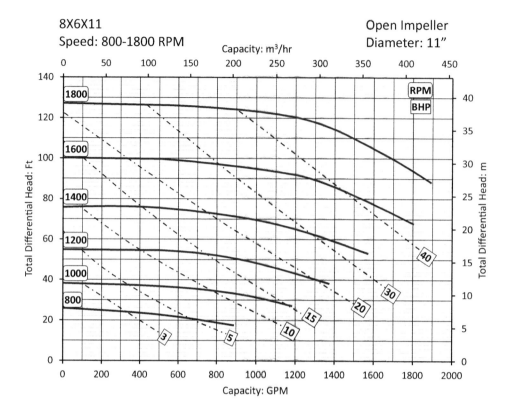

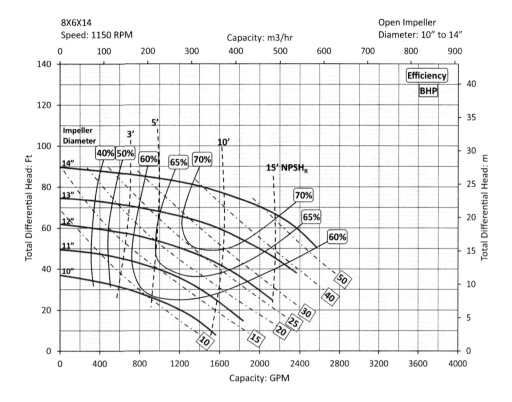

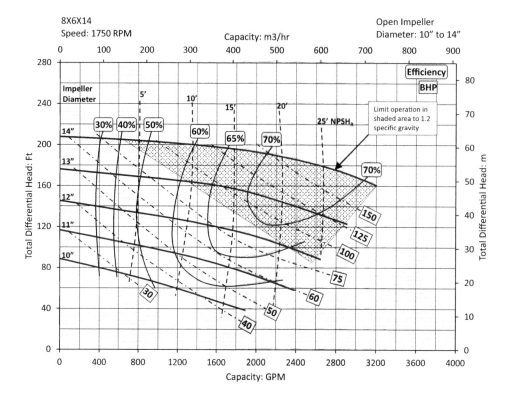

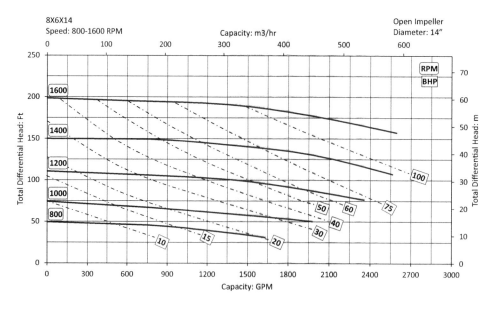

Problems

Problem 10.1

Forty 4" cones are connected to a 8 × G × 14 Magnum Pump. Impeller is rotating at 1750 RPM. The feed manifold is Located 25 ft above the liquid level in the mud tanks. Assume 10 ft of head is lost in the plumbing.

1. What impeller size should be used?
2. What Horsepower pump would be required to pump a 16.6 ppg drilling fluid?
3. What NPSH would be required?

Problem 10.2

A bank of twenty 4" desilter cones are mounted 10 ft above the drilling fluid in the pits. An 8 × 6 × 14, 1750 RPM centrifugal pump is available to supply 50 gpm to each cone. The suction line is 8" in diameter, has two elbows, one butterfly valve, and 10 ft of straight pipe. The centrifugal pump 6" diameter discharge line has four elbows, one butterfly value, and 30' of pipe.

1. What impeller should be used?
2. With this impellor, can two 1" diameter mud guns be used to stir the suction tank if no agitator is available? If not, what impeller is needed? Assume 15 ft of 3" diameter pipe is used to supply fluid to the mud guns.

Problem 10.3

Twenty 4" cones are connected to a 6 × 5 × 14 Magnum Pump. Impeller is rotating at 1750 RPM. The feed manifold is 30 ft above the liquid level in the mud tanks. Assume 15 ft of friction in the plumbing.

1. What impeller size should be used?
2. What Horsepower pump would be required to pump a 16.6 ppg drilling fluid?
3. What NPSH would be required?

Problem 10.4

Ten 4" hydrocyclones are connected to a 5 × 4 × 1750 Magnum Pump. Impeller is rotating at 1750 RPM. The feed manifold is 20 ft above the liquid level in the mud tanks. Assume 10 ft of head loss in the pipe.

1. What impeller should be used?
2. What Horsepower pump would be needed to pump 16.6 ppg drilling fluid?

CHAPTER 11

Fraction of drilling fluid processed

Introduction

To allow all drilled solids removal equipment to perform correctly, drilling fluid must be processed sequentially in compartments that are plumbed correctly. Large solids are removed with shale shakers, smaller solids are removed with hydrocyclones, and the smallest solids are separated with centrifuges. To ensure adequate processing, equipment should be sized to process at least 125% of the anticipated maximum rig circulating rate with the exception of the centrifuge. The total solids removal efficiency depends upon plumbing arrangement necessary to process one hundred percent of the fluid which comes from the well bore through the shakers and hydrocyclones. Generally, the centrifuge processes only a small fraction of the drilling fluid reporting to the surface.

Economics of correct drilling fluid processing is difficult to quantify. One specific event will help capture the general trend of what usually happens when attention is paid to keeping a low concentration of drilled solids. In the early 1990s, a major oil company was planning to drill twelve wells from a specific platform. The contractor installed excellent equipment on the rig to satisfy the contract to drill. Near the end of the sixth well, the Operations Supervisor in the main office was disturbed by the cost of the wells. An investigator was sent to the platform to evaluate the drilling fluid system. The contractor subscribed to the common misconception that every pump should be able to pump from any tank to any other tank in the drilling fluid system. After an initial perusal of the system, the investigator reported that it would be prohibitively expensive to correct the plumbing so that the removal system would operate properly. The Operations Supervisor said "fix-it". After an hour of further explanation and argument, the investigator finally agreed to draw up plans for correct plumbing. After three days of intensive planning, the plans were submitted to the office. At the end of the sixth well, six welders were sent to the platform. The production wells were shut-in and over $250,000 was spent correcting the plumbing. The seventh well cost less than the money spent to properly process the drilling fluid. Even though the cost of modification was high, the changes produced a massive change in well cost.

Frequently, modifications to correct the plumbing cost much less but usually result in great savings on the next wells. For example, on one rig, the overflow from the desilter bank was sent back to the drilling fluid system through a rubber hose. The hose was arranged so that the overflow was returned up-stream from the suction tank. Moving the hose to a down-stream discharge changed a 50% effective system to a 100% effective system. In this case the cost of modification was very low; but the changes can produce a massive change in well cost.

Calculating drilling fluid process efficiency

One of the major problems in drilled solids removal is the inability to process all of the drilling fluid. The fluid processing efficiency can be calculated by dividing the volume of drilling fluid treated by the volume of drilling fluid entering the suction compartment. This equation applies only to compartments where the drilling fluid is well blended and homogeneous. If the drilling fluid is not well mixed, the processing efficiency will be significantly lower than the calculated value. Both of these systems will be discussed.

1) In Fig. 11.1, a desilter processes 400 gpm of drilling fluid. By taking suction from tank #1 and discharging the desilter-cleaned drilling fluid into tank #2, all of the drilling fluid is cleaned. No fluid is entering tank #2 unless it passes through the desilter. This assumes that no cones on the desilter are plugged from solids which could have by-passed the shale shaker. Note, generally, when $4''$ desilters process 50 gpm input, the overflow is only 49 gpm which 1 gpm being discarded. For purposes of estimating the process efficiency of the system, this small discard can be ignored. In this case, the smaller desilter overflow would actually decrease the removal process efficiency to less than 100%.

2) The flow rates in the well may not always be constant. To provide a degree of flexibility, the desilters should process more fluid than is arriving in the suction tank for the desilters. In Fig. 11.2 above, 400 gpm are arriving at the surface. If the desilters process 500 gallons, there will be a back flow between tank #2 and tank #1. This insures that all of the fluid in tank #2 has been processed through the desilters - or 100% processing efficiency.

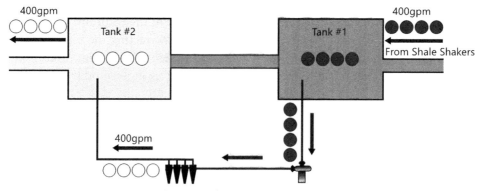

Fig. 11.1 Adequate arrangement.

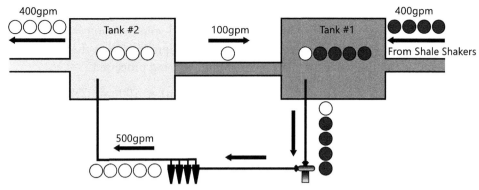

Fig. 11.2 Better tank arrangement with back flow from tank #2 to tank #1.

The fraction of drilling fluid processed, or cleaned, is the volume cleaned by the desilter divided by the volume entering the suction tank of the desilter. In this case, the answer is obvious by observing the 'dirty dots'.

$$\text{Processing Efficiency} = \frac{\text{flow rate entering suction compartment}}{\text{flow rate through desilters}} \times 100$$

$$\text{Processing Efficiency} = \frac{400 \text{ gpm} + 100 \text{ gpm}}{500 \text{ gpm}} \times 100 = 100\%$$

This calculation can be better explained using the next tank arrangement (Fig. 11.3). In this case, the flow entering Tank #1 from the well is 400 gpm; however the desilters are processing only 300 gpm. There will be 100 gpm flowing from Tank #1 to Tank #2. Counting the "dirty dots" in Tank #2, reveals that 300 gpm are clean but 100 gpm have not been cleaned.

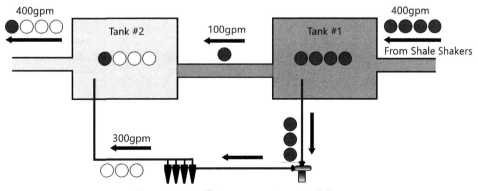

Fig. 11.3 Insufficient processing capability.

3) Insufficient processing:

Cleaning process efficiency is the ratio of the fluid volume being cleaned divided by the volume entering the suction tank of the equipment. From the shale shaker, 400 gpm is entering tank #1 and only 300 gpm is being processed.

$$\text{Cleaning efficiency} = (300 \text{ gpm} / 400 \text{ gpm}) \times 100 = 75\%$$

In Fig. 11.3, tank #2 contains three clean dots and one dirty dot — or three out of every four gallons is being cleaned. This is a 75% cleaning efficiency.

Usually, however, keeping an exact balance is difficult. More fluid is processed by the equipment than is flowing from the well. In tank #2, three cleaned dots and one dirty dot indicate that only 75% of the fluid is being processed through the desilter. The equation predicts the fraction of drilling fluid processed.

Occasionally, someone on the rig will route the overflow from the desilters back into the same tank with the concept that the desilter will process the drilling fluid twice and provide a cleaner drilling fluid (Fig. 11.4). Consider the case where 400 gpm is coming from the well and the desilter is cleaning 500 gpm and discharging into the tank downstream (like the first arrangement). In this case the desilter is processing 500 gpm but 900 gpm (500 gpm + 400 gpm) is entering tank #1. The process efficiency would be:

$$500 \text{ gpm}/900 \text{ gpm} = 0.56 \text{ or } 56\%$$

Instead of the desilter "looking at the mud twice", it only processes about one-half of the fluid entering tank #1.

If the flow from the well was reduced to 350 gpm and processing 400 gpm should obviously provide a good processing plant. But not if the clean fluid from the desilter is put back into the suction tank. In this case, only 56% of the drilling fluid is processed (Fig. 11.4).

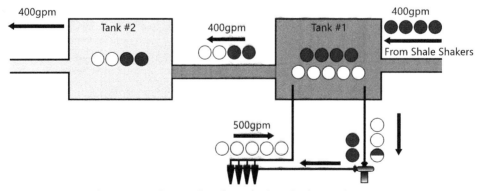

Fig. 11.4 Incorrectly returning the overflow from desilters back into their suction compartment.

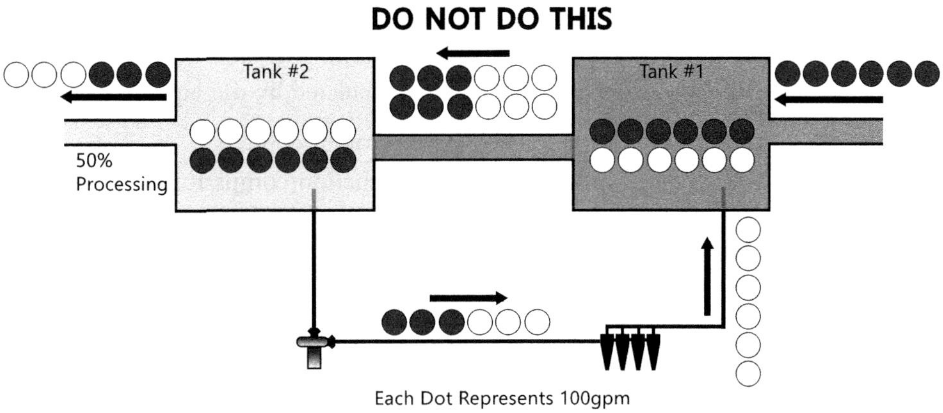

DO NOT DO THIS

Fig. 11.5 Discharge upstream from the hydrocyclone.

Frequently, the desilter discharge is located upstream from the suction. In one case, the discharge was a hose that could easily be moved from one tank to another. In the plumbing arrangement in Fig. 11.5, 600 gpm is entering tank #1 from the well and the desilters are cleaning 600 gpm. The total flow entering the suction tank #2 of the desilters is 1200 gpm. The processing flow rate is 600 gpm. The process efficiency is 50%.

Another incorrect plumbing situation has been observed on several rigs. The desilter and desander suctions are in the same tank and the discharge from both hydrocyclones is into the same tank (Fig. 11.6).

The fluid entering tank #1 from the well is 400 gpm. The desilter is processing 600 gpm and the desander is processing 600 gpm. This probably meets the contract

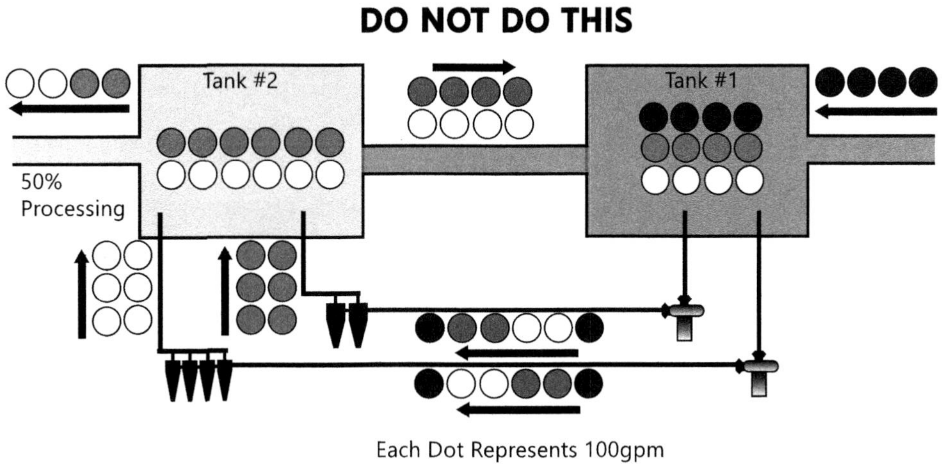

DO NOT DO THIS

Fig. 11.6 Parallel processing of desander and desilter.

agreement when the solids removal equipment is processing much more than the flow rate downhole. However, the flow rate returning to tank #1 from tank #2 is 800 gpm. The desilter processing efficiency is calculated by the equation:

$$\text{Processing Efficiency} = \frac{\text{flow rate through desilters}}{\text{flow rate entering suction compartment}} \times 100$$

$$\text{Processing Efficiency} = \frac{600 \text{ gpm}}{400 \text{ gpm} + 800 \text{ gpm}} \times 100 = 50\%$$

The desander processing efficiency is also 50%. The hydrocyclones should process fluid sequentially from one tank to the next tank. If a fine mesh screen is being used on the main shale shaker, the desander could be eliminated from the processing plant.

4) Field example of poor plumbing:

Notice in Fig. 11.7 that the desander bank (just behind the top of an agitator) is connected to the same line which is feeding the bank of desilters. What is wrong with this system? A sketch of the flow paths reveals a significant problem. The desilter feed and the desander feed lines are from the same pipe. These are in parallel and neither can process 100% of the fluid.

The total number of cones cannot be counted in the picture. Assuming that only 400 gpm of fluid was coming from the well, the desander is processing 600 gpm and the desilter is processing 500 gpm (perhaps by contract), the system looks good. However, using the equation for the fraction cleaned, the desilter is only processing about 45% of the fluid from the well.

Fig. 11.7 Field example of the desilter and desander in parallel with feed from the same pump.

5) Another bad field plumbing job:

The next field example came from a location which wanted to decrease their problems with centrifugal pumps. The Company man connected all the equipment up to one pump. This pump was connected into the suction tank, (Fig. 11.8). But only one centrifugal pump needed to be used. Several spare pumps meant that pumps could be sent in for repair without shutting the system down. As an interesting exercise, note that removing the desander from the system will increase the total fraction of fluid processed by the desilter.

Entering the suction tank#2: 500 gpm from the well, 700 gpm from the degasser jet pump, and 800 gpm from the desander or 2000 gpm and 900 gpm from the desilter bank; for a total of 2900 gpm. (The 600 gpm through the degasser leaves and re-enters the same tank). The desander is processing 800 gpm.

The process efficiency for the desander is:

$$800 \text{ gpm}/2900 \text{ gpm} = 0.28 \text{ or } 28\%$$

The process efficiency for the desilter is:

$$900 \text{ gpm}/2900 \text{ gpm} = 0.31 \text{ or } 31\%$$

Clearly the drilled solids are going to build in this drilling fluid. The contract might read that the hydrocyclones must process at least 100 gpm more than the fluid being pumped downhole. Clearly, in this case, that is insufficient to guarantee good clean drilling fluid.

6) Common incorrect plumbing found on jack-up rigs and platform rigs

Another common plumbing nightmare is found in offshore operations (Fig. 11.9). Contractors are told that they need to pump from anywhere in the system to anywhere in the system. A jigsaw puzzle of valves and pipes attached to the bulk heads challenge anyone to arrange the plumbing so that the flow is correct in the removal system. Valves can leak or accidently be left open or closed which reduce the processing efficiency from the desired 100% to some unknown level. This increases the

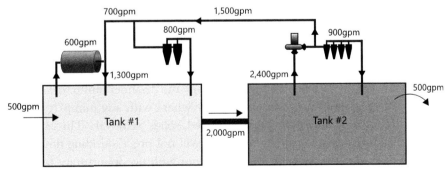

Fig. 11.8 Mud tank arrangement using only one pump.

Fig. 11.9 Typical incorrect plumbing on a jack-up rig.

likelihood that excess quantities of drilling fluid will be necessary to keep the drilled solids concentration to the level needed for drilling fast, trouble-free wells. The cost of poor removal efficiency and dilution will be discussed in the next chapter.

Summary

Subscribe to the concept: one pump, one switch, one function in the drilling fluid processing system. For auxiliary pumping when not drilling, pumps could temporarily be connected to the system to pump from anywhere to anywhere.

Each unit of solids control equipment should have its own dedicated, single purpose pump—with no routing options. When the pump is turned on, there should be only one place for the fluid to go. Hydrocyclones and mud cleaners have only one correct location in tank arrangements and, therefore, should have only one suction location. Routing errors should be corrected and equipment color-coded to eliminate alignment errors. If there is a concern about an inoperable pump suggests allowing other pumps in the system to be used, they generally will not process the drilling fluid in a correct manner. Making an easy access to the pumps and having a standby pump in storage can save money. Common and oft heard justifications for manifolding the pumps are "I want to manifold my pumps so that when my pump goes down, I can use the desander pump to run the desilter, etc.", or "I can pump from anywhere to anywhere with any pump". These statements indicate a poor understanding of drilled solids removal. This arrangement almost automatically guarantees that the system will not process drilling fluid correctly. Having a dedicated pump properly sized and setup with no opportunity for improper operation will give surprisingly long pump life as well as processing the drilling fluid properly. Solids removal section should have a minimum number of valves so that plumbing is correct for processing fluid correctly.

If pumps are needed for completions or for drilling fluid swap-out, they should be added to the drilling fluid processing plant. The drilling fluid processing pumps should not be used or manifolded into that system. Although this may look like a more expensive arrangement, a risk analysis of things that can go wrong should convince the most frugal drilling groups that they are well worth the additional expense. Rule: one pump, one switch, one function.

Problems

Problem 11.1

Determine fraction of fluid processed by desander.

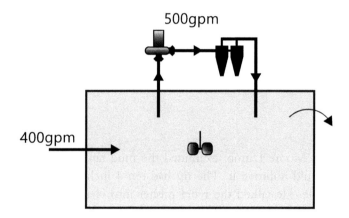

Problem 11.2
Find fraction of fluid processed.

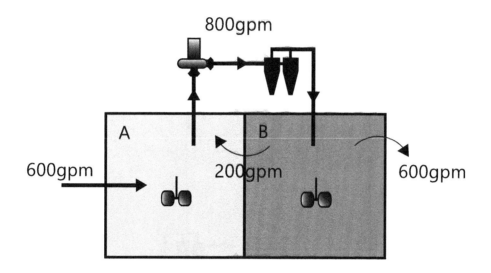

Problem 11.3
Weevil
A weevil engineer, Moore Dumb, examined the mud tank arrangement on a rig and thought that he could improve it. The rig had ten 4 inch desilters and was pumping 450 gpm down hole. He talked the relief pusher into changing the desilter overflow so that the mud went back into the tank upstream (nearer the shale shaker) from the suction of the pump feeding the desilter. His reasoning was that the desilters could "look at the mud" twice before it went back downhole. When Les Eagleye returned from days off, he chewed out the engineer for faulty thinking.

Was Moore or Les right? Why wasn't this a good idea?

Problem 11.4

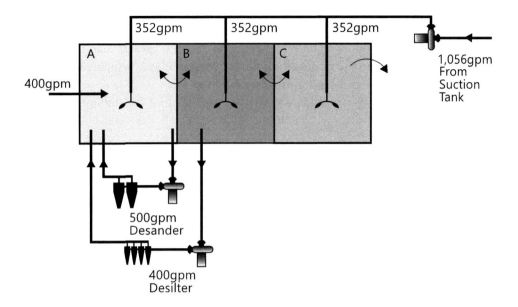

Look at mud twice with desilters.

Cut Comer Drilling Company was trying to do an extraordinary job for an important company hoping that the good drilling performance would result in a long term contract. So the head honcho from the office came to the rig with permission to make the necessary changes to the mud system to insure that the drilling fluid was super cleaned.

The first thing that he checked was the pressure at the manifold of the desander, the desilters, and the mud guns. All gauges indicated 39 psig for the 10.0 ppg drilling fluid. The rig had one 12″ desander and one bank of eight 4″ desilters. The rig was circulating 400 gpm downhole.

He then examined the suction tanks for both the desander and the desilters. They were not well agitated, so he had two mud guns, made from 3″ × 1″ swedges, installed in both tanks. The drilling fluid was supplied to the mud guns from the suction compartment. Only clean drilling fluid agitated the tanks and sufficient fluid was pumped through the mud guns to keep the tanks well-agitated.

Next he examined the solids removal equipment. The solids discharge from the desander and the desilter seemed very low. So he decided that if the hydrocyclones "looked at the mud twice," a cleaner drilling fluid would result. To do this, he routed the overflow from both the desander and the desilters to the compartment upstream from the desander suction.

What results would you expect from these changes?

Problem 11.5

A well flow rate of 400 gpm is entering a desander suction pit that is not agitated. The contractor uses mud guns to stir the pit and justifies this because clean drilling fluid is being used through the mud guns. The desander and desilter are processing 500 gpm each as shown in the figure below. The fluid rate being processed is greater than the fluid rate coming from the well and this satisfies the contract. Each mud gun is supplying 100 gpm.
1. What is the desander and desilter process efficiency?
2. What additional capacity would be needed to process all of the drilling fluid?

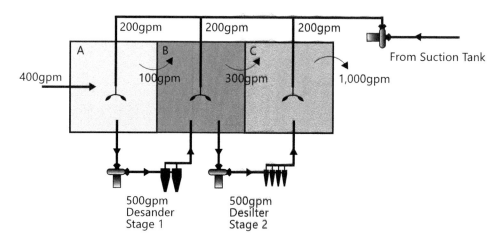

Problem 11.6

Find the fraction of fluid processed by the desanders and desilters using agitators for mixing/blending instead of mud guns for Problem 11.5.

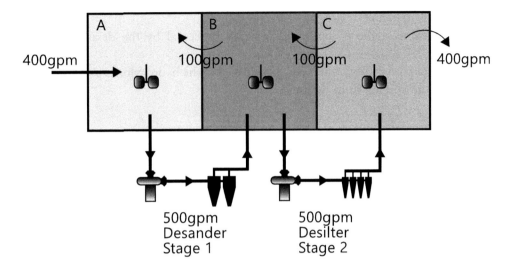

Problem 11.7

Flow rate from well is 450 gpm. Degasser is processing 500 gpm and degasser centrifugal pump is pumping 600 gpm. Desander processing 600 gpm and desilters processing 800 gpm.
1. Calculate the fraction of drilling fluid processed by the desander and desilter.
2. If it is not 100%, how can the plumbing be changed?

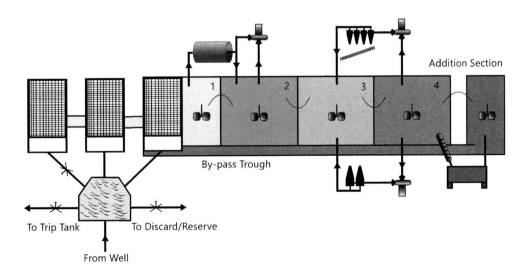

Problem 11.8

Two 1″ diameter mud guns stir each compartment. Fluid through the mud guns is coming from the clean drilling fluid in the suction section. Flow from well is 500 gpm. Degasser is processing 700 gpm with a power-fluid flow rate of 800 gpm. Desander is processing 600 gpm and desilter processing 800 gpm.

1. What fraction of the drilling fluid is being processed by the desanders and the desilters?
2. If this is not correct, how can it be changed so that the hydrocylones process all of the drilling fluid coming from the well?

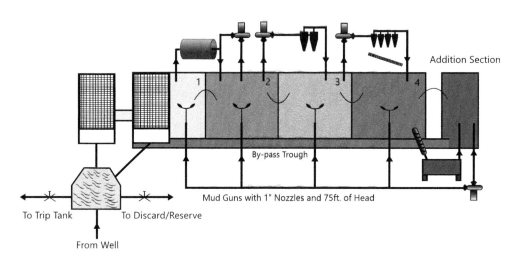

Problem 11.9

1. Find fraction of fluid proceed by desilters.
2. Evaluate processing.

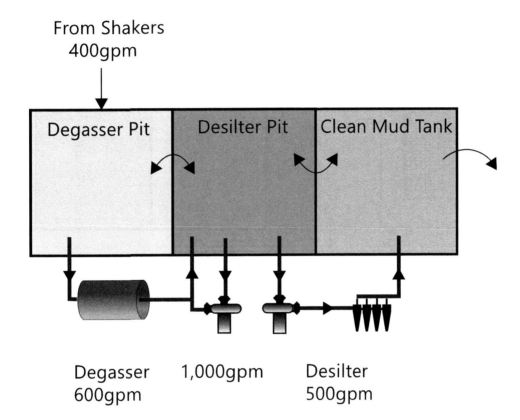

Problem 11.10

Flow from the well into shale shaker is 1000 gpm. Flow into the desanders is 1200 gpm (by contract). What is the processing efficiency of the desanders in each of the two cases below? Sketch the flow rates into and out of each compartment. The process efficiency is the processing rate divided by the flow rate entering the suction tank of the desanders times 100%.

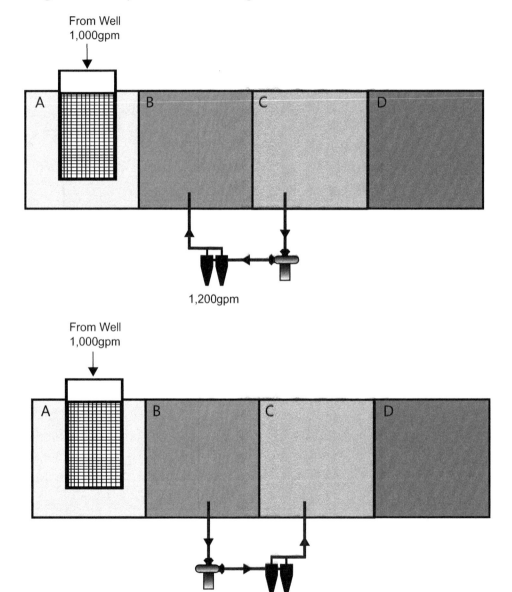

Problem 11.11

The next three field examples of arrangements existing on drilling rigs working in the Gulf of Mexico were reported at the AADE Shale Shaker Conference held in Houston, Texas during November 1998. These problems are not unusual and rig tank arrangements are about the last technology to be utilized by the industry.

Example rig #1

The four compartments all over-flowed from one compartment to the next. Assume that the flow rate down the bore hole was going to be 900 gpm.
1. What problems exist with this mud tank arrangement?
2. Calculate the process efficiency.
3. What changes that could be made to improve solids removal?

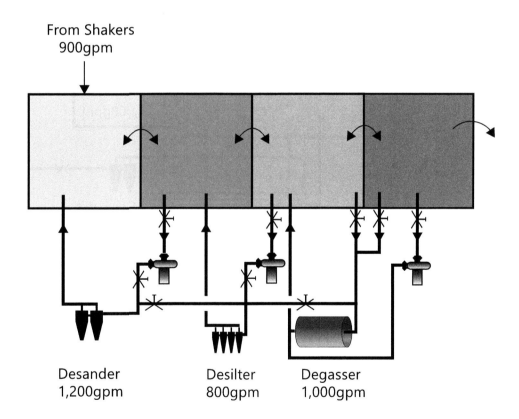

Problem 11.12
Example rig #2
1. What problems exist with this mud tank arrangement?
2. For the best processing with this system, which valves should be open?
3. Find fraction of fluid processed.
4. What modifications should be made?

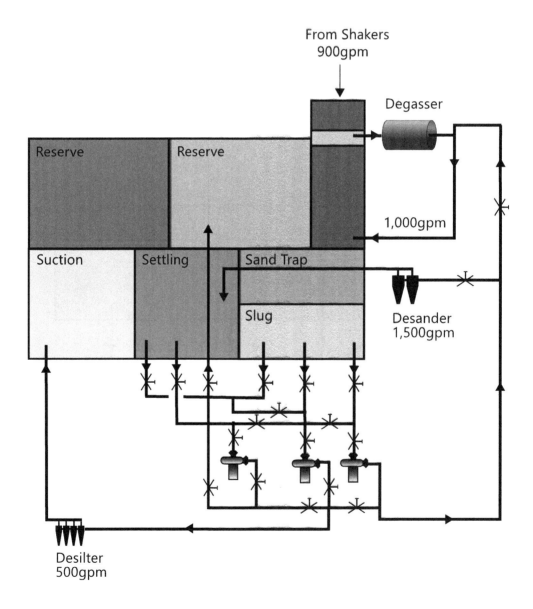

Problem 11.13
Example rig #3
1. Identify which valves should be open.
2. Find fraction of fluid processed by desander and desilter.
3. What is wrong?
4. What changes should be made in this tank arrangement?

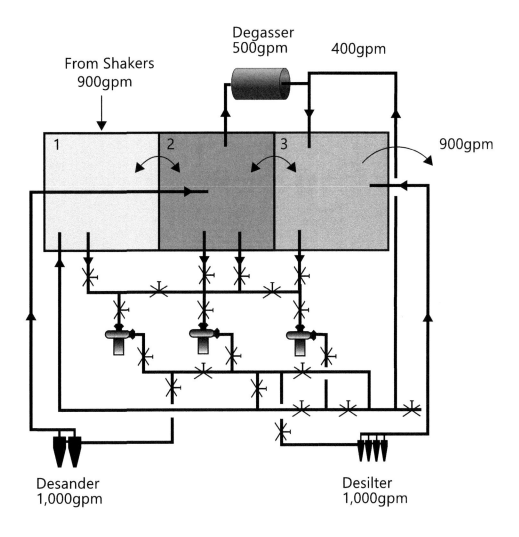

Problem 11.14

Mud guns: 64 ft head, 1″ diameter nozzles, clean drilling fluid from suction tank.
1. Find fractions of fluid processed by desanders and desilters.
2. What changes need to be made?

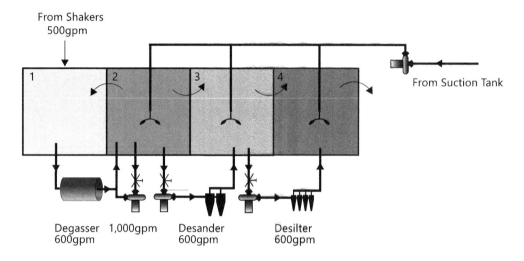

Problem 11.15

1. If the fraction processed is not 100%, what changes need to be made?
2. Find fraction of drilling fluid processed by the desanders and desilters.

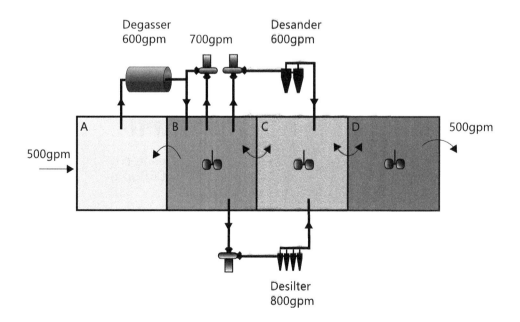

Problem 11.16

A company man had centrifugal pump problems. There were 3 centrifugal pumps available. He was using one, had one on stand-by and the other in the shop for repair. Believe it or note, but he could not control the drilled solids.
1. Determine fraction of drilling fluid processed through desanders and desilters.
2. What needs to be done to make the desilter and desander process 100% of the fluid coming from the well?

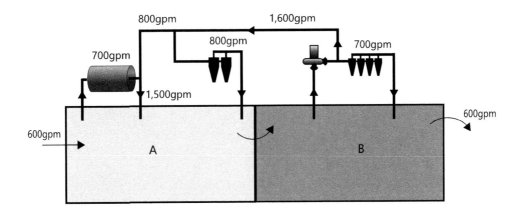

CHAPTER 12

Equipment solids removal efficiency

Effect of equipment removal efficiency

The term "solids removal efficiency" is frequently used to describe solids control equipment performance. This term may be somewhat confusing. The American Petroleum Institute Recommended Practice 13C (API RP 13C) refers to the solids removal process in terms of system performance. Solid control equipment is designed to remove drilled solids. These drilled solids are not dry when they are removed from the system. For example, the underflow discharge from a properly operating 4-inch desilter or hydrocyclone contains around 35% vol solids and 65% vol drilling fluid. The liquid concentration of the discard from a shale shaker depends upon the screens. An API 200 screen discards a much wetter discard stream than an API 20 screen. The liquid drilling fluid that accompanies these separated solids comes from the drilling fluid system. After screening, however, the solids in the discarded drilling fluid cannot be expected to have the same solids distribution as the drilling fluid in the tanks. Quantities of barite in the liquid phase of the discard from a fine screen may not be in the same concentration as barite in the pits. The discard obviously contains more drilled solids than the drilling fluid in the pits. The barite, or weighting agent, will probably also have a different concentration in the liquid phase of the discard than in the drilling fluid in the pit. Measuring the quantity of barite in the discard will not reveal the amount of drilling fluid discarded. The drilling fluid accompanying the drilled solids discard will, however, contain drilled solids that have remained in the system after the fluid was originally processed by the solids control equipment.

Efficiency is defined by the ratio of output to input. For example, if a 100 HP motor drives a rotary table and produces 85 HP to rotate a drill string, the efficiency of the system is 0.85% or 85%. Drilled solids removal efficiency would imply a ratio of output (or discard) to input (circulating volume).

If only the removal of the drilled rock from the drilled rock entering the system is considered in the process calculation, no credit would be taken for the resident drilled solids present in the drilling fluid. The input volume of drilled solids would be the 9 bbl of drilled rock and the output volume would be the 6 bbl. The system performance would be 67% efficiency. This could be called solids removal equipment efficiency or solids removal equipment performance.

Only solids removed which decrease the solids concentration in the drilling fluid are considered in calculating solids removal equipment efficiency. If a valve is opened and 200 bbl of drilling fluid are removed from a system, some drilled solids are obviously removed with the liquid. However, the concentration of drilled solids in the remaining

A Practical Handbook for Drilling Fluids Processing
ISBN 978-0-12-821341-4, https://doi.org/10.1016/B978-0-12-821341-4.00012-9

drilling fluid system does not change. Removal of the 200 bbl of drilling fluid provides space for clean drilling fluid to be added. The addition of the clean drilling fluid will decrease the concentration of drilled solids in the system, but this is a very expensive method, called dilution, of maintaining a low concentration of drilled solids. Solids control equipment removal efficiency calculation only considers the drilled solids removal which decreases the drilled solids concentration in the system compared to the new drilled solids introduced during that interval.

Reasons for drilled solids removal

Many years ago a controversy raged concerning the effect of drilled solids on the cost of a well. Many thought that drilled solids were beneficial as an inexpensive substitute for weighting agents. As oil well drilling encountered more and more difficult conditions, hole problems finally became undeniably associated with excessive drilled solids. Frequently, production horizons near the surface were normally pressured and could be drilled with unweighted drilling fluids. Usually, these drilling conditions were relatively trouble-free and a poor quality drilling fluid was used for drilling. Of course, drilling performances and well productivity could be enhanced with better quality drilling fluids but those effects were difficult to quantify. As these areas graduated from unweighted drilling fluids to weighted drilling fluids, better drilling fluid properties were required to prevent trouble. The primary problem was that large quantities of drilled solids were intolerable. The drilling trouble costs could easily be traced to failure to limit drilled solids concentration. This provided the impetus for most drilling rigs to upgrade their surface systems handling drilling fluids. The benefits of a clean drilling fluid are well stated throughout this book and have been well validated.

Some rigs now process all drilling fluid sequentially in accordance with good practices as discussed in Chapter 11. Drilling fluid type does not affect proper rig plumbing. Dispersed or non-dispersed, fresh or salt water, clay-base or polymer-base, or any other drilling fluid must be treated sequentially to remove smaller and smaller drilled solids.

The cost of solids control equipment was justified initially economically as an insurance policy to prevent catastrophes. Subsequently more expensive drilling fluids required lower drilled solids concentrations. Polymer additives that adhere to active solids require significantly lower concentrations of drilled solids to prevent loss of too much polymer. Environmental concerns also dictate minimization of waste fluid; this requires careful attention to mechanical removal of drilled solids.

Diluting as a means for controlling drilled solids

One way that the drilled solids can be kept at a manageable level is to simply dump some of the drilling fluid containing the drilled solids and replace with clean drilling fluid.

One-half of the drilled solids is eliminated if one-half of the system is dumped and replaced with clean fluid. Generally this is too expensive, so mechanical equipment is used. Traditionally, large volumes of drilling fluid were dumped from the system by aggressively dumping sand traps. This makes room available for clean drilling fluid needed for dilution without calling it "dilution".

Chemical treatment

Chemical treatment of a water-base drilling fluid for solids removal involves adding a "flocculant" to the drilling fluid. This causes extremely small solids to agglomerate together so they can be removed mechanically or allowed to settle by gravity in the mud tanks. Normally, a flocculant is used in conjunction with mechanical treatment. For example, flocculants can be added at the flow line to increase the particle sizes so they can be removed with the shaker screen. Flocculants are also added to drilling fluid being processed by a centrifuge. The low-shear-rate inside of the centrifuge prevents the flocculated particles from separating and this makes an effective tool for decreasing the concentration of very small particles.

Mechanical treatment

This is the method of physically removing solids using shale shakers, desanders, desilters, mud cleaners and centrifuges. Each piece of equipment generally limited to the following range of particle removal:

Scalping Shale Shaker:
- 440 μm and larger

Fine Screen Shaker:
- 75 μm and larger (weighted drilling fluids)
- 44 μm and larger (unweighted drilling fluids).

Mud Cleaner:
- 75 μm and larger (weighted fluids).
- 44 μm and larger (unweighted fluids).

Desanders:
- 100 μm and larger.

Desilters:
- 15 μm and larger.

Centrifuge:
- 5−10 μm and smaller (weighted drilling fluids).
- 5−10 μm and larger (unweighted drilling fluids).

Each piece of mechanical equipment is effective within a certain particle size range. Shale shakers separate by the size of the particles; the other devices which use centrifugal force for separation separate by mass of the particle. Using all of the equipment listed above throughout a drilling program will produce maximum benefits without overloading any one piece of equipment. None of the above items will take the place of another piece of equipment; however no piece of equipment operating at optimum efficiency should cause downstream equipment to become overloaded. In some wells, depending upon the size of the drilled solids, the mud cleaner might not be needed.

Removing solids starting at the initiation of a drilling operation is a first priority in solids control as it is much easier to remove one particle 100 μm in diameter with a fine screen shaker than it is to attempt to remove 125,000 particles of 2 μm size with a centrifuge.

In unweighted drilling fluids, the fine screen shakers, and desilters are generally used until the point of adding barite for weight-up. If only coarse screens (API 80) can be used on the main shaker, the desander is needed to prevent solids overload in the desilters. With fine screens (API 140 and up) the desanders are not needed. Centrifuges can be used to increase drilled solids removal, although this is not common. With weighted drilling fluids, fine screen shakers, mud cleaners and centrifuges are used.

Mechanical separation-basics

Mechanical separation equipment employs mass differences, size differences, or a combination of both to selectively reject undesirable solids and retain desirable solids in a drilling fluid.

Shakers are vital to solids control and should process all of the drilling fluid returning through the flow line. A standard scalping shaker performs adequately for small rigs operating at shallow depths with low solids native drilling fluid, however, fine screen shale shakers are generally more efficient and remove more drilled solids.

The desanders and desilters are located directly downstream from the shale shaker. They should be sized to process at least 125% of the rig circulation rate while discarding undesirable cuttings and solids larger than around 20 μm. The desander removes the majority of the solids down to the 75 μm size range and prevents the desilter from being overloaded. The desilter removes the majority of solids down to around the 15 μm range, in an unweighted drilling fluid.

Liquid loss from desanding and desilting an unweighted drilling fluid is relatively insignificant compared to the amount of drilling fluid that must be removed from the mud tanks to eliminate the same amount of solids. Attempts to recover the liquid phase results in the recovery of very fine colloidal solids and is not recommended.

When drilling with a weighted drilling fluid, the desander and desilter cannot be used economically because they discard too much of the valuable barite. Therefore, fine screen shakers and mud cleaners are used to remove solids down to 75 μm. The mud cleaner will remove solids which have by-passed the main shaker screens and keep all retained solids to sizes less than "sand" (or 75 μm). This is essential to provide the correct ingredients in the drilling fluid to form good, thin, slick, compressible filter cakes.

Colloidal solids will continuously increase in a drilling fluid. This increases the plastic viscosity, decreases filter cake quality, and is detrimental to drilling performance. In a weighted drilling fluid, centrifuges are used to remove solids smaller than 5–10 μm. (It is not used to recover anything but is used like all solids control equipment to eliminate drilled solids without discarding all of the drilling fluid.) Only a fraction of the drilling fluid is processed with each circulation, because the filtration additives and the low-shear-rate modifiers are also removed with the colloidal material. These must be added back to the drilling fluid when centrifuges are used.

Effect of solids removal system performance

This example assumes that the surface system contains 1000 bbl of drilling fluid and the targeted drilled solids level is 4% vol. Assume that 100 bbl of drilled solids report to the surface. For reference, 100 bbl is the volume of 1029 ft of a 10 inch diameter hole.

If this 100 bbl of drilled solids remains in the drilling fluid, the pit levels remain constant. The drilling fluid system increased to 1100 bbl because 100 bbl of new hole was drilled. The volume of the rock represented by the new hole is virtually the same whether it is ground into cuttings or is solid rock. The 4% vol drilled solids concentration before drilling means that the drilling fluid contained [0.04 × 1000 bbl, or 40 bbl] of drilled solids when drilling started. After drilling, these 40 bbls plus the 100 bbl of new drilled solids would be in the 1100 bbl drilling fluid system, or (140 bbl/1100 bbl) × 100 or 12.7% vol drilled solids.

To reduce the concentration of drilled solids to 4% vol by only adding clean drilling fluid would require adding enough clean drilling fluid to reduce the 100 bbl drilled solids to a concentration to 4% vol. Calculation:

$$100 \text{ bbl} = (0.04) \text{ new drilling fluid built}$$

$$\text{New drilling fluid built} = 2500 \text{ bbl}$$

This new drilling fluid built would consist of drilled solids plus clean drilling fluid:

$$2500 \text{ bbl} = \text{clean drilling fluid} + 100 \text{ bbl}$$

$$\text{Clean drilling fluid} = 2400 \text{ bbl}$$

Volume is independent of the original volume of the system.

If the tanks are full when the drilling started, the 2400 bbl of clean drilling fluid could not be contained in the mud tanks. The only volume available for the clean drilling fluid is the volume of discard removed from the system. Since nothing is removed from the system, the volume added must be removed to return the pit levels to the original level. The excess drilling fluid (2400 bbl) would need to be removed from the drilling fluid system to keep the pits from overflowing. Not only would the cost of the clean drilling fluid be prohibitive, but this fluid must also eventually go to a disposal site.

The intent of this analysis is to build the basis for the concept of an appropriate removal efficiency to build the minimum quantity of new drilling fluid and also minimize the volume of discarded fluid. The assumption will be made that the drill pipe will be returned to the position where it started drilling the interval. Although the rock does not change volume, the pit levels will rise because of the volume of the steel added as the drill string enters the well.

Evaluation of the equipment solids removal efficiency of solids control equipment is very difficult to do with only short term tests. Capturing equipment discards for only 15 min to 2 h will not provide sufficient data to calculate equipment solids removal efficiency. The quantity of solids entering the drilled solids removal equipment is usually unknown and impossible or very difficult to determine. Drilled solids do not arrive at the surface in the same order they were drilled. Fluid flow in the annulus is usually laminar. The center part of the annular flow moves faster than the portion of the flow adjacent to the formation or the drill pipe. A 30 min sample from the shale shaker discards may contain samples that were drilled two to 4 h apart even in a well bore drilled to gauge. With rugosity and borehole enlargements lag times are extended. Mud Loggers can sometimes observe formation cuttings that were drilled several days before the sample was taken from the end of the shale shaker. This is the reason that material balances are difficult to obtain with these small snapshots. Circulating a borehole clean before drilling an interval, drilling a known volume of solids, and circulating all of the cuttings from the well bore will permit estimating a known volume of solids arriving to the surface. Another variable in this analysis is the porosity of the formation. Solids analysis requires the initial condition of the quantity, or volume, of solids that arrive at the surface.

Relationship of solids removal efficiency to clean drilling fluid needed

To make a volumetric analysis of the effect of various solids removal efficiencies on the clean drilling fluid needed, an examination of the volumetric balance of the fluid in the well bore and surface system needs to be made. In Fig. 12.1, the bore could be about 5000 ft deep and contain over 500 bbl of drilling fluid. All of the surface drilling fluid will be moved into one tank and could be as much as 2000 bbl of drilling fluid.

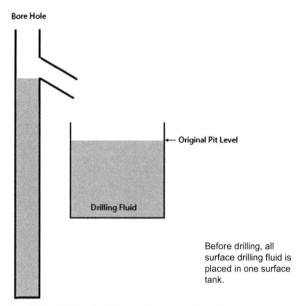

Fig. 12.1 Drilling fluid in well bore and surface equipment.

About 1000 ft of 10″ diameter hole would have a volume of about 100 bbl (Fig. 12.2). If nothing is removed from the fluid system, will the pit levels increase, decrease, or stay the same?

The problem might be examined by simply removing the 100 bbl of hole from the bottom of the bore hole and placing it in the surface tank (Fig. 12.3). Clearly, the hole volume increased by 100 bbl but the rock removed is occupying 100 bbl of volume in the surface tank.

Actually, the rock is, of course, not removed as a single entity but is broken into pieces. Assume the 100 bbl of rock was broken into ten pieces of 10 bbl each (Fig. 12.4). The volume formerly occupied by the rock is now part of the volume of the fluid in the well bore, but the volume of those ten pieces is still 100 bbl (Fig. 12.5).

The pit level does not change if nothing is removed from the system. If a brick is placed at the bottom of a bucket of water, it could be broken into four pieces without changing the volume of the entire system.

The drilled rock does expand when tectonic stresses are removed. When creating tunnels in mountain areas, the stress distribution in the rock is frequently determined with strain gages. A large area on the face of the tunnel is ground smooth and strain gages are affixed to the rock face. A large diameter core is then cut and the strain measured as the rock expands. The rock only changes dimensions by a few thousandth of an inch. For the purpose of the calculations made here, this small expansion will be ignored.

274   A Practical Handbook for Drilling Fluids Processing

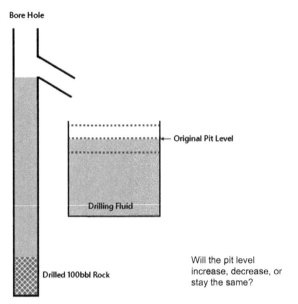

Fig. 12.2 Drill about 100 bbl of hole.

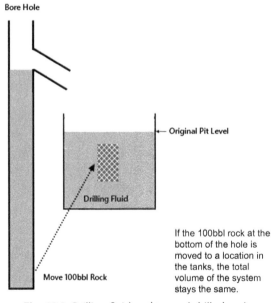

Fig. 12.3 Drilling fluid and moved drilled rock.

Equipment solids removal efficiency 275

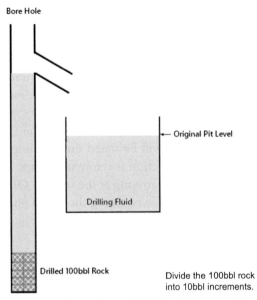

Fig. 12.4 Dividing the 100 bbl of rock into ten pieces.

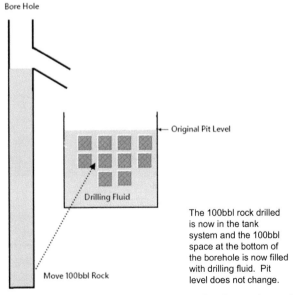

Fig. 12.5 Moving the ten 10 bbl chunks of rock into the tank.

Five examples of the effect of equipment solids removal efficiency

If the mechanical equipment does not remove a significant portion of the drilled solids reporting to the surface, maintaining a reasonable level of undesirable drilled solids can become very expensive. Dilution, then, becomes a major portion of the solids management strategy. Calculations, summarized in Figs. 12.6–12.10, indicate the performance of the solids removal equipment.

This set of calculations is simply a material balance of the volumes added and the volumes discarded. The calculations will be based on a drilling fluid processing plant system in which the solids removal section is removing either 100%, 90%, 80%, 70% or 60% by volume of the drilled solids arriving at the surface. Obviously 100% removal efficiency is not currently possible-while retaining the liquid phase. Removal percentages listed are used simply to demonstrate the method and concepts of solids removal efficiency.

The average drilled solids concentration in the discard stream has been selected as 35% vol. The underflow from a decanting centrifuge and the mud cleaner discharge stream will contain 55% to 63% vol solids. The underflow from desilters and desanders will vary around 35% vol solids. The discharge stream from a shaker can vary from 45% vol solids for very coarse screens to 20% vol for very fine screens. The average for

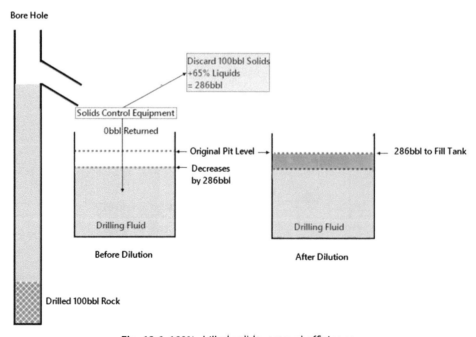

Fig. 12.6 100% drilled solids removal efficiency.

Equipment solids removal efficiency 277

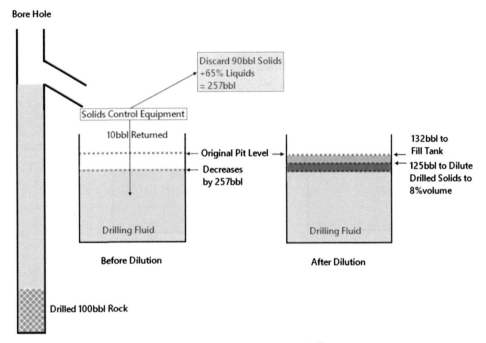

Fig. 12.7 90% drilled solids removal efficiency.

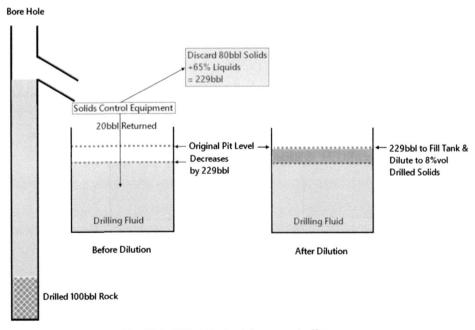

Fig. 12.8 80% drilled solids removal efficiency.

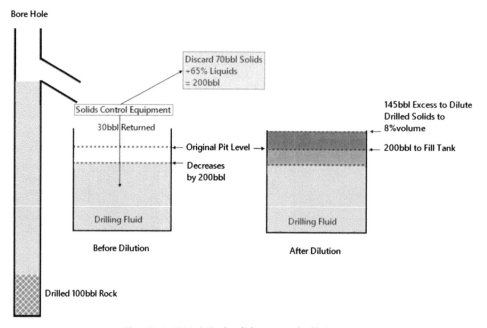

Fig. 12.9 70% drilled solids removal efficiency.

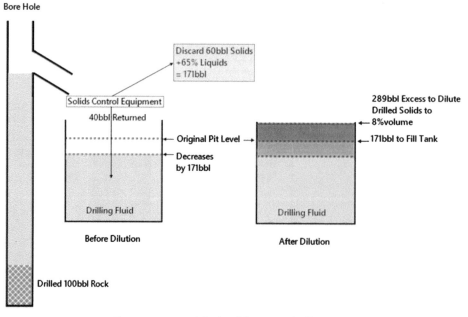

Fig. 12.10 60% drilled solids removal efficiency.

all of these devices is assumed to be 35% for this example. With very coarse screens, most of the liquid is removed from the large solids as they travel down the shaker screen. So the concentration of solids is much higher than it is with screens with smaller openings; although the total volume of solids discarded is usually smaller. Screens with smaller openings remove more solids as well as more liquid. The reason is related to the ratio between the surface area and the volume of the cuttings. For example, a golf ball would retain very little drilling fluid when it was removed from the fluid. Grind the ball into very small pieces and the volume of liquid would increase greatly. The volume of the solid (golf ball) would not change; but the surface area would change greatly. More liquid is required to wet the increased surface area.

To examine the quantity of discards, clean drilling fluid needed to dilute remaining drilled solids, and excess volume of drilling fluid built, four equipment solids removal efficiencies will be discussed: 100%, 90%, 80%, 70% and 60%.

This is a theoretical analysis of the effect of equipment solids removal efficiency and concentration of drilled solids in the discard stream. For these calculations, 100 bbl of drilled solids will report to the surface. The target drilled solids concentration is 8% by volume.

100% removal of drilled solids

If this could be accomplished and the drilled solids were 35% vol of the discard, the discard volume could be calculated:

$$\text{Volume of discarded drilled solids} = (0.35)(\text{volume of total discard})$$

Assume 100 bbls of drilled solids arrive at the surface. If all are discarded, the total volume of discard would be:

$$100 \text{ bbl} = (0.35)(\text{volume of total discard})$$

$$\text{Volume of discard} = 286 \text{ bbl}$$

The pit levels would decrease the volume of the discard. The ratio of discarded volume to volume of drilled solids removed would be 2.86. In other words, for every barrel of drilled solids removed from the drilling fluid system, 1.86 bbl of drilling fluid would accompany the one barrel of drilled solids. The pit levels would drop by 286 bbl during this period and that volume must be added to the active system to keep the pit levels constant. The concentration of drilled solids would decrease from 8% vol to a lower number (depending upon the volume of drilling fluid in the active system).

The addition of 286 bbl of clean drilling fluid will reduce the drilled solids concentration because the 186 bbl of drilling fluid discarded with the 100 bbl of drilled solids would contain 15 bbl of drilled cuttings if the initial concentration of drilled solids in the active system was 8%. This reduces the total drilled solids in the drilling fluid.

90% removal of drilled solids

Again, drill 100 bbl of drilled solids. In this case, 90 bbl of drilled solids would be discarded and 10 bbl of drilled solids would remain in the drilling fluid.

$$\text{Volume of discarded drilled solids} = (0.35)(\text{volume of discard})$$

$$90 \text{ bbl} = (0.35)(\text{volume of discard})$$

$$\text{Volume of discard} = 257 \text{ bbl}$$

$$\text{Ratio of volume of discard to volume drilled solids} = 2.57$$

In this case, 90 bbl of drilled solids and 167 bbl if drilling fluid would be discarded; or a total of 257 bbl would be required to keep the pit levels constant.

The remaining solids would need to be diluted with clean drilling fluid.

$$\text{Drilled solids} = (0.08)(\text{dilution volume})$$

$$\text{Dilution volume} = \frac{10 \text{ bbl}}{0.08} = 125 \text{ bbl}$$

The dilution volume would consist of 10 bbl of drilled solids and 115 bbl of clean drilling fluid. Since 257 bbl would be required to keep the pit volumes constant and only 115 bbl would be needed to keep the drilled solids concentration constant, additional clean drilling fluid would have to be added to the system which would cause the total drilled solids in the active system would decrease.

80% removal of drilled solids

Again, drill 100 bbl of drilled solids. In this case, 80 bbl of drilled solids would be discarded and 20 bbl of drilled solids would remain in the drilling fluid.

$$\text{Volume of discarded drilled solids} = (0.35)(\text{volume of discard})$$

$$80 \text{ bbl} = (0.35)(\text{volume of discard})$$

$$\text{Volume of discard} = 229 \text{ bbl}$$

$$\text{Ratio of volume of discard to volume drilled solids} = 2.29$$

In this case, 80 bbl of drilled solids and 149 bbl of drilling fluid would be discarded; or a total of 229 bbl would be required to keep the pit levels constant.

The remaining solids would need to be diluted with clean drilling fluid.

$$\text{Drilled solids} = (0.08)(\text{dilution volume})$$

$$\text{Dilution volume} = \frac{20 \text{ bbl}}{0.08} = 250 \text{ bbl}$$

The dilution volume would consist of 20 bbl of drilled solids and 230 bbl of clean drilling fluid. Since 229 bbl would be required to keep the pit volumes constant and only 1 bbl would need to be jetted to keep the pit levels constant, the total drilled solids in the active system would be almost balanced. That is the volume of clean drilling fluid needed would be almost exactly the volume which was discarded from the active system.

70% removal of drilled solids

Again, drill 100 bbl of drilled solids. In this case, 70 bbl of drilled solids would be discarded and 30 bbl of drilled solids would remain in the drilling fluid.

$$\text{Volume of discarded drilled solids} = (0.35)(\text{volume of discard})$$

$$70 \text{ bbl} = (0.35)(\text{volume of discard})$$

$$\text{Volume of discard} = 200 \text{ bbl}$$

$$\text{Ratio of volume of equipment discard to volume drilled solids} = 2.0$$

In this case, 70 bbl of drilled solids and 130 bbl of drilling fluid would be discarded; or a total of 200 bbl would be required to keep the pit levels constant.

The remaining solids would need to be diluted with clean drilling fluid.

$$\text{Drilled solids} = (0.08)(\text{dilution volume})$$

$$\text{Dilution volume} = \frac{30 \text{ bbl}}{0.08} = 375 \text{ bbl}$$

The dilution volume would consist of 30 bbl of drilled solids and 345 bbl of clean drilling fluid. Since 200 bbl would be required to keep the pit volumes constant, an additional 145 bbl would be needed to dilute the remaining drilled solids to the targeted value of 8% vol. Only a volume of 200 bbl is available after the solids removal equipment has discarded the 70% vol of solids arriving at the surface and the liquid associated with the cuttings. The actual discard would be the 200 bbl from the equipment and an additional 175 bbl to allow the remaining drilled solids to be diluted to the targeted value of 8% vol. This means that the ratio of actual volume of discard to the volume drilled would be (200 bbl + 145 bbl)/100 bbl or 3.45.

60% removal of drilled solids

Again, drill 100 bbl of drilled solids. In this case, 60 bbl of drilled solids would be discarded and 40 bbl of drilled solids would remain in the drilling fluid.

$$\text{Volume of discarded drilled solids} = (0.35)(\text{volume of discard})$$

$$60 \text{ bbl} = (0.35)(\text{volume of discard})$$

$$\text{Volume of discard} = 171 \text{ bbl}$$

$$\text{Ratio of volume of equipment discard to volume drilled solids} = 1.71$$

In this case, 60 bbl of drilled solids and 111 bbl of drilling fluid would be discarded; or a total of 171 bbl would be required to keep the pit levels constant.

The remaining solids would need to be diluted with clean drilling fluid.

$$\text{Drilled solids} = (0.08)(\text{dilution volume})$$

$$\text{Dilution volume} = \frac{40 \text{ bbl}}{0.08} = 500 \text{ bbl}$$

The dilution volume would consist of 40 bbl of drilled solids and 460 bbl of clean drilling fluid. Since 171 bbl would be required to keep the pit volumes constant, an additional 289 bbl would be needed to dilute the remaining drilled solids to the targeted value of 8% vol. Only a volume of 171 bbl is available after the solids removal equipment has discarded the 70% vol of solids arriving at the surface and the liquid associated with the cuttings. The actual discard would be the 171 bbl from the equipment and an additional 289 bbl to allow the remaining drilled solids to be diluted to the targeted value of 8% vol. This means that the ratio of actual volume of discard to the volume drilled would be (171 bbl + 289 bbl)/100 bbl or 4.6.

The information just calculated for the five different equipment solids removal efficiencies indicates that the volume of discard rises rapidly after it reaches a minimum value. In this case, with 35% vol of drilled solids in the discard and a targeted drilled solids concentration of 8% vol, the optimum solids removal efficiency is around an 80% removal efficiency.

The line designated by the blue diamonds in Fig. 12.11 indicate that the volume of clean drilling fluid required to dilute the drilled solids remaining in the drilling fluid is less than the volume of fluid required to return the pit levels back to their original values. In this case the targeted drilled solids concentration will be decreasing. The line designated by the red squares indicate that more fluid is required to dilute the solids remaining in the pits after processing through the solids control equipment. The pit levels would increase so much that excess drilling fluid would need to be removed from the system. The intersection of these two lines would indicate the smallest quantity of drilling fluid required to maintain a 4% vol concentration of drilled solids in the system. This intersection is predictable and the calculation is presented in the next section.

The volume of discard compared with the volume of formation drilled depends upon the solids removal efficiency, the target drilled solids concentration in the drilling fluid and the concentration of solids in the discard (Fig. 12.12). When the solids are more concentrated, the discard ratio is smaller for any particular targeted drilled solids concentration. As the targeted drilled solids concentration decreases from 12% to 2%, the volume of discarded material increases substantially. Drilling fluids with high drilled solids concentrations do not require as much clean drilling fluid and may appear to be cheaper. If the solids removal efficiency is very low (less than 80%), the cost of a clean drilling

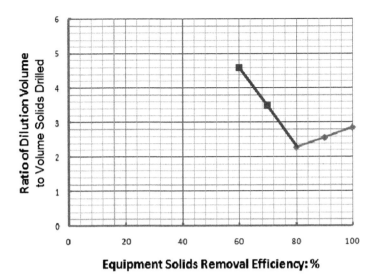

Fig. 12.11 Dilution required to maintain 4% volume drilled solids.

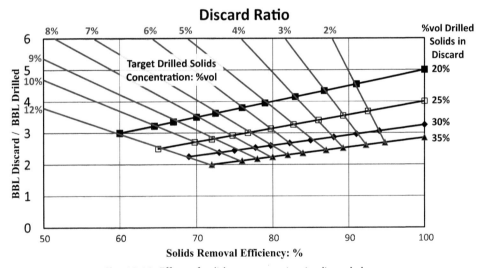

Fig. 12.12 Effect of solids concentration in discard slurry.

fluid may be substantial and not practical. This is the reason that many drilling rigs cannot effectively run a clean drilling fluid.

One note of caution here: these calculations assume that the drilled solids concentration within the drilling fluid remains constant all the time. In real life, this does not happen. With poor solids removal efficiency, the dilution will probably be postponed until the drilled solids have increased significantly. This example was primarily to show the significant effect on drilling fluid cost that can be achieved if the solids removal equipment is planned correctly, plumbed correctly, and operated properly. Of course the solids must be brought to the surface before they grind into particles which are too small to be effectively removed.

Solids removal efficiency for minimum volume of drilling fluid to dilute drilled solids

This optimum value of removal efficiency for various targeted drilled solids concentrations and drilled solids concentration in the discarded slurry should be calculated from the equation:

$$\text{Optimum solids removal efficiency} = \frac{(1 - \text{Target drilled solids conc. in drilling fluid})}{1 - \text{Target drilled solids conc.} + (\text{Target drilled solids conc.})/(\text{Drilled solids conc. in discard})}$$

Assume the drilled solids concentration in the discard is 35% vol and the target drilled solids concentration is 8% vol.

$$\text{Optimum solids removal efficiency} = \frac{(1 - 0.08)}{1 - 0.08 + (0.08/0.35)} = 0.80$$

If the same analysis is performed for other solids removal efficiencies and other values of targeted drilled solids concentrations, a series of curves reveals how rapidly the dilution volumes increase with poor removal efficiencies. As the requirement for a clean drilling fluid decreases (i.e. going from a 4% vol LGS drilling fluid to a 12% vol LGS drilling fluid), the volume of dilution decreases markedly. This, however, simply means that the drilling fluid cost will decrease while the well costs rise rapidly.

The minimum volume required to dilute solids remaining after processing the solids control equipment depends upon the drilled solids concentration in the drilling fluid. If all of the drilled solids are removed from the system, the clean drilling fluid added to return the pit levels back to the original level will dilute the solids already in the drilling fluid. As noted earlier, more clean drilling fluid will be needed to return the pits to the original level with 100% removal than 90% removal of drilled solids. The smallest volume required will occur when the system is "balanced", i.e., no excess drilling fluid is needed to dilute the drilled solids returning to the system. The same solids removal efficiency that provides the minimum quantity of new drilling fluid to be built will also be the removal efficiency that generates the minimum discard volume. This would be a condition where the volume of clean drilling fluid required to dilute the solids remaining after processing through the removal equipment is exactly the volume discarded by the equipment.

Optimum solids removal efficiency
$$= \frac{(1 - \text{Target drilled solids conc. in drilling fluid})}{1 - \text{Target drilled solids conc.} + (\text{Target drilled solids conc.})/(\text{Drilled solids conc. in discard})}$$

The derivation for this equation is presented below. For the case above where the targeted drilled solids concentration is 4% vol, and the drilled solids concentration in the discard is 35% vol, the optimum Solids Removal Efficiency would be:

$$\text{Optimum solids removal efficiency} = \frac{(1 - 0.04)}{1 - 0.04 + (0.35/0.04)} = 89.4\%$$

This agrees with the calculation above which indicates that the 90% removal efficiency almost is an optimum value.

Comment

Targeted drilled solids concentration recommendations are not presented here. The goal of the removal equipment should be to achieve the maximum possible removal of drilled solids without resorting to "dump and dilute". This requires proper selection of solids control equipment and proper plumbing in the removal section. The type of drilling fluid, the type of drilled solids, the transport of solids up the annulus and the proper hydraulic to remove cuttings as they are created will influence the ability to remove drilled solids. The goal should be to improve the removal efficiency with each well drilled in a particular field.

Equation derivation

$$\text{Volume of discard} = (\text{Solids removal efficiency})(\text{Solids to surface}) \times (\text{Drilled solids concentration in discard})$$

Volume of clean drilling fluid:

$$\text{Volume of new drilling fluid built} = \text{Volume of clean drilling fluid} + \text{Volume of retained drilled solids}$$

The volume of new drilling fluid built and the volume of retained drilled solids may be expressed in terms of the discard concentration and the targeted drilled solids concentration in the drilling fluid.

The volume of new drilling fluid built requires that the drilled solid concentration in the new fluid be the targeted concentration:

Drilled solids volume
$$= (\text{Targeted drilled solids concentration})(\text{New drilling fluid built})$$

The second term on the right side of the equation relates to the volume of retained drilled solids, which is determined by the solids removal efficiency:

Volume of retained drilled solids
$$= (1 - \text{Solids removal efficiency})(\text{Drilled solids to the surface})$$

These expressions may now be substituted into the expression for the volume of clean drilling fluid that needs to be added to the drilling fluid system which is:

Volume of new drilling fluid built = Volume of clean drilling fluid needed
+ Volume of retained drilled solids

$$\frac{\text{Volume of drilled solids to surface}}{\text{Target drilled solids concentration}} = \text{Volume of clean drilling fluid needed}$$
$$+ (1 - \text{Solids removal efficiency})(\text{Volume of drilled solids to surface})$$

Solving this equation for the volume of clean drilling fluid that must be added to dilute drilled solids remaining in the system:

$$\frac{\text{Volume of clean drilling fluid}}{\text{Volume of drilled solids to surface}}$$
$$= \frac{1 - \text{Solids removal efficiency} - (1 - \text{SRE})(\text{Target drilled solids concentration})}{\text{Target drilled solids concentration}}$$

Frequently, some rules-of-thumb are discussed in the literature about how much drilling fluid is required per barrel of drilled solids. In the illustration below, a common value of three is shown in Fig. 12.13. This value depends upon the targeted drilled solids concentration and the equipment solids removal efficiency.

Estimating equipment drilled solids removal efficiency for an unweighted drilling fluid from field data

Situation

No-profit prilling company is drilling 100 bbl of hole daily in a formation with 15% porosity. For four consecutive days, 400 bbl of discards and fluid were captured each day in discard tanks. The pit levels remained constant but some drilling fluid was jetted to the

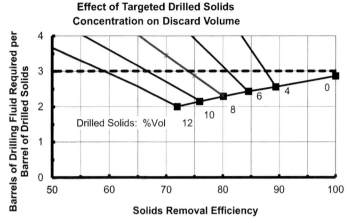

Fig. 12.13 Common rule of thumb.

reserve pits daily to keep the pits from overflowing. The unweighted drilling fluid weighed 9.4 ppg daily and contained 2% by volume bentonite.

Discussion

Since no barite is contained in the drilling fluid, the V_{lg} is the same as the V_s. Assume a specific gravity of the low gravity solids to be 2.6. The equation for determining V_{lg} is:

$$V_{lg} = 62.5 + 2.0V_s - 7.5(\text{MW})$$

or

$$V_s = V_{lg} = 7.5(\text{MW}) - 62.5 = 8\% \text{ volume}$$

Part of this 8% by volume low gravity solids concentration was 2% vol bentonite, so the low gravity drilled solids content was 6% by volume.

The pit levels decrease by the quantity of material removed from them. If no fluid or solids are removed from the system, the pit levels remain constant (except for the increase in volume of the drill pipe entering the hole). In this case the volume decrease is 400 bbl daily or 1600 bbl. This must also be the total volume of clean mud added to the system if the pit levels are returned to their original position.

The volume percent of drilled solids in the mud remained at 6% vol daily. So the drilled solids retained must be 6% vol of the new mud built, or

$$\text{Volume of retained solids} = (0.06)(\text{volume of new mud built})$$

The new drilling fluid built daily comprises the clean mud added and the drilled solids that remain in the drilling fluid after it is circulated through the solids removal equipment.

$$\text{Volume of new drilling fluid built in the drilling fluid system}$$
$$= \text{Drilled solids retained} + \text{Clean drilling fluid added}$$

The quantity of drilled solids retained can be substituted into that equation, resulting in:

Volume drilling fluid mud built = (0.06)(Volume new drilling fluid built)
+ Clean drilling fluid

The volume of clean drilling fluid must be exactly the volume that was discarded or 1600 bbl. This gives one equation with one unknown:

Volume of new mud built = (0.06)(Volume of new mud built) + 1600 bbl

so

$$\text{Volume of new mud built} = \frac{(1600)}{(1-0.06)} = 1702 \text{ bbl}$$

Since the drilled solids retained are 6% vol to the new mud built:

Drilled solids retained = (0.08)(1702 bbl) = 136 bbl

with 15% porosity, the 400 bbl drilled resulted in the addition of 340 bbl of solids to the system. If 136 bbl of drilled solids were retained, 204 bbl were discarded. This gives a ratio (or solids removal efficiency) of 204 bbl/340 bbl or 0.60. The solids removal efficiency is 60%.

For comparison, the procedure in API RP13C is used to calculate the same solids removal efficiency is presented in the Appendix.

Excess drilling fluid built: Normal discards from fine screens on linear motion shale shakers and from hydrocyclones contain about 35% vol solids. If all of these 35% solids are drilled solids, the volume of drilling fluid discarded with the drilled solids can be calculated.

The statement that the volume of discarded solids is equal to 35% vol of the discarded volume could be written:

Discarded drilled solids = 0.35(Volume of fluid discarded with drilled solids)

204 bbl = (0.35)(Volume of total fluid discarded with drilled solids)

Volume discarded with drilled solids = 583 bbl

The 583 bbl of waste discard would contain 204 bbl of drilled solids and 379 bbl of drilling fluid.

Since a total of 1600 bbl was discarded, 1017 bbl of good drilling fluid was removed from the system along with the 583 bbl of waste drilled solids and drilling fluid. The 1017 bbl of good drilling fluid is the excess clean fluid added to dilute the retained drilled solids and could be pumped to a storage pit. However, eventually this excess drilling fluid must go to disposal. Ideally, the amount of clean drilling fluid added to the system will be exactly the volume discarded with the drilled solids (583 bbl).

Consider if the discard from the hydrocyclones and the shale shaker was very wet (meaning a large volume of liquid was discarded with the drilled solids) so that the discard contained 20% vol drilled solids instead of 35% vol.

$$\text{Discarded drilled solids} = 0.20(\text{Volume of fluid discarded with drilled solids})$$

$$50.5 \text{ bbl} = (0.20)(\text{Volume of fluid discarded with drilled solids})$$

or

$$\text{Volume of fluid discarded with drilled solids} = 253 \text{ bbl}$$

Again, since a total of 400 bbl was discarded daily, 400 bbl–253 bbl, or 147 bbl, of good drilling fluid was also discarded.

A conclusion should be obvious at this point. Efforts to eliminate all drilling fluid from dripping from the end of the shale shaker are futile and not needed when the solids removal efficiency is around 60%. This drilling fluid will need to be discarded eventually and shale shakers do a better separation when the effluent is still wet. This will usually point the way to using shaker screens with smaller openings. Coarse screens allow more drilled solids to pass through.

Estimating equipment drilled solids removal efficiency for a weighted drilling fluid

Situation

After drilling 1000 ft of hole with a 12.5 lb/gal drilling fluid circulated at 25 bbl/min, the hole was circulated clean. This required four hole volumes to eliminate all solids in the discard. Assuming the formation averaged about 13% vol porosity, a multi-armed caliper indicated that 97.3 bbl of new hole was drilled. The drilling fluid was a fresh water-base drilling fluid weighted with barite and containing 2 % vol. bentonite, no oil, and 5% drilled solids by volume. While drilling this interval, 1350 sacks (sx) of barite (100 lb/sx) were added to the system, and the drilled solids remaining in the system were diluted as required to control their concentration at the targeted 5% by volume. Some drilling fluid was pumped to the reserve pits and all solids control equipment discards were captured in a container to be shipped back to shore. One drilling fluid technician reported that 200 bbl were hauled to shore and another reported that 180 bbl were captured.

This data is available, more or less in most field operations. Certainly the volume of solids reaching the surface is necessary for any calculation. This data will be assumed to be available from a hole caliper and circulating the hole clean after drilling a particular interval. Next, the volume of clean drilling fluid added to the active system to dilute the drilled solids remaining in the system is needed. With a weighted mud, the number of sacks of barite and an analysis of solids concentrations in the drilling fluid allows a calculation of the clean drilling fluid added. Similarly, if the liquid volume added (water, oil, or synthetics) is known, the volume of clean drilling fluid can be calculated. Finally, if all of the discard volumes are captured in a disposal tank or container, the volume of discard can be measured.

These calculations do not require knowledge of the volume of the circulating system. It is assumed that the system has reached a stable drilled solids concentration and the changes to the system are the primary concern. In actual practice, the system is dynamic with small amounts of drilling fluid ingredients added continuously and continuous discards from the solids removal equipment. At the drilling rig, sand traps are dumped with a variety of quantities of good drilling fluid.

Analysis

1. Volume of new drilling fluid built while drilling interval

Assuming a low gravity solids density of 2.6 g/cc and a barite density of 4.2 g/cc, the low gravity solids content can be calculated from the equation:

$$V_{lgs} = 62.5 + 2V_s - 7.5(\text{MW})$$

where:
V_{lgs} = low gravity solids content, % by volume
V_s = total solids content, % by volume, and
MW = Mud weight, lb/gal

Rearranging to solve for total solids:

$$V_s = \frac{[V_{lgs} + 7.5(\text{MW}) - 62.5]}{2.0}$$

For a 12.5 ppg drilling fluid containing 5% vol drilled solids by volume, 2% vol bentonite (V_{lgs} = 7% vol), the total solids are 19.1% vol. Since bentonite and drilled solids account for 7% vol, the remaining 12.1% vol is barite.

100% low gravity solids weigh 2.6 g/cc[8.34 ppg / g / cc]42 gal/bbl = 911 lb/bbl

With a specific gravity of 4.20, a barrel of barite weighs 1470 lbs [calculated from: (4.2) (8.34 lb/gal) (42 gal/bbl)/100% vol]. The 1350 sx of barite added during the drilling of this interval is equivalent to 91.8 bbls. Assuming that the mud weight and drilled solids were controlled at the stated levels during the drilling of the interval. Barite concentration in the drilling fluid is 12.1% vol of the drilling fluid. Stated in an equation:

Barite volume in the drilling fluid = (12.1%)(volume of new drilling fluid built)

So the volume of new drilling fluid built (91.8 bbl barite/0.121) is the volume of new drilling fluid built or 759 bbl. This new mud volume includes the drilled solids that remain in the system.

2. Drilled solids removal

    The volume of new mud created by adding clean drilling fluid to dilute the drilled solids remaining in the system was 759 bbl. This new fluid contains 5% drilled solids; or 37.9 bbls.

3. Equipment solids removal efficiency

    The new volume of drilled solids added from the drilling process was 97.3 bbls. From the volume of new hole, 97.3 bbl and the retained 37.9 bbls in the mud, the solids removal process has separated the difference of 59.3 bbls, or 61% of the drilled volume. This is the equipment solids removal efficiency.

4. Volume of drilling fluid created from adding the clean drilling fluid

    The volume of new drilling fluid built created was 759 bbl (as calculated from the amount of barite added) and 37.9 bbl of this was drilled solids. The volume of clean drilling fluid added would is 721.1 bbl. This could have been added as a blend of ingredients or, as most common, as individual components during the drilling process.

5. Excess drilling fluid generated

    The pit levels decrease only by the quantity of drilling fluid removed from the system. If nothing is removed, the pit levels would not change (except by the volume of drill pipe added to the system). The 97.3 bbl of drilled solids have been added to the system but 97.3 bbl of new hole means that the pit levels stay constant. The pit levels will drop by the amount of fluid and solids removed from the pits and will rise by the volume of new material added to the system.

    The volume of clean drilling fluid added was 721.11 bbl. The volume of material removed was either 200 bbl or 180 bbl depending upon which drilling fluid technician is correct. This would mean that either 521.1 bbl or 541.1 bbl of new drilling fluid would have to be stored.

6. The inaccuracy in calculating discard volumes

    The volume of discard was either 200 bbl or 180 bbl. In either case the discard tanks had to contain 59.3 bbl of drilled solids that came to the surface from the drilling operation. In one case the drilled solids concentration of newly drilled solids in the discard would be:

    59.3 bbl/200 bbl or 29.7% vol

    or in the other case, 59.3 bbl/180 bbl or 33.9% vol.

In the 200 bbl case 141 bbl of good drilling fluid was discarded; and in the 180 bbl case, 121 bbl of good drilling fluid was discarded. In other words, only 20 bbl of good drilling fluid was the total difference, if the excess 521 bbl or the excess 541 bbl of good drilling fluid would be sent to storage. The difference was relatively insignificant. Note that the 20 bbl error did not in any way affect the calculation of the solids removal efficiency.

Another method of calculating the dilution quantity

The dilution required to compensate for the incorporation of 37.9 bbl of drilled solids was the 759 bbl of new mud, less the volume of the drilled solids in the new drilling fluid built, or, 721 bbl. This is 19 bbl of dilution per bbl of incorporated solids (721 bbl/37.9 bbl).

$$V = (100 - C_{DS})/C_{DS}$$

where:

$V=$ the volume of new mud (dilution) required, bbl/bbl of incorporated solids
and $C_{DS} =$ the concentration of drilled solids

This is a calculation for the dilution; the volume of new mud that must be prepared.

The volume of new mud plus the incorporated cuttings, which is the total volume increase, is simply the volume increase factor (VIF) multiplied by the volume of incorporated solids:

$$\text{VIF} = 100/C_{DS}$$

At 5% drilled solids, V, the dilution volume, is 19 times the volume of incorporated solids, and the volume increase is 20 times that volume.

API method

Using the current API technique (API 13C) in the above problem: the dilution corresponding to total incorporation of the drilled solids (no separation) would be the volume of solids drilled/drilled solids fraction: in this case, 97.3 bbl/0.05 or 1946 bbl. The dilution factor (DF) is the ratio of the volume of new mud actually prepared to that would have been required with no drilled solids removal. In this instance, 759 bbl of new mud has been built and the DF is 759 bbl/1946 bbl, or 0.39. In other words, the required dilution was 39% of what it would have been if none of the drilled solids had been separated.

Drilled solids removal factor

Observe the relationship between the dilution factor (DF) and the drilled solids removal of 61% calculated above. The drilled solids removal factor (DSRF) is defined as follows:

$$\text{DSRF} = 100(1 - \text{DF})$$

It is numerically equal to the drilled solids removal efficiency previously calculated. In this case:

$$\text{DSRF} = 100(1 - 0.39) = 61\%$$

Summary of effective solids control

- Obtain solids removal equipment from reputable manufacturers and size it to process drilling fluid at the manufacturer's recommended capacity. Except for shale shakers and centrifuges, the process rate should be greater than 125% of the flow rate entering the suction tank of the equipment.
- Remove as many drilled solids as possible the first time they are circulated to the surface.
- Do not by-pass the shale shaker or other equipment, if at all possible.
- On scalping shakers, use a very coarse screen.
- Use the smallest screen openings possible on the main shale shakers.
- Maintain an adequate inventory of recommended spare parts.
- Assign rig personnel on each tour to be responsible for equipment operation and maintenance.
- Any drilling fluid brought to a rig should be added to the mud tanks through the shale shaker.
- Sufficient shaker capacity should be available to process the entire top-hole flow rate.

Equipment solids removal efficiency problems

Problem 12.1

Never-wrong Drilling Company is drilling a $12\frac{1}{4}$" hole from a casing seat at 4000 ft to the next casing seat at 10,000 ft. The 12.0 ppg water-base drilling fluid has an 18% vol solids content with an MBT of 18 lb/bbl. After drilling the well, a caliper indicates that the borehole washed-out to an average diameter of 14". While drilling this 6000 ft interval, 7,163 bbl of clean drilling fluid was metered into the system to maintain a constant 18% vol solids concentration

From this data, calculate the removal efficiency of the solids equipment.

Problem 12.2

With this inefficient drilled solids removal system, the decision was made to drill with a polymer drilling fluid system that required a 4% vol drilled solids concentration. The cost of reducing the drilled solids concentration only 2.5% is startling.

Calculate the cost to reduce the drilled solids concentration from 6.5% vol to 4% vol with the removal efficiency indicated in Problem 12.1, if the new drilling fluid costs $90/bbl and the disposal cost is $40/bbl.

Problem 12.3

In Problem 12.2, what would be the cost benefit if the solids removal efficiency was increased to 80% for the 4% vol drilled solids concentration?

Problem 12.4

Demonstration of the effect of a slight increase in the drilled solids concentration:

If the 80% removal efficiency was achieved and 6% vol drilled solids level would not create hole problems, another significant cost reduction is possible. Calculate the cost of dilution and disposal for this case.

(Note greater savings might more easily accrue if the removal efficiency was increased to match the optimum solids removal efficiency for 4% vol solids. The benefits of a clean drilling fluid are immense. The cost of rearranging the plumbing and having the proper solids removal equipment is usually much less than the savings created from the elimination of the visible and invisible non-productive time.)

General comments

Problem 12.1 assumes that the clean drilling fluid volume is known. Frequently, drilling fluid is not metered into the pits but the new fluid is built by adding liquid and solids to the tanks to keep the pit levels constant and to maintain the drilled solids concentration at some target value.

Since the barite concentration in the drilling fluid was also kept at the same value, 9.5% vol, it has been used as a tracer to determine how much clean drilling fluid was added. The dilution fluid, in Problem 12.1, contains 9.5% vol barite or 680 bbl. Barite weighs 1471 lb/bbl, so this would be equivalent to 1,000,000 lb of barite used (or 100,000 sx). The new drilling fluid built contained 7163 bbl of clean fluid and some volume of drilled solids that was returned to the pits by the solids removal equipment.

Barite may be used as a tracer to determine the volume of clean drilling fluid added to the system; however, using measured volumes of barite in the discard is enamored with problems. Do not use barite volumes discarded to calculate barite needed. The liquid removed by fine shale shaker screens does not contain the same concentration of barite as the drilling fluid in the pits. In one field test, between 100 lb/h and 300 lb/h excess barite was discarded from a continuous-cloth screen. During that field tests, a deficiency of as much as 100 lb/h of barite was observed in the discard from panel screens. In this case, barite went through the screen which decreased the barite in the discard below the concentration values of the drilling fluid in the pits.

CHAPTER 13

Cut points

Introduction

Cut point is the ratio of the mass of discarded solids in a particular size range to the mass of solids in the same size range presented to the removal equipment. For example, if 1.0 lb of solids between 1 mm and 2 mm in size are presented to a screening device, and 0.4 lb of solids in that size range are discarded, the mass ratio to discharged solids to feed solids is 40%. This ratio is a cut point and would be labeled D40. A D50 number would the size which would have an equal chance of being in the retained or the discard stream. The chart below (Fig. 13.1), shows typical cut point curves for centrifuges, hydrocyclones and shale shakers. Table 13.1 shows the median size range of solids removed by the

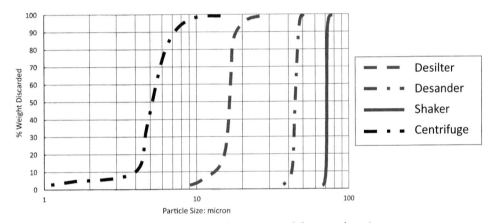

Fig. 13.1 Cut point curves for various solids control equipment.

Table 13.1 Solids removed by solids control equipment.

Equipment	Size	Median-size removed microns
Shale shakers	API 80-screen	Larger than 177
	API 120-screen	Larger than 105
	API 200-screen	Larger than 74
Hydrocyclones	8-inch diameter	Larger than 70
	4-inch diameter	Larger than 25
	3-inch diameter	Larger than 20
Centrifuge (weighted mud)		Larger than 5
Centrifuge (un-weighted mud)		Less than 5

various pieces of solids control equipment. The D50 of a shaker screen is shown to be about 74 µm and the curve is almost vertical. This would indicate that all of the solids larger than 80 µm would be discarded and all solids smaller than about 70 µm would be retained. The "80 µm" number would be called the D100, because 100% of those particles would be discarded. This "sharp" curve might be obtained with a premium square mesh screen and a very low solids load in a low viscosity fluid.

Cut points may be determined for all drilled solids removal equipment. When testing a particular solids control unit, the feed flow rate to the unit and the two discharge flow rates are required. The density of the feed flow multiplied by the volume flow rate provides the mass flow rate into the unit. Discharge mass flow rates are also calculated by multiplying the density of the stream by the volume flow rate. Obviously, the sum of the discharge mass flow rates must be equal to the feed mass flow rate. One of the discharge flow streams is discarded and the other is retained in the drilling fluid. The material balance, both the volume flow rate balance and the mass flow rate balance should be verified before measuring the particle sizes of the various streams.

Solids removal equipment removes only a very small fraction of the total flow into the equipment. For example, a four inch desilter processing 50 gpm of drilling fluid will discard only about 1 gpm of material. Since the discarded material is such a small proportion of the total material processed, the difference between the retained stream flow rate and the feed stream flow rate is difficult to measure. For this reason, more accurate data is acquired by mathematically adding the discarded solid flow rate and concentrations to the retained solids flow rate and concentration to determine the feed solids concentration and flow rate. With shale shakers, the flow rate through the screen is difficult to measure at a rig. Most shale shakers are not installed with easy access to measure the flow rate through the screen. The flow rate through the screen is determined by subtracting the volume flow rate discarded by the screen from the flow rate coming from the well. The mud pumps need to be calibrated to obtain the correct flow rate. A small amount of gas or air in the drilling fluid has a significant effect on the volumetric pump efficiency. To calibrate a pump, take suction from a slug tank while drilling. Do not try to calibrate a pump by pumping into a trip tank with a low pump pressure. Measure the volume of the slug tank by its dimensions. Determine the amount of liquid in the slug tank by subtracting the volume of gas contained in the drilling fluid. The volume fraction of gas can be determined by the equation:

$$\text{Volume fraction of gas} = \frac{\text{Pressurized mud weight} - \text{Unpressurized mud weight}}{\text{Pressurized mud weight}}$$

In one field test, only 6% volume of air reduced the volumetric efficiency of a rig mud pump to 85%.

A check of the volume flow rate balance into and out of the solids control equipment AND the mass flow rate balance into and out of the solids control equipment should be verified before determining the solids distribution in the input and discard streams.

To develop a cut point curve, the quantity of solids discarded from the equipment is compared with the quantity of solids in a series of size ranges that are presented to the equipment. For example, if a screen is discarding 100 g of particles between 40 and 50 μm, and 200 g of particles in that size range are presented to the screen, the cut point is said to be a D50. In other words, the screen separates 50% of the particles in this size range that is in the drilling fluid. If only 60 g of solids are discarded between 20 μm and 25 μm from 300 g presented to the screen in that size range, the screen would be removing 20% of the solids in this size range. The D20 cut point would be between 20 and 25 μm.

To develop a cut point curve, the quantity of solids presented to the equipment in all size ranges must be determined. The quantity of solids discarded in each of those size ranges must be determined. A ratio of the two quantities can then be plotted as a function of the size range of the particles.

Cut points can be obtained for shale shakers, hydrocyclones, and centrifuges. The cut point curves are dependent upon many variables and are usually smoothed to become continuous curves instead of step functions which are dependent upon the size intervals selected. In all cases the size distribution of the input stream and the discard stream must be determined. The flow rate of each of these streams must be measured. As a check of the measurements, the mass flow rate into the device should equal to the mass flow rate out of the device. Mass flow rates are measured by multiplying the volume flow rate times the density of the flow stream. This volume and mass balance needs to be confirmed before measuring the particle sizes in the streams.

A cut point curve graphically displays the fraction of various size particles removed by the solids control equipment compared to the quantity of that size particle presented to the equipment. For example, a D50 cut point is the intersection of the 50% data point on the "Y" axis and the corresponding micron size on the "X" axis on the cut point graph. This cut point indicates the size of the particle in the feed to the solids control equipment that will have a 50% chance of passing through the equipment and 50% chance of discharging off of the equipment. Frequently, solids distribution curves are ERRONEOUSLY displayed as cut point curves. Cut point curves indicate the fraction of solids of various sizes that are separated. They also are greatly dependent upon many drilling fluid factors and only indicate the performance of the complete solids control device at the exact moment in time of the data collection. For example, if no solids larger than 60 μm are presented to a piece of equipment, the cut point will be less than 60 μm. The cut point, or separation, curve also depends upon the particle size distribution in the feed sample. The cut points of the solids control equipment will be determined by the physical condition of the equipment and the properties of the drilling fluid.

Cut points of shale shakers

The initial impression of people unfamiliar with solids control is that the API number is the cut point of the screen. The screen openings are a different size when the screen is dry or wet. Cut point is NOT an invariant function of the screen. A cut point is an indicator of performance that varies with solids loading and drilling fluid properties.

A shaker screen sieves drilling fluid and solids larger and smaller than the openings are discarded. All openings are generally not exactly the same size even in a single layer square mesh screen. The openings in a dry screen may not be the same size as the openings when the screen is wet (Fig. 13.2). For example, when using a non-aqueous drilling fluid (NADF), the screen becomes oil-wet. If a fine screen (like an API 200) becomes water-wet, the NADF may flow off the end of the screen because the openings in the screen are so small. Most screens do not have exactly perfectly uniform opening sizes. The actual cut point varies with the solids loading, rheology of the fluid, the motion of the shaker screen, the stroke length of the screen, and the surface tension of the fluid around the screen wire. In other words, the cut point for a specific screen will not necessarily stay constant during the drilling of a well under dynamic conditions. Cut point is a performance measurement. This is the reason that the API RP13C does not use the cut point as a method of comparing screen performance. The API RP13C simply measures the size of the smallest solid discarded [or the largest solids that could pass through a dry screen] and calls this the API number.

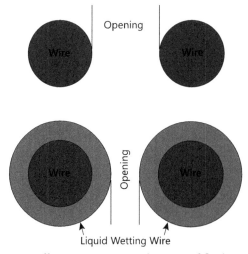

Fig. 13.2 Decrease in effective opening size because of fluid wetting the screen.

Fig. 13.3 A stack of standard screens mounted on a Ro-Tap©.

To determine the solid sizes in a shaker discard and flow stream, standard ASTM square mesh screens are mounted on a vibrating base, such as a RoTap™ (Fig. 13.3).

The fluid collected from flow into the shale shaker or the discard from the shaker screen is poured into the top of a stack of standard sieves (Fig. 13.4).

Fig. 13.4 Pouring the fluid sample into the stack of standard sieves.

Fig. 13.5 Washing the fluid sample through the sieves.

This fluid sample is gently washed through the screens with a light spray of the base fluid from the drilling fluid. In Fig. 13.5, a water-based drilling fluid is being gently washed through the screens.

If ASTM Screens are used for the standard screen, the sieve number indicates the equivalent square mesh size.

After the solids have been sieved, each screen is gently washed with fluid to remove the drilling fluid clinging to the particles and the solids are dried. Each screen with the captured solids is then weighed.

Ratio of weight of solids discarded to weight of solids reporting to shaker screen in specific size ranges is presented in Fig. 13.6. The weight of solids on each size screen is determined and a ratio of these are plotted. This makes a bar graph. Normally cut point curves are presented as continuous curves by drawing a line through the mid points of the bars (Fig. 13.7).

The typical lazy "S" cut point curve is created from the bar graph in Fig. 13.6, as shown in Fig. 13.7. In this case the D50 point would about 128 μm. The D100, which would be probably measured by the procedure in API RP13C, would be about 210 μm.

Cut points of hydrocyclones

The cut point curve for 4″ desilters in the mud cleaner chapter was measured at a drilling rig during the first mud cleaner field test installation. A flow meter was unavailable at the rig site and the flow rate is needed to determine the flow rate through the hydrocyclones. A piece of 30″ casing was mounted on the discharge side of the desilter bank (Fig. 13.8).

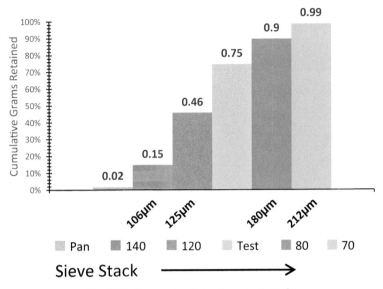

Fig. 13.6 Bar graph of dry sieve test results.

Two lines were painted inside of the casing and the volume between the two lines accurately pre-determined. While pumping drilling fluid through the desilters, the valve was open. To measure the flow rate through the desilters, the valve was closed and the time for fluid to fill the area between the volume lines was measured. Several timed measurements could be made to assure that the flow rate was constant. A sample of

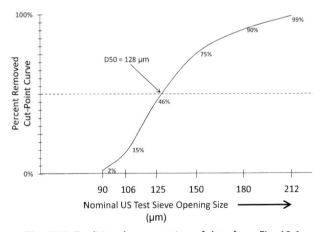

Fig. 13.7 Traditional presentation of data from Fig. 13.6.

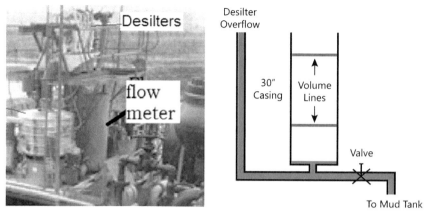

Fig. 13.8 Roughneck proof flow meter to measure flow rate through desilters.

the overflow in the "roughneck-proof" flow meter was taken to determine the solids distribution in the desilter overflow.

The underflow from the desilters was captured in a volume-calibrated bucket under the discharge spout. Normally the desilter underflow would report the mud cleaner screen. This spout could be moved aside and the desilter underflow captured in the bucket to measure the flow rate of the underflow. A sample was captured to measure the particle size distribution in the underflow.

The cut points of 4″ desilters for 11.0 ppg drilling fluid did not appear to be as sharp as the technical brochures had indicated. If the cut point was in the 20–30 μm range, the quantity of barite falling on the mud cleaner screen would have overwhelmed the screen quickly. The shaker screens would not have lasted very long.

The cut point curve indicated that the D50 was a range of solids between 50 μm and 125 μm (Fig. 13.9). The total quantity of solids in this size range is not indicated by the cut point curve. This fluid had been used for many weeks on this rig before the cut point was measured. Most of the barite was smaller than 50 μm.

Cut points of centrifuges

The input to a centrifuge should be with a positive displacement pump. The volume flow rate can be measured by calibrating the pump. The fluid was sampled to provide the density and, eventually, the particle size distribution. The flow rate of the discharge stream (underflow or heavy slurry) from a centrifuge is difficult to measure. Usually, the centrifuge is mounted on top of a mud tank so this slurry falls into a well-agitated location. To measure the underflow rate, a barrel was cut to form a catch pan (Fig. 13.10). Small diameter pipe was welded onto the sides of the barrel so that it appeared to be a "stretcher". A line was painted horizontally on the inside of the barrel near the bottom.

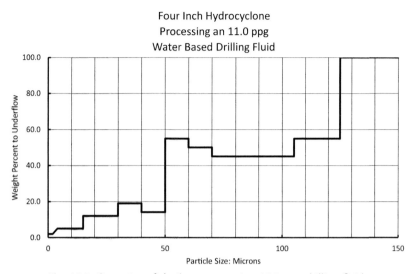

Fig. 13.9 Cut point of desilters processing 11.0 ppg drilling fluid.

Another line was painted near the top of the barrel. Water was placed in the lower part of the barrel below the lowest line. The half-barrel was then placed across the mud tank next to the centrifuge. The barrel was shoved beneath the centrifuge and a stop watch started when the water level reached the first line. The time to move the water level to the top line was measured. Since this volume was known, the volume divided by the time gave the flow rate of the underflow (heavy) slurry coming from the centrifuge. A sample of the underflow was captured in another container for analysis. The density of this slurry was measured. After confirming a volume and a mass flow balance, the particle size distribution was measured.

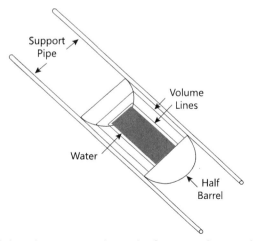

Fig. 13.10 The catch barrel to measure the under flow rate of a centrifuge on a mud tank.

After the mass flow balance and the volume flow balance was confirmed, the particle sizes were determined in the input flow stream and the underflow stream. Before this balance was achieved, these attempts consumed most of one day. After determining the size distribution and the mass flow rates, a cut point curve was generated (Fig. 13.11).

The cut point curve was surprisingly sharp (Fig. 13.11). That means that almost all of the solids above about 10 μm were in the underflow (heavy) slurry and almost all of the solids below 5 μm were in the overflow (light) slurry.

About the time these measurements were being made, another centrifuge was invented and introduced to the industry by Mobil Oil researchers. It consisted of a perforated cylinder rotating inside a chamber, see Chapter 21, Centrifuges. Fluid was introduced into the annulus. The small solids could pass through the holes in the rotating cylinder but centrifugal force kept the larger particles from passing through the perforated cylinder. This centrifuge had an advantage over the decanter centrifuge for a retrofit on a drilling rig. Both discharge streams were fluid and could be pumped (Fig. 13.12).

The underflow (heavy) slurry discharged from a decanter centrifuge must fall into a well agitated spot in the mud tanks. The Mobil centrifuge could be mounted on a trailer and placed next to the mud tanks (Fig. 13.13).

The cut point curve was surprisingly sharp (Fig. 13.14). More solids in the very small size range were discarded and the D100 was somewhat higher than the decanter centrifuge. This was actually surprising and prompted two more tests. Two of the tests were made at a rig site and the third was made in the plant with the equipment. Unfortunately, most people seemed to be skeptical of the effectiveness of this centrifuge and it failed to make a great impact in the industry.

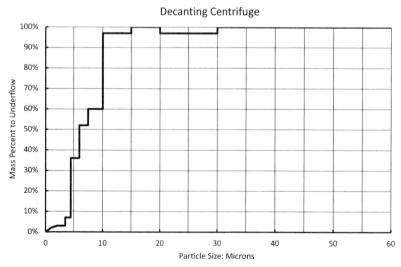

Fig. 13.11 Cut point curve for an 17.2 ppg drilling fluid.

Fig. 13.12 Mobil mud centrifuge discharge.

Fig. 13.13 Mobil mud centrifuge mounted on a trailer.

Of course, this centrifuge would not be useful where attempts are made to decrease the fluid in the effluent—that is, give a dry discharge. It did do a good job of removing the colloidal solids from the mud system.

Comment on particle size presentations

Distributions of particle sizes vary greatly with the manner of presentation. That means that if the size interval chosen is different, the data may be correct but the

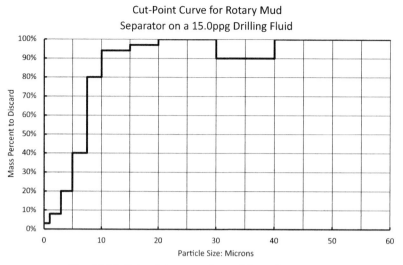

Fig. 13.14 Cut point curve of a mobil mud separator.

implication of the data may be very misleading. For example, consider the two graphs in Fig. 13.15.

Both of the charts in Fig. 13.15 were drawn from the same distribution curve. Different size ranges were used in the charts and this gives a different profile.

The following distribution curves are all drawn from the same data (Figs. 13.16–13.18):

If each interval is divided by the width of the interval, all graphs start appearing alike (Fig. 13.19).

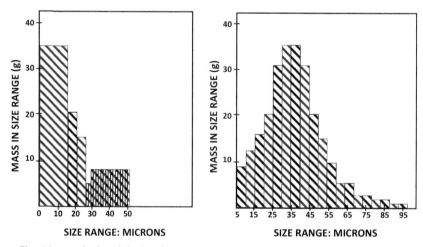

Fig. 13.15 Which solids distribution would you like to see in your drilling fluid?

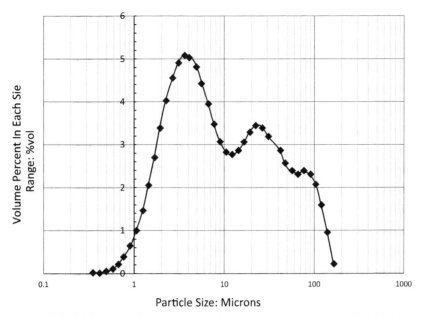

Fig. 13.16 Solids distribution curve #1 using different particle size width bands.

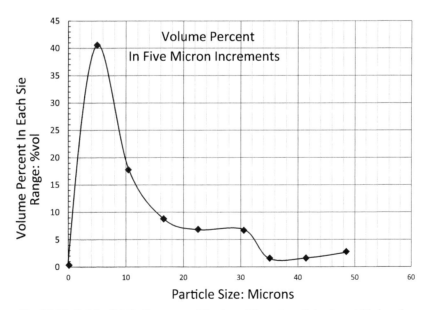

Fig. 13.17 Solids distribution curve #2 using different particle size width bands.

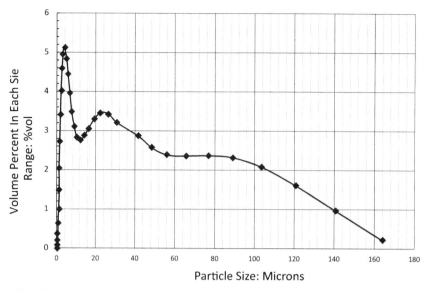

Fig. 13.18 Solids distribution curve #3 using different particle size width bands.

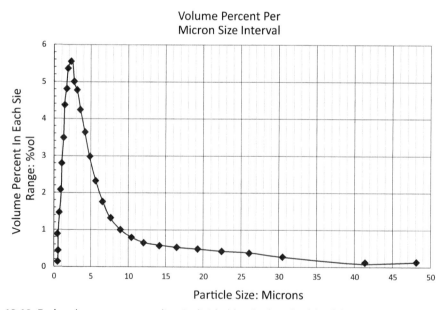

Fig. 13.19 Each volume percent reading is divided by the bandwidth of the size interval selected.

Plotting the volume percent in terms of the logarithm of the interval selected changes the profile greatly (Fig. 13.20). This technique emphases the very small size range. If you are selling centrifuges, this amplifies the interval of solids removal.

If you are selling desilters or shale shakers, a linear presentation of data will magnify the solids interval to be removed with this equipment (Fig. 13.19).

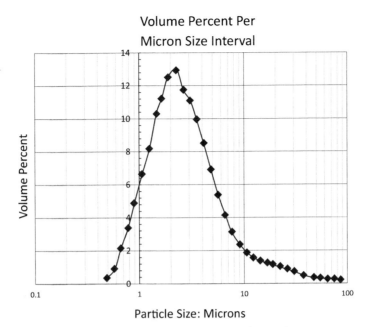

Fig. 13.20 Size plotted on a logarithm scale.

A curve representing the distribution of solids in an active drilling fluid system (Fig. 13.21), may give specific impressions depending upon exactly what intervals were used to determine the percent solids in that interval. Using the same equipment each time can provide some continuity to the analysis but may still not provide the

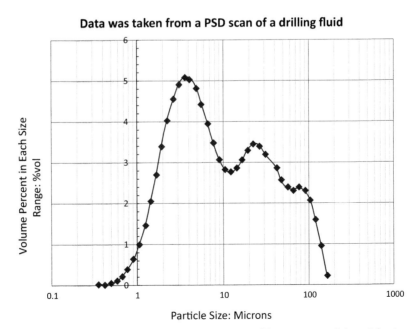

Fig. 13.21 Distribution of solids in a drilling fluid as reported by a commercial particle size measurement device.

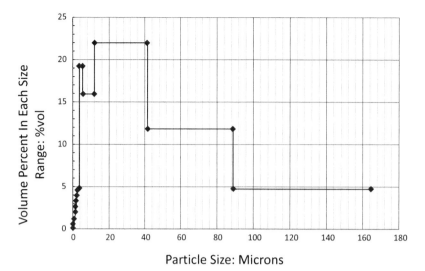

Fig. 13.22 Linear plot of distribution curve.

correct perception of the solids. As noted above, different distributions can be created from the same data by selecting different intervals of particle size. The Malvern scan uses a semi-logarithm presentation. The interval between one and two microns seems to have five weight measurements but the interval between 20 and 30 μm only has three weight measurements.

Plotting the same data in Fig. 13.22 on a logarithm scale changes the shape greatly (Fig. 13.23).

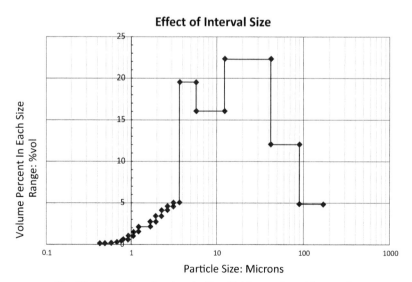

Fig. 13.23 Plotting the data in Fig. 13.22 on a logarithm scale.

CHAPTER 14

Operating guidelines for drilling fluid surface systems

Objective

This chapter presents guidelines for the design and operations of the surface drilling fluid system. These guidelines have been explained in depth in the various chapters and are summarized here for easy access. These recommendations are intended to improve drilling fluid quality and increase drilling performance capability. Many of the comments were made during the API Task Group discussions of API RP13C.

Description of the surface drilling fluid system

The surface system used for processing drilling fluid consists of many components: the flow line, the active tanks, the reserve tanks, the trip tank(s), the agitators, the solids removal devices, the gas removal devices, devices to add drilling fluid components, devices to mix and shear drilling fluid, the piping to correctly plumb the equipment, the pumps to feed the removal devices, and the motors to drive the pumps.

Design of the active surface processing system

Tank depth should be approximately equal to the tank width or diameter. Significantly deeper tanks can present problems with achieving adequate mixing. Special attention is needed to correctly size agitators for the correct blending and mixing of deep compartments. Shallow tanks can lead to formation of vortices and air entrainment. The air entrainment can cause significant corrosion problems as well as create problems for the centrifugal pumps. Centrifugal pumps cannot pump gaseous drilling fluids effectively. To prevent vortices, the removal section should overflow into the suction section so the centrifugal pumps will always have a positive head and not create vortex problems.

The minimum surface area, in square feet, of compartments, should be equal to the maximum circulation rate, in gallons/minute, divided by 40. This facilitates the release of entrained gas. This rule of thumb was developed by George Ormsby (many years ago) and seems very effective.

Piping between tanks and tank bottom equalizers should be 10″, 12″, or 14″ in diameter. These are adequate for circulation rates as high as 935, 1350, and 1830 gpm respectively. The flow velocity through equalizing lines should be between 4 ft/s and 10 ft/s. This rule of thumb was developed to prevent solids settling at the low flow rates and turbulence at the higher flow rates. Generally, if an equalizing line is too large, solids will settle in that line until the velocity increases enough to transport solids.

The active drilling fluids surface system was discussed in depth in Chapter 8. This is a summary of some important aspects of the system; starting with the flow line and continuing through the system through the suction tank. The drilling fluid processing system should have three easily identifiable sections: removal, addition, and suction. The sections could consist of compartments in a tank (for very small drilling rigs) or several tanks for very large drilling rigs.

Flow line

The flow line diameter should be sufficient to handle the maximum circulation rate for a very viscous drilling fluid. Common practice is to slope the flow line down one foot for each ten feet of horizontal distance.

For drilling rigs which have more than one shaker, a flow divider is recommended to allow for each shaker to process the same quantity of drilling fluid. This will allow the same fine screens to be used on all shale shakers.

Removal section

The first of the surface system compartments is located in the removal section (Fig. 14.1). These compartments can be very small. All unwanted ingredients are removed in this section. The shale shakers remove the largest drilled solids from the fluid. Sand traps are designed to remove the larger solids which pass through the shaker screens. The drilling fluid is degassed so that the centrifugal pumps can move the fluid. Centrifugal pump output is very sensitive to air or gas in the fluid. Although the gas cut drilling fluid has very little effect on the down hole pressure, it creates a large problem when feeding the equipment in the surface processing system. Hydrocyclones and centrifuges are

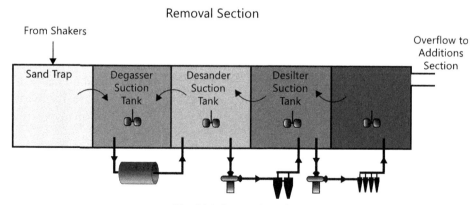

Fig. 14.1 Removal section.

used to remove the finer drilled solids that have passed through the shaker screens. The fluid then overflows into the addition section downstream.

Shale shakers

Four shaker motions are available for drilling rigs: unbalanced elliptical, circular, linear, and balanced elliptical. Circular motion and unbalanced elliptical motion shakers can process rig flow rates through screens as fine as API 80 (180 µm) and occasionally as fine as API 100 (150 µm). Linear motion and balanced elliptical motion shakers can process rig flow rates through screens as fine as API 200 (75 µm) and occasionally as fine as API 325 (44 µm).

On larger drilling rigs, a scalping shaker' is used to remove very large solids which could damage the fine screens on the main shaker. This scalping shaker could be a circular motion or an unbalanced elliptical motion machine since a very coarse screen should be mounted to remove only the very large solids. After the scalping shaker, the fluid could flow directly into the main shakers back tank (called a 'possum belly') or into a fluid distribution chamber for multiple shakers to receive the same flow rate of drilling fluid.

Rig up of shale shakers
- The shaker skids should be level.
- Space, walkways, and handrails should be provided to properly service the shaker.
- Flow lines should be connected to the lower part of the back tank (a.k.a. Possum belly) of the shale shaker.
- If the flow line must enter the top of the back tank, an elbow should direct the fluid to the bottom of the back tank.
- The screens should be cleaned of drilling fluid before tripping pipe.
- The back tank should NOT be dumped into the active system before cleaning the screens.
- A by-pass port should be available to permit the disposal of cement, pills and spotting fluids, contaminated fluids, bottoms-up (if needed), and dumping the shaker back tank.
- Screens should have an API label indicating the maximum screen opening size, the conductance, and the open area.

Operational guidelines for shale shakers
- Shale shaker screens should not be by-passed while drilling or tripping pipe.
- If lost circulation material is required, circulate a pill first and attempt to stop the lost circulation problem before contaminating the entire drilling fluid system.
- Lost circulation is less likely if the drilled solids content is very low and the filter cake is very thin.
- If lost circulation is not resolved with a pill, by-passing the shaker may be necessary to retain the lost circulation material. In this case, remove all lost circulation material after this interval of hole is cased.

- Inspect the shaker screens for holes whenever circulation ceases.
- Any fluid arriving on location from any source should be added to the drilling fluid system through the shale shaker.
- On trips into the hole, any fluid from the trip tanks should be returned to the active system through the main shaker.
- Spray bars may be necessary to handle sticky clays. The fluid should be delivered in a mist, not a cohesive jet of fluid.
- The shaker screen should be as fine as possible without excessive drilling fluid loss.
- Do not keep the shaker screens on linear, or balanced elliptical motion shakers elevated to the maximum position while drilling. Elevate during bottoms-up and then return to a lower position. This will prevent solids degradation on the screens as the solids are transported off the screen.
- Follow the manufacturer's guidelines on screen installation and tensioning, as well as maintenance.
- When using shaker screens that need tensioning, tension should be checked 15–30 min after installation, again after 2–3 h, and regularly thereafter. (The screen 'relaxes' and becomes loose after installation. A loose screen has a very short life.)

Sand traps

Sand traps were very useful when very coarse shaker screens were used on the main shakers. Solids would settle and could be dumped into a waste tank. They are usually eliminated from the tank system when fine screens are used on the main shaker.

Rig up of sand traps

Generally the sides of the sand trap must have an angle steeper than the angle of repose—or larger than about 40 degrees (Fig. 14.2).

The discharge valves (dump gates) on tanks should be large, non-plugging, and quick opening and closing. The settled solids slide down the slope to the valve when it is opened. The valve is closed when drilling fluid begins to flow through the valve. The valve should be operable from the level of the tanks and flow from the valve should be visible from the position from which the valve is operated.

Fluid should enter the sand trap at its upstream end, and flow from it over a high, tank width, weir at its downstream end. The weir height should be about six inches below the top of the tank.

Operating guidelines for sand traps

When fine screen shakers (API 140 or finer) are used, the solids passing through the screen would settle so slowly that the sand trap became ineffective. A seventy micron

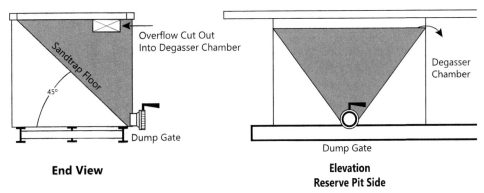

Fig. 14.2 Sand trap.

barite particle settles much faster than a seventy micron drilled solid. The majority of the settled solids will be barite.

Currently, the carrying capacity of drilling fluids is generally much higher than it was before linear motion shakers were developed. This means that the low-shear-rate viscosity of the drilling fluid is much larger. This high viscosity hinders good settling. Most large drilling rigs, which are outfitted with sufficient shaker capacity, have eliminated the sand trap from the surface tank system.

Degassers

Centrifugal pumps do not pump gaseous fluids very well. The centrifugal force applied to the fluid causes the fluid to move outward and the gas to collect in the center of the impeller. This causes the fluid to cease moving through the pump housing. Since down stream equipment requires the use of centrifugal pumps, the gas should be removed as soon as it enters the active surface system.

Rig up of degassers
- Suction for a degasser should be in the compartment down stream from the sand trap.
- Fluid from the sand trap should overflow into the degasser suction compartment.
- If no sand trap is used, fluid passing through the shale shaker screen should enter the degasser suction compartment.
- The degassed fluid should enter the next compartment down stream from the degasser suction and over flow back into the degasser suction.
- The lower end of the degasser suction line should be about one foot above the bottom of the suction compartment.
- The degasser suction compartment should be well-agitated.

- The degasser should process more fluid than is entering the degasser suction compartment.
- For vacuum degassers, the centrifugal pump feeding the eductor jet must supply sufficient head to remove the fluid from the vacuum compartment.
- Atmospheric degassers are positioned in the same manner as vacuum degassers.
- Verify the degasser is performing correctly by comparing the mud weight in the suction tank with the mud weight in the discharge compartment.

Operational guidelines for degassers
- Degassers are needed whenever gas bearing formations are drilled.
- After trips, the bottoms-up fluid should be processed through the degassers.
- Gas can enter the drilling fluid from the in-situ gas in the formations as the rock is drilled.
- Increasing background gas does not mean a kick is imminent. This gas needs to be removed.

Hydrocyclones

Hydrocyclones are available in a variety of sizes. The 4″ desilter cone is the most common. When it was introduced to the industry, coarse screens were normally used on shakers. This created a problem with solids overload in these small cones. Desanders were introduced to decrease the solids load on the small cones. Desanders are hydrocyclones with inside diameters of six inches or larger. When fine screens (such as API 140 and finer) became common on shakers, the desanders are no longer needed to reduce the solids overload in the desilters.

Rig up of hydrocyclones
- If desanders are used, the desander suction tank should be the degasser discharge tank.
- If desanders are not used, the desilter suction tank should be the degasser discharge tank.
- If desanders are used, the desander overflow should be into a compartment down stream.
- If desanders are used, the desilter suction is in the desander discharge compartment.
- The desilter discharge is directed into the next compartment down stream.
- An underflow equalizer line should permit back flow from the discharge compartments into the compartment up stream.
- Each bank of hydrocyclones should have dedicated centrifugal pumps to provide the proper head at the manifold.
- One pump, one switch—do not use the feed pumps for other functions.
- The piping between the pump and the feed manifolds for hydrocyclones should be as short and straight as possible

- No elbows or swages should be within three pipe diameters of the entrance to the hydrocyclone manifold.
- Pressure gauges should be mounted on the hydrocyclone manifold.
- The hydrocyclone input manifold must be above the liquid level in the suction tank.
- If the hydrocyclone discharge manifold is more than about five feet above the drilling fluid in the discharge tank, a vacuum breaker should be installed.[1]
- Sufficient walk ways, ladders (if necessary) and hand rails should be provided to permit safe and easy servicing of the hydrocyclones.

Operational guidelines for hydrocyclones
- Follow manufacturer's guide lines for the proper head on the manifold.
  - Head, in feet, can be calculated from the equation: Head = 19.23 divided by the mud weight, in ppg.
  - From the "well control" equation: Pressure = 0.052 (mud weight) (depth).
- Maintain the flow rate into and out of the hydrocyclones between four and eight feet per second. See Centrifugal Pumps Chapter 10.
- Check for a spray discharge. A 'rope' discharge means poor separation of drilled solids and could be caused by too many solids or a blockage in the feed line.
- If too many solids are in the drilling fluid, more cones may be needed and/or the shaker screen has a hole in it.

Mud cleaners

Mud cleaners are used to remove drilled solids larger than barite. Even though API 200 screens are mounted on the main shakers, the mud cleaner acts as insurance for removal of solids which by-pass the shaker screens or solids which pass through holes in the screen.

Rig up of mud cleaners
- Mud cleaners are installed in the same location on the removal section as hydrocyclones.
- Usually, the hydrocyclones on the mud cleaner are used in the unweighted part of the hole with the screen replaced with a solid plate.
- The underflow from a hydrocyclone has no carrying capacity and neither does the screen throughput. The fluid flowing through the screen must be returned to a well-agitated location so that it will be blended back into the drilling fluid.

---

[1] Make certain that the end of the vacuum breaker pipe does not have threads on it. Rig crews will connect hoses or needle valves instead of leaving the pipe open (voice of experience).

Operational guidelines for mud cleaners
- The mud cleaner should process all of the fluid flowing into the mud cleaner suction compartment.
- If the solids that fall on the screen from the hydrocyclones become too dry and clump together, spray drilling fluid on the screens not the liquid phase of the drilling fluid.
- The head on the hydrocyclones can be lower than required for unweighted drilling fluids because the excess fluid will assist cuttings transport on the mud cleaner screen.

Centrifuges
A centrifuge is usually used to remove small drilled solids (and small barite) from the active drilling fluid system. These solids significantly increase the plastic viscosity and decrease drilling performance. The heavy slurry (or underflow) from a centrifuge contains both barite and drilled solids. A centrifuge separates particles by mass. A barite particle and a drilled solid that has the same mass are found together. Do not use the quantity of barite recovered as an economic guideline for a centrifuge performance. A centrifuge is a solids removal piece of equipment which can be very effective in removing the damaging very small solids from the drilling fluid.

Rig up of centrifuges
- Centrifuges must be mounted on the mud tanks so the heavy (or underflow) slurry falls into a well-agitated location. The heavy slurry has no carrying capacity.
- A centrifuge should process the drilling fluid after it has been processed through a mud cleaner.
- A centrifuge may discard the retained fluid into the additions tank down stream from the removal section.

Operational guidelines for centrifuges
- When a centrifuge is used, it should be used continuously for at least one complete circulation of the fluid.
- A centrifuge is used to control the plastic viscosity of a weighted drilling fluid.
- A centrifuge is more efficient if the input flow rate is lower than the rated value. Running the lower flow rate allows the centrifuge to continuously process drilling fluid efficiently to create a homogeneous drilling fluid with a low plastic viscosity.
- The centrifuge should not be used to "recover" barite from a stored drilling fluid and add to an active system.
- A centrifuge can be used as a 'super' hydrocyclone on an unweighted drilling fluid by discarding the heavy (or underflow) slurry and retaining the light (or overflow) slurry.
- Centrifuges processing weighted drilling fluids will discard the small detrimental colloidal drilled solids as well as the beneficial low-shear-rate viscosifiers and filtration

control additives. The filter cake will become thicker and incompressible. These additives must be continuously returned to the drilling fluid when using a centrifuge.
- While processing weighted fluids, the feed fluid should be diluted as necessary to control the overflow viscosity at no more than 37 sec/qt with water based fluids, and under 40 sec/qt for oil-based fluids.

Addition section

After all of the detrimental things have been removed from the drilling fluid, the volume removed must be replaced to keep the drilling fluid level in the mud tanks. This volume is added through the addition section. Additional material may be needed to replace additives which have been destroyed during the downhole journey. All additions of new material to the drilling fluid system should be made into the addition section, which is the compartment down stream from the removal section. Additions made from reserve pits, fluid stored on location, or fluid arriving from another location should be made through the shakers. The addition section can be small but the additives should be well-blended with the drilling fluid. The addition section should also be blended with the suction section.

Rig up of the addition section
- The removal section overflows into the addition section. (This keeps the pit levels high in the removal section to prevent air from vortexing into the centrifugal pumps and minimizes the volume which responds to lost circulation or a kick.)
- Mud hoppers should be plumbed into the addition section.
- A venturi head recovery should be mounted on the discharge side of the mud hopper.
- A simple cylinder at the end of the mud hopper addition line can be used to eliminate much of the air that is frequently entrained by the mud hopper.
- The addition section should be well-agitated.
- The addition section should have mud guns with their suction in the suction tank.
- Liquid additives, such as base fluid, caustic, should be added in the addition section.
- A prehydration tank is required for bentonite additions. Clay should be soaked for 24 hours before using.
- Deflocculants should not be added to a prehydration tank. Lignosulfonate is an inhibitor for dispersion of bentonite.
- Some polymers and clays may need to be sheared through a shearing device to accomplish dispersion before adding to the addition tank.
- Do not circulate the drilling fluid through a shearing device. (This will create detrimental colloidal particles from the drill solids.)
- Frequently, a premix tank is filled with new drilling fluid to be added continuously to replace drilling fluid volumes discarded by the solids removal equipment.

- The premix tanks should have their own mud hoppers and agitators to facilitate treatment of the liquid in the premix tanks.

Operational guidelines for additions compartment
- Dry bentonite will not disperse properly when added to a gel/lignosulfonate drilling fluid. Use the prehydration tank.
- Use mud guns to blend the additives with the drilling fluid and to mix with the suction tank.
- The quantity of material added should be extended during the period of one circulation.

Suction section

The final section of the active surface drilling fluid processing system should be large enough to provide a homogeneous drilling fluid to the drill string all the time drilling is taking place. Chapter 2 contains the necessary guidelines for the correct volume. The suction section must have enough agitators and mud guns to guarantee that the fluid will have the same density and rheology in all compartments. The suction section should be well blended with the additions section.

Rig up of the suction section
- The suction section should be much larger than the maximum volume expected in the drill string.
- The suction section needs to be large enough while drilling the surface hole that the fluid has a residence time long enough to be properly treated.
- The suction section needs to have effective agitators and submerged mud guns positioned to keep the suction section well blended with the addition compartment.
- Baffles around each agitator should prevent vortexing and prevent aeration of the drilling fluid.
- The suction section should have two small well agitated pits to use for pills and for slugs.
- The pill and slug tanks need to have a mud hopper arranged so that the fluid can be circulated as material is added.
- The plumbing should be arranged so that the mud pumps can take suction from the last compartment in the suction section, or the pill tank, or the slug tank.

Operating guidelines for the suction section
- Keep the suction section well stirred and blended.
- Check pressurized mud weight periodically to determine the amount of air or gas in the drilling fluid. (As much as 6% air in the drilling fluid has reduced the pump volumetric efficiency to 85% and the fluid did not appear to be foamy.)

- Check the mud weight in all compartments regularly. This validates the homogeneity of the fluid in the suction section.

Mud guns

Mud guns are excellent for blending ingredients into the active system. They are only used in the addition and suction sections. Mud guns will distribute the drilling fluid into the different compartments to create a homogeneous drilling fluid.

Rig up of mud guns

- A centrifugal pump usually provides the fluid flowing through the mud guns.
- The mud gun pump takes suction from the far end of the suction tank and distributes it into all compartments after the removal section.
- Mud guns are not recommended for the removal section unless it is the suction and discharge to the centrifugal pump is in the same tank.
- Fluid provided from the suction compartment to a compartment in the removal section will require more removal equipment capacity. (The clean drilling fluid from the suction section will blend with the dirty drilling fluid and must be reprocessed.)
- Mud guns can be made from pipe swedges or purchased with venturi nozzles which will entrain more drilling fluid within the compartment.

Operating guidelines for mud guns

- The flow rate through the mud guns should be sufficient to blend the compartments thoroughly before the fluid disappears down hole.
- At least one mud gun should also be mounted in the addition compartment to help blend the fresh drilling fluid with the suction tank.
- The mud gun line will also be used to fill the trip tanks.

Design of the trip tank system

Trip tanks are used to keep the well bore filled with drilling fluid while removing the drill string. Some trip tanks systems are mounted next to the BOP stack and gravity feed fluid into the well bore as drill string volume is removed. Caution must be used with these systems to monitor drilling fluid gelation with prevents fluid from flowing into the well bore. More commonly now, a small centrifugal pump is used to keep the well bore full. Fluid flows into the well bore and overflows into back into the trip tank. As pipe is removed, the trip tank level drops. This allows calculation of the volume of fluid added to the well bore.

If no fluid is added when a stand of pipe is removed, the formations may be supplying the fluid to replace the volume of the drill string. A kick might be imminent. The prudent course of action is to return the bit to the bottom of the hole and circulate bottoms-up. The mud weight can then be adjusted to prevent further influx during a trip.

Most trip tanks have two compartments: one is used to measure the volume of fluid required to replace the volume of drill string removed; the other compartment is refilled while the other compartment is being emptied. The fluid to fill the tanks is supplied from the suction section by the mud gun line.

On the trip into the hole, the fluid from the trip tanks should be added back to the active system through the shale shaker. Usually, some large solids that have been captured in the BOP are entrained in the fluid coming from the hole as the drill string is run back in the hole. These solids plug hydrocyclones and can be removed with shakers.

Design of the prehydration tank

When bentonite (or gel) is used in water-based drilling fluids, or some polymers are used, they must first be dispersed. Bentonite should be soaked in water for at least 12 hours, preferably for 24 hours. This allows the clay platelets to separate and make the addition of the gel much more effective. With a slurry of 30–40 lb/bbl, the gel strength and low-shear-rate viscosity will become very high. Do not add a deflocculant (or so-called dispersant) like lignosulfonate to try to incorporate more gel in the fluid.

The prehydration tank should be equipped with a mud hopper and a centrifugal pump. Solids are added through the hopper and blended with a mixer in the prehydration tank. After the solids are added, the centrifugal pump can then be connected to a mud gun line to assist in blending the gel and the hydration of the clay.

If polymers, like HP007, are used, they need to be processed through a shear pump to assist the dispersion of the polymer.

Design of the clean drilling fluid addition system

In many instances, clean drilling fluid is blended in a separate tank and used to keep the pit levels constant. This system keeps the fluid in the suction section more homogeneous. Instead of adding liquid and all of the ingredients required to keep the drilling fluid within specifications, the added fluid has the proper properties. This type of system needs two tanks for blending clean drilling fluid. While one tank is being used to keep the pit levels constant, drilling fluid can be blended in the second tank and allowed to reach equilibrium. A small centrifugal pump can be used to transfer fluid from the mixed tank into the active mud system. The fluid should be added into the addition section of the active system (just as it would be if the ingredients were blended into the suction section as described in Chapter 3).

SECTION III

Solids control equipment

CHAPTER 15

Drilled solids removal

Introduction

Drilled solids should be considered EVIL because they affect both the trouble costs and drilling performance. Trouble costs are easy to identify and down time on a drilling rig stands as a monument to the viciousness of drilled solids. Problems with drilling performance are seldom recognized and more difficult to quantify. Failure to achieve maximum drilling rates, failure to locate all productive zones, difficulty with log interpretation, and loss of production because of cement placement allows flow behind casing are hidden with other excuses. This discussion will present information about both trouble costs and failure to achieve the best drilling performance. This could be called visible and invisible non-productive time (NPT).

Control of drilled solids begins at the bottom of the bore hole. Cuttings should be removed before they are reground by the drill bit. This requires operating with the maximum possible hydraulics and keeping the bit loading below the founder point. Solids are much easier to remove if they are large when they reach the surface. Cuttings should be transported to the surface without tumbling and grinding into smaller particles.

Economics of solids control

Very few drilling programs are written with scheduled Non-Productive Time (NPT). Although stuck pipe may be frequently experienced, it is not usually scheduled. Trouble costs are related to many factors. Excessive quantities of low-gravity-solids (drilled solids) is one major factor. These costs are difficult to quantify. Some field examples can be recited but the exact cost and savings are usually considered classified by operators.

Field example #1

An intensive solids control evaluation was recently conducted on an offshore platform rig in which the existing primary solids removal devices on the flowline were changed to higher capacity solids removal devices. The particular well that was analyzed was the longest extended reach well ever drilled for the operator. The intent of the evaluation was to compare previous well's overall costs to the costs after the change was made. Not only did the study include cost savings associated with drilling fluid and products costs, it also included cost savings related to reduced NPT, downhole tool failures, and time on well. With the ability to run finer screens on the high capacity solids removal devices, the operator was able to reduce dilution requirements significantly while at the same time reducing the sand content to less than 1% throughout the well. The tangible savings to being able to run finer screens resulted in over $900 K USD per well savings in drilling fluid cost alone. Other cost savings related to downhole tool failures, equipment maintenance, time on well, and NPT reduction (both visible and invisible) resulted in over $800 K USD per well savings bringing the overall cost savings of over $1.7 M USD per well. A return on investment in this case was realized over the course of two wells.

Field example #2

A Gulf of Mexico platform was programmed to drill 12 production wells. While drilling the sixth well, the Operation Supervisor called for an evaluation of the drilling fluids system. The equipment on the mud tanks was first class and appeared to be operating as it should be. A complete evaluation of the plumbing revealed that the contractor had been under the erroneous impression that flexibility was essential. The drilling fluid could be pumped from any compartment to any other compartment. The plumbing manifold had an enormous number of valves. The contractor had spent a lot of money needlessly because of the misunderstanding about how drilling fluid should be processed. The cost to correct the plumbing was so large that the evaluation recommendation was to do nothing. The Operation Supervisor insisted that plans be drawn to correct the system. After three days of examination and tracing flow patterns, a new plan was created. When the sixth well was finished, the production was shut-in. Six welders arrived at the rig and changed (simplified) the mud tank arrangement at the operator's expense. The seventh well cost less than any of the first six. The savings as indicated by the AFE of the seventh well was more than the cost of shut-in and modification of the tank system. The next five wells all cost significantly less than any of the first six wells.

The plumbing was correctly arranged as presented in this material. Each pump in the solids removal section had only one function. Turn on the pump and process 100% of the flow entering the pump's suction tank. This is true no matter what drilling fluid systems are used. One hundred percent processing is required of simple clay-based fluids, polymer fluids, potassium chloride fluids, salt systems, and Non-Aqueous Fluids. In other words, a contractor should not have to change the plumbing regardless of the operator's preference in drilling fluids.

Field example #3

One field foreman believed that he could save the company a lot of money by using drilled solids as a weighting agent instead of barite. The cost savings were easy to calculate. For a 10 ppg water-based drilling fluid, no barite was permitted to be added except for slugs for trips. The foreman was extremely proud of his frugality. He was also proud of the fact that he could recover stuck pipe faster than any of his colleagues. He had much more experience. The fisherman was on his speed dial. The explanation he used for this was that the company knew he was the best at saving money and drilling troublesome wells, so they always assigned him to the most difficult wells. It appeared to be true except he was having stuck pipe problems in fields that had a very minimum number of those kind of problems. He used drilling fluids with low fluid loss; consequently he claimed that it was not a filter cake problem down hole. Unfortunately, he did not believe that fluid loss could be lowered with drilled solids and the filter cake thickness increase.

The lesson here is that minimizing individual costs (like the cost of barite) will not always reduce the cost of a well. The consequences of the minimization needs to be evaluated. What was not considered in this case was the effectiveness of a cement job. Cement may seal the bottom of the hole but fail to provide barriers where the filter cakes prevent the cement from sealing. A large quantity of hydrocarbons can be flowing behind the casing and entering barren formations.

Example #4
One of the drilling divisions in major operator company decided to assign someone to watch drilling fluid properties daily in all of the wells. This person had a directive to visit any rig that seemed to be having problems. The plastic viscosity and solids content of the drilling fluid was monitored in each well daily. Frequently, visits were required to work with the Company Man to make minor changes in processes on the rig. For eighteen months, this program worked so well that there was no stuck pipe in the drilling division. The process was terminated because "there were no problems for the mud watcher to solve". The division had stuck pipe the following month.

Roles of the drilling fluid processing system

Non-aqueous drilling fluids (NADFs) are used frequently because of the drilling performance advantages they offer over water-based drilling fluids (WBDFs). But several differences are apparent when using an NADF which affect how the fluid is prepared, transported downhole, returned to the surface, and reconditioned for recycling back into the well. These steps constitute what is called the "drilling fluid processing system". However, the equipment and the plumbing for removing solids from the drilling fluid is the same no matter what fluid is used. The difference between water-based and NADF is usually related to the tendency to not want to lose so much of the expensive liquid phase.

Effective and efficient solids control, i.e. removal of drilled cuttings from the drilling fluid, is arguably the most important task of the drilling processing system. This is necessary to maintain the target mud properties of the mud program and avoid potential problems such as lost circulation and stuck pipe.

Nevertheless, it is equally important for the drilling fluid processing system to provide a drilling fluid with uniform mud weight (density) from the top to the bottom of the well. Safe control of the well depends on that. Management of a kick requires determining the formation pressure, which depends on knowing the mud weight of the fluid throughout the mud column. If the mud weight is uniform, it is also relatively easy to compute the required volumes of mud that must be manipulated to ensure that a blow-out does not occur. The section on mud tank arrangements addresses this issue.

Another important role of the drilling fluid processing system is preparation of the drilling fluid for cementing. Proper cementing of casing requires that the drilling fluid be conditioned sufficiently, which in the case of NADFs requires a multi-step procedure. Cement will not adhere to a hydrophobic surface. (Hydrophobic means "oil-wet" and is the opposite of hydrophilic.) To achieve a good bond and develop the proper barriers in a casing annulus, the cement must be preceded by spacers that displace and remove NADF from the wellbore. (A spacer is a volume of fluid which separates one fluid from another.) Generally this requires a large volume of strong water-wetting agent in the lead spacer fluid.

Drilled solids sizes

Drilled cuttings created by drill bits are usually small. Large chunks of rock which appear on the shale shaker screens usually come from the side of a borehole. The tendency is to think of a well bore wall as being a smooth surface. Experience does not support that concept except when drilling unconsolidated sands and when the balanced activity NADF allows a stable well bore in shales. (Balanced-activity means that the chemical activity, or fugacity, of the NADF matches the activity of the shale and prevents water from leaving the shale or the fluid.) In one instance, several sensors were mounted in a drill collar immediately above a drill bit. The bit was tripped in the hole and then removed without drilling. The sensors indicated that the bit had experienced a 50 g impact during the trip in the hole. The weight indicator did not indicate that the bit experienced such an impact. Clearly, the bit had bounced off of a ledge in the well bore. When running casing, the casing shoe is rounded to allow the casing to get to bottom. If the casing does not have the rounded shoe at the bottom, the flat end will stop on a ledge and the driller will have to "work" the pipe to move it into the hole. Most logging tools have rounded lower shapes to facilitate them getting to the bottom a hole. Indications also exist that the shales in the well frequently have micro-cracks created either by the drill bit, the drill string, or naturally. Casing drilling has found that wells that experienced lost circulation before can be drilled with no lost circulation even though the annular pressure is much higher than conventional drilling strings. The "Smear" effect has been postulated to be the effective cause of the failure to fracture the rock and lose circulation. The casing rubs against the well bore wall, and squeezes solids into the mouth of the small micro-cracks. This seals the crack from fluid entry and prevents fractures from forming. Even though most well bores are considered "stable", a significant number of solids can enter the drilling fluid system from the side of the well bore.

Drilled solid sizes can range from large chunks of rock about 5″–12″ round to very small solids less than one micron in size. One micron is 10^{-4} cm or 0.00003937 in Barite ranges from about 2–75 μm in size. API specifies that no more than 3% weight of barite can be larger than 75 μm and no more than 30% weight can be smaller than 6 μm. For size comparison, talcum powder ranges from about 5–50 μm.

Both the solids which are drilled and those from the side of the well bore are both considered "drilled solids". These solids need to be removed to enjoy the benefits of a clean drilling fluid. The degradation of these solids as they are being transported from the well bore can create large surface areas which must be wetted with the continuous liquid phase of the drilling fluid. To illustrate the large increase in solids surface area, a one inch cube is sliced so that each face is cut in half. This would make eight smaller cubes and the surface area would increase from 6 in^2 to 12 in^2—or double. If each of the eight smaller cubes of rock is split in the same manner, sixty four cubes will be created and the surface area would increase from 12 in^2 to 24 in^2. With only eight cuts, the solids now decrease from a one inch cube to 0.003906 in or 99.2 μm (Table 15.1) and have 1,536 in^2 of surface area.

Table 15.1 The Increase in surface area as a cube of rock is sliced.

No. cuts, n	No. pieces, 8^n	Size (in), $1/(2^n)$	Surface area (in^2), $6(2^n)$	Size (μm)	API screen designation	Size (μm)
0	1	1	6	25,400.05		
1	8	0.5	12	12,700.03		
2	64	0.25	24	6350.01		
3	512	0.125	48	3175.01		
4	4096	0.0625	96	1587.50	12	1700
5	32,768	0.03125	192	793.75	25	710
6	262,144	0.015625	384	396.88	45	355
7	2,097,152	0.007813	768	198.44	80	180
8	16,777,216	0.003906	1536	99.22	200	75
9	1.34E+08	0.001953	3072	49.61	325	45
10	1.07E+09	0.000977	6144	24.80	400	38
11	8.59E+09	0.000488	12,288	12.40	450	32
12	6.87E+10	0.000244	24,576	6.20		
13	5.5E+11	0.000122	49,152	3.10		
14	4.4E+12	6.1E-05	98,304	1.55		
15	3.52E+13	3.05E-05	196,608	0.78		

For comparison of these small sizes, the API Screen Designation for shaker screens in that general size is included in Table 15.1. Of course, the solids created in the well bore are not cubes of rock, but they are subjected to grinding in the annulus in both vertical wells and horizontal wells. If the cuttings are being brought to the surface by very high annular velocities, turbulence will create the opportunity for solids degradation. In horizontal wells, the cuttings also be abraded by the drill string rotating on them at the low side of the bore hole. Additional treatizes on comminution theory are available.[1] The mining industry is interested in rapid degradation of ore, however the drilling industry wants to prevent this.

To illustrate the effect of this increase in wetted-surface area, consider drilling one foot of a $9^7/_8''$ hole in a 20% volume porosity shale.

The volume of rock drilled will be:

$$= (.80)\frac{9.875^2}{1029} = 0.0758 \text{ bbl} = 3.18 \text{ gallons} = 735.5 \text{ in}^3$$

If this volume of rock is considered to be one inch cubes of rock, 735 pieces of rock would start from the bottom of the hole. If each rock splits just 11 times, the surface area of each one inch cube of rock will increase from 6 in^2 to 12,288 in^2. The total surface area for all of the 735 pieces would be 735 times 12,288 or 9,032,000 in^2(An American Football field has 8,242,560 in^2 of area). If a 10 μm film of liquid wets the surface of these small cuttings:

$$\text{Volume} = (9,032,000 \text{ in}^2)(10 \text{ μm})\left(\frac{0.00003937 \text{ in}}{\text{μm}}\right)$$

$$= 3555.8 \text{ in}^3 \text{ or } 15.4 \text{ gallons}$$

If the drilling rate is just 50 ft/h, the free liquid lost from the drilling fluid to wet the new surfaces would be 770 gallons (or 18.3 bbl) every hour. This would double in the cuttings that were ground twelve times.

This is the reason that a "solids problem" may arise in either a water based (WBDF) or non-aqueous drilling fluid (NADF) without an increase in solids content. As the solids are ground into smaller pieces as they traverse the annulus, the total volume of solids does not change but, the surface area requiring absorption of the liquid phase increases greatly. Smaller particles are much more difficult to separate from the surrounding fluid phase. The drag due to the viscosity of the fluid (proportional to surface area) can overwhelm the momentum induced into the particle by the application of an external force designed to push the particle toward one of the two exit streams of a piece of solids control

---

[1] Fuerstenau, Han, editors. Principles of Mineral Processing, SME, 2009.Specifically Chapter 3 by Herbst, Lo, Flintoff. Size Reduction and Liberation, within the same handbook.

equipment. The plastic viscosity of the fluid significantly increases by particle—particle interactions in the fluid simply due to the large number of particles added.

Why control drilled solids?

When drilled solids accumulate in a WBDF, the low-shear-rate viscosity and gel structure of the drilling fluid increases. This is a very visible effect and, consequently, there is a general acceptance that drilled solids affect WBDF properties and are detrimental. NADFs are frequently considered "drilled solids tolerant" because the build-up of drilled solids does not cause an obvious change in the drilling fluid properties in the mud pits. Most of the detrimental effects of drilled solids in NADF are much more subtle but they are still there.

Visible NPT or trouble costs

Lost circulation and stuck pipe are common problems while drilling and many of these situations can be avoided by removal of drilled solids from the drilling fluid. Both of these problems can be related to the filter cake and the particle size distribution in the drilling fluid. If the solids removal equipment is not used properly, drilled solids build in the drilling fluid and can only be reduced by dilution. Dilution increases the waste disposal problem and can be very costly-making it a very visible trouble cost. Drilled solids, specifically sand or large abrasive particles, also create a potential for rapid wear of expendable components in the drilling fluid system.

The most significant trouble created by drilled solids is caused by thick poor filter cakes. They are responsible for poor cement placement, difficulty with log interpretation, excessive surge and swab pressures, formation damage, and stuck pipe. These problems will be described briefly here and discussed in greater detail in other chapters.

Stuck pipe

To examine the effects of stuck pipe, a little review is appropriate about friction forces. If a 200 lb weight is pulled along a surface, the force to pull can be calculated by multiplying the force holding the surfaces together times the coefficient of friction. The coefficient of friction of steel on rock is around 0.3–0.4 (Fig. 15.1).

$$\text{Force to pull} = (\text{Coefficient of friction})(\text{Force to hold surfaces together})$$
$$= (0.3)(200 \text{ lb}) = 60 \text{ lb}$$

In this case, a force of 60 lb would cause the block to slide on the surface. The coefficient of friction between steel and rock is about 0.3 and 0.4. Obviously, if the surface was lubricated with a grease, or slick clay, the coefficient of friction would be lower. For

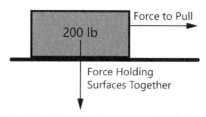

Fig. 15.1 Calculating the force to pull a sled on rock.

example, if the coefficient of friction was reduced to 0.09 with clay, the force to pull the block would be only 18 pounds.

Apply these concepts to a drill collar in a well bore. Most drilling operations attempt to keep about 200 psi over-balance at the bottom of the hole. Generally, it is larger than that. Consider the BHA (bottom hole assembly) touching 100 ft of formation with a filter cake deposited around the collars. The sealed area around the collars would expose about 4 in of the formation to the drill collar (Fig. 15.2).

The force holding the surfaces together (collar and formation) would be the pressure times the area of contact. For an estimate of this area, the 100 ft by 4 in will be considered to be a rectangle.

Force holding surfaces together:

$$= (\text{Pressure})(\text{Area})$$
$$= (200 \text{ lb}/\text{in}^2)(100 \text{ ft} \times 12 \text{ in}/\text{ft} \times 5 \text{ in})$$
$$= 1,200,000 \text{ lb}$$

The force required to pull the drill collar upward would be the weight of the drill collar plus the frictional force. Assuming the coefficient of friction is 0.3,

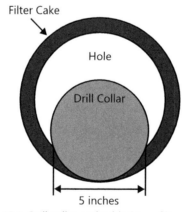

Fig. 15.2 Drill collar embedded in a filter cake.

$$\text{Force to pull} = (0.3)(1,200,000 \text{ lb})$$
$$= 360,000 \text{ lb}$$

This is a very large force and would probably indicate that the drill collar would not come free.

If the filter cake was changed so that the drilled solids were removed to provide a thin slick, compressible filter cake. This would reduce the area of contact to 100 ft by 1 in and decrease the coefficient of friction to 0.08. The force to pull free could now be calculated by the procedure presented above (Fig. 15.3).

Force holding surfaces together:

$$= (\text{Pressure})(\text{Area})$$
$$= (200 \text{ lb/in}^2)(100 \text{ ft} \times 12 \text{ in/ft} \times 1 \text{ in})$$
$$= 240,000 \text{ lb}$$
$$\text{Force to pull} = (0.08)(240,000 \text{ lb})$$
$$= 19,200 \text{ lb}$$

Compare the values:
With the correct filter cake, the force is considerably lower.

$$\text{Force to pull} = (0.09)(240,000 \text{ lb})$$
$$= 19,200 \text{ lb}$$
$$\text{Force to pull} = (0.3)(960,000 \text{ lb})$$
$$= 360,000 \text{ lb}$$

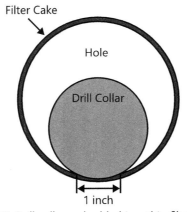

Fig. 15.3 Drill collar embedded in a thin filter cake.

Thin, slick, compressible filter cakes require removal of drilled solids. When drilled solids remain in the drilling fluid, they grind into much smaller solids and absorb much of the free liquid. Even though the fluid loss may be low, the filter cake gets thicker.

Filter cakes in NADF

NADF is reported to be very tolerant of drilled solids. This is usually based on the concept that the yield point does not change significantly in NADF as it does in a WBDF. A change in yield point makes a very visible effect when looking at the fluid in the mud tanks. The addition of drilled solids also changes the plastic viscosity of both WDF and NADF. This change makes not visible change in the appearance of the fluid in the mud tanks. Plastic viscosity is the viscosity that a fluid has at a very high shear rate, such as flowing through a bit nozzle. This viscosity should be as low as possible to allow the fluid to remove the new cuttings from beneath the drill bit teeth as soon as they are made.

When first blended, NADF filter cakes are usually very thin and slick. The filter cake in Fig. 15.4 was created in a high temperature high pressure (HTHP) filtration cell at room temperature and 500 psi differential pressure.

After the NADF was used in three wells, the filter cake formed at room temperature and 500 psi was much thicker (Fig. 15.5).

If the filter cake is thin, slick, and compressible, stuck pipe is not likely. This requires that drilled solids be removed from the drilling fluid. For visualization: consider the difference in pulling force between a sled on ice and a sled on coarse sand paper.

Many think that differential pressure sticking is not possible with NADF even though the filter cakes down hole may look like those shown in Fig. 15.5. In one incidence in Lake Parentis, France with NADF, the pipe was stuck and would not move up or down. The drill pipe was backed off, and a drill stem test tool was run in the hole. The drill stem test tool was screwed into the fish, and a lower mud weight was pumped into the drill

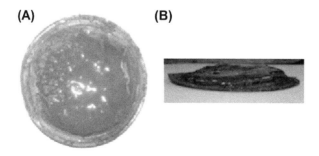

Fig. 15.4 Filter cake for freshly mixed NADF.

Fig. 15.5 Filter cake of NADF being used in a third well.

pipe. The packer was set and the drill pipe opened for flow back. Shortly thereafter, the weight indicator read the full string weight. The driller helped the crew clean the floor in preparation for tripping pipe and did not move the drill pipe. When the floor was clean, the pipe was stuck again. The unbalanced condition was again applied and the full string weight appeared on the weight indicator. This time the driller let the rig crew clean the floor and disconnect the surface equipment while he moved the drill pipe. The fish was recovered without incidence but represented a confirmation of differential pressure sticking with NADF.

Extra note

If the decision is made to double centrifuge NADF to "save the expensive liquid phase", the very small detrimental particles remain with the liquid. Spacer fluids and cement will not displace these thick filter cakes. The cement may bond with the casing but a space will be left between the cement sheath and the well bore wall. If this decision is made, consider a new method of completing the well. About two hundred feet above the production interval, run the production long string into the well and cement. Test the casing, drill out, and run a pressure integrity test (PIT). Verify that a barrier exists between the shoe and the annulus. Then drill the production interval and run a liner. With a rotating liner hanger, pump sufficient cement to fill the annulus plus some extra. Reverse the excess cement out while the drill pipe is still in the hole. Test the cement job with high and low pressure to guarantee a seal. Complete the well with tubing set above the top of the liner. Any cement failure in the production zone will simply move fluid into the production tubing.

Cement placement

If filter cakes are thick, cement will not displace them. After a filter cake is exposed to cement filtrate, it loses some of its ability to seal the formation. This provides a continuous path for flow behind the casing.

Casing should be moved [preferably rotated] while cement is flowing up the annulus. With thick filter cakes, the drag and torque forces are so large that this maybe impossible. Without casing movement, successful cementing barriers are difficult—or even impossible—to develop.

Surge and swab pressure

Thick filter cakes make good seals around a drill bit. As the bit passes through the thick filter cake, it acts like a piston. As the bit moves into the hole, the fluid from beneath the bit must be displaced up the hole. With a thick filter cake sealing or restricting the flow, the pressure beneath bit increase and can cause lost circulation. As the bit is withdrawn from the bottom of the hole, this same seal will cause the bottom hole pressure to decrease which may swab formation fluid into the bore hole. An influx of gas can be dangerous. As the gas expands as it is pumped or migrates up the hole, the heavier weight drilling fluid is displaced. This reduces pressure on the formations which can cause more gas to enter the well bore and create a kick.

Wear

Expendables (like seals, pump parts, swivels, etc.) all wear much faster when impacted with abrasive solids. Large solids act like sandpaper and erode metal.

Disposal costs

Poor solids removal equipment, practices, or plumbing arrangement results in excessive quantities of dilution. This will increase drilling fluid costs and disposal costs. Dilution and the cost consequences of poor solids removal efficiency is discussed in depth in Chapter 5.

Lost circulation

Shale formations frequently have micro-cracks created either by the drill bit, the drill string, or naturally. When a well bore fractures because of pressure, the pressure to open the fracture is the sum of the tectonic stress and the fluid pore pressure. If the drilling fluid can enter the crack and apply this pressure, a fracture can be propagated. The fracture will propagate as long as the pressure reaches the end of the crack. To prevent a fracture from propagating, the fluid must be prevented from entering the small micro-cracks around the well bore wall.

The strength of porous material is a function of the pressure differential or the difference between the confining pressure and the pore pressure. This means when the well bore pressure is much higher than the interstitial pore pressure, the rock is much stronger in compression. However, if fluid can enter the crack, the crack propagation depends on the pressure in the crack exceeding the stress holding the rock together (which is the sum of the pore pressure and tectonic stress). The first mud cleaner test was a great example of this phenomena. The gas sand at 11,000 ft had never been produced and had 11.0 ppg pore pressure. The sands down to 16,000 ft had been produced, some to pore pressures as low as 2.3 ppg equivalent. These sands were drilled with no lost circulation and no stuck pipe because the drilling fluid had virtually no drilled solids. Both the mud cleaner and a centrifuge were used to remove drilled solids. The mud cleaner test was terminated about 200 ft above TD and the rig had problems logging the well and high torque and drag on the BHA.

Invisible NPT or trouble costs
Invisible NPT
Visible non-productive time (NPT) is easily observed. For example, with stuck pipe, the rig is obviously experiencing NPT. Invisible NPT probably cost more than the visible NPT. If a drill bit is drilling in a foundered condition, money is being wasted. One example of invisible NPT could be described in an actual incident. In one field in Texas, a major operator had drilled over one thousand wells. The drilling program was well established. The time to drill, to run casing, and cement was very reproducible. Adjusting for inflation, the cost of the new wells was very predictable. However, in the next well, drill-off tests indicated that the drillers were applying too much bit loading and the bit was floundering. Cuttings were not being removed from beneath the bit. Drill-off tests determined that the bits would drill twice as fast if the bit loading was decreased. In this well, the bits not only drilled twice as fast but lasted three times longer. By eliminating bit trips and drilling faster, this well cost 30% less than predicted from the previous wells. That clearly is invisible NPT and has been an expensive NPT during the past few wells.

Log interpretation
Thick filter cakes may mean that a large quantity of filtrate has entered the formation or the logging sonde is sticking and not moving up the well bore with a uniform velocity. If the formation is filled with filtrate near the well bore, any resistivity measurement must be made through this invaded zone. This may hide the true concentration of fluids in the formation. Some production might be lost.

Formation damage

Thick filter cakes are an indicator of large quantities of filtrate entering the formation. If the filtrate interacts with the formation fluid some insoluble products may be formed and precipitate in the pore space. This will plug the formation. If the damaged region next to the well bore is thicker than the penetration expected from a perforator, production may be permanently diminished. A thin, slick, compressible filter cake would decrease the quantity of fluid entering the formation. This requires eliminating drilled solids from the drilling fluid.

Cementing

The noise log has revealed many, many wells have flow behind casing. Although this is not "drilling NPT", it still is caused by the inability to form a cement barrier next to thick filter cakes. This represents a large expense as hydrocarbons flow into barren formations. The noise log was invented by accident. A researcher was assigned the task of developing a method to locate faults in formations around a well bore. The method would involve creating a shock wave and listening for the reflection from faults in the near vicinity. A small explosive was developed to be used in the well bore for the shock wave. The electronics were developed to provide a receiver to listen for the reflected waves. The electronics were tested in a well that was to be plugged because production had declined and the bottom part of the well was full of sand. When the receiver reached the top of the sand, there was a lot of noise; so much that the researcher thought the electronics had failed because of the pressure or the temperature. The receiver was slowly retrieved and suddenly all of the background noise ceased. The receiver was lowered back down the hole and the noise reappeared. An electric log indicated that the noise ceased next to a very thick water sand. Thus the noise log was invented to indicate flow behind casing. (The researcher was never able to return to the original project.)

Another invisible NPT is created because of thick filter cakes deposited on the well bore wall create drag as the casing is run into the hole. Over thirty years ago, measurements indicated that failure to move the casing while cementing would prevent good barriers from the cement. When a driller has to "work the casing to bottom" because of the drag on the casing, movement while cementing is very unlikely. The driller suffers consequences if the casing gets stuck far off bottom. The casing may be moved while the cement is in transit but is placed on bottom when the cement arrives at the casing shoe. The driller can truthfully answer "yes" to the question: "Did you move the casing while cementing?" The question should have been: "Did you move the casing when the cement was flowing up the annulus?" (This situation has been observed many times in the field.)

Drilling performance

Problems with drilling performance are seldom recognized and difficult to quantify. Good solids control starts at the bottom of the bore hole. Cuttings should be removed from the hole bottom before the next row of cutters reach them and grind them into finer solids. These cuttings then need to be transported to the surface without further grinding or size reduction. If they are tumbled severally in the annulus, the smaller debris increases the colloidal content of the drilling fluid and also makes the drilled solids more difficult to remove.

The hydraulics required to remove the most cuttings can be determined at the rig site. Every drilling rig has a hydraulic limit. This is the maximum pressure that can be produced by the mud pumps and the horsepower available from those pumps. The goal is to have the fluid strike the bottom of the well with the maximum force possible or expend the maximum power possible. As flow rate increases, pressure loss in the drill pipe increases and leaves a smaller pressure available for the bit nozzle pressure losses. Therefore, increasing the flow rate is not the solution. This is addressed in API RP 13D.

After increasing the bit loading to a certain point, suddenly the drilling rate no longer increases when more weight is applied to the bit. Either the bit is vibrating or the bit is starting to grind up cuttings that were not removed from the bottom of the hole. The bit loading which ceases to increase penetration rate is called "founder point".

The founder point of a drill bit was described by some great Shell drilling engineers—Grant Bingham introduced the concept; in the 60s, Rubin Feenstra and Van Leuven published one of the most important graphs and concepts that drillers need to understand. As weight on the bit increases, the drilling rate goes up because the bit teeth are being pushed further into the rock. However, at some point, the drilling fluid ceases to remove all of the cuttings before the next row of cutters arrive. Now the bit teeth must regrind cuttings that were already made. An increase in bit weight does not create the same increase in penetration rate. In some cases, the cuttings can accumulate to significantly decrease the drilling rate at the higher bit weights.

For a roller cone bit, drilling rate depends upon weight on bit and rotary speed until the founder point is reached. As the weight on bit increases, drilling rate increases as a square of the weight on the bit until the bit teeth start regrinding cuttings that were not removed expeditiously. The drilling rate will then cease to increase as a square of the weight on the bit. The founder point depends upon hydraulics—the ability of the drilling fluid to remove cuttings from beneath the drill bit. These tests were initially performed by placing the roller cone bit on bottom with the maximum weight possible and drilling a few inches. The brake was then locked to prevent the traveling block from moving. As the bit drills, the hook load increases. As the hook load increases by 2000 lb, the drill pipe stretches the same amount for each interval. The time interval for the hook load to increase by 2000 lb is measured. With the stretch constant for the

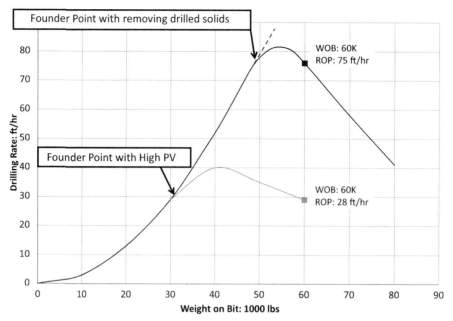

Fig. 15.6 Effect of drilled solids on the founder point of a bit.

drill string, the distance the bit moves can be calculated. The distance moved divided by the time interval provides the drilling rate. One such curve (Fig. 15.6), shows the founder point at about 36,000 lb weight on bit. Cuttings can be removed more effectively if the viscosity of the fluid through the nozzles is lower. In this case, the plastic viscosity was decreased by removing drilled solids and the founder point moved up to 52,000 lb weight on bit. The drilling rate at the founder points increased from 40 ft/h to 80 ft/h. If the bit weight recommended by the bit manufacturer had been used (60,000 lb), the dirty drilling fluid would have been drilling at 28 ft/h and the clean drilling fluid drilling at 75 ft/h.

Plastic viscosity depends upon four things:
1. Liquid Phase Viscosity
2. Size
3. Shape
4. Number of solids

Once the liquid phase has been decided and the drilling fluid created, very little can be changed with the liquid phase viscosity. The size, shape, and number of solids must be controlled by the proper procedures at the rig. Cleaning cuttings from the bottom of the hole without regrinding will provide the largest solids possible entering the annulus. This means not exceeding the founder point. The solids must be brought to the surface as quickly as possible without tumbling and grinding in the annulus. Failure to keep the cuttings moving upward as they are brought to the surface, will result in smaller solids and a larger number. Both of these effects will increase plastic viscosity. After the cuttings arrive at the surface, they should be removed as expediently as possible.

Polycrystalline diamond compact bits

The same effect can be observed with Polycrystalline Diamond Compact (PDC) bits. A PDC bit drill-off test cannot be performed the same way as is done with the roller cone bit. The amount of energy input into the bit is compared to the drilling rate achieved with the bit as the bit weight is slowly increased (Fig. 15.7). Drilling rates with PDC bits do not correlate with weight on bit but rather correlate with torque at the bit. The energy input is the torque times the rotary speed. As the weight on the PDC bit is increased, the ratio of the energy input to the drilling rate remains constant until the bit starts regrinding cuttings. This is the founder point of a PDC bit. It, too, is a function of the hydraulics—or more specifically, the viscosity of the fluid removing the cuttings (the plastic viscosity). Reducing the plastic viscosity allows the PDC bit to drill faster.

Since these drill-off tests are not common on drilling rigs (yet), the drilling rates are frequently much lower than they could be because too much bit weight is applied in accordance with the bit manufacturer's guidelines. If the drilling rig does not have adequate hydraulics to remove the cuttings, the drilling rates are much lower than they could be and the bits are wearing out faster than they should. Eliminate a few bit trips and save much money.

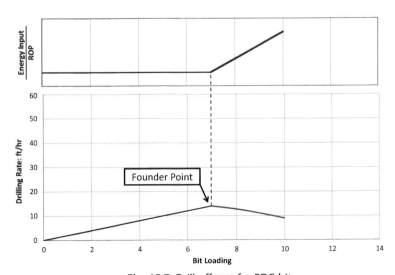

Fig. 15.7 Drill-off test for PDC bit.

Carrying capacity

Good carrying capacity requires the drilling fluid bring drilled solids to the surface. Cuttings transport is a very complex subject and has many very academic, elegant papers describing the interactions of all of the variables. Because these variables are not always known in a well, an empirical solution has been published and accepted in API RP13D. It relies on data available in a normal morning report from the drilling rig to give an indication of whether the cuttings are being transported to the surface without excessive tumbling. The guidelines are based on cuttings having sharp edges when they reach the surface. Three things control the ability of a drilling fluid to transport cuttings: Annular velocity, mud weight, and fluid viscosity in the annulus. Fluid viscosity can be identified with the "K" value for viscosity. The "K" viscosity can be calculated from the plastic viscosity and the yield point (or more accurately, the shear stress at 600 RPM and 300 RPM). This was discussed in Chapter 1.

The K viscosity from the Power Law Model: (Shear Stress = K (Shear Rate)n) is the viscosity of the fluid at one reciprocal second. This viscosity must be large enough to bring cuttings up a vertical hole. Decreasing the plastic viscosity from 15 cp to 5 cp while the yield point remains at 15 lb/100 sq.ft. the K viscosity will increase from 306 eff cp. to 1373 eff cp (Fig. 15.8).

Interestingly, drilled solids are removed to decrease plastic viscosity and decreasing plastic viscosity will bring more solids to the surface (from an increase in K). The system also works the other way. Failure to bring cuttings to the surface while they are large enough to be removed by the equipment will increase the plastic viscosity. Increasing plastic viscosity will decrease the ability to bring cuttings to the surface and allow them to grind into smaller, more numerous particles. This increases the plastic viscosity which decreases the carrying capacity. When this happens, the normal reaction is to

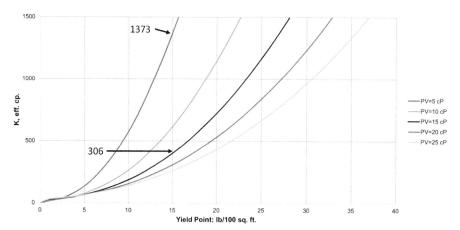

Fig. 15.8 Effect of plastic viscosity on carrying capacity of a fluid.

increase the yield point. When the yield point increases significantly, the shale shaker screens may be too fine to handle the more gelled fluid. The screens will need to be changed to a coarser screen. This decreases the quantity of drilled solids that can be removed which again increases the plastic viscosity. This cyclic change requires careful attention and some drastic action to reverse the cycle. This is a 'solids problem" which arises without always having an increase in total solids content.

Summary

When using water-based drilling fluids many years ago, drillers believed that they could save a significant amount of money by increasing the mud weight with drilled solids until the mud weight reached 10 to 11 ppg. The cost of the barite saved was easily calculated and drilled solids were not considered undesirable. Barite was added only after the drilling fluid properties could not be achieved because of the drilled solids. The savings involved from not buying barite was "obvious". As the mud weight requirements continued to increase because of higher pore pressures, stuck pipe and lost circulation became "normal" events. Finally, most drillers made the connection between trouble costs and allowing the drilled solids to increase. The NPT became very obvious after improved methods to eliminate drilled solids from the water-based drilling fluid were established. The detrimental effects of drilled solids became understood and the NPT costs decreased. Processing NADF has now reached that point in history. Even though the liquid phase is expensive, the very large visible and invisible NPT costs are now being recognized. This chapter is designed to provide guidance for properly processing NADF and that includes mud tank arrangements, and plumbing, as well as removal of drilled solids.

Problems
Problem 15.1

Lost hole Drilling Company is drilling a turnkey wildcat through an old gas field and expects to see several hundred feet of drawn-down sands. The mud spec sheet calls for a very low fluid loss to prevent stuck pipe. The drilling fluid now has a fluid loss of 8 cc/30 min. the "OUT" mud properties were: MW 10.6 ppg, Temp 155°F, FV 46, PV 14, YP 11, Gels 3/12, API 8.0, Solids 10, Ca 12O, Cl 2000, pH 10.0, MBT 20.

The drilling foreman found several holes in the linear motion shaker screen. He noticed that during the past five days, the solids content had increased but the fluid loss was decreasing. The mud properties are now: MW 10.6 ppg, Temp 160°F, FV 45, PV 20, YP 15, Gels 8/25, API5.5, Solids 12, Ca 120, Cl 2500, pH 10.0, MBT 25.

The mud program called for rather expensive filtration products since the bottom hole temperature was now over 300°F. Discussion with a mud engineer convinced him that he could achieve the 3—5 cc/30 min, if he would let the drilled solids increase only a little more. The rationale was that the filter cake needed a range of solids sizes to insure the maximum packing density. By allowing the drilled solids to provide this wide range of solid, all of the holes in the filter cake would be filled with solids. This maximum packing would decrease the fluid loss.

Two days later, the fluid-loss was down to 3 cc/min. The mud properties were: MW 10.6 ppg, Temp 160°F, FV 55, PV 25, YP 20, Gels 10/35, API 4.5, Solids 15, Ca 140, Cl 3000, pH 10.0, MBT 28.

One more day of drilling remained before the well reached the first depleted sand with low pore pressure. He figured that he saved his company several tens of thousands of dollars in chemical costs. He also observed that he did not have to add much barite during the past 4 days. With all of the savings generated by this concept, he was expecting a big raise and promotion. He petitioned the office to allow him to implement this technique throughout the company. It could easily save a lot of barite and chemicals and therefore, a lot of money. His turnkey bids would always be lower than the competition. When he called his boss, what do you think he was told? Did the drilling foreman solve the problem cheaply? Can this technique be used to significantly decrease costs? What should be done, if anything, before that sand is reached?

Problem 15.2

While drilling a $9^7/_8''$ hole, 100 bbl of drilled solids reported to the surface during one interval. The hole was circulated clean before the interval and after drilling. (Note: this would be about 1000 ft of 10″ diameter hole) While drilling this interval, 1000 bbl of clean drilling fluid was added to keep the drilled solids concentration at 3% volume. The discarded drilled solids from the solids control equipment captured in tanks was a volume of 200 bbl.

Find:
1) Solids Removal efficiency of the system
2) How much excess drilling fluid was discarded to keep the pit levels constant,
3) What should be the solids removal efficiency to eliminate generating the excess drilling fluid (i.e. the lowest cost removal efficiency)?

CHAPTER 16

Shale shakers

Introduction

One method of removing solids from drilling fluid is to pass the mud over a screen surface. Particles smaller than the openings in the screen pass through the holes of the screen along with the liquid phase of the mud. Particles too large to pass through the screen are thereby separated from the fluid and discarded. The shale shaker, in various forms, has played a prominent role in the oilfield solids control schemes for several decades. Shakers have evolved from small, relatively simple devices capable of running only very coarse screens to the models of today which are able to sieve the fluid through very fine screens.

Shale shakers are the most important piece of solids removal equipment. In almost all cases, shakers are the most inexpensive and effective means of controlling the very undesirable drilled solids. If the shale shaker is not being used efficiently, the remaining solids control equipment will not perform properly. Care must be taken when selecting a shaker system that will provide the optimal solids removal at the most critical phase of the solids removal system—the flow line. The criticality of removing the solids at the flow line is amplified when solids are allowed to remain in the system and break down over time after multiple circulations and ultimately causing issues with the drilling fluid rheology and filter cake deposition. The effects of solids retained in the system and their effects on viscosity, filter cake, stuck pipe, cementing, and ROP was discussed in Chapter 15.

In this chapter, screens will be discussed first and then the shale shakers which use the screens to separate the undesirable drilled solids from the drilling fluid.

Screens

The shaker should be designed to use the finest screen possible with the least amount of whole mud loss. Selection of a screen is a compromise between solids removal, circulating rate and the life expectancy of a screen. New technology has improved screen life by making screens repairable. Generally, two types of repairable screens are available for most shakers. The first is a perforated metal plate to which one or more layers of fine screen cloth are bonded. The second type is similar to the first except the metal backing plate is corrugated to increase the square feet of available screen area in the same footprint as a flat type screen. The particle size that a shaker will separate from the feed particles is largely determined by the screen opening sizes. The actual separation sizes will be determined by many other factors, including particle shape, fluid viscosity, screen deck angle, shape of vibration, vibration frequency and amplitude, solids and liquid feed rates, and particle cohesiveness. Shale shakers use fine mesh screens to remove solids larger than the opening sizes. Some relatively new screen designations have been introduced by API RP13C and will be discussed before discussing shale shakers.

For many years, screens have been described by "mesh" size. "Mesh" is defined as the number of openings per unit length (Fig. 16.1).

Size of solids moving through the screen depends mesh and on the diameter of the wire. The openings (and solids that pass through) are larger if the wire has a small diameter (Fig. 16.2).

The wire mesh on the left will have a greater flow capacity but will also return more solids to the drilling fluid. Specifying "mesh" does not identify the ability of a shaker screen to remove solids.

The screen's mesh is the number of openings per linear inch, counting from the center of a wire. A mesh count of 50 × 50 indicates a square mesh having 50 openings per inch in both axis directions. A 60 × 40 mesh indicates a rectangular opening having 60 openings per inch in one direction and 40 openings per inch in the other. For a single layer screen cloth with a known mesh count and wire diameter, the opening size can be calculated as follows:

$$D = 25,400\{(1/n) - d\}$$

where

D = opening size in micron
n = mesh count, in number of wires per inch (1 per inch)
d = wire diameter, inches

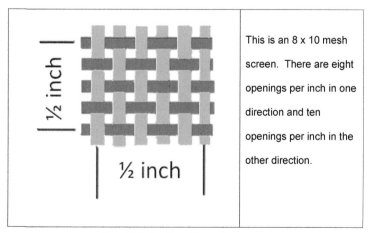

Fig. 16.1 An 8 × 10 mesh screen.

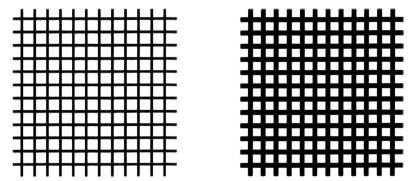

Fig. 16.2 Same mesh but with different wire diameters.

The maximum flow rate of a screen is directly related to the open area that the mud can flow through. For this reason, wire thickness is very important. Two screens with the same opening but different wire thickness will have different flow capacities. The screen with the smaller wire will have a greater flow capacity due to the larger opening area. It will also have a shorter screen life.

Over the past several years, many new screen designs and types have created much confusion in the drilling industry. This is understandable when we consider the variety of screens available. For example, screens come with one, two, three, or, in some cases four layers of cloth, with or without a repairable plate, and with or without bonding material. Layered or "sandwich" screens have contributed to this confusion. The sandwich screen is a combination of two fine mesh screens supported by a calendered backing screen held together in such a way that the cross wires of the bottom screen create an interference in the free openings of the upper screen. This prevents near size or oversize particles from embedding in the top screen. The size separation is thus made by the combination of cross wires from both fine screen surfaces.

Panel screens

The screens in Fig. 16.3 are continuous cloth screens. They are mounted on a shale shaker and tensioned. When the tension decreases because the screen wire stretches or "relaxes", the screen must be retensioned for long life. If the fine screens are not retensioned, they tend to fail in a very short time. When a continuous cloth screen does fail, or rip, the entire screen must be replaced. This became very expensive. Shaker manufacturers started stretching the wire over a panel (Fig. 16.4A), and fixing the wire to the panel. The panel segments (Fig. 16.4B), are relatively small and can be plugged if a wire in the opening does fail.

A close-up of the screens mounted on a panel (Fig. 16.4B), reveals how some of the open area available for sieving drilling fluid is now blocked by the metal frame. The

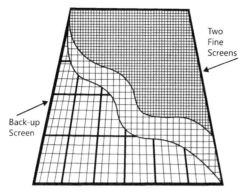

Fig. 16.3 Continuous cloth three layered screen.

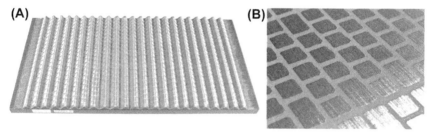

Fig. 16.4 Panel screens.

advantage of this construction, however, is that if a hole does develop in the screen, it can be easily plugged. This reduces the screen area very slightly but makes this screen much more economical than the continuous cloth screen.

Currently, triple layer screens are used which seem to improve flow capacity while removing a large quantity of solids. Even a two layer screen is difficult to describe with the term "mesh" (Fig. 16.5).

If the screens on the left are placed on top of the screen on the right, the resulting screen will have a variety of opening sizes. This cannot be described by a simple

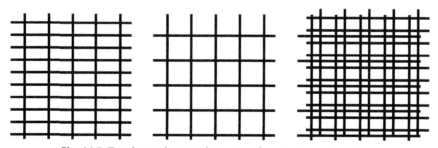

Fig. 16.5 Two layered screen has no uniformity in opening sizes.

"mesh" description. A new system is now offered by the API RP 13C and ISO 13501. This new system was adopted in December 2004. Screens are now compared with standard ASTM screens in their ability to capture aluminum oxide grit. Aluminum oxide grit has a variety of shapes and closely resembles shale cuttings that arrive at the surface in a well bore which has good cuttings transport. The grit is sieved through screens that meet the ASTM E-11 requirements.

When drilled cuttings exit the well, they should be removed as quickly and as efficiently as possible. Shakers are used first to remove particles larger than the openings in the shaker screen. Particles smaller than the openings in the screen pass through the holes of the screen along with the liquid phase of the drilling fluid. Particles too large to pass through the screen are separated from the drilling fluid for disposal. Basically, a screen acts as a "go-no-go" gauge: Either a particle is small enough to pass through the screen opening or it is not. The drilled solids which are removed are not dry, of course. Consequently some drilling fluid is lost with the cuttings.

Screen design

Most shale shaker screens and all fine mesh screens are manufactured with a backing cloth. The backing cloth gives support to the top sizing screen to prevent sagging and premature screen failure. All high quality screens are manufactured with a calendered backing cloth. A calendered backing cloth has all of the wire intersection high points flattened by passing the wire cloth between heavy rollers, giving the material a smooth surface. Backing cloth is calendered to reduce deflection on the fine mesh sizing cloth thus reducing its susceptibility to metal fatigue (Fig. 16.6).

Generally, two types of bonded screens are available. The first is a perforated metal plate to which one or more layers of fine screen cloth are bonded. A second type is a screen or screens bonded to the backing cloth with either adhesive or heat-set plastics. This second type has poorer screen life and has been replaced by the more durable perforated metal backed screens.

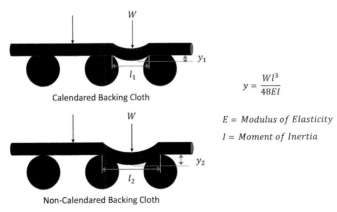

Fig. 16.6 Calendaring.

The perforated metal backed screens exhibit a vastly improved life span, in part because tears are contained and can be easily plugged or repaired with quick-drying adhesives. However, a reduction in throughput is realized because of the blanked-off bonded area. Further, the non-plugging characteristic of the layered screens is partially lost. In summary, bonded screens have resulted in vastly improved screen life, but with a sacrifice in throughput and plugging resistance.

Screening surfaces

Screening surfaces used in solids control equipment are generally made of woven wire screen cloth, in many different sizes and shapes. The screen cloth used on shale shakers has changed significantly during the past several years. At one time, the screens were defined by the mesh size. When all of the screens had square openings and were all made from the same diameter wire, this was a very efficient way of describing screen cloth. When different diameter wires were used, this description failed to indicate what size particles would pass through the screen.

Two screen wires with twenty openings per inch in each direction would be called 20 mesh screens. The performance on a shale shaker, however, would be significantly different. The screen on the left of Fig. 16.2 would be able to handle a larger flow rate of drilling fluid than the screen on the right. The screen on the right would remove more drilled solids than the screen on the left. Clearly, designation by screen mesh would not be descriptive of actual performance here.

The screening industry started making screens where the openings were not the same dimension in each direction. These screens with oblong openings and were designated with one number which was the sum of the openings in each direction. This screen has 20 openings per inch in one direction and 40 openings per inch in the other direction and was labeled as a B60 (Fig. 16.7).

The screen designation then became even more confusing when one screen was placed on top of another screen. The opening sizes were no longer uniform in either direction. If the oblong screen was placed on top of the square mesh screen in drawings

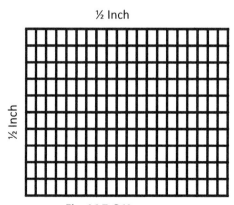

Fig. 16.7 B60 screen.

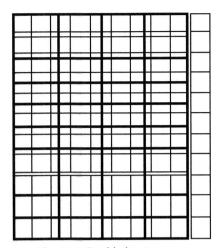

Fig. 16.8 Double layer screen.

above, the opening sizes of this layered screen cannot be described by using the "mesh" concept (Fig. 16.8).

The API formed a committee of experts to address this problem and attempt to describe shaker screens. At first, this committee wanted to develop a performance test that could be used to predict behavior of these screens on a drilling rig. This was soon deemed impractical because too many variables affect performance. The next quest was to provide some method of describing screens that would be capable of providing a fair comparison between different vendors. Finally, the decision was made to describe the largest particle that would be returned to the drilling fluid. A distribution of openings was not definitive enough. The distribution of opening sizes had been used earlier because the curves resembled a "cut-point" curve. However, solids did not select openings which were exactly their size. Small solids pass through the large openings along with the larger solids. After several tests, the committee finally evolved a test method that could give repeatable results in several laboratories. A small amount of aluminum oxide grit in a variety of sizes is placed on a screen sample. The screen sample is shaken with a vibrator for 10 min and the largest particle which passes through the screen is determined. This became the recommendations in API RP13C and is now used widely to compare shaker screens.

Screen weaves

Plain square weave

The plain square weave design is one of the most common weaves available in the petroleum drilling industry and provides a straight through flow path with the same diameter warp and shute wires in an over-and-under pattern. The top and side views of a square weave design is shown in Fig. 16.9.

Fig. 16.9 Plain square weave wirecloth.

Rectangular weave

The rectangular opening design provides maximum percent open area and tends to prevent blinding or clogging of material. The longer openings prevent material build-up and the smaller dimension controls the sizing. The top and side views of a rectangular opening design is shown in Fig. 16.10.

Screen openings

To determine the size of the maximum opening in a screen, it is mounted in a disc and place in a Ro-Tap® with two ASTM standard screens above it and two standard screens below it (Fig. 16.11).

The Ro-Tap® in Fig. 16.11 shows an unknown screen mounted below three standard ASTM screens (70, 80, and 100 mesh) and above two standard ASTM screens (120 and 140 mesh).

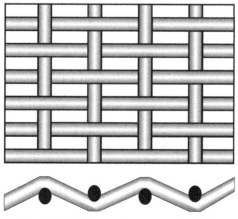

Fig. 16.10 Rectangular weave wirecloth.

Fig. 16.11 Ro-Tap® with unknown screen mounted between three standard screens above and two standard screens below.

Another Ro-Tap® arrangement (Fig. 16.12), shows another arrangement of ASTM sieves mounted with the test screen in the center. Sieves used for this test range from 70 to 140. The screen's opening size is determined by comparing quantity of test particles trapped by test screen with quantities in ASTM sieves above and below test screen.

A known quantity of various size grit is placed on the top screen and the screens shaken for 10 min. By weighing the grit on each screen, the size of the maximum opening can be determined. This is explained in great detail in API RP13C. Several grits were tested by the API committee and the grit which gave reproducible results was aluminum oxide.

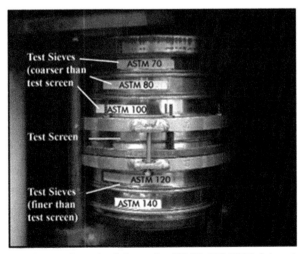

Fig. 16.12 Ro-Tap® stack of sieves for API RP 13C D100 determination.

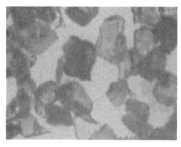 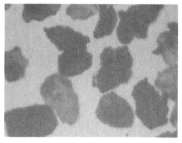

Aluminum Oxide   Shale

Fig. 16.13 Microscope pictures of shale cuttings and aluminum oxide grit with the same magnification.

Under a microscope, the aluminum oxide grit looks almost like shale cuttings which arrive at the surface when the cuttings are being properly transported to the surface (Fig. 16.13).

To sieve the aluminum oxide grit, the ASTM E-11 Wire Cloth Standard was adopted. This lists the opening size for screens in metric units—either millimeters or microns. Since this was going to be an international standard and metric units would be used, the concept of openings per inch could not be used. However, the standard had the "mesh" designation listed as an alternative designation. API RP13C used this number as the "API Number". This meant that the labels on the screens would have familiar units for the rig hands—even though they were no longer "mesh". The only screen sizes will therefore be only those sizes listed in the E-11 Specification. There will be an API 170 or an API 200 screen but there can be no API 175 or API 210 screen if they are labeled according to the procedures listed in API RP13C. Perhaps not surprisingly, several vendors had drastic changes that needed to be made in their labeling. But, the procedure outlined in API RP13C levels the playing field for vendors. This procedure will clearly describe the maximum opening sizes of the screens but it does not, nor is it intended to, predict performance of the screen.

For screen designation, API RP13C also describes the non-blocked area of the screen. Attempts were made to try to identify the open area of a screen that is not blocked by wire. This would be a very difficult problem with the fine screens currently used. Instead, API RP13C recommends reporting the area of the screen which is not blocked with panels or adhesives. All openings in a panel or a screen are measured and summed to provide the area of the total screen available for sieving.

Conductance

API RP13C recommends that the conductance of the screen be included in the screen description. Screen conductance describes the flow capacity of a screen. Conductance is defined as the permeability of the screen divided by the screen thickness. Darcy's Law, Eq. (16.1), is used to determine the permeability of screens.

$$q = \frac{K(\Delta p\, A)}{\mu\, L} \tag{16.1}$$

The constant of proportionality (K) is called the permeability. If the flow rate is measured in cubic centimeters per second, the cross sectional area in square centimeters, the fluid viscosity in centipoise, the pressure differential in atmospheres, and the length in centimeters, the permeability will have the units of Darcy.

Solving Darcy's law for the permeability per unit length, or conductance, C:

$$C = \frac{K}{L} = \frac{\mu\, q}{\Delta p\, A} \tag{16.2}$$

where:
  C is the conductance usually reported in kilodarcys per millimeter,
  K is the permeability of the screen, in kilodarcies,
  L is the thickness of the screen, in millimetres,
  μ is the viscosity of the fluid, centipoises,
  q is the flow rate through the screen, in gpm,
  Δp is the pressure drop across the screen, in psi, and
  A is the area of the screen, inches^2.

Conductance of a screen is determined by measuring the flow rate of a Newtonian fluid with a known viscosity, flowing through a shaker screen, with a measured area perpendicular to the flow, and a known pressure drop.

Motor oil was selected as the fluid to be used for conductance measurements because it was viscous enough to flow at a slow rate and it also contains ingredients which would make the screen oil-wet. The screen wires must be wet with the fluid used in the test. The flow rate through the screen must be laminar to provide a reproducible number. Therefore, the velocity should be maintained below a range of 2 cm per second to 3 cm per second (around 1 inch per second).

The screen is mounted between two plastic cylinders. API RP13C offers three different heights of the cylinders above the screen. The intent is to use the smallest height with very coarse screens and the largest height with very fine screens to help insure laminar flow through the screen.

A large volume of motor oil is place in a large container above the sample screen. The screen is mounted in a cylinder of PVC pipe that has an inside diameter of 5.75 inches and extends one, two, or three inches above the screen. The screen is placed under a container of about 50 gallons of motor oil. The motor oil flows onto the screen and overflows over the edges into overflow containers. The fluid which flows through the screen is captured in a container mounted on an electronic balance. When the weight in the catch pan starts increasing at a uniform rate, the increase in weight can be timed. The volume flow rate can be determined from charts indicating the density of the oil as a function of temperature.

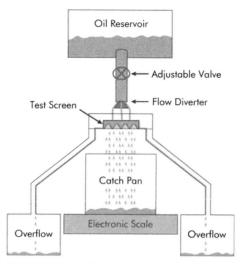

Fig. 16.14 Schematic of equipment used to determine conductance.

When the weight increase in the catch pan becomes constant, the flow rate through the test screen is constant (Fig. 16.14). The overflow height of oil above the rim of the screen is used to determine the pressure difference across the screen.

One laboratory arrangement to measure screen conductance.

The flow rate (Q) of motor oil is measured through a screen mounted between two short sections of 6″ diameter PVC pipe. From previously determined values of the density and viscosity (μ) of the oil as a function of the oil temperature, the height of the oil overflowing the screen determines Δp. The thickness of the screen (L) does not have to be measured. The right side of the above equation calculates the conductance in kilodarcies per millimeter screen thickness.

Screen labeling

Screens that have been tested with the procedure in API RP13C, should have a label that provides the information about the largest opening size (in microns) from the test, the API number, the conductance, and the unblocked screen area. The API information appears in the section on the left side of the label and the manufacturer's part number and other information appears on the right side of the label.

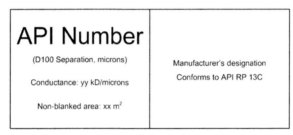

Screen Label—Style #1

The label may also be arranged in a vertical manner. The API information will appear in the top section and the manufacturer's information will appear in the bottom section.

Screen Label—Style #2

ASTM screens are created with a specific wire diameter and opening sizes. The "alternative designation" will be called the "API number". For example, API 170 would have a 90 μm equivalent opening. The specifications on opening sizes provide a large latitude to meet specifications. An API 170 standard screen could have openings from 85 to 95 μm.

Equivalent aperture opening size

Known weights of aluminum oxide grit previously sized on a stack of ASTM screens are placed on the top screen and vibrated for 10 min. The grit weight on the test screen is used to compare with the apertures of known ASTM square mesh screens.

Fig. 16.15 illustrates a cut point determination test in which a screen labeled "XYZ 210" was tested to determine the API D100 cut point. A cumulative weight of 13.40 g was captured on the test screen and was inserted between the ASTM 140 and 170 sieves in a stack of 6 sieves ranging from 100 to 230. This is equivalent to an opening size of 101.1 μm. This cut point is on the finer spectrum of the range for an API 140 screen (>98 to 116.5 μm). This screen would have a designation of API 140 (101.1 μm).

How a shale shaker screens fluid

Shale shakers should remove as many drilled solids and as little drilling fluid as possible. These dual objectives require that cuttings (or drilled solids) move off the screen while simultaneously separating and removing most of the drilling fluid from the cuttings. Frequently, the objective of a shale shaker is stated as "remove the maximum quantity of drilled solids. Stopping a shale shaker is simplest way to remove the largest quantity of drilled solids. Of course, this will also remove most of the drilling fluid. For this reason, shakers should be viewed as having two objectives.

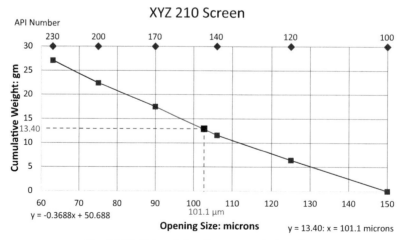

Fig. 16.15 Determining the screen designation.

Cutting sizes greatly influence the quantity of drilling fluid that tends to cling to the solids. As an extreme example, consider a golf-ball size drilled solid coated with drilling fluid. Even with a viscous fluid, the volume of fluid would be very small compared with the volume of the solid. If the solids are sand-sized, the fluid film volume increases as the solids surface area increases. For silt-size or ultra fine solids, the volume of adhered liquid coating the solids may even be larger than the solids volume. More drilling fluid returns to the system when very coarse screens are used than when screens as fine as API 200 are used.

Drilling fluid is a rheologically complex system. At the bottom of the hole, faster drilling is possible if the fluid has a low viscosity. In the annulus, drilled solids are transported better if the fluid has a high viscosity. When the flow stops, a gel structure slowly builds to prevent cuttings or weighting agents from settling. Drilling fluid is usually constructed to perform these functions. This means that the fluid viscosity depends upon the history and the shear within the fluid. Typically, the low-shear-rate viscosities of drilling fluids range from 300 to 400 cP up to 1000 to 1500 cP. As the shear rate (or usually the velocity) increases, drilling fluid viscosity decreases. Even with a low-shear-rate viscosity of 1500 cp, the plastic viscosity (or high-shear-rate viscosity) could be as low as 10 cp.

A significant head would need to be applied to the drilling fluid to force it through the screen.

Imagine pouring honey onto an API 200 screen (Fig. 16.16). Honey at room temperature has a viscosity around 100 to 200 cP. Flow through the screen would be very slow. If the screen is moved rapidly upward through the honey (Fig. 16.17), more fluid would

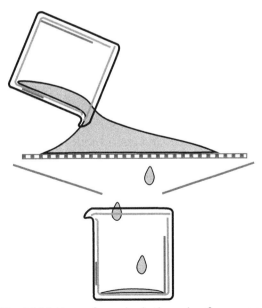

Fig. 16.16 Honey flows slowly through a fine screen.

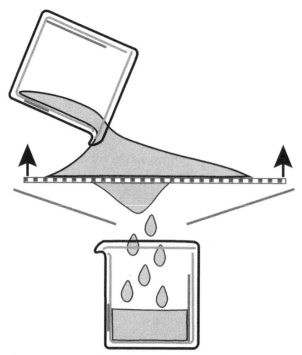

Fig. 16.17 More honey flows through a moving screen.

flow in a given period of time. The introduction of vibration to this process applies upward and downward forces to the honey. The upward stroke moves the screen rapidly through the honey. These same forces of vibration affect drilling fluid in a similar manner. The upward stroke moves drilling fluid through the screen. Solids do not follow the screens on the downward stroke, so they can be propelled from the screen surface.

The upward motion of the shaker screen forces fluid downward through the shaker openings and moves solids upward. When the screen moves on the downward stroke, solids do not follow the screen. They are propelled forward along the screen. This is the reason that the elliptical, circular, and linear motion screens transport solids.

Screens are moved upward through the fluid with the elliptical, the circular, and the linear motion shale shakers. The linear motion shaker has an advantage because solids can be transported out of a pool of liquid and discharged from the system. The pool of liquid creates two advantages: not only does it provide an additional head to the fluid but it also provides inertia or resistance to the fluid as the screen moves upward. This significantly increases the flow capacity of the shaker.

The movement of the shaker screen through the drilling fluid causes the screen to shear the fluid. This decreases the effective viscosity and is an effective component to allow the shaker to process drilling fluid.

The shaker screen movement upward through the fluid is similar to pumping the drilling fluid through the screen opening. If the fluid gels on the screen wires, the effective opening is decreased. This would be the same as pumping drilling fluid through a smaller diameter pipe. With the same head applied, less fluid flows through a smaller pipe in a given period of time than a larger pipe. If a shaker screen becomes water-wet while processing an oil-base drilling fluid, the water ring around the screen opening effectively decreases the opening size available to pass the fluid. This too would reduce the flow capacity of the shaker.

When screens are used on a drilling fluid, a high surface tension film will form on the stainless steel wires on the screen cloth. This film reduces the effective opening to some value smaller than the aperture opening size would indicate. This film effect is increased as drilling fluid viscosity increases. This effect plus piggybacking (where particles smaller than the screen opening adhere to larger particles that are removed by the screen) account for the fact that the screens remove some solid particles smaller than the aperture opening size.

Discussion of API RP13C

The API Task Group that wrote the new API RP13C document wrote a tutorial to describe the intent of the new way to designate screens. That tutorial is duplicated here to explain the procedure. When the document was first published, much confusion existed because some companies were trying to use this designation process to predict screen performance. The API RP13C simply provides a method of measuring the largest particles which will pass through the screen. A super fine screen (one with 600 openings per inch in each direction) would remove many solids unless there were two openings that were two inches by two inches. A plot of the opening distribution would make this screen look superior to many coarser screens but the particles do not seek to go through holes which are the size of the particle. The new API RP13C document provides a method of comparing screen openings and does not indicate anything about performance. The following section is the tutorial that the API task group published to explain the new concepts.

API RP13C tutorial

Summary

API has revised the shale shaker screen testing procedures and numbering convention. By using the new API Screen Number, confusion among screen types is reduced and comparison between screen types can made fairer. Some screens which may previously have been named "200 mesh" may now have an API Screen Number of only 100 to 140. However, ALL screens which are tested according to RPI3C and have the same API Screen Number will remove solids of a similar size.

Note: The change from using D_{50}–D_{100} (50% of a specific particle size removed vs. 100%) will change the rating of most screens—the extent of the change being dependent upon screen type/design.

The new number describes the size at which particles will be rejected (removed) under laboratory test conditions. The new API number is NOT intended to describe how the screen (or indeed the shaker) will operate in the field. This will depend upon several other parameters such as fluid type & properties, shaker design, operating parameters, ROP, bit type, etc.

The greatest value of the new numbering system is that ALL conforming screens are measured using the same process which will allow cross-comparison of screen designs/types based on a uniform solids size removal value.

Any manufacturer labeling their screens as "conforming to API RP13C" must supply the test data for that screen upon request to the end user/purchaser

Definitions
Mesh
Mesh, as it relates to a piece of woven wire cloth, is a measure of the number of holes in a linear inch (such as 100 mesh) or in a linear inch in each direction (such as 100 × 60 mesh).

D_{50} cut point
The D_{50} cut point of a screen is the particle size at which half of those particles reporting to the screen will pass through the screen and half will be retained.

D_{100} cut point
The D_{100} cut point of a screen is the particle size at which no particles that size or larger will pass through the screen.

API screen number
The API Screen Number is determined using a specific test procedure (as set out in API RP13C). The test uses a specifically graded sample of aluminum oxide which is passed through a stack of sieves—including the test sieve—and mounted on a Tyler Ro-Tap®.

The method determines the D_{100} cut point of the test screen and relates it to the D_{100} of an "equivalent" standard ASTM test sieve.

Screen conductance
Conductance, measured in kilodarcies per millimeter (kD/mm), defines a Newtonian fluid's ability to flow through a unit area of screen in a laminar flow regime under prescribed test conditions. All other factors being equal the screen with the higher conductance number should process more flow.

Non-blanked area

The non-blanked area of a screen describes the net unblocked area in square feet (ft^2) or square meters (m^2) available to permit the passage of fluid.

Background

The use of the term "mesh" (when considering the capabilities of shake screens) was made obsolete by the introduction of oblong mesh and multi-layer screens which resulted in variations in "aperture sizes".

The following photomicrographs (60×) show four different screens from four different manufacturers (Fig. 16.18).

Rig-based personnel continued to rely on the "mesh number" indicated by the manufacturer. This may have borne little relationship to the actual separation potential of the screen being used. Hence, comparison of screens from different manufacturers (or even across one manufacturer's series) could be difficult or inaccurate.

In December 2004, the API passed RP13C—titled "Recommended Practice on Drilling Fluids Processing Systems Evaluation". The practice combined and updated two previous separate documents RP13C and RP13E. This document was also passed by ISO as ISO 13501.

RP13C covers a number of subjects relating to fluids treatment systems. However, this tutorial only addresses screen testing (cup point and conductance), classification and labeling.

API screen number determination

RP13C describes a method to define and compare the absolute (or D_{100}) separation potential of any shale shaker screen to an equivalent standard ASTM test sieve. A representative section of the screen is mounted in a holder (see below) and is placed in the middle

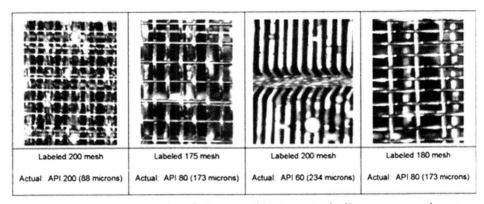

Fig. 16.18 Demonstration of confusion caused by "non-standard" screen nomenclature.

Fig. 16.19 Test screen mounted in holder and Ro Tap® with test screen and sieve stack.

of a stack of ASTM test sieves (calibrated according to ASTM E-11). Using a Ro-Tap® (see below) a defined amount of dry aluminum oxide is sieved and the results collated and graphed. The test is repeated three times and the results from each test are then averaged to determine the D_{100} cut point (Fig. 16.19).

This cut point is the *absolute* or D_{100} cut point. Any particle larger than this value will not pass through the screen. The cut point is cross-referred to Table 5 in the RP13C document (Table 16.1), and compared to standard ASTM sieves of know separation ability.

Table 16.1 D_{100} cup point and API screen number.

D_{100} (μ)	API screen number
>462.5–550.0	API 35
>390.0–462.5	API 40
>327.5–390.0	API 45
>275.0–327.5	API 50
>231.0–275.0	API 60
>196.0–231.0	API 70
>165.0–196.0	API 80
>137.5–165.0	API 100
>116.5–137.5	API 120
>98.0–116.5	API 140
>82.5–98.0	API 170
>69.0–82.5	API 200
>58.0–69.0	API 230
>49.0–58.0	API 270
>41.5–49.0	API 325
>35.0–41.5	API 400
>28.5–35.0	API 450

Important Note: RP123C states that this test describes the openings of the screen and does not predict the performance of the screen in the field. However, if all other variables are equal, a screen with a higher API Screen Number (smaller holes) should remove more and finer solids.

For example, if the measured D_{100} cut point of the test screen is 114.88 μm (114.88 μ), the table indicates that it compares to the ASTM 140 sieve. The test screen would then be classified as an API 140 screen.

Comparison with RP13E

The predecessor to the new PR13C was called RP13E. This (now obsolete) procedure used a light source, a microscope and a computer system to measure the size and distribution of apertures within a screen. Values for D_{16}, D_{50} and D_{84} were calculated from the results.

The D_{50} was called the "mean" value and was generally considered the primary cut point used to "name" the screen. Fig. 16.20 illustrates how screens classified by their D_{50} (75 μ) could have very different D_{100} values. The graph also shows a comparison with standard ASTM 200 sieve.

Fig. 16.21 shows a significant difference between the D_{50} (RP13E) and the D_{100} used in the new procedure.

The change from RP13E to RP13C is a positive step because the new procedure moves away from measuring openings in the screen to an actual physical test using real solids.

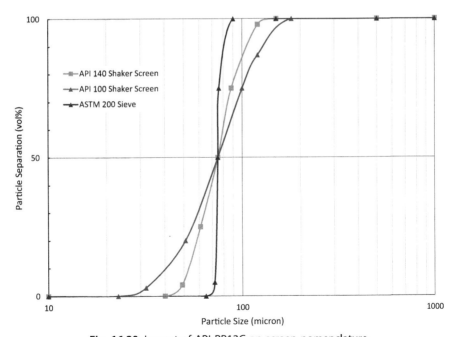

Fig. 16.20 Impact of API RP13C on screen nomenclature.

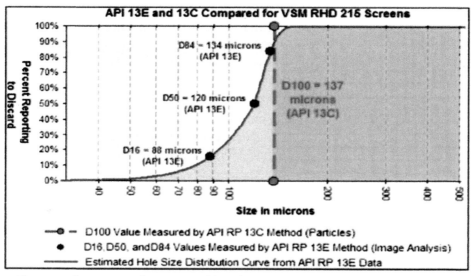

Fig. 16.21 Comparison of API RP13E (old) and API RP13C (new). *(Courtesy of Brandt NOV.)*

One further advantage of RP13C is that any screen with any aperture shape can be tested using the same procedures.

Rig site performance

RP13C does not predict rig site performance given the myriad combinations of screens, shakers, fluids, flow-rate, solids loadings, etc. Performance will depend upon various factors including the properties of the fluid, the operational parameters of the shaker and the particle size distribution of the drilled solids presented to the screens.

Labeling requirements

RP13C states that the designation system (labeling) will consist of no fewer than the following minimum elements.
1. API screen designation or API number (this must be 2× larger than any other information);
2. D_{100} (Equivalent aperture) in microns (μ);
3. Conductance in kilodarcies per millimeter (kD/mm);
4. Non-blanked screen area in square meters (m^2) or square feet (ft^2);

5. Manufacturer's designation and/or part number (although not currently required to confirm to API13C, API recommends manufacturers use the API screen designation in the part number)
6. Conforms to API 13C

The following information is optional (but recommended);

1. Manufacturer's name
2. Application or description
3. Country of origin
4. Lot number
5. Date of manufacture
6. Order number
7. Bar code

The label/tag must be permanently attached to the screen in a visible place. Examples are shown in below.

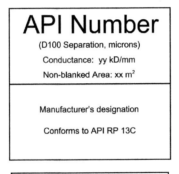

Sample #1—Vertical labeling format and example.

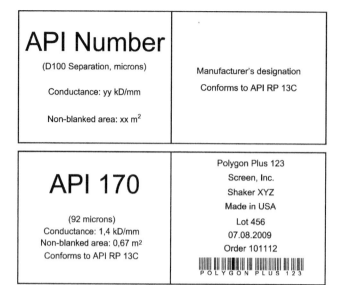

Sample #2—Horizontal labeling format and examples.

Putting the label to use

Because screens that conform to RP13C have all been tested using the same procedure, the labels are very helpful when it comes to comparing different screens.

For example, if it is determined that there is a need for the cut point provided by API 170 screen then a screen labeled API 170 can be selected regardless of the manufacturer of the screen.

Alternatively, if there is excess shaker capacity but longer screen life is desired, selecting a screen with larger diameter wires and perhaps more bonding material should provide increased screen life.

Using screens conforming to RP13C can help the operator make a more informed decision.

Conclusion

By specifying screens that conform to RP13C, much of the previous confusion can be eliminated and screen selection/comparison simplified.

Frequently asked questions regarding RP13C

1. **What is the D_{100}?**
   a. The D_{100} cut point of a screen is the particle size at which no particles will pass through the screen.
   b. D_{100} is a single number determined from a prescribed laboratory procedure—the results of the procedure should yield the same value for any given screen.
   c. The D_{100} should not be compared in any way to the D_{100} value used in RP13E.
   d. The method used in RP13C provides only the D_{100} value with no provision for D_{16}, D_{50}, or D_{84} values.
2. **What does the API Screen Number tell us?**
   a. The API Screen Number corresponds to the API defined range of sizes into which the D_{100} value falls.
3. **What does the API Screen Number NOT tell us?**
   a. The API Screen Number is a single number which defines solids separation potential under specific test conditions. It does NOT define how a screen will operate on a shaker in the field as this will depend upon several other parameters such as fluid type & properties, shaker design, operating parameters, ROP, bit type, etc.
4. **Will a screen with a D_{100} of 172μ remove solids finer than 172μ?**
   a. By definition, every particle of the D_{100} size or coarser will be discarded.
   b. In practice, some portion of solids finer than 172μ may be removed.
5. **Why is there such a large variance between the old and new labeling values?**
   a. The biggest variation is due to the shift from using D_{50} to D_{100}.
   b. The previous test procedure (RP13E) measured the distribution of the apertures in the screen. RP13C uses a physical and repeatable test using actual solids (dry aluminum oxide) which measures the coarsest particle that can pass through the screen.
6. **Should I use the old screen number or the new API Screen Number when ordering replacement screens?**
   a. Although some companies are changing their part numbers to reflect their conformance to RP13C, others are not. It is therefore best to specify the RP13C value you want.
7. **Should I screen with a finer screen now if what used to be called a "200 mesh" screen is now labeled API 120?**
   a. Users are advised to determine the screens that work best in their applications regardless of what the new API Screen Number happens to be.
8. **What is the practical value of RPC13C to the end user?**
   a. RP13C provides an unequivocal procedure and benchmark for comparing different screens.
   b. The primary intent of RP13C is to provide a standard measuring system for screens.

Screen types and tensioning systems

Most screens on the market are flat. They are either mounted on a panel or tensioned on the shaker. Failure to tension screens properly usually results in premature failure. The screens need to be re-tensioned frequently during the first two days of use. Panel screens are pre-tensioned and fastened to the frame. This allows longer use of the screen because if one small opening screen wire tears or wears out, a plug can be placed in the opening and the screen use continued. Other screens are made with a three dimensional form to provide more screening surface area. The three dimensional screens provide additional non-blanked area in the same footprint as a flat panel screen (Fig. 16.22).

Screen types

1. **Pre-Tensioned Screen Panels (supported):** The term pre-tensioned refers to the fact that the screen cloth is tensioned across a support structure and no further tensioning of the screen cloth is required to prevent the wire cloth to move under vibration. Pre-tensioned screen panels consist of either a single layer or a multitude of layers of screen cloth that are bonded to each other and a perforated rigid structure that has openings to allow fluid to pass through. The shapes of the openings in the perforated plate vary between manufacturers and there is a compromise between the size of the opening, the API cut point of the screen cloth arrangement, and the backing cloth wire diameter. The larger the opening in the perforated plate, more fluid will be allowed to pass through (increase in conductance), however, the larger the opening, the more the wire cloth is allowed to flex under vibration subjecting it to fatigue and ultimately failure. This system of perforated plate and wire cloth is attached to a support frame made from either metal or a composite type material.

2. **Pre-Tensioned Screen Panels (unsupported):** Similar to the pre-tensioned screen panels (supported), the unsupported pre-tensioned screen panels consist of either a single layer or a multitude of layers of screen cloth that are bonded to each other and a perforated rigid structure that has openings to allow fluid and undersized solids to pass through, however in this case, there is no additional structure required (metal or composite frame) to support the system under vibration. The shaker design provides the support structure for this type of screen system.

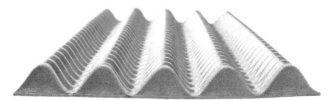

Fig. 16.22 Three dimensional shaker screen.

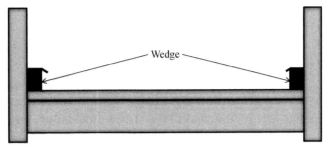

Fig. 16.23 Wedge screen tensioning system (front view of shaker).

3. **Non-Pre-Tensioned Screens:** This type of screen system consists of either a single layer or a multitude of layers of screen cloth that remain un-tensioned until they are placed on the shaker. Also known as a hook strip type of screen, this type of system is difficult to repair and is not used on most shaker systems today.

Screen Retention systems:
1. **Wedge Retention System:** The wedge system illustrated in Fig. 16.23 consists of a pre-tensioned screen panel that is held in place on a flat shaker deck with two wedges, one on each side of the screen. Typically, the wedge is hammered into place with a large hammer, however there are current shakers on the market with alternative methods for securing the wedge in place such as a notched wedge system with pry bar and a cam lock style system. With this type of tensioning system, a significant support structure is required for the screen panel to prevent it from moving independent of the shaker bed under vibration depending on the "g" factor of the shaker. The higher the "g" factor, the thicker the screen frame required. Through the mechanical advantage of the wedge being hammered in place, a large amount of force is applied to the screen being held in place preventing it from moving under vibration.
2. **Overslung Crowned Deck Screen Retention System:** The overslung crowned deck system illustrated in Fig. 16.24 consists of a crowned shaker deck design in which a screen or layers of screen bonded typically to a supporting perforated plate is stretched over the crowned deck to create tension along the entire span of the screen system. There are two basic methods for pulling the screen over the crowned shaker deck. The first method involves the use of a draw bar with holes for bolts in which the

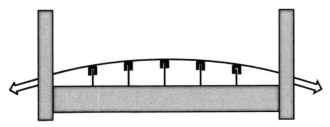

Fig. 16.24 Overslung crowned deck screen tensioning system (front view of shaker).

bolt is placed through the sidewall and drawbar and the drawbar grabs the screen and pulls it toward the sidewall as the bolt is tightened. The other method for tensioning a screen in an overslung deck system is the use of fingers to grab the screen from underneath and pull it toward the sidewall through the mechanical action of a ramp action bolt.

3. **Air Bladder Screen Retention System:** Similar to the wedge system, the air bladder tensioning system illustrated in Fig. 16.25 provides pressure to the edges of a pre-tensioned screen surface that is bonded to a support frame. The air bladder system requires consistent positive air pressure typically in the 80-90psi range to hold the screen panel in place.

4. **Compression Screen Retention System:** A new method for screen tensioning has been developed in which the screen panel is compressed in place through the use of pins located on the sidewall(s) of the shaker. With this tensioning system illustrated in Fig. 16.26, the panel is pressed into place either with a manual or pneumatic compression type system.

Gumbo and water-wet solids problems

Screening gumbo and water-wet solids poses major problems for screening devices such as shale shakers. Gumbo in a water-based drilling fluid tends to stick to the screen and is difficult to convey off the screen. Gumbo will not convey out of a pool of liquid and

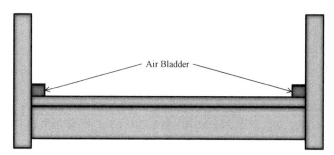

Fig. 16.25 Air bladder screen tensioning system (front view of shaker).

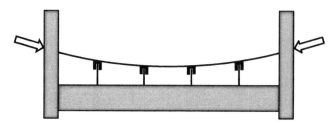

Fig. 16.26 Compression screen tensioning system (front view of shaker).

most "high performance" screening machines will have to be tilted down to improve the conveyance. The shaker should be cleaned and initially tilted downhill to 5 degrees; flow should then be fed to the unit. The deck angle should be tilted downhill until the gumbo moves continuously from the feed and discharges over the end of the unit. A change in API screen size may be necessary. Some rigs have devices called Gumbo Busters built into their flow lines to remove the gumbo. To minimize solids loading and gumbo effects, a "scalping" shale shaker may also be used. The scalping shaker does not need to be a high performance shaker and should be run with API 10 to 35 screens. Running any finer screens on the scalping shaker will usually require more than one scalping shaker to meet screen area demands for fluid throughput, and may actually hinder the performance of the fine screen linear motion shakers below them. No shaker removes gumbo well, but to obtain reasonable results the bed has to be tilted downward. This is true for machines using any type of motion (linear, circular, elliptical). A similar phenomenon can be experienced with water-wet solids in a non-aqueous fluid (NAF). When the electrical stability of a NAF, and the emulsion of water in oil and the oil-water ratio is high, then any clay present in the drilling fluid can become hydrated by the un-emulsified water and ultimately become sticky. When this happens, the sticky solids tend to stick to the shaker screens causing them to blind and deteriorate prematurely.

Blinding and plugging of screens

The blinding and plugging of shaker screens can have a significant effect on the performance of a shaker. There are several factors that come into play that can cause a screen to plug or blind. Blinding refers the inability for a shaker screen to "see" the process fluid. It typically occurs when a polymer has been introduced to a mud system that has not sheared entirely and the conglomerate sticks to a screen surface and "blinds" it off. Plugging (Fig. 16.27), on the other hand, refers to the solids in the mud becoming wedged either inside the wire cloth openings or between the various layers of a multi-layered screen. Near sized particles are defined by Taggert[1] as particles ranging in size from 25% smaller to 50% larger than the pore openings in the screen.[2] When a significant (>60%) amount of the particles in the feed stream particle size distribution fall within this range, plugging of the screens occurs causing a large amount of fluid to flow off the end of the shaker (Fig. 16.28).

---

[1] Taggert AF. Handbook of mineral dressing-ores and industrial minerals. New York City: John Wiley & Sons Inc., 1953. p. 7–72.
[2] Manohar L, Hoberock LL. Solids-conveyance dynamics and shaker performance. SPE Drilling Engineering December 1988;385.

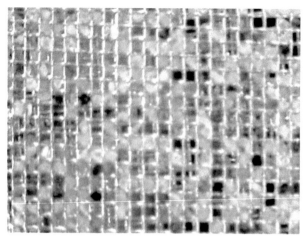

Fig. 16.27 Example of near sized particle plugging.

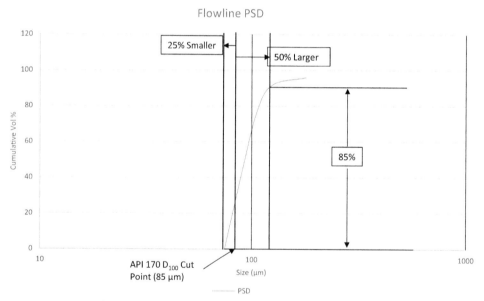

Fig. 16.28 Example of near sized particle plugging with API 170 screen.

Shale shaker history

The first shale shakers did not actually "shake". A wire-mesh drum was placed in the flow line so that the flow of fluid rotated the drum. Liquid easily passed through the large openings and very large drilled cuttings were rejected from the system (Figs. 16. 29–16.31).

Shale shakers 375

Fig. 16.29 Linda "*K*" rotary shale extractor.

Fig. 16.30 Showing tailing blades.

Fig. 16.31 Hudson-Boucher automatic shale separator.

On the job photograph of a Thompson shale separator and sample machine (Note the full flow of the reconditioned mud pouring back into the pit. The Thompson separator is considered standard equipment by many leading oil companies).

Shale shaker design

There are several various designs and types of shakers that have been employed over the years and the consistent theme across all of the shaker designs is the fact that they are vibrating due to a force being imparted on a structure at a certain frequency causing the vibrating structure to behave in a particular manner. The primary factors incorporated in the design of a shale shaker are the type of motion, the stroke length, and the frequency (RPM). The shape and axial direction of the vibration motion along the deck is controlled by the position and orientation of the vibrator(s) relative to the deck and the direction of rotation of the vibrators.

Motion types

If a single vibrator is mounted close to the screens and center of gravity, the motion is circular, as shown in Fig. 16.32B. Cuttings travel direction and speed on a several factors to include horizontal deck depends on the direction of rotation, the frequency of vibration, and the amplitude of motion. Amplitude of motion is the distance from the mean position of the motion to the point of maximum displacement. For a circular motion, the amplitude is the radius of a point on the screen deck side. The stroke is the total movement or twice the amplitude.

If a vibrator is mounted above the deck, the motion is elliptical at the ends of the deck and circular below the vibrator (Fig. 16.32C). The rate of travel of the solids

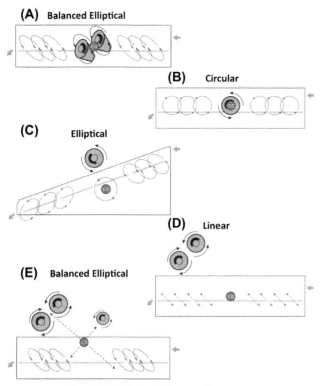

Fig. 16.32 Shale shaker motion types.

(conveyance) is controlled by the axis of the ellipse, slope of the screens, and direction of rotation. Shale shakers that use an elliptical motion usually have divided decks with screens placed at different slopes in order to provide proper discharge of cuttings, or the screen deck is tilted in a downhill position.

There are two methods for achieving a balanced elliptical type motion on a shaker as illustrated in Fig. 16.32A and E. The first method involves orienting two motors on the side of the basket to create the elliptical motion that is "balanced" along the entire length of the shaker deck as illustrated in Fig. 16.32A. Another method for achieving elliptical motion is to include a third smaller motor added to a linear motion machine that is perpendicular to the center of gravity and when energized, the third motor creates the elliptical motion along the length of the shaker basket as illustrated in Fig. 16.32E.

Straight line or linear motion is also used, as in Fig. 16.32D. The rate of travel of cuttings depends on the slope of the motion axis (vibration angle), the slope of the screens (screen deck angle), the length of stroke (twice the amplitude), and the vibration frequency (RPM of the motors). Linear motion machines can be run uphill, which allows greater deck coverage by the fluid, a deeper pool depth that increases the hydraulic head and pressure differential across the screening surface, and provides a higher solids load capacity due to the increased efficiency of conveyance. Usually vibrator mounting position and the vibrator motion provided is more complex than the examples shown. For any linear motion shale shaker with dual vibrating motors, it is critical to assure the proper direction of rotation (motors turning in opposite direction from each other), and to ensure that the unit is mounted level. The stroke length for a given design depends on the eccentric weights employed and the weight of the vibrating structure or screen frame. The greater the eccentricity of the weights relative to the live screen frame weight, the greater the thrust. It is important to note that the horsepower rating of the vibrator motors is not related to the performance of the shaker. In other words, an increase the horsepower rating of a vibrator motor while the eccentric counterweights remain the same as well as the vibrating screen frame weight, there will be no change in the acceleration of the screen (or G-force) or ultimately in the performance of the shaker. The horsepower rating of the vibrator motors is solely related to the internal reactions within the motor itself and is required to turn that motor/shaft/eccentric weights at a particular speed.

Scalping shakers

Several important factors control whether a shale shaker/scalping shaker can effectively remove gumbo. If the screen deck can be tilted downward (Fig. 16.33), with an articulated deck, gravity will assist gumbo removal. If the older style unbalanced elliptical motion shaker is used, the fixed downward angle will usually satisfactorily convey gumbo. If the deck angle is flat, like most circular motion machines, the shaker has to generate a sufficient negative, or downward, force vector, normal to screen, to overcome the

Fig. 16.33 Unbalanced elliptical shaker.

adhesion factor (or stickiness) of the gumbo so that the screen separates from the solids. If it doesn't separate, the gumbo is effectively glued to that spot on the screen.

The industry then selected a shaker from the mining industry. The vibrator was located above the screen and provided an elliptical motion at the ends of the shaker screen and a circular motion near the middle of the screen. The fluid at the discharge end was rotated back toward the feed end of the screen. Advertisements touted this feature as a method of better sieving the material. The screen had to be tilted downward to cause the solids to be discarded. Even then, when drilling a very sticky clay, the clay would roll back up the screen and gather more material. Usually it had to be removed manually.

If the vibrator is placed at the center of gravity, the screen motion is circular (Fig. 16.34). These screens required less surface area and most of the shakers were double-deck shakers. The intent was to place a coarse screen on the top deck to spread the fluid across the fine screen on the lower deck. Unfortunately, it was difficult to see when a hole appeared in the lower deck. These shakers could handle screens with openings around 177 μm but no smaller.

Scalping shakers are generally adequate for top hole drilling and for shallow holes when used with other solids control equipment, such as hydrocyclones. For deeper holes and when using weighted drilling fluids or an expensive liquid phase, a scalping shaker might be used with fine screen shakers. Tests have indicated that when processing drilling fluid through a scalping shaker in front of a fine screen shaker, a very coarse mesh screen (such as an API 10 TO API 20) should be used on the scalping shaker to remove the largest quantity of drilled solids.

To prevent damage to fine screens mounted on the linear or balanced elliptical motion shakers, a very coarse screen is mounted on a shaker to treat the drilling fluid as soon as it returns to the surface. This is a good application for the circular motion or the unbalanced elliptical motion shakers. Tests have indicated that more solids are removed

Fig. 16.34 Circular motion shaker.

from the system if a very coarse screen is mounted on the scalping shaker before the fluid is processed through the fine screens on the main shaker. The main shaker would be a linear motion or a balanced elliptical motion shaker. If a fine screen is mounted on the scalping shaker, some of the solids seem to deteriorate because of the multiple impacts. The purpose of the scalping shaker is to remove the very large solids which frequently arrive at the surface. These solids usually come from the borehole wall and can be very large. They need to be removed to prevent damage to the fine screens which should be mounted on the main shakers (linear motion or balanced elliptical motion).

When linear motion shale shakers were introduced to oil well drilling operations, drilling fluid could routinely be sieved through API 200 screens for the first time. This goal was desirable because it allows the removal of drilled solids sizes down to the top of the size range for barite which met API specifications. Circular motion and unbalanced elliptical motion shale shakers were usually limited to screens in the range of an API 80. Drillers soon found, however, that gumbo could not be conveyed up hill on a linear motion shale shaker. The material adhered to the screen. To prevent this, the circular and unbalanced elliptical motion shakers were used as scalping shakers if the gumbo buster was not available.

The "rig" shakers remained attached to the flow line to remove very large cuttings and gumbo. These were called scalping shakers. Even in places where gumbo might not be present, scalpers were used to prevent very large cuttings or large chunks of shale from damaging the API 200 + screens. These screens have finer wires and are much more fragile than wire used on an API 10 or API 12 screen. The scalping shakers also had the advantage of removing some of the larger solids that would enter the mud tanks when a hole appeared in the fine screen.

Tests indicated that fitting the scalping shakers with the finest screen possible, an API 80 or API 100, did not result in the removal of more solids when combined with the API 200 on the main shaker. Apparently, shale in that size range would break apart on the scalping shakers into pieces smaller than 74 μm or it would damage the cuttings enough so that the linear motion shaker screen broke the cuttings. This action results in fewer total solids rejected from the system. Use the scalping shakers as an insurance package to prevent very large cuttings from hitting the fine screens. The scalpers, even with an API 20 screen, will still convey gumbo.

Scalping shakers should be used with either the linear motion or the balanced elliptical motion shale shakers. Gumbo must be removed before any screen can convey drilled solids uphill out of a pool of liquid. These motions may be used on shakers to remove gumbo if the screen slopes downward from the back tank to the discharge end of the shaker.

Combination shakers are available that mount a downward sloping screen on a linear motion shaker above an upward sloping screen with a linear motion. Gumbo will move down the top screen and be removed before the fluid arrives at the lower screen. Again, the scalping screen should be a very coarse screen.

Several important factors control whether a shale shaker/scalping shaker to effectively remove gumbo. If the screen deck can be tilted downward, with an articulated deck, gravity will assist gumbo removal. If the older style unbalanced elliptical motion shaker is used, the fixed downward angle will usually satisfactorily convey gumbo. If the deck angle is flat, like most circular motion machines, the shaker has to generate a sufficient negative, or downward, force vector, normal to screen, to overcome the adhesion factor (or stickiness) of the gumbo so that the screen separates from the solids. If it doesn't separate, the gumbo is effectively glued to that spot on the screen.

Some designs use dual screens, dual decks and dual units in parallel to provide more efficient solids separation and greater throughput. Depending on the particular unit and screen openings used, capacity of scalping shakers can vary from 100−1600 gpm or more. Screen sizes commonly used with scalping shakers range from API10 to API80.

Scalping shakers normally require minimal maintenance. Other than periodic greasing, the following check list should be implemented while making a trip:
- Wash down screens.
- Check screens for proper tension.
- Shut down shaker when not drilling in order to extend screen life.
- Dump and clean possum belly. But, do not empty the possum belly into the active system.
- Clean the tension rails.
- Inspect rubber screen supports for wear.
- Most important—Install replacement screens properly, square on the deck, with even tension according to the manufacturer's recommendations.

Fine screen shakers

When the linear motion shaker was introduced to the industry, a great change occurred with processing drilling fluid. Removal of the large solids had a great impact on creating thin, slick filter cakes. Now, the balanced elliptical shakers and the linear motion shakers can process drilling fluid through screens as fine as API 325 in some cases. API barite specifications allow 3 wt% of the barite to be larger than 75 μm (or API200). Usually, this is the finest screen typically used on these shakers.

Linear motion on the screen was created by placing two vibrators oriented at an angle to the screen. The vibrators would reinforce the motion up or down but cancel each other as they moved toward or away from each other (Fig. 16.35). These screens could handle screens as fine as 75 μm. Barite, which meets API specifications, can only have three weight percent of solids larger than 75 μm. The screen could be tilted upward so that a pool of liquid was formed in the upstream end of the shaker and the solids moved up the screen and off the end.

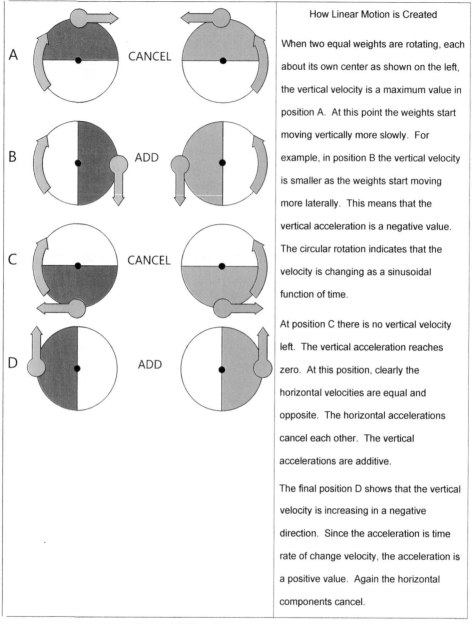

How Linear Motion is Created

When two equal weights are rotating, each about its own center as shown on the left, the vertical velocity is a maximum value in position A. At this point the weights start moving vertically more slowly. For example, in position B the vertical velocity is smaller as the weights start moving more laterally. This means that the vertical acceleration is a negative value. The circular rotation indicates that the velocity is changing as a sinusoidal function of time.

At position C there is no vertical velocity left. The vertical acceleration reaches zero. At this position, clearly the horizontal velocities are equal and opposite. The horizontal accelerations cancel each other. The vertical accelerations are additive.

The final position D shows that the vertical velocity is increasing in a negative direction. Since the acceleration is time rate of change velocity, the acceleration is a positive value. Again the horizontal components cancel.

Fig. 16.35 How linear motion is created.

On a linear motion shaker with a 0.13″ stroke at 1500 RPM, the maximum acceleration is at an angle of 45 degrees to the shale shaker deck. The "G" factor would be 4.15. The acceleration is measured in the direction of the stroke. If the shale shaker deck is tilted at an upward angle from the horizontal, the stroke remains the same. The component of the stroke parallel to the screen moves the solids up the 5 degree incline.

On a linear motion shaker, the motion is generally at an angle to the screen. Usually the two rotary weights are aligned so that the acceleration is 45 degrees to the screen surface. The higher liquid capacity of linear motion shale shakers for the same mesh screens on unbalanced elliptical or circular motion shakers is primarily related to the fact that a pool of drilling fluid is created at the entry end of the shale shaker. The linear motion moves the solids out of the pool, across the screen, and off the end of the screen.

Concern about the rapid reversal of motion on the linear shaker caused one additional development: the balanced elliptical motion. The vibrators are rotated away from the direction of movement of cuttings. This causes the linear motion to develop a slightly wider orbit. Solids are removed from a pool of liquid in the upstream end and move up the screen in the same manner as with the linear motion.

In unbalanced elliptical and circular vibration motion designs, only a portion of the energy transports the cuttings in the proper direction. The remainder is lost due to the peculiar shape of the screen bed orbit and is manifested as solids becoming nondirectional or traveling the wrong direction on the screen surface. Linear motion designs provide positive conveyance of solids throughout the vibratory cycle because the motion is straight-line rather than elliptical or circular.

Generally, the acceleration forces perpendicular to the screen surface are responsible for the liquid and solids passing through the screen, or the liquid capacity. The acceleration forces parallel to the screen surface are responsible for the solids transport, or the solids capacity.

Shaker capacity

Shale shakers have **capacity limits**. Exceeding a capacity limit means excessive mud will be discharged over the ends along with the solids. Capacity limits can only be defined when the screens are not blinded.

There are two capacity limits on a shale shaker:
1. The **solids capacity limit** is the maximum amount of solids that a device will convey.
2. The **liquid capacity limit** is the maximum gallon per minute capacity for various drilling fluids.

Usually the solids capacity limit is encountered only when drilling soft, gummy formations or drilling large diameter holes with high penetration rates. The overall capacity of the shaker is a combination of the solids capacity limit and the liquid capacity limit.

Solids conveyance

Conveying of solids is defined as the movement of the oversized solids on a vibrating screen bed toward the discard end of the shaker. There are several factors that can affect the conveyance of solids on a vibrating screen bed. The primary physical factor that affects the conveyance of a solid is the dynamics of the screen surface and the interaction of that surface with the solids it is trying to remove. With several motion type shakers available on the market today, and each of them with their theories, which motion shaker is the best? First, the application must be considered. If the condition yields a predominantly high solids loading situation in which you need to remove a large amount of solids in a relatively short amount of time, higher G-forces will move the solids off the screen bed in a faster, more efficient manner. If those solids have a tendency to stick to the screening surface, the type of motion could have an effect on the conveyance of those solids. In regards to the shaker motion type, the major axis of the ellipses formed across the length of the basket has a major impact on solids conveyance. Specifically, it is desirable for the major axis of the ellipsoidal motion to be directed toward the solids discharge end as in both the linear and balanced elliptical motion shakers. The aspect ratio of the major to minor axis of the ellipse in a balanced elliptical motion shaker plays an important role in conveyance as well. As the aspect ratio approaches 1:1, the motion becomes more circular which, in turn, reduces the conveyance to a minimum state.

Generally speaking, the conveyance of these solids should be considered throughout the entire process of screening i.e. through the pool and continuing on across the drying panel until they are ultimately discarded. Once equilibrium of the system is obtained, these conveyance values both in the pool and on the drying panel will be the same. The surface tension of the fluid plays a large role in the adhesiveness of both the solids to each other and the solids to the screen surface.

"G" factor determination

The "G" factor refers to a ratio of an acceleration to the earth's gravitational acceleration. For example, a person on Earth that weighs 200lb would weigh 600lb on Jupiter because the planet Jupiter is so much larger than Earth. A person's mass remains the same on Earth or Jupiter but weight is a force and depends upon the acceleration of gravity. The gravitational acceleration on Jupiter is three times the gravitational acceleration on earth. The "G" factor would be 3.

The term "G-Force" is sometimes used incorrectly to describe a "G-Factor". In the above example, the "G-force" on earth would be 200 lbs and the "G-force" on Jupiter would be 600 lb. This is because the acceleration of gravity on Jupiter would be 3 times the acceleration of gravity on Earth.

One empirical method used to compare screening devices is the "G" factor, which is approximately proportional to thrust where:

$$G - \text{Factor} = \frac{\text{Stoke(inches)} \times \text{RPM}^2}{70,490}$$

Accelerations are experienced by an object or mass rotating horizontally at the end of a string. A mass rotating around a point with a constant speed has a centripetal acceleration (C_a) which can be calculated from the equation:

$$C_a = r\omega^2,$$

where r is the radius of rotation, and
ω is the angular velocity in radians per second.

This equation can be applied to the motion of a rotating weight on a shale shaker to calculate an acceleration. The centripetal acceleration of a rotating weight in a circular motion with a diameter (or stroke) of 2r in inches rotating at a certain RPM (or ω can be calculated from the above equation).

$$C_a = \left(\frac{1}{2}\right)(\text{diameter})(\omega)^2$$

$$C_a = \frac{1}{2}(\text{Stoke, inches})\left(\frac{1 \text{ ft}}{12 \text{ inches}}\right)\left[\text{RPM}\left(\frac{2\pi \text{ radians}}{\text{revolution}}\right)\left(\frac{1 \text{ min}}{60 \text{ s}}\right)\right]^2$$

Combining all of the conversion factors to change the units to feet per second squared:

$$C_a\left(\text{in } \frac{\text{ft}^2}{\text{sec}}\right) = \frac{(\text{stroke, inches})}{(\text{RPM})^2}$$

Normally this centripetal acceleration is expressed as a ratio of the value to the acceleration of gravity.

$$\# \text{ of G's} = \frac{C_a}{32.2 \text{ft/s}^2} = \frac{(\text{Stroke, inches}) \times (\text{RPM})^2}{70,490}$$

Shale shakers are vibrated by rotating eccentric mass(es). A tennis ball rotating at the end of a 3 ft string and a 20 lb weight rotated at the same RPM at the end of a 3 ft string will have the same centripetal acceleration and the same "G" factor. Obviously, the centripetal force, or the tension in the string, will be significantly higher for the 20 lb weight.

The rotating eccentric weight on a shale shaker is used to vibrate the screen surface. The vibrating screen surface must transport solids across the surface to discard and allow fluid and solids smaller than the screen openings to pass through to the mud tanks. If the rotating weights rotated at a speed or vibration frequency that matched the natural frequency of the basket holding the screen surface, the amplitude of the basket's vibration will continue to increase and the shaker will be destroyed. This will happen even with a very small rotating eccentric weight. Consider a child in a swing on a playground. Application of a small force every time the swing returns to full height (amplitude) soon results in a very large amplitude. This is a case where the "forcing function" (the push every time the swing returns) is applied at the natural frequency of the swing.

When the "forcing function" is applied at a frequency much larger than the natural frequency, the vibration amplitude depends upon the ratio of the product of the unbalanced weight (w) and the eccentricity (e) to the weight of the shaker's vibrating member (W); or.

$$\text{Vibration amplitude} = \frac{w\,e}{W}$$

The vibration amplitude is one half of the total stroke length.

The peak force, or maximum force, on a shaker screen can be calculated from Newton's second law of motion:

$$\text{Force} = \frac{W}{g}a,$$

where a is the acceleration of the screen.

For a circular motion where the displacement is described by the equation:

$$x = X \sin\frac{Nt\pi}{30}$$

The velocity is the first derivative of the displacement dx/dt, and the acceleration is the second derivative of the displacement d^2x/dt^2.

This means that the acceleration would be

$$a = \left(\frac{N}{30}\right)^2 (\pi)^2 X \sin\left[(\pi)^2 \frac{Nt}{30}\right]$$

The maximum value of this acceleration occurs when the sine function is one. Since the displacement (X) is proportional to the ratio of we/W for high vibration speeds, the peak force, in pounds (from the peak acceleration) can be calculated from the equation:

$$F = \frac{we\,N^2}{35,200}$$

So the force available on the screen surface is a function of the unbalanced weight (w), the eccentricity (e), and the rotation speed (N).

Stroke length for a given design depends upon the amount of eccentric weight and its distance from the center of rotation. Increasing the weight eccentricity and/or the RPM increases the "G" FACTOR. The "G" factor is only an indication of the acceleration of the vibrating basket and is not necessarily an indicator of performance. Every shaker design has a practical "G" factor limit. Most shaker baskets are vibrated with a five horsepower or smaller motor and produce two to seven "G"s of thrust to the vibrating basket.

Conventional shale shakers usually provide a "g" factor of between 4 and 6. This means the acceleration of the screen is 4–6 times the acceleration of gravity. There are currently shakers on the market providing a "g" factor larger than 8. The higher the "g" factor, the greater the solids separation possible.

A "g" factor that is too high for a shaker's screen supports may reduce the life of a screen as well as the shaker basket components if the two systems are allowed to move independent of each other under vibration. Proper screen tension is critical to assure both good screen and shaker life. It is important that the acceleration (or "G's") exerted by the vibrating screen frame is directly transmitted to the screening surface and that there is little to no difference in amplitude between the two.

There are two basic methods for determining the G factor of a shale shaker:

1. The first involves either a dual or single axis accelerometer that is either attached to a magnet that can be placed on the shaker or bolted to the shaker. The angle of vibration and maximum amplitude can usually be determined when using a dual axis accelerometer in that both the x and y components of the reading are determined. A single axis accelerometer only takes the amplitude in one direction so the placement and orientation of the sensor is crucial in obtaining an accurate reading. For linear motion shakers, the maximum amplitude can be determined through the use of a single axis accelerometer as long as the x and y components can be measured. To determine the maximum amplitude and vibration angle, measurements must be taken both parallel and perpendicular to the plane in which the screens are oriented as illustrated in Fig. 16.36.

Where the resulting G-force from the determined x and y component values is (Fig. 16.36):

$$G's = \sqrt{(y - \text{component}^2) + (x - \text{component}^2)}$$

and the resulting vibration angle is:

$$\phi = \tan^{-1}\left(\frac{y - \text{component}}{x - \text{component}}\right)$$

For example if you were to place the single axis accelerometer on the top of the vibrating basket and obtained a y-component value of 5.32 G's then placed the accelerometer perpendicular to that reading and obtained a x-component value of 4.76 G's, the resulting G's would be 7.14 with a resulting vibration angle of 48.18 degrees from the screen plane.

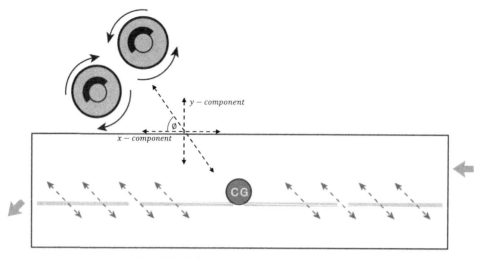

Fig. 16.36 G-force and angle determination.

2. The second method commonly used to determine the G factor of a shale shaker is either a sticker or magnet consisting of a series of circles called a stroke detector in combination with a reed tachometer. Stroke detectors can be obtained from most shaker manufacturers. Fig. 16.37 is an example of Derrick's stroke detector mounted on the motor mount of a shaker (Fig. 16.38).

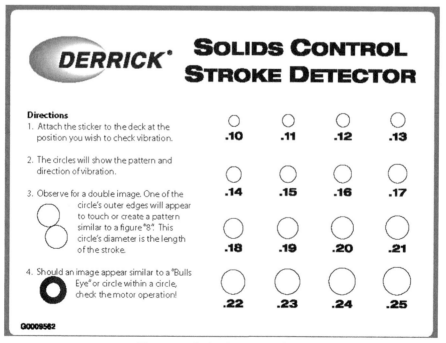

Fig. 16.37 Stoke detector sticker.

Fig. 16.38 Stroke detector sticker mounted on shaker.

Pick the circle that corresponds to the shakers stroke length. Run the shaker unloaded (no mud or cuttings) and look at the stroke detector. The possibilities and diagnosis are as follows (Table.16.2):

Relationship of "G" factor to stroke and speed of rotation

An unbalanced rotating weight vibrates the screen deck. The amount of unbalanced weight combined with the speed of rotation will give the "G" factor imparted to the screen deck (See preceding section). The stroke is determined by the amount of unbalanced weight and its distance from the center of rotation and the weight of the shale shaker deck. This assumes that the vibrator frequency is much larger than the natural frequency of the shaker deck. Stroke is independent of the rotary speed.

"G" factor can be increased by increasing the stroke, or the RPM, or both. The "G" factor can be decreased by decreasing the stroke, or the RPM, or both. The stroke must be increased by the inverse square of the RPM reduction to hold the "G" factor constant (Table 16.3).

Some shakers have one deck, some have two decks, and some have three decks with screens mounted on them.

Dryer shakers

The dryer shaker, or dryer, is a linear motion shaker used to minimize the volume of liquid associated with drilled cuttings discharged from the main rig shakers and

Table 16.2 Stroke detector readings and troubleshooting.

The shaker is operating normally. Proper G force and stroke length	The shaker is operating at a lower than normal G force. Probable cause, a vibrator was replaced with one that doesn't have enough eccentric weight.	The shaker is operating at a higher than normal G force. Probable cause, a vibrator was replaced with one that has too much eccentric weight.	One of the vibrators is not running for some reason and the shaker is operating in elliptical motion. Also the cuttings near the discard end will move backwards.	Both vibrators are rotating in the same direction. Basically there is little or no basket movement
The shaker is shaking correctly	Solution, replace with the correct vibrator or weights	Solution, replace with the correct vibrator, or weights.	Make sure switches for both vibrators are turned on. Check switches, wiring, and finally the vibrator. Replace the faulty item(s).	Reverse the rotation of one of the vibrators so the shaker can operate in linear motion.

hydrocyclones. The liquid removed by the dryers is returned back to the active system. Dryers were introduced with the closed loop systems and the environmental efforts to reduce the liquid waste haul off. Two methods are available to minimize liquid discharge: chemical and mechanical. The chemical system, called a dewatering unit. Linear motion shakers are used to mechanically diminish the liquid discard. These systems may be used separately or together.

The dryer is used to deliquify drilled cuttings initially separated by another piece of solids separation equipment. These drilled solids can be the discharge from a main shaker

Table 16.3 Examples to hold 5 "G"s constant while varying stroke length at different values of RPM.

5 G's	@ 0.44" stroke at 900 RPM	4 G's	@ 0.35" stroke at 900 RPM
5 G's	@ 0.24" stroke at 1200 RPM	4 G's	@ 0.20" stroke at 1200 RPM
5 G's	@ 0.16" stroke at 1500 RPM	4 G's	@ 0.13" stroke at 1500 RPM
5 G's	@ 0.11" stroke at 1800 RPM	4 G's	@ 0.09" stroke at 1800 RPM

or a bank of hydrocyclones. Dryers recover liquid discharged with solids in normal liquid-solids separation that would have been previously discarded from the mud system.

The dryer family incorporates equipment long used as independent units: the main linear motion shaker, the desander, and the desilter. They are combined in several configurations to discharge their discard across the fine mesh screens of a linear motion shaker to capture the associated liquid. These units, formerly identified with mud cleaners, are mounted on the mud tanks, usually in line with the primary linear motion shaker. They can be tied into the now line to assist with fine screening when not being used as dryers. Their pumps take suction from the same compartments as desanders and desilters and discharge their overflow (effluent) into the proper downstream compartments.

A linear motion dryer may be used to remove the excess liquid from the main shaker discharge. The flow rate across a linear motion dryer is substantially smaller than the flow rate across the main shaker. The lower flow rate permits removal of the excess fluid by the linear motion dryer. This dryer is usually mounded at a lower level than the other solids separation equipment to use gravity to feed the solids to it. Whether using a slide or conveyor, the cuttings dump into a large hopper, located above the screen and in place of the back tank or possum belly. As the cuttings convey along the screen, they are again deliquified of more of the liquid associated with drilled cuttings. This excess fluid, with the fine solids that passed through the screens, is collected in a shallow tank that takes the places of a normal sump. This liquid is pumped to a catch tank that acts as the feed for a centrifuge or back to the active system.

A dryer unit can be used to remove the excess fluid from the underflow of a bank of hydrocyclones (desanders or desilters). This arrangement resembles a mud cleaner system. When used in this configuration, the dryer unit may be used on either a weighted or unweighted mud system. The liquid recovered by the linear motion shaker under the hydrocyclones can be processed by a centrifuge as previously described.

In most configurations, the dryers use the same style of screens, motors, and/or motor/vibration combinations as other linear motion shakers by the same manufacturer.

Depending on the fluid, it may pose a very large financial advantage by saving previously lost liquid from the discard. It can also provide a relatively dry discharge that can be handled by back hoe and dump truck rather than a vacuum truck.

Drilling fluid properties must be monitored properly when the recovered liquid is returned to the active system. Large quantities of colloidal solids may be recovered with the liquid. This could affect plastic viscosity, the yield point, and the gel strengths of a drilling fluid.

Triple deck shakers

A relatively new procedure is being used now in areas where lost circulation is prevalent. The concept is based on the observation that the hoop stress around a well bore can be increased if a fracture is propped open with large solids. This effect was observed initially in PIT tests. When the fracture is propped open, the well bore is "strengthened" or a "stress cage" is developed which increases the pressure required to open the fracture again. To form this stress cage, large solids must be present in the drilling fluid, consequently, large particles (usually calcium carbonate or limestone) is added to the drilling fluid. To prevent loss of this material, triple deck shakers are being used.

A triple deck shaker was introduced in the early 1960s. This was a circular motion shaker and never became very popular. Louis Brandt was a design engineer and resigned about the time these shakers were placed on the market. He formed The Brandt Company and made double deck, circular motion shakers instead of a triple deck. The design was intended to use the top deck as a scalping deck with a very coarse screen and the finer screen (usually an API 80 or coarser) was mounted on the lower deck. In use, many drilling superintendents mounted the finer screen on top, because they had trouble observing when the screen failed.

The top deck removes the very large particles. The middle deck screen is sized to capture the propping particles and return them to the drilling fluid. The lower deck is designed to remove particles from the drilling fluid that are larger than barite but smaller than the added proppants. The drilling fluid rheology cannot be accurately measured with the traditional rheometers, because the large particles must be removed for the measurement. The gap between the bob and the outer rotating cylinder is usually smaller than the added solids.

The triple deck shakers and well-bore strengthening effects have been proven to work. When circulation is lost, the fractures are sealed quickly. However, the side-effects of adding solids to the drilling fluid does not seem to have been explored with a dedicated engineering study. Visible and invisible non-productive time has been shown to be affected by excessive low-gravity-solids in the drilling fluid (see Chapter 15).

There are indications that fewer lost circulation events happen when the drilling fluid is kept clean. However, this hypothesis has also not been validated with an engineering study Fig. 16.39.

Fig. 16.39 The first triple deck shaker.

Gumbo removal

Gumbo is formed in the annulus from the adherence of sticky particles to each other. It is usually a wet, sticky mass of clay, but finely ground limestone can also act as gumbo. The most common occurrence occurs while drilling recent sediments in the ocean. Enough gumbo can arrive at the surface to lift a rotary bushing from a rotary table. This sticky mass is difficult to screen. In areas where gumbo is prevalent, it should be removed before it reaches the main shale shakers.

Many gumbo removal devices are fabricated at the rig site, frequently in emergency response to a "gumbo attack". These devices have many different shapes but are usually in the form a slide at the upper end of a flow line. One of the most common designs involves a slide formed from steeply sloped rods spaced one to three inches apart and about six to eight feet long. The angle of repose of cuttings is around 42°, so the slides have a slope greater than 45°. Gumbo, or clay, does not stick to stainless steel very well; consequently, some of the devices are made with stainless steel rods (clay does not adhere to stainless steel as well as it does to iron). The drilling fluid easily passes through the relatively wide spacing in the rods and the sticky gumbo mass slides down to disposal.

Several manufacturers have now built gumbo removal devices for rig installation. One of these units consists of a series of steel bars formed into an endless belt. The bars are spaced one to two inches apart and disposed perpendicular to the drilling fluid flow. The bars move parallel to the flow. Gumbo is transported to the discharge end of the belt and the drilling fluid easily flows through the spacing between the bars.

Another machine (Fig. 16.40), uses a synthetic mesh belt with large openings, like an API 5 to an API 10 screen. The belt runs uphill and conveys gumbo from a pool of drilling fluid. A counter rotating brush is used to clean gumbo from the underside of the belt. The belt speed is variable so it can be adjusted for the solids loading and fluid properties.

Fig. 16.40 Gumbo remover.

A shaker that doesn't shake

Recently another option has been marketed to screen drilling fluid without shaking the screen. A continuous belt of screen cloth receives the fluid from the flow line. A vacuum beneath the screen helps the drilling fluid to flow through the screen. An air jet removes the solids from the belt after the vacuum has pulled most of the fluid through the screen. This concept was first introduced in the 1980s but failed to develop into a commercial product.

Maintenance

Because of their greater complexity and use of finer screens, fine screen shakers generally require more attention than scalping shakers. Nonetheless, their more effective screening capabilities more than justify the higher operating cost. This is especially true when rig rates are high and/or expensive drilling fluid systems are used.

Besides periodic lubrication, fine screen shakers require the same minimum maintenance as scalping shakers while making a trip:

1. Wash screens. Drilling fluid which dries on a shaker screen during a trip will plug the small screen openings and is very difficult to remove.
2. Check screens for proper tension.
3. When using panel screens, plug any openings which have ruptured screens.
4. Shut down shaker when not drilling to extend screen life. Vibrating a dry screen drastically shortens screen life of fine screens.
5. Dump the back tank into a disposal tank. Solids dumped from the back tank do not settle in sand traps. They tend to stay suspended and very quickly plug desilters as soon as circulation resumes. These solids also eventually grind into smaller pieces and are detrimental to drilling performance.

In addition, frequent checks must be made for screen plugging or blinding and broken screens. All will occur more frequently on fine screen shakers than on the coarser screens on scalping shakers. Specifically, the screens should be checked while making a connection when all of the fluid has drained from the back section of the shaker.

Screen blinding, while present to some degree on scalping shakers, is more frequent with fine screen shakers. If the openings become coated over, the throughput capacity of the screen can be drastically reduced and flooding of the screen may occur. Screen blinding can be caused by sticky particles coating over the screen openings by sticky clay, the evaporation of water from dissolved solids, or from grease. Linear motion or elliptical motions shakers do not transport sticky clays efficiently. A scalping shaker is necessary to remove most of these particles before they reach the main shakers. Most of the time, a screen wash-down is needed to cure the problem. This wash-down may simply be a high pressure water wash, a solvent (in the case of grease, pipe dope or asphalt blinding), or a mild acid soak (in the case of blinding caused by hard water). Stiff brushes should not be used to clean fine screens because of the fragile nature of fine wire in the screen cloth.

Screen capacity, or the volume of drilling fluid which will pass through a screen without flooding, varies widely depending on shaker model and drilling conditions. Screen opening size, drilling rate, bit type, formation type, and drilling fluid type, weight, drilling fluid surface tension, thickness of the wetting ring of liquid around the wires, and viscosity affect throughput to some degree. When drilling with non-aqueous drilling fluid, the screens must become oil-wet. If a fine screen becomes water-wet, the water ring around the wires closes the openings and drilling fluid floods off the end of the screen. The surface tension of a drilling fluid is seldom measured but has a great impact on a screen capacity, no matter what brand is used. The shaker capacity is directly related to the opening sizes in the screen and the smaller the opening sizes, the lower the screen capacity. Drilling rate affects screen capacity because increases in drilled solids loading reduce the effective screen are available for drilling fluid to flow though.

Increased high-shear-rate viscosity (called PV), is usually associated with an increase in percent solids by volume and/or increase in mud weight, has a markedly adverse effect on screen capacity. As a general rule, for every 10% increase Plastic Viscosity (PV), there is a 2–5% decrease in throughput capacity. Other factors which decrease the screen capacity are: screen motion, screen velocity, the low-shear-rate viscosity of the drilling fluid, total solids loading (amount of oversize and the quantity of solids which pass through the screen), the thickness of the ring of liquid adhering to the screen wires, the thickness of the layer of drilling fluid adhering to the solids, and the surface tension of the fluid (Fig. 16.41).

Mud type also has an effect on screen capacity. Higher viscosities generally associated with non-aqueous drilling fluid (NADF) results in lower screen throughput than would

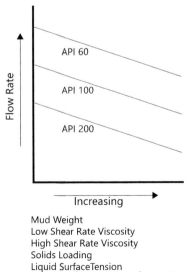

Fig. 16.41 Screen capacity decreases with a variety of variables.

be possible with a water-base drilling fluid of the same mud weight. Some drilling fluid components, such as synthetic polymers, also have an adverse effect on screen capacity. Starch, for example, is large enough to plug an API200 screen. As a result, no manufacturer can offer a standard throughput for all operating conditions. This is the reason the API RP13C committee elected to simply try to describe the screens instead of trying to develop a procedure which would predict performance. The capacities of shakers can vary from 50 to 800 gpm.

Power systems

The most common power source for shale shakers is the rig electrical power generator system. The rig power supply should provide constant voltage and frequency to all electrical components on the rig. Most drilling rigs generate 460 VAC, 60 Hz, 3-phase power or 380 VAC, 50 Hz, 3-phase power. Other common voltages are 230 VAC, 190 VAC, and 575 VAC. Through transformers and other controls, a single power source can supply a variety of electrical power to match the requirements of different rig components.

Shale shakers should be provided with motors and starters to match the rig generator output. Most motors are dual-wound motors. These may be wired to use either of two voltages and starter configurations. For example, some are 230/460 VAC motors and some are 190/380 VAC motors. Dual-wind motors allow the shaker to be operated properly with either power supply after relatively simple rewiring. Care must be taken, however, to make certain that the proper voltage is used. Electrical motors are designed to rotate with a specific speed. Typically the rotational speed is 1800RPM for 60 Hz applications and 1500RPM for 50 Hz application.

Shale shakers use a vibrating screen surface to conserve the drilling fluid and reject drilled solids. The effects of this vibration is described in terms of the "G" factor, or the function of the angular displacement of a screen surface and the square of the rotational speed. (For a detailed discussion, see preceding section on "G" factor.) Angular displacement is achieved by rotating an eccentric mass. Most shale shakers are designed to be operated at a specific, fixed "G" factor by matching the stroke to a given machine rotational speed. It follows that any deviation in speed will affect the "G" factor and influences the shaker performance.

Deviations in speed may be caused by one or more factors, but typically caused by fluctuations in voltage or the frequency of the alternating current. If the voltage drops, the motor cannot produce the rated horsepower, and may not be able to sustain the velocity need to keep the eccentric mass moving correctly. Low voltage also reduces the life of electrical components. Deviations in frequency result in the motor turning faster (frequencies higher than normal) or slower (frequencies lower than normal). This directly influences RPM and shaker performance.

Slower RPM reduces the "G" factor, and causes poor separation and poor conveyance. Faster RPM increases the "G" factor, and may improve conveyance and separation, but often increase screen fatigue failures. In extreme cases, higher RPM may cause structural damage to the shale shaker. Thus, it is important to provide proper power to the shale shaker.

For example, a particular shale shaker is designed to operate at 4 G's. In this example, the shaker has an angular displacement, or stroke of 0.09 inches. This shaker must vibrate at 1750RPM to produce 4.1 G's. At 60 Hz, the motor turns at 1750RPM so the G factor is 4.1, just as designed. If frequency drops to 55 Hz, the motor speed reduces to 1650RPM, which results in a G factor of 3.5. Further reduction of frequency to 50 Hz results in 1500RPM and a G factor of 2.9.

Most rigs provide 460 VAC, 60 Hz power, so most shale shakers are designed to operate with this power supply. However, many drilling rigs are designed for 380 VAC, 50 Hz electrical systems. To provide proper G factors for 50 Hz operations, shale shaker manufacturers rely on two methods, increasing stroke length, or through voltage/frequency inverters (transformers).

A motor designed for 50 Hz applications rotates at 1500RPM. At 0.09 inch stroke a shale shaker will produce 2.9 G's. Increasing the stroke length to 0.13 inches provides 4.1 G's, similar to the original 60 Hz design. However, the longer stroke length and slower speed will produce different solids separation and conveyance performance. At the longer stroke lengths, shakers will probably convey more solids and have a higher fluid capacity. If the stroke length is not increased, some manufacturers use voltage inverters to provide 460 VAC, 60 Hz output power from a 380 VAC, 50 Hz supply.

Constant electrical power is necessary for good, constant shale shaker performance. The tables below assist in designing a satisfactory electrical distribution system. Alternating Current (AC) motors are common on most shale shakers. The motor rating indicates the amount of electric current required to operate the motor. The values in Table 16.4 provide some guidelines for various motors. The manufacturer recommendation should always take precedence over the generalized values in these tables. The

Table 16.4 Electric current required by motors running at full load.

Motor rating	Single phase		Three phase		Three phase	
HP	115 V	230 V	190 V	230 V	460 V	575 V
1	16	8	8	3.6	1.8	1.4
1½	20	10	10	5.2	2.6	2.1
2	24	12	12	5.8	3.4	2.7
3	34	17	17	9.6	4.8	3.9
5	56	28	28	15.2	7.6	6.1
7½	80	40	40	22	11	9.0
10	100	50	50	28	14	11
15			42	21	17	
20			54	27	22	

Table 16.5 Maximum allowable electric current in various wire sizes.

Current amps	35	50	70	80	90	100	125	150	200	225	275
Wire size AWG	10	8	6	5	4	3	2	1	0	00	000

amount of electric current that a conductor (or wire) can carry increases as the diameter of the wire increases. Common approximate values for currents are presented with the corresponding wire size designation in Table 16.5. Conductors, or even relatively large diameter wire, still have some resistance to the flow of electric current. This resistance to flow results in a line voltage drop. When an electric motor is located in an area remote from the generator, the line voltage drop may decrease the motor voltage to unacceptably low values. Some guidelines of wire diameter needed to keep the voltage drop to 3% are presented in Table 16.6.

> **WARNING**
>
> Electrical Hazard - Follow ALL National Electrical Codes, Local Electrical Codes, and Manufacturer's Safety and Installation Instructions. Always conform to regulatory codes, as apply, regionally and internationally

Sample #3.

These values are to be used as general guidelines only. Many factors, including insulating material and temperature control the values.

Cascade systems

The first Cascade system was introduced in the mid-1970s. A scalpel shaker received fluid from the flow line and removed gumbo or large drilled solids before the fluid passed through the main shaker with a fine mesh screen. The first unit combined a single-deck, elliptical motion shaker mounted directly over a double-deck, circular motion shaker. This combination was especially successful offshore where space is at a premium. It was, however, subject to the technology limitations of that time period that made 80–120 mesh screens the practical limit.

One advantage of multiple deck shale shakers is their ability to reduce solids loading on the lower, fine-mesh screen deck. This increases both shaker capacity and screen life. However, capacity may still be exceeded under many drilling conditions. The screen mesh, and thus the size solids returned to the active system, is often increased to prevent loss of whole mud over the end of the shaker screens.

Processing drilling fluid through shale shaker screens, centrifugal pumps, hydrocyclones, and drill bit nozzles can cause degradation of solids and aggravated problems associated with fine solids in the drilling fluid. To remove drilled solids as soon as possible,

Table 16.6 Copper wire size required to limit line voltage drop to 3%.

Current, amps	120 V single phase			
	Wire length, feet			
	50	100	150	200
10	12	12	10	8
20	12	8	6	6
40	8	6	4	3
60	6	4	2	1
80	6	3	1	0
100	4	2	0	0

Current, amps	190 V three phase			
	Wire length, ft			
	50	100	150	200
10	12	12	12	12
20	10	10	8	8
40	8	6	6	4
60	6	4	4	3
80	4	4	3	2
100	4	3	2	1

Current, amps	230 V three phase			
	Wire length, ft			
	150	200	250	300
10	12	12	12	12
20	10	10	8	8
40	8	6	6	4
60	6	4	4	3
80	4	4	3	2
100	4	3	2	1

additional shakers are installed at the flow line so that the finest mesh screen may be used. Sometimes as many as six to ten parallel shakers are used. Downstream equipment is often erroneously eliminated. The improved shale shaker still remains only one component (a very important component) of the drilled solids removal system.

A system of cascading shale shakers, using one set of screens (or shakers) to scalp large solids and gumbo from the drilling fluid, and another set of screens (or shakers) to receive the fluid for removal of fine solids, increases the solids removal efficiency of high performance shakers, especially during fast, top-hole drilling or in gumbo-producing formations. The cascade system is used where the solids loading exceeds the capacity of the fine API screen.

The advantages of the cascade arrangement are:
1. Higher overall solids loading on the system
2. Reduced solids loading on fine API screens.
3. Finer screen separations
4. Longer screen life
5. Lower fluid well costs

There are three basic designs of cascade shaker systems:
1. The separate unit concept.
2. The integral unit with multiple vibratory motions, and
3. The integral unit with a single vibratory motion.

The choice of which design to use depends on many factors, including space limitations, performance objectives, and overall cost.

1. *Separate unit*—The first of these, the separate unit system, mounts useable rig shakers (elliptical or circular motion) on stands above newly-installed linear motion shakers. Fluid from the rig shakers (or Scalping shakers) is routed to the back tank of a linear motion shaker. Line size and potential head losses must be considered with this arrangement to avoid overflow, and loss of drilling fluid. This design may reduce overall cost by utilizing existing equipment. This concept, where space is available, has the advantages of highly visible screening surfaces and ease of access for repairs.
2. *The Integral Unit with Multiple Vibratory Motions*—The second design type combines the two units of the first design into a single, integral unit mounted on a single skid. Commonly, a circular motion shaker is mounted above a linear motion shaker on a common skid. The main advantages of this design are reduced installation costs, and reduced space requirements. The internal flow line eliminates the manifold and piping needed for the two separate units. This design reduces screen visibility and accessibility to the drive components.
3. *The Integral Unit with a Single Vibratory Motion*—The third design consists of an integral unit with a single vibratory motion. Typically, the units use linear motion shakers. Visibility of, and access to, the fine screen deck can be limited by the slope of the upper scalping deck.

Summary

Cascading systems use one set of shakers to scalp large solids and/or gumbo from the drilling fluid, and another set of shakers to remove fine solids. Their application is primarily during fast, top-hole drilling, or in gumbo formations. This system was designed to handle high solids loading. High solids loading occurs when rapidly drilling large diameter hole or when gumbo arrives at the surface.

The introduction of high-performance linear motion shale shakers has allowed development of fine screen cascade systems capable of 200-mesh separations at the flow line. This is particularly important in areas where high circulating rates and large amounts of drilled solids are encountered. After either the flow rate or solids loading is reduced in deeper parts of the borehole, the scalping shaker should be used only as an insurance device. Screens as coarse as API 10 may be used to avoid dispersing solids before they arrive at the linear motion shaker. When the linear motion shaker, with the finest mesh screen available, can handle all of the flow and the solids arriving at the surface, the need for the cascade system disappears. Even when the fine mesh screen can process all of the fluid, screens should be maintained on the scalpel shaker. These screens can be a relatively coarse screen (API 10 to 40) but they will protect the finer screens on the main shaker.

Shaker users guide

An evaluation should be conducted to determine the drilling fluid processing system needed to minimize drilling and disposal costs. On the basis of this evaluation, the number, type, and configuration of shaker(s) can be chosen. Any solids removal system should have enough shale shakers to process 100% of the drilling fluid circulating rate.

Installation

1. Consult the owner's manual for correct installation procedures. If unavailable, the general guidelines below may be helpful in installing shale shakers
2. Low places in the flow line will trap cuttings. Flow line angle should be such that solids settling does not occur, i.e. a 1-inch drop for every 10-feet of flow line seems to be a good rule of thumb.
3. When a back tank, also known as a possum belly, is used, the flow line should enter at the bottom to prevent solids from settling and building up. If the flow line enters over the top of the back tank, the flow line should be extended to within one pipe diameter of the flow line from the bottom.
4. Rig up with sufficient space and approved walkways around the shaker(s) to permit easy service.
5. When more than one shaker is required, use a flow distribution module.
6. A cement bypass that discharges outside the active system is desirable.
7. Mount and operate the shale shaker where it is level. Both the solids and fluid limits will be reduced if this rule is not followed.
8. Motors and starters should be explosion-proof. Local electrical codes must be met. Be sure the proper sized starter heaters are used.
9. Provide the proper electrical voltage and frequency. Low line voltages reduce the life of the electrical system. Low frequency reduces the motion and lowers the capacity of the shale shaker.

10. Check for correct motor rotation.
11. Check for correct motion of the shale shaker deck.
12. Check drive belts for proper tension according to manufacturer's instructions.
13. Screens should be installed according to manufacturer's instructions.
14. Provide a wash-down system for cleaning.
15. Water-spray bars, if installed, should provide only a mist of water—not a stream of water.
16. Linear motion screens should have liquid covering of one half of the deck area.

Operation

1. For double-deck shale shakers, run a coarser mesh screen on the top deck and a finer mesh screen on the bottom—the coarser screen should be at least two standard mesh sizes coarser. Watch for a torn bottom screen. Replace or patch torn screens at once. During normal drilling operations, cover at least 75%–80% of the bottom screen with drilling fluid to maximize utilization of available screen area. Properly designed FLOW-BACK PANS may improve shaker performance. (Gumbo shakers mounted above as an integral part of linear shale shakers are not called double deck shale shakers—although the operation guidelines above still apply.)
2. For single deck shale shakers with multiple screens on the deck, try to run all the same mesh screens. If coarser screens are necessary to prevent drilling fluid loss, run the finer mesh screens closest to the back tank. All screens should have approximately the same size openings. For example, use a combination of MARKET GRADE (MG) 100 mesh (140 µm) and MG 80 mesh (177 µm), but not MG 100 mesh (140 µm) and MG 50 mesh (279 µm). Under normal drilling conditions, cover at least 75%–80% of the screen area with drilling fluid to properly utilize the screen surface area.
3. Spray bars (mist only) may be used for sticky clay to aid conveyance which reduces whole drilling fluid loss. High pressure washers should not be used on the screen(s) while circulating. They disperse and force solids through the screen openings. Spray bars are not recommended for weighted or oil-based drilling fluids.
4. Do not bypass the shale shaker screens or operate with torn screens; these are the main causes of plugged hydrocyclones. This results in a build-up of drilled solids in the drilling fluid. Dumping the back tank into the pits (to clean the screen or for whatever reason) is a form of bypassing the shale shaker and should not be done.
5. All drilling fluids that have not been processed by solids removal equipment and intended to be added to the active system should be screened by the shale shakers to remove undesirable solids. This specifically includes drilling fluid delivered to a location from remote sources.

Maintenance

1. For improved screen life with non-tensioned screens, make certain the components of the screen tensioning system, including any rubber supports, nuts, bolts, springs, etc. are in place and in good shape. Install screens according to the manufacturer's recommended installation procedure.
2. Lubricate and maintain the unit according to the manufacturer's instructions. (Some units are SELF-LUBRICATING and should not be "re-lubricated").
3. Check the tension of screens one, three, and 8 h after installation and hourly thereafter.
4. Check the tension of and adjust drive belts according to manufacturer's instructions.
5. If only one deck of a multiple deck shaker is used, be sure other tension rails are secured.
6. Wash screens at the beginning of a trip so as not to allow fluid to dry on the screen(s). (Do not dump the back tank into the active system, or the sand trap below the shaker. This results in plugging of hydrocyclones downstream and/or an increase in drilled solids concentration in the drilling fluid.
7. Inspect screens frequently for holes and repair or replace when torn.
8. Check condition of VIBRATION ISOLATORS members and SCREEN SUPPORT RUBBERS and replace if they show signs of deterioration or wear.
9. Check the fluid bypass valve and other places for leaks around the shaker screens.
10. Remove drilling fluid build-up from vibrating bed, vibrators, and motors. Caution: Do not spray electrical equipment or motors with oil or water.

Operating guidelines

1. Shale shakers should run continuously while circulating. Cuttings cannot be removed if the shaker is not in motion.
2. Drilling fluid should cover most of the screen. If the drilling fluid covers only 1/4 or 1/3 of the screen, the screen mesh is too coarse.
3. A screen with a hole in it should be repaired or replaced at once. Holes in panel screens can be plugged. Install screens according to Manufacturer's recommended installation procedures. Cuttings are not removed from the drilling fluid flowing through the hole.
4. Shaker screen replacements should be made as quickly as possible. This will decrease the amount of cuttings remaining in the drilling fluid because the shale shaker is not running.
5. Locate and arrange tools and screens before starting to make the replacement. If possible, get help.
6. If possible, change the screen during a connection. In critical situations, the driller may want to stop (or slow) the pumps and stop drilling while the screen is being replaced.

7. For improved screen life with non-pretensioned screens, make certain the components of the screen tensioning system, including any rubber supports, nuts, bolts, springs, etc. are in place and in good shape.
8. Check condition of vibration isolators members and screen support rubbers and replace if they show signs of deterioration or wear.
9. Water should not be added in the possum belly (or back tank) or onto the shale shaker screen. Water should be added downstream.
10. Except for cases of lost circulation, the shale shaker should not be by-passed, even for a short time.
11. Wash screen(s) at the beginning of a trip so fluid will not dry on the screen(s).
12. The possum belly (or back tank), should not be dumped into the sand trap or mud tank system just before making a trip. If this is done, cuttings will move down the tank system and plug desilters as the next drill bit starts drilling.

Summary

Shale shakers are the most important and easiest to use solids removal equipment. In most cases, they are highly cost-effective. If shale shakers are not being used correctly, the remaining solids removal equipment will not perform properly.

A shale shaker can be used in all drilling applications where liquid is used as the drilling fluid. Screen selection is controlled by the circulation rate, shaker design, and drilling fluid properties. Variation in drilling fluid properties control screen through put to such a large degree that no shaker capacities are included in this handbook.

Most operations involved in drilling a well can be planned in advance because of experience and engineering designs for well construction. Well planners expect to be able to look at a chart or graph and determine the size and number of shale shakers required to drill a particular well anywhere in the world. They expect to be able to determine the mesh size of the shaker screens used for any portion of any well. There are too many variables involved to allow those charts to exist. Many shale shaker manufacturers, because of customer demand, publish approximate flow charts that indicate their shakers can process a certain flow rate of drilling fluid through certain size screens. These charts are usually based on general field experience with a lightly-treated water-base drilling fluid, usually from Gulf of Mexico drilling. These charts should be used to provide only very inaccurate guesses about screen meshes that will handle flow rates for a particular situation.

Rheological factors, fluid type, solids type and quantity, temperature, drilling rates, solids/liquid interaction, hole diameters, hole erosion, and other variables dictate actual flow rates that can be processed by a particular screen. Drilling fluid without any drilled solids can pose screening problems. Polymers that are not completely sheared tend to blind screens and/or appear in the screen discard. Polymers that increase the low-shear-rate viscosity or gel strength of the drilling fluid also pose screening problems. Polymers, like starch, that are used for fluid-loss control are also difficult to screen through a fine mesh screen (like a API 200). Oil-base drilling fluids without adequate shear and proper mixing are difficult to screen. Oil-base drilling fluid without sufficient oil-wetting additives are very difficult to screen through screens finer than API 100.

Screen selection for shale shakers is very much dependent upon geographical and geological location. Screen combinations that will handle a specific flow rates in the Middle East or Far East will not necessarily handle the same flow rates in Norway or the Rocky Mountains. The best method to select shale shaker screens and/or number of shale shakers for a particular drilling site is to first use the recommendations of a qualified solids control advisor from the area. Screen use records should be established to guide further use.

CHAPTER 17

Sand traps

Introduction

Drilling fluid enters the removal tank section after it passes through the main shale shaker. Immediately below the main shaker, the first pit can be a settling pit or sand trap. Fluid passing through the shaker screen flows directly into this small compartment. The fluid in this compartment is not agitated. This allows solids to settle. The fluid overflows from the sand trap into the next compartment, which should be the degasser suction pit. Sand traps are rarely used on many rigs because of the small size of the solids that remain in the fluid after passing through API 170 or API 200 shaker screens.

The sides of a sand trap should slope at 45 degrees or more to a small area in front of a large opening discharge valve. The angle of repose of drilled solids is around 42 degrees. The solids are dumped after the mud pumps are shut-off. In many cases, during periods of fast drilling with coarse or damaged shaker screens in use, the sand trap will fill several times per day.

An effective sand trap requires an overflow weir of maximum length to create a liquid column as deep as possible. A common and recommended practice is to utilize the full length of the partition between the sand trap and the degasser suction pit.

Settling rates

Most particles settling in a fluid will reach a terminal velocity. For example, a person jumping out of a balloon at 20,000 ft would achieve a velocity of 1135 ft/s if there was no drag force. The speed of sound in air (at 70 °F) is 1130 ft/s. In other words, with no drag force on the object falling from 20,000 ft, the object would break the sound barrier. When a parachute opens, the drag force is significantly increased. The object quickly reaches a terminal velocity—which hopefully will allow a "soft landing". If the particles are very small, they may not settle at all. For example, a fog is water vapor that is suspended in the atmosphere. Water will pour from a vessel, but if the water is in very small droplets, it does not settle. The particles are then said to be "colloidal". Solids in drilling fluid below 2–6 μm may be colloidal and will not settle.

The rate at which larger solids settle depends upon the force causing the settling, the dimensions of the solid, and the fluid viscosity in which the solid is settling. Analysis of forces acting on irregular shaped objects is extremely complicated. Analysis of forces acting on spheres is not as complicated but is addressed here. For simplicity, the solid will be assumed to be spherical and settling in a quiescent fluid. The forces acting on the sphere would be the gravitational force causing it to fall and the buoyant force tending to prevent settling. The force causing settling could also be a centrifugal force in a device such as a hydrocyclone or a centrifuge. This section will develop the equation relating to solids settling through a drilling fluid in the sand trap.

The drag force on a spherical particle when it reaches a terminal velocity while settling in liquid can be calculated from Stokes' Law:

$$F = 6\pi \mu v R \qquad (17.1)$$

where F is the force applied to the sphere by the liquid, in dynes;
μ is the fluid viscosity, in poise;
R is the radius of the sphere, in cm;
v is the particle velocity, in cm/s.

Stokes' Law was developed when the centimeter/grams/second (cgs) unit system was popular with scientist. Viscosity is defined as the ratio of shear stress in a liquid to the shear rate. One poise has the units of dynes-s/cm², in absolute units, or (g/cm s), in cgs units. The unit of dyne has the units of g cm/s²

A sphere falling through a liquid experiences a downward force of gravity, and an upward force of the buoyancy effect of the liquid. The buoyancy force is equal to the weight of the displaced fluid:

$$\text{Buoyant force} = \frac{4\pi}{3}(R^3)\rho_l \qquad (17.2)$$

The downward force is mass times acceleration or the weight for gravity settling. The mass of the sphere is the volume of the sphere times the density of the sphere:

$$\text{Mass of sphere} = \frac{4\pi}{3}(R^3)\rho_s \qquad (17.3)$$

The total downward force on the particle would be:

$$F_{\text{down}} = \frac{4\pi}{3}(R^3)\rho_s - \frac{4\pi}{3}(R^3)\rho_l \qquad (17.4)$$

When the particle reaches the terminal velocity, the upward force can be calculated from Stokes law.

$$F_{\text{up}} = 6\pi \mu v R \qquad (17.5)$$

When terminal velocity is achieved, the downward force is equal to the upward force:

$$\frac{4\pi}{3}(R^3)\rho_s - \frac{4\pi}{3}(R^3)P_l = 6\pi \mu v R \qquad (17.6)$$

$$\frac{4}{18}(R^2)(\rho_s - \rho_l)_s = \mu v \qquad (17.7)$$

Solving this equation for velocity, and changing the radius R to diameter d in centimeters:

$$v = \frac{d^2}{18\mu}(\rho_s - \rho_l)g \tag{17.8}$$

$$v : \text{cm}/\text{s} = \frac{(d : \text{cm})^2}{18(\mu : (\text{poise}))}\left[(\rho_s - \rho_l) : \text{g}/\text{cm}^3\right]\left(g : \text{cm}/\text{s}^2\right) \tag{17.9}$$

$$v : \text{cm}/\text{s} = \frac{(d : \mu\text{m} \times 10^{-4}\text{cm}/\mu\text{m})^2}{18(\mu : (\text{cp}/100))}\left[(\rho_s - \rho_l) : \text{g}/\text{cm}^3\right]\left(980 \text{ cm}/\text{s}^2\right) \tag{17.10}$$

$$v : \text{ft}/\text{min} = \frac{(d : \mu\text{m} \times 10^{-4}\text{cm}/\mu\text{m})^2}{18(\mu : (\text{cp}/100))}\left[(\rho_s - \rho_l)\right]$$
$$: \text{g}/\text{cm}^3\right]\left(980 \text{ cm}/\text{s}^2\right)(\text{ft}/30.48 \text{ cm})(60 \text{ s}/\text{min}) \tag{17.11}$$

$$v : \text{ft}/\text{min} = \frac{1.07 \times 10^{-4} (d : \mu\text{m})^2}{\mu : (\text{cp})}\left[(\rho_s - \rho_l) : \text{g}/\text{cm}^3\right] \tag{17.12}$$

where
v = Settling or terminal velocity, ft/min,
d = Particle equivalent diameter, μm,
ρ_s = Solid density, g/cm^3,
ρ_l = Liquid density, g/cm^3,
μ = Viscosity of liquid, centipoises.

A 2.6 g/cm^3 drilled solid passing through an API 20 screen (850 μm diameter) would fall through a 9.0 ppg, 100 cp liquid with a terminal velocity of 1.17 ft/min. This could be calculated from Eq. (17.12):

$$v : \text{ft}/\text{min} = \frac{1.07 \times 10^{-4} (d)^2}{\mu}(\rho_s - \rho_l)$$

$$v : \text{ft}/\text{min} = \frac{1.07 \times 10^{-4} (850 \text{ μm})^2}{100 \text{ cp}}(2.6 - [9.0 \text{ ppg}/8.34 \text{ ppg}]) \tag{17.13}$$

$$= 1.17 \times 10^{-2} \text{ ft}/\text{min}$$

If the rig is circulating 500 gpm through a 50 bbl settling tank or sand trap, the fluid remains in this tank for a maximum of 4.2 min. If the sand trap holds 100 bbl of drilling fluid, the retention time is 8.4 min. Solids can settle about 5 feet during the 4.2 min retention time or ten feet during the 8.4 min retention time.

The selection of a viscosity to use in the equation is complicated. On drilling rigs, normally the lowest viscosity measurement made is with the 3 RPM viscometer reading. Some drilling rigs using polymer drilling fluids are using Brookfield viscometers which measure very low shear rate viscosities. Drilling fluid viscosity is a function of shear rate. As a particle settles, the fluid viscosity impeding the settling depends upon the settling rate. As the velocity decreases, the viscosity of the fluid increases. The "K" viscosity is the viscosity of a fluid at one reciprocal second—which is within the shear rate range of a small solid falling through a drilling fluid, and can be determined on most drilling rigs. Some drilling fluids are constructed to have very large low-shear-rate viscosities to facilitate carrying capacity as the solids are moved up the bore hole. Many drilling fluid systems have K viscosities in the range of 1000 eff cp instead of the 100 cp used in the example above. Solids settling will be greatly hindered in these fluids because they are designed to prevent settling.

Consider the comparison of the settling rate of barite and that of low gravity drilled solids. Stokes' Law can be used to describe the anticipated settling rate for spheres of barite or low gravity drilled solids:

$$V_B = \frac{d_B^2}{18\mu}(\rho_B - \rho_l)g \tag{17.14}$$

$$V_{LG} = \frac{d_{LG}^2}{18\mu}(\rho_{LG} - \rho_l)g \tag{17.15}$$

Eqs. (17.14) and (17.15) can be used to solve for the ratio of diameters that will cause the settling velocity of barite to be equal to the settling velocity of low gravity solids:

$$D_B = 0.65\, D_{LG} \tag{17.16}$$

Eq. (17.16) indicates that a 20 μm barite settles at the same rate 30 μm low gravity solid; or a 48 μm barite settles at the same rate as a 74 μm low gravity drilled solid. Note this is true regardless of the viscosity of the fluid in which these particles are settling.

Comment: Linear motion and balanced elliptical motion shale shakers permit the use of finer screens than were used in the past. Consequently, sand traps are frequently ignored in a system. Considering the inescapable fact that screens regularly tear and wear out, sand traps offer the ability to capture some of the solids that would normally be left in the drilling fluid.

When API 80 screens were used on shale shakers and represented the smallest openings possible for processing drilling fluid, sand traps were a very important component of the surface drilling fluid system. Normally, screens as coarse as API 20 to API 40 (850 μm to 425 μm), were used in the upper part of a bore hole. The solids that passed through these screens settled quite rapidly. When API 200 screens are installed on the main shakers, the largest solid presented to the fluid in the tank is 74 μm. These solids settle much more slowly than the larger 850 μm (API 20) solids that were separated earlier.

The sand trap is still used in a system to provide backup for failures in the main shaker screen. These screens sometimes break and the failure may go unnoticed for a long period of time. The sand trap offers the possibility of capturing some of the solids that pass through the torn screen.

Although not intended to be used as an insurance shield, the scalping shakers also provide the opportunity to remove solids larger than API 20 to API 40 before the fluid reaches the main shaker. This provides some relief from large solids reaching the sand trap if the finer wires of the main shaker break.

Bypassing shale shaker

One rule cited frequently is "Do not by-pass the shale shaker." Cracking the by-pass valve at the bottom of the shaker back tank allows a rig hand to screen most of the fluid on a finer screen. However, the solids that are not presented to the shaker screen are not removed and cause great damage to the drilling fluid. Sand traps provide some insurance against this activity; but they do not capture all of the larger solids that by-pass the screen—so it is still a very bad practice.

Another activity common on drilling rigs by-passes the shaker screen more subtly. Before making a trip, the possum belly, or back tank, is dumped into the sand trap to clean the shaker. Drilling fluid left on a shaker screen dries during a trip and causes the screen to flood. In an effort to prevent any screen plugging, the back tank is also cleaned and all of the settled solids are dumped into the sand trap. All of the dumped solids, however, do not settle. When circulation ceases and there is no fluid flowing through the sand trap, the gel structure of the drilling fluid starts increasing. The barite in the drilling fluid must remain suspended to prevent a kick while tripping pipe. The drilled solids dumped into the sand trap also remain suspended. When circulation is restored, these suspended solids migrate down the removal system until they reach the apex of a hydrocyclone. These solids plug many cones on drilling rigs. Back tanks, or possum bellies, should be dumped to a waste pit NOT into the drilling fluid system.

Comment: The Stokes Law calculation only identifies the terminal settling velocity of spheres. If round balls are reporting to the surface on a drilling rig, there is a bigger problem than calculating settling rates. The bore hole is not being circulated clean. Solids are tumbling in the annulus and creating an increase in plastic viscosity. In holes with angles less than about 35–40 degrees, the carrying capacity index needs to be adjusted. In high angles and horizontal holes, the only solution is to rotate the pipe and keep the drilling fluid moving as fast as possible. If this is a water-based drilling fluid, additional XC polymer can provide an elastic modulus to clean the hole. No additive is currently available to create an elastic modulus in non-aqueous drilling fluids (NADF). No guidelines have been developed, yet, for the value needed for the elastic modulus to insure cleaning.

CHAPTER 18

Degassers and mud gas separators

Introduction

When drilling subsurface formations, the fluid inside the formations is released into the drilling fluid system. If gas is contained in the rock being drilled, this gas is circulated out of the hole with the drilling fluid. This is called "back-ground" gas. As gas rises up the hole and the pressure is decreased, a gas bubble will expand. A large amount of gas at the surface could be a very small amount at the drill bit.

The degassers are not specifically used to remove gas from the drilling fluid before it is pumped back down hole because the down hole pressure will decrease. Centrifugal pumps will not pump gaseous drilling fluid very efficiently. The gas collects in the center of the impeller and eventually blocks liquid flow from entering the pump. The chart illustrates the minimal effect that gas has on mud weight at various hole depths. At 20,000 ft the decrease in mud weight is so small that the bottom hole pressure changes very little (Fig. 18.1).

The background gas needs to be removed from the drilling fluid so that centrifugal pumps can be used to process the drilling fluid. This background gas does not indicate an impending kick. There is no need to try to increase the mud weight to eliminate background gas.

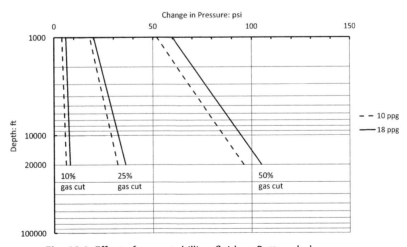

Fig. 18.1 Effect of gas cut drilling fluid on Bottom hole pressure.

Air can also be introduced into the drilling fluid system through the mud hopper. When the mud hopper is left running, air is pulled into the flow stream in the additions section. Some contractors place a short piece of 20″ or 26″ casing at the end of the mud hopper line (As illustrated in Chapter 3 Fig. 3.6). The discharge line from the mud hopper enters tangentially into the short piece casing that is positioned vertically. A top, with a large (10″ to 12″ diameter) hole, is welded to the upper end of the casing. The bottom is left open at the top of the drilling fluid in the tank. When the mud hopper is left running, the air entrained in the drilling fluid is removed with the centrifugal force of the drilling fluid swirling inside of the casing.

Degassers are the most effective way to remove unwanted gas. They are designed to rapidly bring gas bubbles to the surface of the drilling fluid, break them and remove them to a safe location away from the rig.

Vacuum degassers use a combination of turbulent flow and reduced internal tank pressure to move gas-cut drilling fluid and release gas bubbles. Several designs are available; the most common types are the horizontal tank/jet pump design, the vertical tank/jet pump design, and the vertical tank/self-priming pump design.

The horizontal tank/jet pump design has a long horizontal tank with long down-sloping baffles inside. Drilling fluid flows down these baffles in a thin layer, releasing the gas bubbles. A vacuum pump is used to remove the gas from the tank and dispose of it at a safe distance from the rig. The vacuum pump also reduces the internal tank pressure, drawing fluid into the tank and increasing the gas bubble sizes, improving removal efficiency. Most of the time, the volume of gas removed is small compared to the capacity of the vacuum pumps. A 3-way valve is installed in the gas piping to let air in and prevent too much vacuum in the tank. The fluid level inside the tank and the operation of the 3-way valve is controlled automatically by a float inside the tank.

The jet pump discharges the degassed drilling fluid from the tank and returns it to the next downstream compartment (Fig. 18.2). There is no re-mixing of released gas and fluid. The jet pump is used because there is still a small amount of gas left in the drilling fluid—but it may be enough to gas-lock a direct feed centrifugal pump. The gas passes easily through the jet pump, floats to the surface of the discharge compartment and breaks out from surface.

The vertical tank/jet pump design has two variations. The first of these is similar to the horizontal/jet pump design. Instead of a long horizontal tank with a single series of baffles, this design has several conical baffles stacked inside a vertical cylindrical tank (Fig. 18.3).

This design increases baffle surface area in a compact footprint. The vacuum pump and jet pump arrangement are the same as for the horizontal design, although some vertical designs have been used with self-priming feed pumps (Fig. 18.4).

An atmospheric degasser was invented by Walter Liljestrand and developed in the early 1970s. A submerged centrifugal pump sprays drilling fluid in a thin sheet of drilling fluid

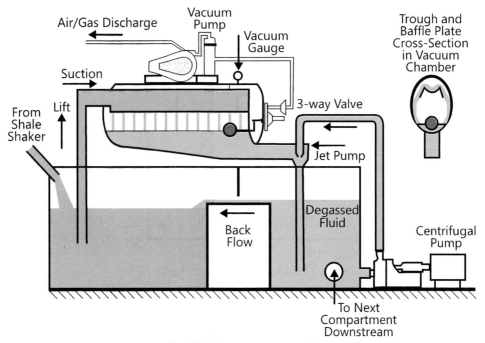

Fig. 18.2 Vacuum degasser #1.

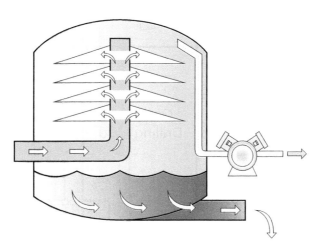

Fig. 18.3 Vertical degasser.

against the wall of a tank. Gas leaves the thin layer of the of drilling fluid and the impact causes the remainder of the gas to separate from the drilling fluid. Comparison of mud weight before and after processing indicates this effectively removed gas. The degassed drilling fluid drains from the spray tank through a trough or pipe to the next downstream compartment. The released gas flows away from degassed fluid. This gas could be piped

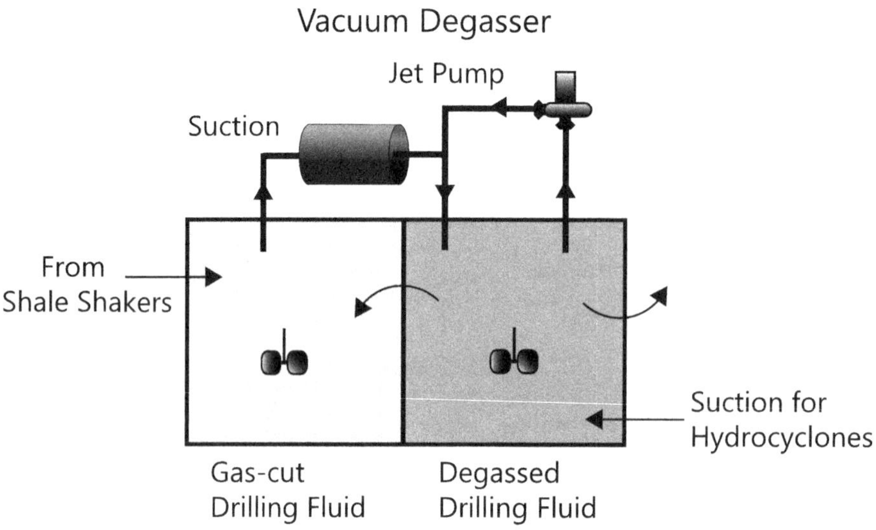

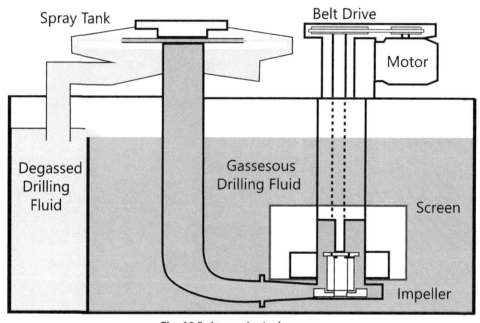

Fig. 18.4 Vacuum degasser #3.

Fig. 18.5 Atmospheric degasser.

away from the rig by covering the trough with a vent hood and flexible hose. If required, a small blower can be mounted on the vent hood to aid with gas removal. Generally, the gas escapes without using a blower. This piece of equipment looks so much different from the conventional vacuum degassers that acceptance in the field is frequently difficult (Fig. 18.5).

Effects of gas-cut drilling fluid

Gas-cut drilling fluid has several effects. Some of these are obvious and others are not. Wrong action in a gas-cut drilling fluid can cause higher drilling costs, lost circulation or a blowout. It is important to recognize both the source (gas or air) and effects of "bubbles in the drilling fluid".

In conventional drilling fluids, air in the fluid is usually a result of the drilling fluid flowing down the flow line and through processing equipment. The main damage from air is corrosion.

Air in the fluid:
- Makes foam on the surface of the mud tanks.
- Reduces measured mud weight.
- Usually makes larger bubbles than hydrocarbon gas.
- Corrodes the drill string.
- Will not be detected by the mud logger.
- May reduce centrifugal pump performance.
- May significantly reduce the mud pump volumetric efficiency.

Gas-cut drilling fluid reduces the mud weight measured with a mud balance. It does not change the true mud weight, but it creates a wrong, urgent feeling to weight up. This can result in great harm.

For example

The well profile calls for a 10.0 ppg drilling fluid to maintain pressure control at 10,000 ft. The mud engineer mixes the right ingredients to make a 10.0 ppg fluid. While drilling ahead the mud is gas-cut by 0.6 ppg but this is not realized. So, even though the actual mud weight is 10.0 ppg, the measured mud weight in the mud balance is 9.4 ppg. Barite is added to bring measured weight up to 10.0 ppg, but this causes the true mud weight to be 10.6 ppg.

Three things happen:
1. The increased mud weight reduces drilling rate with the roller-cone bits.
2. The gas in the mud reduces pump volume efficiency and the fluid flow rate down the drill pipe.
3. The risk of losing circulation and/or stuck pipe due to greater hydrostatic pressure is increased if the formation is pressure sensitive. At 10,000 ft, this increases bottom hole pressure by 312 psi.

When weighing mud samples:
- Use a clean and calibrated mud balance.
- Be sure the place the sample is taken is well stirred.
- Be sure the sample is the same as the fluid being circulated.

- Fill the mud balance cup completely.
- If gas-cut mud is suspected, use a pressurized mud balance (see API RP13B) or hand vacuum to degas the sample thoroughly before weighing.

Another technique that seems to give mud weights within 0.05 ppg of the value measured with a pressurized mud balance is to use a defoamer on the sample. Add some defoamer to a mud cup full of drilling fluid and pour it through the funnel two or three times to agitate, and then weigh in a regular mud balance.

If the true mud weight shows a low reading, it still may not be due to gas or air. Oil or water flows will also reduce mud weight as will weighting material dropping out of poorly agitated drilling fluid systems. Inadequate suspension properties in a drilling fluid may also result in barite leaving the drilling fluid on the way out of the hole. A degasser cannot restore mud weight caused by these problems.

Main mud pumps are positive displacement pumps. They are designed to pump gas-free drilling fluid with about 95%–97% volumetric efficiency. Gas-cut drilling fluid reduces pump flow rate because the positive displacement cylinders are not filled with liquid. Measurements indicated in one well that 6% volume gas/air in the water-based drilling fluid reduced the volumetric efficiency of a triplex pump to 85%. This makes it difficult to maximize the hydraulic impact or hydraulic power of the fluid passing through the nozzles of the drill bit.

Removing gas bubbles

Mud/gas separators are designed to remove large amounts of large bubbles from the drilling fluid. Sometimes called "poor-boy" degassers or "gas-busters", mud/gas separators receive severely gas-cut drilling fluid from a rotating control device (i.e. rotating head) or a choke manifold during a kick.

Mud/gas separators flow the gas-cut drilling fluid in thin sheets over a series of baffles arranged inside a vertical tank. The resulting turbulent flow breaks out large gas bubbles which then rise through a vertical vent line and are released a safe distance from the rig. Caution should be used to make the discharge line for the gas effluent very large to decrease the pressure required to dispose of the gas. If the gas discharge line is too small, the back pressure may eliminate the liquid seal at the bottom of the tank and dump gas onto the drilling fluid tanks. The return drilling fluid flows into the back tank of the shale shakers for further processing.

Gas discharge lines offshore are typically $8''-12''$ in diameter. Onshore, the discharge lines may be only $6''$ in diameter depending upon the drilling area. When a gas bubble reaches the surface during a well control event, the velocity of the gas can be very large. The pressure loss through the discharge line varies as the fifth power of the vent line inside dimension.

Installation

Actual placement of the degasser and related pump will vary with the design of the degasser, but these recommendations may be used as a general rule:
- Install a screen in the inlet pipe to the degasser to keep large objects from being drawn into the degassing chamber.
- Locate the screen about one foot above the pit bottom and in a well-agitated spot.
- There should be a high equalizer line between the suction and discharge compartment. This allows the remaining gas at the surface of the downstream compartment to flow back into the degasser compartment for further gas removal.
- The equalizer should be kept open to allow backflow of processed drilling fluid to the suction side of the degasser.
- Route the liquid discharge pipe to enter the next compartment or pit below the liquid level to prevent aeration.
- Install the gas discharge line to safely vent the separated gas to atmosphere or to a flare line.

Maintenance of degassers varies considerably depending on make and model. In general, the following guidelines apply:
- Check to make sure the suction screen is not plugged.
- Routinely lubricate any pumps and other moving parts and check for wear.
- Keep all discharge lines open and free from restrictions, such as solids build-up around valves.
- If the degasser uses a vacuum, keep it at the proper operating level, according to the manufacturer's recommended range for the mud weight and process rate.
- Check all fittings for air leaks.
- If the unit uses a hydraulic system, check it for leaks, proper oil level and absence of air in the system.

Procedure for installation of degassers:
- The degasser shall draw suction from the compartment immediately downstream from the sand trap.
- When the sand trap is in use, flow to the degasser compartment shall be over a long, high weir.
- While the degasser is in use, there shall be no tank bottom equalization between the degasser compartment and those adjacent to it.
- Fluid shall be degassed before it reaches the pumps feeding the downstream equipment.
- The pump used to power the jet on vacuum degassers should take suction from the same compartment into which the vacuum degasser discharges.
- The degasser suction should be positioned 30 cm (12″) above the tank bottom.
- The degasser compartment should be well agitated.

- The centrifugal pump feeding the eductor jet of vacuum degassers should provide the feed head recommended by the manufacturer. A pressure gauge should be installed to permit the head to be verified.
- The degasser capacity should be at least equal to the planned circulation rate in all of the hole intervals in which gas intrusion is considered to be a hazard.

Procedure for operation of degassers

- Degassers should be operated when receiving "bottoms up" after trips. This instructs the crews with the start up procedure and provides regular checks to confirm that the equipment is working properly.
- The volume percent gas or air in a drilling fluid is calculated by dividing the difference in pressurize drilling fluid weight and the unpressurised drilling fluid weight by the pressurized drilling fluid weight, and multiplying this fraction by 100. Just 9% volume of gas in a drilling fluid can reduce the mud pump volumetric efficiency down to 85% (measured in the field).

Mud/gas separators

Degassers are designed to remove small quantities of gas from the drilling fluid. When pumping out a kick, the quantity of gas arriving at the surface will be too large to admit to the mud tanks. Even a small kick, like 5 bbls of gas, will expand greatly when it arrives at the choke manifold and expands to atmospheric pressure. A gas bubble this size will overload the flow line and blow through the rotary table—usually burning the rig down. This fluid passes through the choke manifold and is processed through a separate tank, called a mud/gas separator, or gas buster, or poor-boy degasser (Fig. 18.6). These separators are frequently "home-made" and may be inadequate for handling large kicks.

In mud/gas separators, the gas-cut drilling fluid flows in thin sheets over a series of baffles arranged inside a vertical tank. The relatively gas-free drilling fluid flows into the back tank of the shale shakers for further processing. The resulting turbulent flow inside of the mud/gas separator allows large gas bubbles to leave the liquid. The gas will then rise through a vertical vent line and be released a safe distance from the rig. Caution should be used to make the discharge line for the gas effluent very large to decrease the pressure required to dispose of the gas.

Gas discharge lines offshore are typically 8″−12″ in diameter. Onshore, the discharge lines may be only 6″ in diameter depending upon the drilling area. When a gas bubble reaches the surface during a well control event, the velocity of the gas can be very large. The pressure loss through the discharge line varies as the fifth power of the vent line inside dimension. Frequently, small pipes, like 4″ diameter, are mounted on the mud/gas separator. A large kick will cause the gas

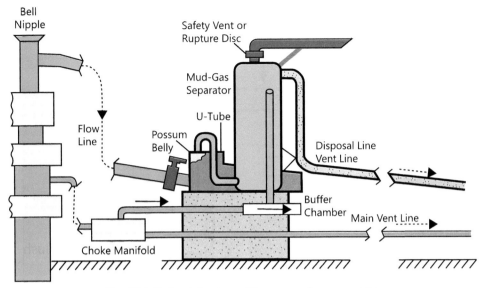

Fig. 18.6 Basic piping to and from a mud-gas separator.

to expel fluid from the bottom of the mud/gas separator and flow over the top of the mud tanks. The size of the kick that can be handled is determined by the size of the gas discharge line on top of the mud/gas separator. If the gas discharge line is too small, the back pressure may eliminate the liquid seal at the bottom of the tank and dump gas onto the drilling fluid tanks. An air gap at the possum belly breaks siphon. Minimum head liquid seal is measured to the open end of the pipe.

Vent lines should be large enough so that with the maximum kick, the pressure to flow gas through the line will not be sufficient to eliminate the head of the liquid seal at the bottom of the separator. Drains should be place at low spots in the plumbing because the lines could become plugged with gelled drilling fluid or ice (in cold climates) (Fig. 18.7). Vent lines should move the gas far enough away from the drilling rig that no hazard exists for the rig crew or equipment. Small quantities of methane or natural gas can be discharged without burning. Usually, however, the vent lines should go to a flare stack and the discharge burned (or flared). This is absolutely required if H_2S is anticipated or even has the potential for being in the gas stream. Very small quantities of H_2S are lethal and H_2S is heavier than air. It will settle around a rig and envelope a large area with toxic fumes.

Because of the possibility of unforeseen events happening, a by-pass line should also be installed. This by-pass line will take fluid directly from a choke manifold and dump it far away from the drilling rig.

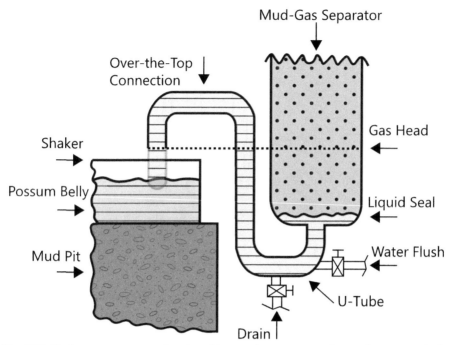

Fig. 18.7 Mud-gas separator mud outlet with over-the-top connection to the possum belly.

CHAPTER 19

Hydrocyclones

Introduction

Hydrocyclones (also referred to as cyclones or cones) are simple mechanical devices without moving parts, designed to speed up the settling process. Feed pressure is transformed into centrifugal force inside the cyclone to accelerate particle settling. In essence, a cyclone is a miniature settling pit which allows very rapid settling of solids under controlled conditions.

Hydrocyclones have become important in solids control systems because of their ability to efficiently remove particles smaller than the finest shaker screens. They are also uncomplicated devices which make them easy to use and maintain.

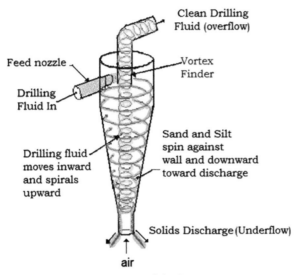

Diagram of desilter.

A hydrocyclone consists of a conical shell with a small opening at the bottom for under flow discharge, a larger opening at the top for liquid discharge through an internal "vortex finder", and a tangential feed nozzle on the side of the body near the wide (top) end of the cone (Fig. 19.1).

Drilling fluid enters the cyclone under pressure from a centrifugal feed pump. The velocity of the fluid causes the particles to rotate rapidly within the main chamber of the cyclone. Small solids and the liquid phase of the drilling fluid tend to spiral inward and upward for discharge through the liquid outlet (overflow). Heavy coarse solids

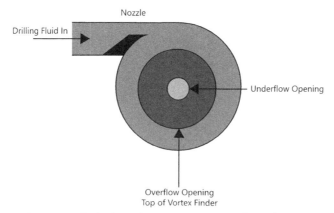

Fig. 19.1 Top view of a hydrocyclone—cross section through upper cylinder.

and the liquid film around them tend to spiral outward and downward for discharge through the solids outlet (underflow).

The centrifugal force of the swirling liquid moves the solids to the outside wall. In drilling operations, hydrocyclones use these centrifugal forces to separate solids in the 15–80 μm size range from the drilling fluid. This solids-laden fluid is discharged from the lower apex of the cone and the cleaned drilling fluid is discharged from the overflow discharge.

Fluid from a centrifugal pump enters the hydrocyclone tangentially at high velocity through a feed nozzle on the side of the top cylinder. As drilling fluid enters the hydrocyclone, centrifugal force on the swirling slurry accelerates the solids to the cone wall. In the center of the rotation, an air cylinder is formed, (Figs. 19.2 and 19.3). The low pressure is created by the rotation of the fluid just as rotation of air creates a cyclone or tornado.

Fig. 19.2 Transparent desilter.

Fig. 19.3 Air cylinder collapsing.

Drilling fluid, a mixture of liquid and solids, rotates rapidly while spiraling downward toward the apex. The higher mass solids move more rapidly toward the cone wall than solids with a lower mass. Movement progresses to the apex opening at the cone bottom. At the apex opening, the solids along the cone wall, along with some drilling fluid, exit the cone. The discharge is restricted by the size of the apex. Most of the drilling fluid and smaller mass particles are forced to reverse flow direction into an upward spiraling path at the center of the cone to exit through the vortex finder.

There are two countercurrent spiraling streams of fluid in a hydrocyclone; one spiraling downward along the cone surface, and the second spiraling upward along the cone center axis. The counter current directions together with turbulent eddy currents concomitant with the extremely high velocities result in an inefficient separation of particles. The two streams tend to co-mingle within the contact regions, and particles are incorporated into the wrong streams. Hydrocyclones, therefore, do not make a sharp separation of solid sizes.

Hydrocyclones are designated arbitrarily by cone diameter at the inlet. Desanders (by convention, cone sizes larger than $6''$) and desilters (by convention, cone sizes less than $6''$) are normally used on unweighted drilling fluids. This is primarily because the cut point of the devices is in the size range of weighting material. Prolonged use of hydrocyclones on a weighted mud will result in a reduction in density and loss of significant weighing material.

The size of oilfield cyclones commonly varies from $4''$ to $12''$ inside diameter. This measurement refers to the inside diameter of the largest cylindrical section of the cyclone. In general, the larger cones have higher cut points and a greater through put. Typical cyclone capacities and feed pressures are shown in the table below. The cut points shown

Table 19.1 Hydrocyclone cut points.

Cone size (I.D.)	4″	5″	6″	8″	10″	12″
Capacity (gpm)	50–75	70–80	100–150	150–250	400–500	500–600
Feed pressure (psi)	30–40	30–40	30–40	25–35	20–30	20–30
Cut point (microns)	15–20	20–25	25–30	30–40	30–40	40–60

are for very light slurries of drilling fluid. The cut points for weighted drilling fluids are much higher (Table 19.1).

Most balanced cones are designed to provide maximum separation efficiency when the head at the inlet is 75 ft. Fluid will always have the same velocity within the cone if the same head is delivered to the hydrocyclone inlet. Pressure can be converted to feet of head with the equation frequently used in well-control calculations:

$$\text{Head, ft} = \frac{\text{Pressure, psi}}{0.052 \times \text{Mud weight, ppg}}$$

The relationship between manifold gauge pressure and drilling fluid weight at constant 75 ft feed head is summarized in Table 19.2.

A centrifugal pump is the ideal choice to supply fluid to the hydrocyclones since it is a constant head device. Once the impeller is selected, the head will remain constant. The input pressure will automatically increase when the mud weight is increased. This is discussed in depth in Chapter 10, Centrifugal Pumps.

Solids, in hydrocyclones, separate according to mass, a function of both density and particle size. However, in unweighted drilling fluids the solids density is a comparatively narrow range. Size has the greatest influence on their settling. Centrifugal forces act on the suspended solids particles, so those with the largest mass (or largest size) are the first moved outward to the wall of the hydrocyclone. Large solids, with accompanying

Table 19.2 Pressure for 75 ft of head for various mud weights.

Pressure, psig	Feed head, ft	Mud weight, ppg
32.5	75	8.34
35	75	9.0
37	75	9.5
39	75	10.0
41	75	10.5
43	75	11.0
45	75	11.5
47	75	12.0
49	75	12.5
51	75	13.0

Table 19.3 Flow rates through hydrocyclones.

Designation	Cone diameter, in	Flow rate thru each cone, gpm
Desilter	2	10–30
Desilter	4	50–75
Desilter	5	70–80
Desander	6	100–150
Desander	8	**150–250**
Desander	10	400–500
Desander	12	500–600

drilling fluid, are concentrated at the cone wall. Smaller particles and most of the drilling fluid concentrate in the inner cylinder of flow.

Larger size (higher mass) particles, upon reaching the conical section, are exposed to the greatest centrifugal force and remain in their downward spiral path. The solids sliding down the wall of the cone, along with the bound liquid, exit through the apex orifice. This creates the underflow of the hydrocyclone.

Smaller particles are concentrated in the middle of the cone with most of the drilling fluid. As the cone narrows, the downward spiraling path of the innermost layers is restricted by the reduced cross-sectional area. A second, upward, vortex forms within the hydrocyclone, the center fluid layers with smaller solids particles turn toward the overflow. At the point of maximum shear, the shear stress within a 4″ desilter is on the order and magnitude of 1000 reciprocal seconds.

Since most hydrocyclones are designed to operate with 75 ft of head at the input manifold, the flow rate through the cones is constant and predictable from the diameter of the cone, (Table 19.3). The "bold" numbers in the table indicate the most common flow rate. Obviously, manufacturers may select different orifice sizes at the inlet of the cone. The orifice size determines the flow rate through the cone at 75 ft of head.

A bank of hydrocyclones (Fig. 19.4) oriented in a straight line experience a head reduction along the manifold. Refer back to Fig. 3.3 and the head loss along the length of the manifold is linear. Therefore, the actual processing rate of each cone along the length of the manifold will reduce proportional to the square root of the head at each cone based on the flow through of a nozzle calculation:

$$\text{Flow rate, gpm} = 19.6 \left(\sqrt{\text{head, ft}} \right) (\text{nozzle diameter, in})^2$$

For example, if the head at the inlet of a desilter manifold is 75 ft and the head at the last cone on the manifold is 70 ft of head, the flow through 0.75 in nozzles would be 95 and 92 gpm respectively.

Fig. 19.5 illustrates and example of a circular style desilter manifold. In this setup, a near-equal head is presented to each cone.

Design features of cyclone units vary widely from supplier to supplier and no two manufacturers' cyclones have identical operating efficiency, capacity or maintenance characteristics. Earlier hydrocyclones were commonly made of cast iron with replaceable liners and other wear parts made of rubber or polyurethane to resist abrasion. Most of the current hydrocyclones are made entirely of polyurethane and are less expensive, last longer, and weigh less.

Manifolding multiple cyclones in parallel can provide sufficient capacity to handle the required circulating volume plus some reserve as necessary. Manifolding may orient the cyclones in a vertical position or nearly horizontal -the choice is one of convenience, as it does not affect cyclone performance

Fig. 19.4 Bank of hydrocyclones.

Fig. 19.5 Circular style desilter package.

The internal geometry of a cyclone seriously affects its operating efficiency. The length and angle of the conical section, the size and shape of the feed inlet, the size of the vortex finder, and adjusting the underflow opening all play important roles in a cyclone's effective separation of solids particles.

Operating efficiencies of cyclones may be measured in several different ways, but since the purpose of a cyclone is to discard drilled solids with minimum fluid loss, both aspects must be considered. In a cyclone, larger particles have a higher probability of reporting to the bottom (underflow) opening, while smaller particles are more likely to report to the top (overflow) opening. The most common method of illustrating particle separation in cyclones is through a cut point curve. The data for the cut point curves below were for processing an unweighted, relatively thin water-base drilling fluids and operated with the proper head applied.

Particle separation in cyclones can vary considerably depending on such factors as feed pressure, mud weight, percent solids and properties of the liquid phase of the drilling fluid. Generally increasing any of these factors will increase the size of solids actually separated by the cyclone and decrease the volume of solids removed.

Most of the hydrocyclones currently used are balanced cones. Balanced cones mean that only a small amount of liquid exits the lower apex because the air cylinder in the middle of the cone is about the diameter of the orifice. To set a cone to balance, slowly open the apex discharge while circulating water through the cone. When a small amount of water is discharged, and the center air core is almost the same diameter as the opening, the cone is "balanced."

A hydrocyclone operating with a spray discharge will remove a significantly greater amount of solids than a cone in "rope" discharge. The initial impression for rope discharge is that the cone is removing more solids; but the plugging of the lower apex simply means that some of the solids which would be separated are being swept back into the active system. Hydrocyclones should not be operated in rope discharge because it will drastically reduce the cone separating efficiency. In a rope discharge, the solids become crowded at the apex, cannot exit freely from the underflow and become caught by the inner spiral reporting to the overflow. Solids which otherwise would be separated are forced into the overflow stream and returned to the mud system. This type of discharge can also lead to plugged cones and much higher cyclone wear (Fig. 19.6).

Rope discharge is a situation where material pours from the cone apex as a slow moving cylinder (or rope). The hydrocyclone effects only inefficient solids-liquid separations. The apex velocity in rope discharge is far less than that in spray discharge. Separations are less efficient and, because of the lower velocity, fewer solids are discarded.

Rope discharge should be immediately corrected to re-establish the higher volumetric flow and greater solids separation of spray discharge. A rope discharge indicates equipment is overloaded, and additional hydrocyclones may be needed. This is the reason that desanders are usually used to process drilling fluid which has passed through API 100

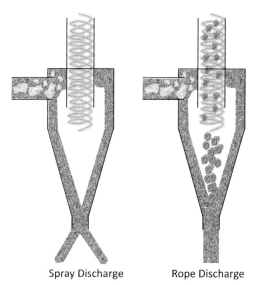

Fig. 19.6 Illustration of rope discharge.

and coarser screens. Fine screens, such as API 140 and finer, will decrease the normal solids load so that desilters have a spray discharge. Although a spraying underflow will also discharge more fluid, the benefits of more efficient solids removal and less cone wear outweigh cost of the additional fluid loss.

Cones with longer lower sections (Fig. 19.7), reportedly provide a sharper cut-point hen separating drilled solids from a drilling fluid.

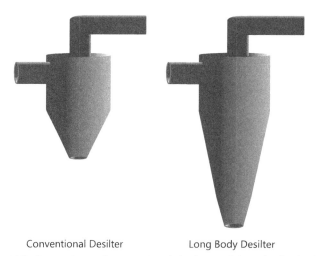

Fig. 19.7 Comparison of conventional desilter with long body desilter.

Several conditions restrict good separation of solids with a hydrocyclone. They are:
- Excessive solids concentration
- Excessive volumetric feed rate per cone
- Excessive fluid viscosity
- Restricted (too small) apex
- Inadequate feed pressure

If too many larger solids are entrained within the central vortex stream to exit with the overflow, the discharge pattern changes from spray to "rope discharge"—characterized by a cylindrical or "ropy" appearance. There is no air core sucking through the center of the cone. Here the apex acts as a choke restricting flow, rather than a weir. The shear rate at the point of most separation in spray discharge is around 1000 s^{-1}. This would indicate that the viscosity of the drilling fluid with the 600 RPM viscometer reading would be an indicator to separation performance.

Discharges from the apex of these cones are discarded when normally used on unweighted drilling fluids. Prolonged use of these cones on a weighted drilling fluid results in a significant mud weight reduction caused by discard of weighting material. When these cones are used as part of a mud cleaner configuration (Chapter 20), the cone underflow is presented to a shaker screen. The shaker screen returns most of the barite and liquid to the drilling fluid system; rejecting solids larger than the API screen size. This is a common application of unbalanced cones, since the cut point is determined by the shaker screen and not the cone.

Cut points

In spray discharge, for any set of cone diameters, feed slurry compositions, flow properties, volumetric flow rates, and pressure conditions, some particle size (mass) which will be 100% discarded out of the apex. Conversely, there will also be some particle size (mass) which will be 100% returned, through the vortex finder, to the drilling fluid.

The D50 cut point of a solids separation device is defined as the particle size at which one-half of weight of the particles goes to the underflow and one-half of the weight of the particles goes to the overflow. The cut point is related to the inside diameter of the hydrocyclone (Fig. 19.8). For example, a 12 inch cone is capable of a D50 cut point around 60–80 µm, a 6 inch cone around 40–60 µm, or 4 inch cone around 20–40 µm. These cut points are representative for a fluid that contains a low solids content. The cut point will vary according to the size and quantity of solids in the feed along with the liquid phase viscosity of the fluid in the feed.

For every size and design of cone operating at a given pressure with feed slurry of given viscosity, density and solids distribution, there will be a certain size (mass) of particle that shows no preference for either top or bottom discharge. As a result 50% of that particular size leaves through the vortex and 50% leaves through the apex.

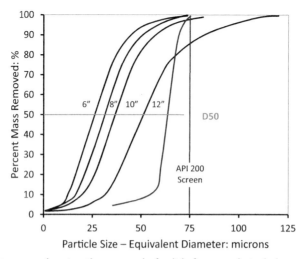

Fig. 19.8 Cut point curves showing the removal of solids from a relatively low-weight drilling fluid.

The particle size which exits 50% through the apex and 50% with the overflow is variously termed the "median cut", or "median size particle", or more frequently in drilling operations, the D50 cut point.

The median cut, or D50 cut point, does not mean that all larger particles exit at the apex and smaller particles exit at the vortex. The D50 cut point of a solids separation device is defined as that particle size at which one-half of the weight of those particles goes to the underflow and one-half of the weight of those particles goes to the overflow. A D30 cut refers to that particle size which is concentrated 30% in the underflow and 70% in the overflow. A D80 cut refers to that particle size found 80% in the underflow, and 20% in the overflow.

The cut point is related to the inside diameter of the hydrocyclone. For example, a 12 inch cone has a D50 cut point for low gravity solids in water around 60–80 μm, a 6 inch cone around 30–60 μm and a 4 inch cone around 15–20 μm.

Desanders

Desanders are hydrocyclones larger than 5″ in diameter (6″, 8″, 10″ or 12″ I.D.). Generally, the smaller the cone, the smaller size particles the cone will separate. Desanders are primarily used to remove the high volumes of solids associated with extremely fast drilling of a large diameter hole, especially when a fine screen shaker is not available.

Desanders are installed downstream from the shale shaker and degasser. The desander removes sand-sized particles and larger drilled solids which have passed through the shaker screen and discards them along with some liquid. The partially-clean drilling fluid is discharged into the next compartment downstream.

Summary

Use desanders in unweighted mud when shakers are unable to screen down to API 140 (100 μm). The role of desanders is to reduce loading downstream on desilters. A desander ahead of the desilter relieves much of the solids loading on the desilter and improves its efficiency. High rates of penetration, especially unconsolidated "surface hole" where the largest diameter bits are used, result in generation of larger concentrations of drilled solids. This may place desilters in rope discharge. So desanders which have greater volumetric capacity and can make separations of coarser drilled solids are placed upstream of desilters. Desanders remove higher mass (coarser drilled solids) during such periods of high solids loading. Desilters can then efficiently process the reduced solids content overflow of the desanders.

When installing a desander, follow these general recommendations:
- Size the desander to process 100%–125% of the flow rate entering the suction tank of the desander.
- Keep all lines as short and straight as possible with a minimum of pipe fittings. This will reduce loss of pressure head on the feed line and minimize back-pressure on the overflow line.
- Do not reduce the diameter of the overflow line from that of the overflow discharge manifold.
- Direct the overflow line downward into the next downstream compartment at an angle of approximately 45 degree. The overflow discharge line should never be installed in a vertical position, doing so may cause excessive vacuum on the discharge header and pull solids through the cyclone overflow thus reducing the cyclone's efficiency.
- Install a vacuum breaker in the overflow line if the desander is over 8–10 ft above the drilling fluid level in the mud tanks.
- Install adequate walkways and hand rails around the desander to allow proper maintenance.
- Keep the end of the discharge line above the surface of the liquid level in the pits to avoid creating a vacuum in the line.
- Install a low equalizer line to permit back-flow into the desander suction.

Operating the desanders at peak efficiency is a simple matter, since desanders are relatively uncomplicated devices. Here are a few fundamental principles to keep in mind:
- Operate the desander unit at the supplier's recommended head (or feed manifold pressure, usually around 30–35 psi). A feed pressure that is too low decreases the separation efficiency, while too high a pressure shortens the life of cyclone wear parts.
- Check cones regularly to ensure the discharge orifice is not plugged.
- Run the desander continuously while drilling and shortly after beginning a trip for "catch-up" cleaning.

- Operate the desander with a spray rather than a rope discharge to maintain peak efficiency.

Use of desander is normally discontinued when expensive materials such as barite or some polymers are added to a drilling fluid because a desander will discard a high proportion of these materials along with the drilled solids. Similarly, desanders are not generally cost effective when a non-aqueous drilling fluid (NADF) is used because the cones also discard a significant amount of the liquid phase.

Desilters

A desilter uses smaller hydrocyclones (usually 4″ or 5″ I.D.) than a desander and therefore, generally removes smaller particles. The smaller cones enable a desilter to make the finest particle size separation of any full flow solids control equipment—removing solids in the range of 15 μm and larger. This makes it an important device for reducing average particle sizes and removing abrasive grit from unweighted drilling fluids.

The cyclones in desilter units operate on the same principle as the cyclones used as desanders. They simply make a finer cut and the individual cone throughput capacities are less than desander cones. Multiple cones are usually manifolded in a single desilter unit to meet throughput requirements. Desilters should be sized to process 100%–125% of the flow rate entering the suction tank for the desilters. (Note that this does not say 100%–125% of the flow rate down the hole.)

Installation of the desilters is normally downstream from the shale shaker, degasser, and desander and should allow ample space for maintenance. Here are some fundamentals for installing desilters:

- Take the desilter suction from the compartment receiving fluid processed by the desander.
- Do NOT use the same pump to feed both the desander and desilter. If both pieces of equipment are to be operated at the same time, they should be installed in series and each should have its own centrifugal pump.
- Keep all lines as short and straight as possible.
- Install a guard screen with approximately $1/4''$ openings at the suction to the desilter pump to prevent large trash or drilled solids from entering the unit and plugging the cones.
- Position the desilter on the pit high enough so the overflow manifold will gravity-feed fluid into the next downstream compartment at an angle of approximately 45 degree. REMEMBER-no vertical overflow discharge lines.
- Keep the end of the discharge line above the surface of the liquid in the tanks to avoid creating a vacuum in the line.

- Install a low equalizer line for back flow to the desilter suction compartment.
- Running a desander ahead of a desilter is required if coarse screens are used on the shale shakers. Desanders take a big load off the desilters and improve their efficiency.
- Operate the cones with a spray discharge. Never operate the desilter cones deliberately with a rope discharge since a rope underflow cuts cone efficiency in half or worse, causes cone plugging, and increases wear on cones. Use enough cones and adjust the cone underflow openings to maintain a spray pattern.
- Operate the desilter unit at the supplier's recommended feed manifold pressure. This is generally between 70 and 80 ft of head. Too much pressure will result in excessive cone wear. As mud weight increases, feed pressure will also increase. As a rule of thumb, desilter cones should operate at a feed pressure of 4 times mud weight. [Calculate this with the equation used in well control: Pressure, psi = 0.052 (mud weight, ppg) (head, ft)].
- A centrifugal pump is a constant head device so the pressure will automatically increase as the mud weight increases.
- As mud weight increases, the cone bottoms can be opened slightly to help increase solids removal efficiency.
- Check cones regularly for bottom-plugging or flooding, since a plugged cone allows solids to remain in the active system. If a cone bottom is plugged, unplug it with a welding rod or similar tool. If a cone is flooding, the feed maybe partially plugged or the bottom of the cone maybe worn out.
- Run the desilter continuously while drilling and also for a short time during a trip. The extra cleaning during the trip can reduce overload conditions during the period of high solids loading immediately after a trip.
- When hydrocyclones are mounted more than 8–10 ft above the liquid level in the mud tanks, a siphon breaker should be installed in the overflow manifold from the cones.

Maintenance

The smaller cones of a desilter are more likely than desander cones to become plugged with oversized solids, so it is important to inspect them often for wear and plugging. This may generally be done between wells unless a malfunction occurs while drilling. The feed manifold should be flushed between wells to remove trash. Keep the shale shaker well maintained-never by-pass the shaker or allow large pieces of material to get into the active system. (Note: the fact that some solids can plug the bottom of a cone means that all of the fluid from the well did not pass through the shaker screens.)

A desilter will discard an appreciable amount of barite because most barite particles fall within the silt size range. Desilters are therefore not recommended for use with weighted drilling fluids. Similarly, since hydrocyclones discard some liquid along with the drilled solids, desilters are not normally used with NADF unless another device (centrifuge or mud cleaner) is used to decrease the liquid discard in the cone underflow.

Hydrocyclone tanks and arrangements

Hydrocyclones are arranged with the larger cone size unit (desanders) upstream of the smaller cone size unit (desilters). A separate tank is needed for each size unit. Generally a desander size and a desilter size are available as part of the rig equipment. The tank arrangements are discussed in detail in Chapter 8.

Suction is taken from the tank immediately upstream of that tank into which discharge is made. The number of cones in use should process 120% of the flow rate of all fluids entering the suction tank for the hydrocyclone. A back flow between the hydrocyclone discharge and suction compartment insures adequate processing. Rules of thumb based on rig circulation rates are usually inadequate if the plumbing is not arranged correctly. For example, if a 500 gpm hydrocyclone over-flow is returned to the suction compartment and a 400 gpm rig flow rate enters the suction compartment, adequate processing is not achieved even though more fluid is processed than is pumped downhole. In this case, the flow entering the hydrocyclone suction compartment is 900 gpm. The fraction of drilling fluid processed would be 500 gpm/900 gpm or 56%. This is discussed in depth in Chapter 11, Fraction of Drilling Fluid Processed.

Hydrocyclone operating tips

- Other than cone and manifold plugging, improperly sized or operated centrifugal pumps is the greatest source of problems encountered with hydrocyclones. Centrifugal pump and piping sizing are critical to efficient hydrocyclone operation. A pressure gauge should be mounted on the inlet manifold.
- Hydrocyclones should always have a pressure gauge installed on the inlet to quickly determine if proper feed head is supplied by the centrifugal pump.
- Hydrocyclones are usually mounted in the vertical position, but may even be mounted horizontally. Cone orientation does not matter as the separating force is that supplied by the centrifugal pump.
- Feed slurry must be distributed to a number of hydrocyclones operating in parallel. A "radial manifold" provides each cyclone with the same slurry regarding feed solids concentration and particle size distribution at the same pressure. An "in line" manifold sees the higher mass (larger diameter) particles tend to pass by the first cyclones and enter the last cones. These particles have a higher energy, and resist entering the first cones. Thus, the last cones in an in-line manifold receive a higher concentration of coarse feed particles. Performance from cones in an in-line manifold will not be identical, as feed concentrations and particles size distributions are different for the

various cones. Further, if the last cyclone(s) in an in-line manifold are taken off line, the end of the manifold has a tendency to plug.
- To minimize loss of head along the feed line and back pressure on the overflow (top) discharge line, keep all lines as short and as straight as possible with minimum of pipe fittings, turns and changes in elevations. Pipe diameters should be 6 or 8 in.
- Operate balanced cones in spray discharge with a central air suction, and check cones regularly to insure apex discharge is not plugged. Operate unbalanced cones according to manufacturer's instructions.
- When balanced cones no longer spray-discharge, either too many solids are being presented for design processing, or large solids have plugged the manifold or apex, or the feed pressure is not correct. If feed pressure is the correct 75 ft of hydraulic head, often the inability to maintain spray discharge can be traced to the shale shakers. Check for torn screens, open by-pass, or improperly mounted screens. Otherwise assure that there are sufficient hydrocyclones to process the total fluid being circulated by the mud pumps.
- Install a low ("bottom") equalizer to permit backflow from the discharge tank into the suction compartment. Removable centrifugal pump suction screens reduce plugging problems.
- Operate hydrocyclones at recommended feed head, usually around 75 ft. Efficiency is decreased by too low a feed head. Too high a head wears hydrocyclone parts and provides insignificant separation improvements.
- Do NOT use the same pump to feed desilter and desander. Each unit should have its own centrifugal pump(s).
- Run desilter continuously while drilling with unweighted mud, and process at least 100% of total surface pit volume after beginning a trip.
- Run desander when shale shakers will not handle API140 (100 μm) or finer screens.
- Install a suction screen with approximately $1/4$ inch openings to prevent large trash from entering and plugging the inlets and/or cone apexes.
- Regularly check cones for bottom plugging or flooding. Desilter cones will plug more often than desander cones. Plugged cones may be cleaned with a welding rod. A flooded cone indicates a partially plugged feed or a worn cone bottom section.
- Between wells or in periods when drilling is interrupted, flush manifolds with water and examine the inside surfaces of the cones.
- Keep shale shaker well maintained, and never by-pass it—This includes dumping the possum belly into the sand trap before a trip.
- Hydrocyclones discard absorbed liquid with the drilled solids. Discharge solids dryness is a function of the apex opening relative to the diameter of the vortex finder.

Appendix 19.A

In a lighter moment, as a committee was writing the AADE Mud Equipment Manual, someone suggested that it would be appropriate to write the ten commandments for each piece of solids removal equipment. George Ormsby was the only one who wrote one.

The ten commandments for hydrocyclones

(Modified from private communication from George Stonewall Ormsby).
 1. Thou shalt forthwith remove from thy unit any cone that is plugged, yea, that is plugged even partially. For a cone that is plugged worketh not for thee, and a cone that worketh not for thee worketh iniquity against thee.
 2. Thou shalt not operate any other device in parallel with the hydrocyclone unit, for the hydrocyclone unit is a jealous unit. The results will be displeasing for all to see.
 3. Thou shalt apply thyself to centrifugal pumps with diligence, for it is written that without good centrifugal pumping there can be no cleaning.
 4. Thou shalt not hide the hydrocyclones in remote and cursed places. For hydrocyclones need to be observed by day and by night; lest they fall by the wayside and become laggard and thou knowest not.
 5. Thou shalt not place the cones on the highest mountain nor in the lowest valley, nor in any other inaccessible place, for maintenance is the staff of life for the cyclone. For this, it will repay thee forever.
 6. Thou shalt not close the discharge of the hydrocyclones from the suction of the centrifugal pump. Instead, thou shalt permit the discharge to equalize in the mud tank with the suction compartment. Yea, by a lower route thou shalt permit it and this will save thee and thy seed many evils.
 7. Thou shalt not close the underflow of the hydrocyclone nor make it heavy. This shall liken it to a serpent and thou processing shalt come to naught. Instead thou shalt open the underflow like unto the rain; for this is the essence of fine drilled solids removal.
 8. Thou shalt not cast away drilling fluid in the wholeness thereof to rid thyself of unwanted solids, but when barites are not therein, thou shalt instead release the cone underflow from all bondage, for by this shall your drilling fluid be freed of unwanted solids and your purse will remain heavy and your family will know their parents.
 9. Thou shalt know and love thy hydrocyclone units and care for it as thine own, and it will cause thee and thy seed to prosper forever.
 10. Let not thy new man servants, nor even thy maid servants, remain in darkness but instead teach them of the many bountiful blessings which accrue from following the wise teachings of the hydrocyclones.

CHAPTER 20

Mud cleaners

Introduction

Mud cleaners are a combination of hydrocyclones mounted above shaker screens with small openings (Fig. 20.1). Mud cleaners can be leased, rented, purchased as independent units, or assembled on a drilling location.

Frequently, on very large drilling rigs, several shakers are available to process the drilling fluid. When the well gets deeper and the mud weight increases, one of the shakers is not needed to process all of the fluid coming from the well. A bank of desilters can be mounted over this shaker and it used as a mud cleaner for the weighted drilling fluid, as shown in Fig. 20.2.

The flow from the flow distributor goes to the three main shakers and no flow from the flow line goes to the shaker used for the mud cleaner. A solid plate is used below the desilters when unweighted drilling fluid is in the hole. All of the desilter underflow is discarded. When the barite goes in, the plate is removed and the desilter underflow reports to the shale shaker screen. The shaker discards the solids larger than 74 μm and returns the solids which pass through the screen. The desilter underflow does not have any low-shear-rate viscosifiers and will not support solids. The desilter overflow is then routed so that it blends with the fluid going through the shaker and sends this fluid to a tank downstream of the desilter suction tank.

A history of the mud cleaner development is presented in the appendix to this chapter. When mud cleaners were invented in 1971, main shale shakers on drilling rigs were either unbalanced elliptical or circular motion machines. The finest screen at that time could only separate solids larger than about 177 μm (API 80) from a normal drilling rig circulating rate. The solids larger than 75 μm and smaller than 177 μm remained in the drilling fluid. These solids are very detrimental to the drilling process.

Most 4-in hydrocyclones discard around 1–3 gpm for every 50 gpm input. With a 1000 gpm flow rate input, the underflow from twenty 4″ hydrocyclones would only be 20–60 gpm. Most shale shakers at that time could process this small flow rate through 74 μm (API 200) screens. Barite smaller than the openings, as well as drilled solids smaller than the openings, returns to the drilling fluid. Solids larger than the screen opening and the solids clinging to those larger solids are discarded. Solids larger than 74 μm make filter cakes incompressible and also increase the coefficient of friction between the drill string and the well bore. This is not unlike making a coarse sandpaper filter cake; so stuck pipe was common.

Fig. 20.1 Commercial mud cleaner.

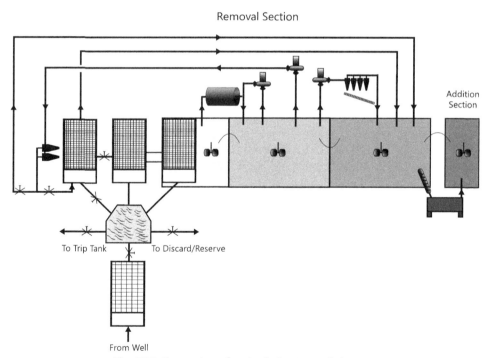

Fig. 20.2 Conversion of main shaker to mud cleaner.

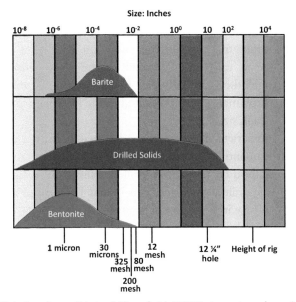

Relative size solids in drilling fluid (ASTM sieve sizes shown).

The D50 cut point of a 4″ hydrocyclone is usually reported to be around 10–20 μm for an unweighted drilling fluid (Fig. 20.3).

With the same cut point in a drilling fluid weighted with barite, an enormous amount of barite would be contained in the underflow even in an 11–12 ppg drilling fluid. The cut point curve of a weighted drilling fluid processed through a desilter is neither sharp nor is the D50 cut point as low as 20 μm.

The barite in an 11.0 ppg drilling fluid does not have a constant size distribution after several days of circulation. The separation curve in Fig. 20.3 indicates that the "cut point"

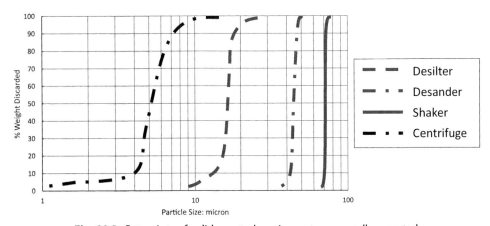

Fig. 20.3 Cut points of solids control equipment as normally reported.

curve for 4″ hydrocyclones is not a sharp separation as might exist for a shale shaker screen or even a 4″ hydrocyclone with an unweighted drilling fluid.

Note about cut points: traditionally the cut point curve is a smooth line drawn through the flat spots on each interval used in the calculations. For example, in Fig. 20.4, about 13% of all the solids between 40 and 50 μm reported to the underflow. This does not indicate how many were presented to the hydrocyclone—only the ratio of the solids discarded in each size range.

Barite specifications restrict solids to sizes predominately smaller than 74 μm. Hydrocyclones remove large quantities of barite from a weighted drilling fluid and for this reason were not generally used to remove drilled solids from weighted drilling fluids. This means that all of the drilled solids between 74 and 177 μm are available for removal but could not be removed with equipment available at that time. The mud cleaner was invented to remove those drilled solids (Fig. 20.5).

Very little barite has a size above 50 μm in this circulating drilling fluid. In the separation curve, about 50% of the solids between 50 μm and 120 μm are separated and 50% remain in the drilling fluid. Perhaps this is surprising until the barite distribution in the 11.0 ppg drilling fluid is considered (Fig. 20.4). If no solids larger than 10 μm were in a drilling fluid processed through a desilter cone, the cut point for this cone would obviously be lower than 10 μm.

Not much barite is in the large size range that will report to the underflow of the hydrocyclones. If this was not true and the cut point of a 4″ hydrocyclone was in the 10–20 μm range as reported for unweighted drilling fluid, the mud cleaner screens would fail from a weight overload during the first few minutes of operation. The 20 μm cut point normally reported for 4″ hydrocyclones should also contain some

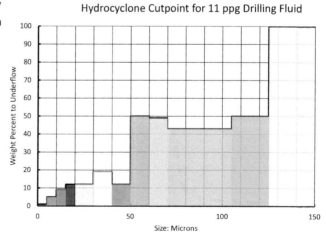

Fig. 20.4 Separation curve for 4″ hydrocyclones processing an 11 ppg drilling fluid.

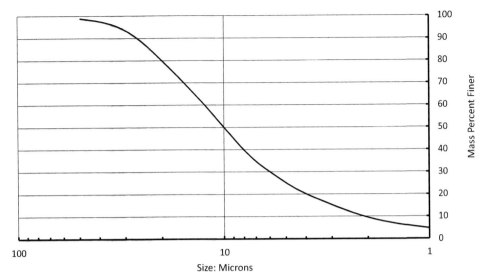

Fig. 20.5 Barite size distribution in an 11.0 ppg drilling fluid.

statement about the size distribution injected into the cones. If there were no solids larger than 12 μm, the cut point would obviously be lower than the 12 μm.

Since the mud cleaner screen separates solids larger than the opening size, a low and sharp cut point is not needed from the hydrocyclones. The underflow of the hydrocyclones should be very wet. A better separation is made on the shaker screens if ample fluid is available to allow the screen to separate the solids from the slurry. To insure sufficient liquid is available for successful screening, some mud cleaners have a small hose attached to the desilter overflow manifold so that some additional fluid can be used to wet the solids on the screen.

Some mud cleaners are designed to return some of the cone overflow to the screen to assist solids separation. This is preferable to spraying water or oil onto the screen to enhance the solids separation. Water or oil sprayed on the screen dilutes the drilling fluid and requires the addition of more barite and chemicals. For example, if only 2 gal/min are sprayed onto the mud cleaner screen, 2880 gallons would be added to the system in one day. This 68.6 bbl of liquid phase will require the addition of all of the solids and chemicals required to create clean drilling fluid. This would be acceptable if about 18 bbl of drilled solids are discarded from the solids control equipment in a slurry of about 35% volume drilled solids in a drilling fluid slurry. This quantity of drilled solids could be generated if only 180 ft of a 10″ diameter hole is drilled during the 24 h period. This, of course, may not be the total quantity of drilled solids reaching the surface, because many solids fall into the well bore from the side of the hole.

When the linear motion shale shakers were introduced, mud cleaner utility appeared obsolete. Linear motion shale shakers could separate drilled solids larger than the barite while handling all of a drilling rig circulation. However, many times desilters are plugged with solids too large to pass through a shale shaker screen in unweighted drilling fluids. These drilled solids reached the desilters and plugged the bottom apex. Their method of entry into the drilling fluid system varied from holes in shaker screens, carry over from the discard end of the shale shakers, to improper mounting of shaker screens, and even gaps left between the upper end of shaker screens and the back tanks. Plugged desilter cones are a common sight around drilling rigs if no one is assigned the task of unplugging. Because this is such a common occurrence, it is not surprising that mud cleaners downstream of API 200 (74 μm) shaker screens are usually loaded with solids. Sometimes solids over load an API 200 screen on a mud cleaner even though an API 200 screens are mounted on the main shale shakers.

With the advent of 200 mesh screens on linear-motion shale shakers in the late 70s and early 80s, mud cleaner usage decreased drastically. Slowly, they have returned to the removal system as trial usage indicates that they still remove a large quantity of solids. Plugging of the lower apex of hydrocyclones is a chronic problem on almost all rigs. This indicates that very large solids frequently bypass shale shaker screens. These large solids enter the mud pits. If solids large enough to plug a 4″ cone are in the system, a wide range of other solids are present also. A mud cleaner is usually needed and their 4″ desilter cones also plug frequently. Even when linear motion shale shakers are operating properly, mud cleaners remove a significant quantity of drilled solids.

Mud cleaners do not compete with the centrifuges. Centrifuges remove colloidal size particles, usually smaller than the weighing agent, from drilling fluid. Mud cleaners remove drilled solids larger than the weighing agent. Improperly used, mud cleaners will also discard large quantities of barite. The slurry on the screen must not "dewater" too soon or the barite will "clump" together and be discharged. Frequently, a small stream of drilling fluid from the desilter overflow line will prevent loss of too much barite.

Mud cleaner hydrocyclones typically receive drilling fluid through a centrifugal pump, just as desanders and desilters do. There have been reports of solids degradation because of the centrifugal pump. Particle size measurements at a variety of rig sites fail to confirm this. Solids do eventually degrade in size but slowly and in other places as well. Some of those places include the turbulent zone of decanting centrifuges, mud guns, crushing by the drill pipe, and the high-shear drill bit nozzles.

Most cut point curves of 4″ desilters show that only about 5% of the underflow from the cones is less than 5 μm. Viscosifiers are in this size range. The underflow that passes through the mud cleaner screen has such a small *amount of viscosifier and colloidal material that it will* not support the other solids. The mud cleaner should be viewed as another method of removing drilled solids larger than the fine mesh size on the shaker screen.

If this material remains in the drilling fluid, they will also eventually become undesirable ultra-fine solids. If a mud cleaner is not used, colloidal material will increase even faster because of the deterioration of larger solids that the mud cleaner could remove.

In unweighted drilling fluids, the mud cleaner can be used to salvage an expensive liquid phase, such as oil or a high concentration of KCl. Usually better results are obtained if the hydrocyclone underflow is fed to a decanting centrifuge instead of being screened. A centrifuge processing the underflow from hydrocyclones would be a better choice to return the liquid phase, if one is available. A mud cleaner can be used but not as effectively.

It is best not to justify any solids control equipment on the basis of "saving barite". Solids control equipment is designed to remove drilled solids. There is no piece of equipment currently available which will separate barite from low gravity solids. The fake sales pitch of "a barite recovery centrifuge" is common in the oil patch. The underflow of a centrifuge does contain an appreciable amount of barite. However, it may also contain an appreciable quantity of drilled solids. A centrifuge separates by mass not species.

Currently the hydrocyclones and screens are used to diminish the liquid discharge from a system. This combination is called a "dryer" when one of the high "G" machines is used to process the hydrocyclone underflow.

Uses of mud cleaners

The principal use of mud cleaners has always been the removal of drilled solids larger than barite. Sufficient drilling fluid by-pass shale shakers so that even with 74 μm (API 200) screens on the shakers, many drilled solids are removed from a weighted drilling fluid. When linear motion shakers permitted API 200 screens to process all of the rig flow, mud cleaner usage declined rapidly. However, whenever a mud cleaner was used downstream, the screens were loaded with solids. Verification of larger solids by-passing the shale shaker is evident from the prevalence of plugged desilters on a rig. Plugged desilters are very common. Usually, no one is assigned to unplug desilters, so they do not get unplugged. (This has a great effect on the solids removal efficiency.)

These larger solids can come from holes in the main shaker screen or a variety of other places. One of the most prevalent is from the back tank (possum-belly) of a shale shaker. Derrickmen are told to wash the screens off before a trip to prevent dried drilling fluid from plugging the shale shaker screens. The procedure usually involves opening the shale shaker by-pass valve and dumping the accumulated solids into the tank immediately below the shaker. This compartment is usually called a sand trap or settling tank. Unfortunately, all of these larger solids do not settle in this tank. Within minutes of starting circulation with a new bit on bottom, hydrocyclones plug. Another prevalent method of getting large solids downstream of a shale shaker is dumping the trip tank into the compartment below the shale shaker. Again, solids do not settle but do plug desilters.

When solids plug the discharge port of a desilter, that desilter no longer removes solids. The solids removal efficiency decreases accordingly. If over half of the hydrocyclones are plugged, a properly designed removal system will suffer significantly; costs and trouble will increase accordingly.

A secondary use of mud cleaners is the removal of drilled solids from unweighted drilling fluids with a very expensive liquid phase—such as the initial application with the potassium chloride drilling fluid. In this case, the underflow from desilters is screened to remove solids larger than the screen opening. Solids and liquids pass through the screen and remain in the system. This is beneficial for non-aqueous drilling fluids as well as saline water-based drilling fluids.

In unweighted drilling fluids, the desilter underflow could be directed to a holding tank. A centrifuge could separate the larger solids for discard and return the smaller solids and most of the liquid phase to the drilling fluid system. This method is easy to apply if a centrifuge is already available on a drilling rig. More drilled solids would be rejected by the centrifuge than by the mud cleaner screen; however, renting a centrifuge for this purpose may be more expensive. Both techniques have been used extensively in the field.

The cost of the liquid phase provides an incentive to recover as much as possible: But, this liquid phase from either a centrifuge or a mud cleaner screen through-put also recovers all of the small solids. In some wells these solids do not have a significant impact on drilling or trouble costs. In general, drilled solids in weighted drilling fluids have much more obvious impact on drilling costs than in unweighted drilling fluids. The impact on unweighted drilling fluids may be less obvious. In an unweighted drilling fluid, mud weights can be maintained around 8.8–8.9 ppg with proper solids control. Poor solids control, equipment or arrangement, may make it almost impossible to keep the mud weight below 9.5–10 ppg. A one pound per gallon difference will make 520 psi difference in bottom hole pressure at a depth of 10,000 ft. This reduces the drilling rate of roller cone bits from chip hold-down and from rock strengthening. Not only does it prevent achieving a good drilling rate but it also affects hole cleaning. With unweighted drilling fluids, the plastic viscosity can be reduced from 12 cp–6 cp when the mud weight is decreased by removing solids. For a yield point of 10 lb/100 sqft, this would increase the "K" value from 224 to 466 eff.cp. This means the bore cleaning would be significantly improved (See discussion in Chapter 6 on cuttings transport.). Good solids control requires generating large cuttings and bringing them to the surface without deterioration.

In some areas, mud cleaner discards do contain almost all barite. This is an indication that the formations being drilled are dispersing into the drilling fluid. Formations containing large quantities of smectite drilled with a fresh water drilling fluid will tend to disperse into very small particles. This is an indication that the drilling fluid should be more inhibitive. Generally, the mud cleaner should be shut down and a centrifuge used to remove the very small particles.

Most operations go through serious expensive, agonizing learning experiences when switching from unweighted drilling fluids to weighted drilling fluids. Decisions to retain the liquid phase and entrained solids may save some money with liquid recovery but may also impact the final drilling cost because it more directly affects trouble costs. Trouble costs, or failure to make hole because of problems is obvious to anyone watching operations. The increase in drilling cost because of poor performance with unweighted drilling fluids may be more subtle and not detected in a review of operations. For this reason, more attention seems to be directed toward solids control while drilling with weighted than with unweighted drilling fluids.

Avoid the common mistake of evaluating systems with immediate temporary cost evaluations instead of total impact on drilling costs. For example, an analysis of replacing barite with drilled solids appears to be an easy calculation. If drilled solids are allowed to accumulate and increase mud weight, the barite additive cost is avoided until the mud weight approaches 11 pounds per gallon. The amount of barite "saved" is easily calculated. Even at five cents per pound, this barite savings cost is significant; However, this will result in a much more expensive well. Review the problems associated with not removing drilled solids discussed earlier and expect to see most of those problems arise from weighting a drilling fluid with too many drilled solids.

Situations where mud cleaners may not be economical

1. When drilling highly dispersible shales with fresh water-based fluids, most of the shales cuttings which reach the surface may be much smaller than the screen openings on the mud cleaner. A centrifuge is highly recommended in this case.
2. When transporting solids in long intervals of horizontal holes with high flow rates and pipe rotation, the solids are frequently ground so small that they easily pass through the mud cleaner screens. One such situation involved several thousand feet of horizontal holes and almost all of the solids reaching the surface were smaller than 44 μm (API 325). If the water based drilling fluid carrying capacity is modified with the addition of excess XC polymer to create an elastic modulus, larger solids would report to the surface in which case a mud cleaner would be beneficial.

Location of mud cleaners in a drilling fluid system

Mud cleaners are normally positioned in the same location as desilters in a drilling fluid system. Frequently, the desilters, or hydrocyclones, are used in the unweighted portion of a borehole by diverting the underflow away from the mud tanks. When a weighting agent, barite or hematite, is added to the system, screens are placed on the mud cleaner shakers. Solids discarded from the hydrocyclones are sieved to discard solids mostly larger than barite and return solids smaller than the screen size with most of the liquid phase. [Practical tip: barite goes in; screens go on.]

Another method has also been used to create a mud cleaner using one of the main shale shakers, mostly on offshore rigs. When several linear, or balanced elliptical, shale shakers are needed to handle flow in the upper part of a well bore, fewer shakers can handle the flow after the hole size decreases and the mud weight is increased. Rigs have been modified so that as many as twenty 4″ hydrocyclones have been mounted above one of the main shakers. The feed and overflow (light slurry) from the hydrocyclones are plumbed as usual. (See Chapters 8 and 9 on mud tank arrangements). All of the desilter underflow is discarded in the unweighted part of the hole as usual, while the shale shaker is screening drilling fluid from the flow line. Usually, this is the largest flow rate expected while drilling that well. As the well gets deeper, a weighted drilling fluid is required and usually the flow rate is lower. When barite goes in, a valve prevents flow line drilling fluid from going to one of the main shale shakers. The desilter underflow is diverted onto the shale shaker screen so that the shale shaker becomes a mud cleaner.

Operating mud cleaners

When the first mud cleaners were introduced into the field, they had to be shut off during weight-up. A significant amount of barite was discarded during the first circulation. Actually, this revealed that the mud tanks were plumbed improperly. Drilling fluid was frequently pumped through mud guns from the additions or suction section back upstream to the removal tank. Barite can meet API specifications and still have a large amount that will be removed with an API 200 (74 µm) screen. API allows 3 wt% of the barite to be larger than 74 µm. This means each 100 lb sack of barite could contain 3 lb of solids which will be removed by the fine screens. If the barite has passed through a drill bit nozzle, the many particles will split and not be removed with such a screen.

One comment frequently heard when a weighted drilling fluid is initially passed through a mud cleaner is: "It is throwing away all my barite". What creates such a comment? The mud weight is decreasing and more barite than normal is required to maintain mud weight. When any solids are removed from a drilling fluid (barite or drilled solids), mud weight will decrease. Actually, removal of solids larger than 74 µm is beneficial to drilling a trouble-free hole—whether those solids are drilled solids, barite, gold, silver, or diamonds. Those solids make a poor, incompressible filter cake and lead to stuck drill strings and poor cement barriers in the annulus. The appearance of the screen discard from a mud cleaner is similar to the underflow, or heavy slurry, from a centrifuge. Although visually, it appears to have mostly barite, tests will reveal that is not true.

API specifications for barite state that 3% by weight may be larger than 74 µm. If 100,000 lb of barite is added to a drilling fluid during a weight-up, 3000 lbs of barite could be removed by an API 200 screen. This is one reason that fluid from the addition compartment should not be circulated upstream. The main shale shaker will also discard most of this size barite from an API 200 screen. The barite is not as noticeable because of the quantity of drilling fluid normally clinging to the shaker discard.

The screen discard from a mud cleaner looks a lot like the underflow from a centrifuge. The solids concentration is usually around 60% volume and 40% volume liquid. This initially appears to be an irrational large quantity of liquid. Researchers frequently pack columns with loose sand to examine various oil recovery procedures. Dry sand is poured into a cylinder while vibrating the pack. If a porosity of the sand pack is 33%–35% volume, the packing is about as tight as can be achieved. Loose sand on the beach, immediately after a wave has washed back out to sea, has about 40% volume water in it. It can be scooped up without water draining from the sand pile. Mud cleaner screen discards and the heavy, or underflow, slurry from a centrifuge has about the same volume percent liquid.

The mud cleaner is designed to continuously process drilling fluid just like the main shale shakers. The mud cleaner screen keeps larger particles from entering the system. Operating the equipment for only part of the time allows solids to remain in the drilling fluid system. These solids grind into smaller particles that become more difficult to remove. Centrifuges will be able to remove these solids from a weighted drilling fluid but they generally do not process all of the rig flow. The mud cleaner can remove these solids before they grind into smaller particles if the mud cleaner is used continuously.

Again, note that the mud cleaner and the centrifuge are complimentary to each other—not competition for each other. The mud cleaner removes solids larger than barite; centrifuges remove solids smaller than most barite.

Estimating ratio of low gravity solid volume and barite volume in mud cleaner screen discard

An estimate of the low-gravity-solids content of the mud cleaner screen discard can be made by weighing the discard. Since the solids concentration will be around 60% volume, the mud weight will be a reasonable predictor of the low gravity concentration. For low gravity solids with a specific gravity of 2.6 and a barite specific gravity of 4.2, the equation to determine the low gravity concentration (V_{lg}) is:

$$V_{lg} = 6.25 + 2.0 V_S - 7.5 \text{MW}$$

where V_{lg} is the volume percent of total suspended solids, and MW is the mud weight in ppg.

Refer to Chapter 5, Drilled Solids Calculations for the derivation of the equation.

Assume that the V_{lg} is 60% volume for a mud weight of 19.0 ppg and the equation will calculate that the volume concentration of low gravity solids is 40% volume. This means that 20% of the volume is barite. So, twice the volume of low gravity solids is being discarded as barite. Even if the actual V_{lg} was 57% instead of 60%, the low gravity concentration would be 34% volume. In most cases the decision to continue running the mud cleaner would not be affected by this inaccuracy. Even if the barite in the discard exceeded the low gravity concentration, the benefits of removing those larger solids will be evident. Accurate results, of course, can be obtained by retorting the solids; but this is a tedious process because the solids are difficult to handle. Care must be taken to obtain a representative sample and pack it into the retort cup without leaving void spaces. A much more accurate method is to use the gravimetric procedure where larger quantities can be used and no volume measurements are made.

Generally the discard from a mud cleaner screen is relatively dry and contains around 60% volume solids. The density of this slurry can be measured with a mud balance but the solids concentration is difficult to measure with a retort. Generally decisions about the performance of a mud cleaner or a centrifuge can be made by weighing the heavy, or underflow, discard from a centrifuge or the screen discard. Very accurate measurements are not really needed. The chart in Fig. 20.6 allows an estimate of the concentration of low gravity solids. The chart is accurate for 2.6 SG low gravity solids and 4.2 SG barite. It is intended to be used to estimate the amount of low gravity solids in a mud cleaner screen discard. Using a retort on the high concentration of solids is a very tedious task. Estimating a concentration of around 60% volume of total solids is a reasonable estimate. When performing reservoir flooding experiments by packing sand columns, using a vibrator for maximum packing, a porosity greater than about 35% is difficult to achieve. The decision to continue to use a mud cleaner usually will not require very accurate measurements of the total percent of solids in the discard. Frequently, when drilling very dispersible clay formations with fresh water drilling fluids, very few large solids remain in the drilling fluid when it reaches the surface.

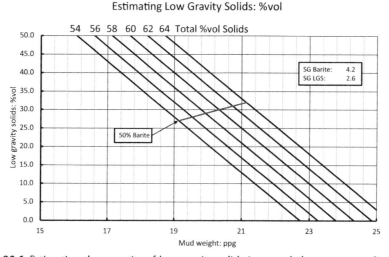

Fig. 20.6 Estimating the quantity of low gravity solids in a mud cleaner screen discard.

The concentration of low gravity drilled solids may indicate a centrifuge is required to remove the small particles. This chart assists in making that determination.

For example, if the mud weight of a mud cleaner screen discard weighs 18.0 ppg, the low gravity solids concentration would be about 42% volume if the solids concentration is 58% volume. Barite would be 16% volume. Low gravity solids concentration would be about 48% volume if the discard total solids concentration is 60% volume. Barite would be only 12% volume for this condition. Note that in either case, the mud cleaner is doing a great job of removing drilled solids or low gravity solids from the drilling fluid. An accurate measurement is not needed to make the decision to continue running the mud cleaner.

Mud cleaner performance

Drilling soft dispersible shales with a fresh-water drilling fluid usually results in drilled solids that cannot be removed with mud cleaners. In these cases centrifuges should be planned for use in weighted drilling fluids. Usually, if solids are being removed with shale shakers, a mud cleaner will probably be beneficial. Solids removed by mud cleaners will cover a wide range of quantities depending on formations drilled, bore hole stability, dispersion of solids as they move up the bore hole, type of drill bit, type of drilling fluid, and other variables.

During one of the first field tests of the mud cleaner, the pressure at the entrance to desilters was varied and the mud cleaner discard examined for drilled solids and barite. With an unweighted drilling fluid, a head of 75 ft was recommended for this brand of desilter. This head creates a balanced hydrocyclone with good separation of low gravity solids; however, this may not necessarily be true for weighted drilling fluids. This well was drilling with a $9^7/_8''$ drill bit between 9300 ft and 9400 ft. Six 4″ hydrocyclones were mounted above a 200 square mesh screen. Samples of the screen discharge were captured during three circulation sequences. The samples were analyzed for the low-gravity and barite solids concentrations (Figs. 20.7—20.9). The centrifugal pump was powered by a diesel engine and the impeller rotation speed could be controlled. This allowed the test to evaluate the flow rate effect on the removal of solids. The flow rate of the mud cleaner screen discard was measured for each test and the concentration of low gravity solids and barite were calculated for each different manifold pressure (Table 20.1).

During the first circulation after new drill bit reached bottom, higher quantities of drilled solids are discarded by the mud cleaner screen as the head is increased on the desilter feed. One method of analysis is to compare the concentration of barite lost with the drilled solids discarded. In the three charts above, the total solids discarded and the quantity of drilled solids are shown as a function of the desilter manifold pressure. The lowest ratio of barite to drilled solids occurs when the manifold head (or pressure) is low. This might be misleading, however. Larger quantities of drilled solids

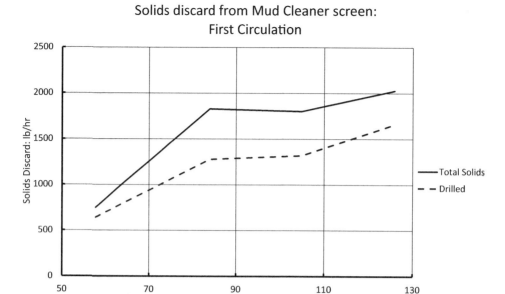

Fig. 20.7 Solids discarded from mud cleaner during first circulation.

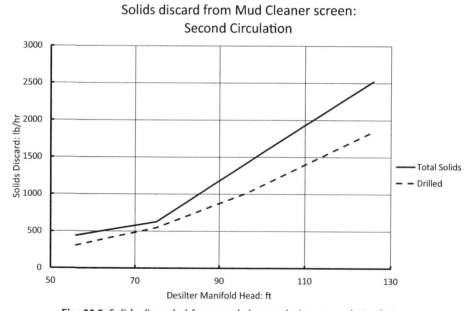

Fig. 20.8 Solids discarded from mud cleaner during second circulation.

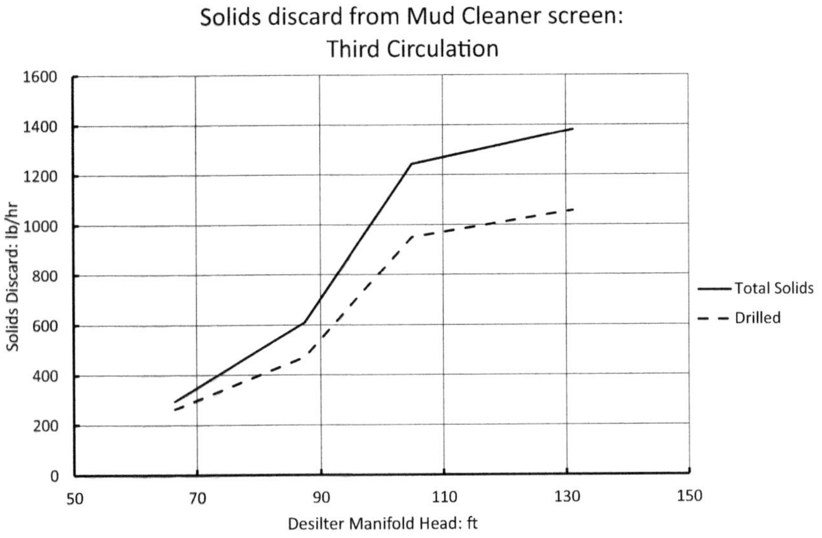

Solids discard from Mud Cleaner screen: Third Circulation

Fig. 20.9 Solids discarded from mud cleaner during.

Table 20.1 Comparison of low gravity solids and barite discarded by a mud cleaner as the manifold pressure changes.

Cyclone manifold pressure, psi	Manifold head, ft	Screen discharge, sec/qt	Discharge density, ppg	Volume percent, solids, %	Drilled solids removed, lb/h	Barite discarded, lb/h
First circulation—bottoms up after TIH with new bit						
33	57.7	16	16.8	58	640	108
48	83.9	7	17.7	58	1275	552
60	104.9	7	17.5	58	1317	484
72	125.9	6	17	58	1659	368
Second circulation						
32	55.9	25	16.4	50	308	132
43	75.2	19	16.7	58	547	79
55	96.2	9	17.6	58	1008	403
72	125.9	5	17.5	58	1844	678
Third circulation						
38	66.4	40	16.6	58	263	32
50	87.4	20	17.1	57	471	138
60	104.9	10	17.3	58	951	292
75	131.1	9	17.3	58	1057	324

discards are much more desirable, even if some additional barite is lost. Higher manifold pressures are preferred to eliminate the largest quantity of drilled solids from the drilling fluid.

A word of caution is appropriate here. The purpose of solids control equipment is to remove drilled solids. Economics certainly justify sacrificing a small additional amount of barite for a good removal of drilled solids. If solids concentrations are reduced by dilution only, the cost would be many times higher than sacrificing some barite to remove these drilled solids. So, evaluating performance by comparing discard ratios can be very misleading.

Mud cleaner economics

When the first two commercial units were used in field operations, the company saved enough money on those two wells to pay for all of the research required to develop the mud cleaners. The two wells (one a production well and the other was an exploration well) in the Pecan Island Field in Louisiana had no stuck pipe problems, fast penetrations rates, and were able to reciprocate the long casing strings for the first time in this field. Since many wells had been drilled in this field, the cost estimate of the wells, or money approved for expenditure (AFE), was relatively accurate in most wells before this.

One major operator's drilling division kept track of stuck pipe issues in all wells for the year following the introduction of the first commercial mud cleaner. There were no incidences of stuck pipe on any well which was using a mud cleaner for the 12 months.[1] At the 1974 IADC Conference on View Points on Solids Control, Jack Warner, President of Goldrus Drilling Company made the comment that all of their turnkey rigs were equipped with mud cleaners. This is a clear indication that the system was economically viable.

There was one incidence during that time reported by an off-set operator. They had installed a mud cleaner (not a Sweco) on a well and had been fishing for stuck pipe for a long time. A research technician, who was working with the major operator, was sent to that rig to find out why the mud cleaner did not prevent stuck pipe. He went up on the rig floor and asked the driller about the mud cleaner. The driller said that they never turned that they never turned on that piece of "@#$@%*##" because they knew it wasn't going to help. The driller said that the mud cleaner would remove the good solids that was giving his mud the body that it needed to do its work.

The question frequently arises "Would it be cheaper to simply jet drilling fluid from the system instead of using solids control equipment?" This is called controlling solids

---

[1] Lou Carlton's comments at the 1974 IADC Conference on View Points on Solids Control, March 17–18, 1974, in New Orleans, LA.

with dilution and is discussed in depth in Chapter 5. However, to answer this question, examine the above discard rate and compare with the volume of drilling fluid which must be discarded to eliminate the same volume of drilled solids. Cost of drilling fluid ingredients vary from company to company and from contract to contract. Rather than actually calculate a cost, the comparison will be made between methods of eliminating low gravity solids from the system with either the mud cleaner or pumping drilling fluid from the system to eliminate the same quantity of low gravity solids using the measurements in the above tables.

An 11 ppg drilling fluid with 13% volume total solids would contain 6.4% volume low gravity and 6.6% volume barite. With the first circulation and 60 psi manifold pressure, 1317 lb/h drilled solids and 484 lb/h of barite are discarded. Since the solids in the drilling fluid are presented in terms of volumes instead of weights, the mass flow rate needs to be translated into volumes.

With a specific gravity of 4.2, the density of barite would be 4.2 times the density of water (8.34 lb/gal), or 35.03 lb/gal. With a specific gravity of 2.6, the density of low gravity solids is 2.6 times the density of water, or 21.68 lb/gal. The volume flow rate of solids would be the mass flow rate divided by the density. The volume flow rate of barite would be (484 lb/h)/(35.03 lb/gal) or 13.8 gal/h and the volume flow rate of low gravity solids would be (1317 lb/h)/(21.68 lb/gal) or 60.7 gal/h.

In this drilling fluid there are 6.4% volume drilled solids. If the discarded low gravity solids are 6.4% of the mud volume and 60.7 gal/h drilled solids are discarded, the total volume discarded to contain that amount of solids would be:

Volume of low gravity solids discarded = 6.6% (volume of 11 ppg drilling fluid).

Or, 60.7 gal/h = (0.066) (volume of 11 ppg drilling fluid).

Volume of 11 ppg drilling fluid discarded is 920 gal/h.

This quantity of drilling fluid contains 6.6% volume barite or (0.066) (920 gal/h) = 60.7 gal/h of barite. Converting barite volume to barite weight: (60.7 gal/h) (35.03 lb/gal), means that 2128 lb/h of barite would be lost to eliminate the 1317 lb/h of low gravity solids. Compare this with the measured loss of only 484 lb/h of barite loss from the mud cleaner screen. In addition to the cost of the excess barite, the pit levels would decrease by 920 gallons every hour. This is equivalent to losing:

(920 gal/h) (24 h)/(42 gal/bbl) or 525.7 bbl of drilling fluid daily

Clearly, concentrating the low gravity solids with equipment is preferable to dumping drilling fluid to eliminate drilled solids.

The bottoms up sample indicates that the smallest discard rate had the highest ratio of drilled solids to barite. However, the largest flow rate of drilled solids, 1659 lb/h, was still a flow rate 4.5 times as large as the barite flow rate.

Accuracy required for specific gravity of solids

Determination of drilled solids in a drilling fluid depends on an accurate determination of the mud weight, the total solids in the drilling fluid, and the density of the drilling fluid ingredients. For example, with a fresh water 11 ppg drilling fluid containing 2.6 specific gravity low gravity solids and 4.2 specific gravity barite, a change in only 1% volume measured solids concentration makes a 2% volume change in calculated low gravity solids. In Table 20.2, for an 11 ppg drilling fluid, a 13% volume solids concentration would indicate 6% volume and a 12% volume solids concentration would indicate 4% volume.

Accurate solids determination is needed to properly identify mud cleaner performance. A small change in the density of the low gravity solids and the barite affects the calculated low gravity solids calculation. Table 20.3 indicates that 14% volume total non-soluble solids and 11 ppg drilling fluid could have between 6.3% and 9.8% volume low gravity solids, depending upon the density of the barite and the low gravity solids.

Obviously the density of the low gravity solids and the barite must be known for an accurate determination of the concentration of solids in the drilling fluid or the discard from any solids control equipment. The most common error in viewing the discard from a mud cleaner screen is the erroneous conclusion that large quantities of barite are contained in the discard. This is the reason that the new API solids analysis recommends that the density of the barite and solids on the shaker screen be determined on location.

One frequent comment while running a mud cleaner is that the mud cleaner is discarding so much barite that large quantities of barite must be added to keep the mud weight constant. Actually, discarding any solids from a drilling fluid decreases density. Replacing low gravity solids with barite will decrease the total solids concentration;

Table 20.2 Comparison of change in total solids to change low gravity solids by 2% vol.

V_{lg}, %vol	V_s, %vol	MW, ppg
8	14	11
6	13	11
4	12	11
2	11	11

Table 20.3 Comparison of fraction of solids calculated depending upon the density of low gravity solids used.

For an 11 ppg drilling fluid with 14% vol total solids		
Density of low gravity solids	Density of barite	Volume fraction of low gravity solids
2.4	4.2	7.1
2.9	4.2	9.8
2.4	4	6.3
2.9	4	9.1
2.6	4.2	8.0

make filter cakes more compressible; decrease the propensity for stuck pipe and lost circulation; and improve the possibility of faster drilling (by increasing the founder point). However, some formations that disperse significantly in a fresh-water drilling fluid and very few large solids arrive at the surface. Solids discarded from the mud cleaner screen should be examined to determine if drilled solids are being removed. If they are not, centrifuges should definitely be considered to remove low-gravity solids.

Actually, mud cleaners frequently do not discard as much barite as the main shale shakers. Visual observations tend to give an erroneous view. The discard from a mud cleaner screen (and the underflow from a decanting centrifuge) contains around 60% volume solids while the shale shaker discard contains around 35%—40% volume solids. The more liquid (shale shaker) discard does not appear to contain as much barite. However, the shale shaker discard may concentrate barite or deplete barite from the flow line drilling fluid. The way drilling fluid discarded from a shale shaker appears does not reveal the concentration of barite.

Heavy drilling fluids

Various arbitrary procedures seem to be developed by operating personnel in the field. Often a centrifuge is run for a specific number of hours per day, or a hydrocyclone bank is used only for a short period. Mud cleaners seem to also attract a variety of erroneous rules of thumb.

In heavily weighted drilling fluids, above 13—14 ppg, mud cleaners are frequently shut off because of excessive barite discards. When a mud tank system is plumbed incorrectly, specifically when mud guns transport freshly added barite fluid into the removal system, a significant amount of barite will be discarded. Barite that meets API specifications may have as much as three weight percent larger than 74 µm. If a thousand 100 lb sacks of barite are added to a drilling fluid system while drilling an interval (used as clean drilling fluid to dilute remaining drilled solids), 3000 lb of the 100,000 lb added will be larger than 74 µm. Now the question becomes—"What damage will these large particles create?" Any particles larger than 74 μm, whether they are drilled solids, barite solids, diamonds, or pieces of gold, will make a poor quality filter cake and enhance the probability of incurring all of the problems created by a large quantity of drilled solids in the drilling fluid.

In a drilling fluid system where the system is arranged properly, barite will have traveled down the drill string, passed through the bit nozzles and then traversed the borehole before reaching the mud cleaner. Any particle larger than 74 μm should be removed from the drilling fluid system. Barite larger than 74 μm should be removed from the system. In situations where the system is not arranged properly, the mud cleaner might be shut down for two or three circulations after a significant weight-up. This is a good test for correct mud tank arrangement. If the mud cleaner starts discarding barite immediately upon weight-up, the system has serious flaws. The objective is to remove large particles from the drilling fluid so the filter cake will be thin, slick, impermeable, and compressible. With solids larger than 74 μm in the cake, this objective will not be achieved.

One common problem with sieving hydrocyclone underflow with a mud cleaner is the tendency of the material to dewater, or de-liquefy, before reaching the discard end of the screen. Clumps of material traveling down a screen with too little liquid will not separate solids properly. A small reflux of drilling fluid from the hydrocyclone overflow should be sprayed on the screen to break these clumps into material that can be separated. A spray of water or oil (depending on the liquid phase) could be used but will generally dilute the system too much. The drilling fluid reflux enhances the screen capability to separate solids from the slurry. If the slurry remains liquid until all separation has been completed, the mud cleaner screen should not remove any more barite than would be removed on the main shaker with the same size screen.

There are situations where the solids disperse as they travel up the borehole. Usually only slivers and cavings from the borehole wall are removed by the main shaker. When this occurs, a centrifuge is needed to remove the very small particles. A change in drilling fluid systems might be contemplated before drilling the next well.

Operation guidelines

After removal of large cuttings with a shaker, drilling fluid is pumped into the mud cleaner's hydrocyclones with a centrifugal pump. The overflow from the cyclones is returned to the active system. Instead of discarding the underflow, the solids and liquid exiting the bottom of the cyclones are directed onto a fine screen. Drilled solids larger than the screen openings are discarded: the remaining solids, including most of the barite in a weighted system, pass through the screen and are returned to the active drilling fluid system.

The cut point and amount of solids removed by a mud cleaner depends primarily on the fine shaker screen used. Since there are many designs of mud cleaners available, performance and economics will vary with machine and drilling variables.

Most mud cleaners use multiple 4″ or 5″ cyclones, processing 400–850 gpm. The liquid through-put is only one measure of mud cleaner capacity; more important is the capacity of the vibrating screen to remove drilled solids.

Some field data of a mud cleaner processing an 11.2 ppg drilling fluid shows the mud cleaner was discarding 46,800 lbs of drilled solids each 24 h, along with 2,925 lbs (29 sacks) of barite. The fine screen under the hydrocyclones salvaged 71,955 lbs (720 sacks) of weighting material per day. From this, it is obvious why mud cleaners should be used in place of desilters alone in weighted drilling fluid applications. Comparing the drilling fluid content of the cone underflow (8 bbl/h) to the fluid content of the mud cleaner discard (1.4 bbl/h) shows another benefit of mud cleaners over desilters in non-aqueous drilling fluid (NADF) and other drilling fluids which have an expensive liquid phase. The primary purpose of solids control equipment is to remove drilled solids, *not* recover barite. Salvaging barite is a great by-product of the device but the removal of the drilled solids is the most important aspect.

Mud cleaners should be considered in these applications:
1. Whenever the application requires finer screens than the existing shaker can handle
2. Unweighted NADF
3. Expensive polymer systems
4. Whenever the cost of water is high
5. Unweighted water-base drilling fluids with high disposal costs and/or environmental restrictions
6. When use of coarse lost circulation material forces by-passing of the shale shaker
7. Work over and completion fluid cleanup
8. As a back-up insurance for solids that are not removed by the main shakers

Installation

Mud cleaners are installed downstream of the shale shaker and the degasser. The same pumps used to feed the rig's desander or desilter are often reconnected to feed the mud cleaner when weight material is added. (Most mud cleaners are designed to also function as desilters on unweighted drilling fluid by rerouting the cone underflow or by removing or blanking off the screen portion of the unit. The mud cleaner may then be used to replace or augment the rig's desilter during top hole drilling.) Frequently, a bank of desilters is mounted over a main shaker if it can use an API 170 or an API 200 screen.

In the upper part of the hole (unweighted drilling fluid), the shaker will process fluid from the flow line and the desilters will discard all of the underflow. Down deeper in the hole, where the flow rate in the well does not require as many main shakers, one shaker can be converted into a mud cleaner. The flow from the well bore no longer goes to one of the main shakers; instead it will process the underflow from the desilters.

Follow these guidelines when installing mud cleaners to allow peak efficiency:
- Size the mud cleaner to process 110%–125% of the flow rate entering the desilter suction tank.
- Take the mud cleaner suction from the compartment receiving fluid processed by the degasser.
- If the mud cleaner has both a desander and a desilter bank of cones, the suction and discharge for each set of cones is the same as it would be in an unweighted drilling fluid system.
- Confirm that the mud cleaner can process over 100% of the flow entering the suction compartment of the desilters.
- Keep all lines as short and straight as possible.
- Install a guard screen with approximately $1/4''$ openings at the suction to the desilter to prevent large trash from entering the unit and plugging the cones. The open area of the screen should be at least twice the pipe area.
- Position the mud cleaner on the pit high enough so the overflow manifold will gravity-feed fluid into the next downstream compartment at an angle of approximately 45 degree. Remember—no vertical overflow discharge lines.
- Provide walk-way and sufficient space for routine maintenance.
- Provide a vacuum breaker in the desilter overflow manifold to avoid creating a vacuum in the desilter overflow manifold.
- Install a low equalizer line for back-flow to the mud cleaner suction compartment.
- Return the fluid underflow from the mud cleaner screen in a well-agitated spot. This will prevent concentrated barite from settling in the mud tank. (The screen underflow will have no carrying capacity.)

To operate mud cleaners at maximum efficiency, remember these fundamentals:
- Operate mud cleaners continuously on the full circulating volume to achieve maximum drilled solids removal.
- Operate mud cleaners within the limits of the screen capacity. A mud cleaner with a cyclone throughput of 800 gpm is of little value if the cone underflow exceeds the screen capacity resulting in flooding and high drilling fluid losses.
- Do *not* judge screen efficiency simply on the basis of cuttings dryness or color. The total amount of drilled solids in the discarded material along with the ratio of barite to drilled solids must be determined to evaluate economic performance.
- Select the number of cones to be operated so that all of the drilling fluid entering the desilter suction tank can be processed and use the finest screen possible, preferably an API 170 or an API 200.

General guidelines for correct mud cleaner operation:

- Run the mud cleaner continuously while drilling and for a short period of time while making a trip for "catch-up" cleaning.
- Start up the shaker before engaging the feed pump.
- Install a pressure gauge on the desilter input manifold to verify the correct head is applied to the hydrocyclones.
- Shut down the feed pump before turning off the vibrating screen. Permit the screen to clear itself, then rinse the screen with water or oil spray before shutting down the screen vibrator.
- For peak efficiency, operate the cones with a spray rather than a rope discharge. This is just as important, or maybe more so, with a mud cleaner as when operating the desilters and desanders in unweighted drilling fluids.
- Check cones regularly for bottom plugging or flooding, since a plugged cone allows solids to return to the active system. If a cone bottom is plugged, unplug it with a welding rod or similar tool. If a cone is flooding, the feed is partially plugged or the bottom of the cone may be worn out.
- When a significant amount of barite is added to increase mud weight, incorrectly plumbed surface systems will require that the mud cleaners be shut down for one or two full circulations. The 3% by weight of API barite larger than 75 μm will result in a significant quantity of barite being removed from the drilling fluid system. Circulating through the bit nozzles tends to decrease the barite size.
- If the quantity of liquid exiting the desilters is insufficient to allow the screen to properly separate solids, a small spray of drilling fluid has proven to be effective in allowing better screening of the underflow. Sprays of water or oil generally will increase the dilution of the drilling fluid and can be costly. Frequently, one of the desilters can be removed from the manifold and a short hose with valve can be used to provide the small amount of drilling fluid needed to prevent "piggy-backing" of the solids.

Maintenance of mud cleaners generally combines the requirements of desilters and fine screen shakers:

- Periodic lubrication
- Check screen for proper tension.
- Inspect the screen to ensure it is free of tears, holes, and dried drilling fluid before startup.
- Shut down unit when not drilling to extend screen life.
- Check feed manifold for plugging of cyclone feed inlets; clean each as necessary.
- Check cyclones for excessive wear and replace parts as necessary

Non-oilfield usage of mud cleaners

One use of mud cleaners, that has been very profitable, has been the use of the equipment with micro-tunneling. Tunneling under roadways lakes, or streams for pipelines, or optical fiber conduits, or other installations require drilling with a circulating fluid. Frequently, the liquid is difficult to acquire and disposal is a problem. So these small drilling systems install a mud cleaner on top of a one-tank circulating system. The solids from the mud cleaner screen are removed and the excess liquid is returned to the tank. The screens used in this application usually have larger openings and larger wire diameters than the screens used for oil-well applications. The purpose of these screens is to dewater the cuttings and retain as much fluid as possible. Drilled solids in this application are not as detrimental as drilled solids are in oil well drilling so the openings in the screens do not need to be as small as with oil well drilling.

Appendix 20.A
History of the development of mud cleaners

The first mud cleaner was a combination of two 12″ diameter and twenty 4″ hydrocyclones mounted above a specially-built, five-foot diameter, cast iron, round Sweco shaker. Even though the mud cleaner was invented for use with a weighted drilling fluid, the first application was on a well drilled using one of the first unweighted, potassium chloride drilling fluids while drilling a shallow 2200 ft research well near Houston, Texas in 1971. Since this was a research well, the drilling fluid was new, the mud cleaner was new, and the pick-up and lay down device for the drill pipe was new. The fluid contained visbestos which made the yield point very high but the plastic viscosity was around 3 cp. An API 80 screen was mounted on an unbalanced elliptical motion Linkbelt shale shaker to process the fluid as it left the wellbore. The fluid was then pumped to the hydrocyclones. The well was to be used, initially, to evaluate the use of air injection into risers to reduce annular pressure at the sea floor and, subsequently, used as a research test well.

The first mud cleaner used in the field was a stainless steel, round, 5′ diameter Sweco shaker (Fig. 20.A.1). It was mounted on a platform at the end of the removal section on a drilling rig in the Bayou Sale Field near Franklin, LA.

Ten Pioneer 4″ desilter cones were fed with a centrifugal pump driven by a diesel motor. Having the centrifugal pump driven by a motor allowed investigation of the desilter output for a variety of manifold pressures. The large "pond" in the background was the "reserve pit" and the small pond just behind the mud cleaner was a "duck's nest" used to store excess drilling fluid when it was removed from the system for dilution.

The large vertical section of casing just beside the shiny mud cleaner was a "roughneck proof" flow meter (Fig. 20.A.2). To obtain cut points, the flow rates from the desilters needed to be measured. The overflow from the desilters was routed beneath the vertical casing with a valve downstream of the casing. The inside of the casing was calibrated to measure gallons. When the valve was closed, the overflow from the desilters filled the casing. By timing the fill, the flow rate could be established.

Fig. 20.A.1 The first mud cleaner used in the field.

The system was so new that it had to be mounted on a platform at the end of the degasser tank because the operator did not want to "clutter up" the mud tank system. The 4″ cones were from Pioneer Centrifuge Co.—each weighed about 40 lbs with a rubber insert in a cast iron body. The mud cleaner was 5′ in diameter and had two decks in it. Ten cones put about 10–50 gpm on the 200 square mesh screens. Plastic ring "sliders" mounted beneath the single layer screen prevented near-size blinding and gave support to the fine wire (Fig. 20.A.3).

The vibration motor was mounted underneath the screens and rotating a vertical shaft. The unbalanced weight on top and another one on the bottom of the motor controlled the height of the screen motion and the rotation speed of the slurry as it

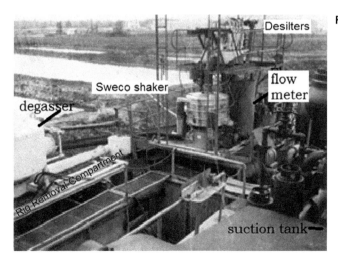

Fig. 20.A.2 Bayou Sale installation.

Fig. 20.A.3 Replacing the plastic sliders beneath the fine screen on the bottom screen.

rotated in an increasing diameter until it reached the discard port. Both ports (clean fluid through screen and discard solids off the screen) had rubber sleeve down spouts (Figs. 20.A.4 and 20.A.5).

Surface casing was set around 3000′ and the hole was drilled with a gel/lignosulfonate drilling fluid. The mud cleaner was installed just before the bit reached 11,000′. Measurements on discarded solids were made every 2 h from 11,000′–16,000′ during most of the months of November and December. Just before Christmas, the unit was shut down because they thought they only had 80′ to drill and "we were not helping because they were having no problems". The interval was being drilled with an 11 ppg gel/lignosulfonate fluid through about a dozen or more drawn-down Miocene sands. One formation at

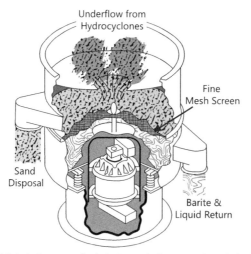

Fig. 20.A.4 Sweco called their mud cleaner a "sand shaker".

Fig. 20.A.5 Bayou Sale Shaker was a double deck Sweco Shaker that caused the solids to traverse a spiral toward the discharge spout.

11,000′ had the original pore pressure. The produced formations had pressure differentials in the 2000′–6000 psi range. No stuck pipe or lost circulation was experienced during the drilling of this 5000′ interval. They actually had to drill 200′ more and they called just before new years to come back over and turn on our "robot". They had to make wiper trips between every logging run. Logging tools were sticking and the torque and drag was significant. Several circulations and a wiper trip were required before it was safe to run and cement the protective casing string. The casing was run to bottom of the hole, reciprocated, and cemented with no difficulty. The 200′ drilled after the mud cleaner was shut-off, would have contributed only about 31 bbl of drilled solids to the drilling fluid if nothing was removed. Most of the cuttings coming to the surface were coming from the side of the hole. Many times, measurements of solids discarded from the mud cleaner indicated more solids were being discarded by the mud cleaner that day than were drilled that day. The shale shaker was also removing many solids (Fig. 20.A.6).

Note this lucky event of turning off the mud cleaner was probably the reason that mud cleaners became commercial. No drilling program schedules stuck pipe. None was programmed here. The research plan should have included a procedure to validate the mud cleaner performance. Since no problems were encountered and none were expected (although stuck pipe and lost circulation are common with 6000 psi overbalance in a well bore), no comparative data was acquired to prove that the mud cleaner was performing properly—until the machine was luckily shut off. This was also a great lesson in planning research for the drilling processes. The research team concentrated on keeping the drilling fluid in good shape and minimizing the impact of drilled solids, unfortunately the primary function should have been to prove machine beneficial. At that time, not all drilling personnel believed that drilled solids were detrimental or evil.

The mud cleaner U. S. patent 3,766,997 (October 23, 1973), granted to J.K. Heilhecker and L.H. Robinson and assigned to Exxon Production Research Company,

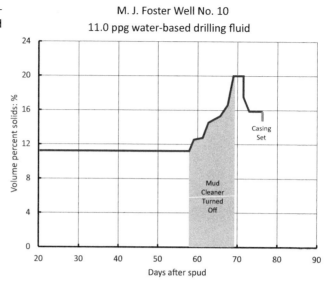

Fig. 20.A.6 History of solids increase when Bayou Sale mud cleaner was stopped.

was found to be invalid because of prior art later discovered in the British Patent office. A German inventor was granted a patent on a similar device in the late 1800s. Although his invention had never been reduced to practice, or used in the oil patch, and screens were much coarser in those days, the existence of the information in the public domain prevented collection of royalty for application of this technology. As an interesting note, all of the companies providing mud cleaner service offered to pay a nominal $5.00 per day per unit for a technology transfer fee for help in developing the product. This offer was rejected and the service companies were told that a much larger royalty payment would be required when the patent issued. With an invalid patent, the service companies never had to pay a royalty and certainly did not pay a technology transfer fee. This actually assisted in the rapid deployment of the mud cleaners into field operations.

The first commercial application, less than a year after the initial patent was submitted involved two wells, one production and one exploration, in the Pecan Island field in Louisiana. For the first time in that field, the intermediate casing could be reciprocated during the cement job because of the lower drilled solids concentration in the drilling fluid. The torque and drag in these wells were spectacularly lower than previous wells.

When the mud cleaner was first introduced, many would try to decide whether to use a mud cleaner or a centrifuge. The problem with this decision is that mud cleaners do not compete with centrifuges in solids removal. In weighted drilling fluids, mud cleaners are designed to remove drilled solids larger than barite (larger than 74 μm). Centrifuges remove solids mostly smaller than most barite (less than 5–7 μm).

Field test #2, in the Katy, Texas, tested a lighter, smaller version of the shaker used in the first field test. Since the size of the shakers was being designed while the platform was being erected, the desilter bank was installed at a high elevation to make certain that the new shaker would fit. Again, the option to place this new piece of equipment on top of the mud tanks was not available. Two single deck shakers were installed below 12 Magcobar (MI-Swaco) 4″ desilters.

Since a vacuum breaker was not an option on the rental desilter bank, a 4″ × 4″ piece of lumber was inserted into the 4″ hose that was used for an overflow line to provide a back pressure to siphoning (Fig. 20.A.7).

The platform was located at the upper end of the mud tank system. The drilling fluid was pumped from a compartment downstream from a Baroid circular motion shale shaker to the desilters. The long line to the compartment downstream of the desilter suction compartment and the high elevation of the desilter overflow caused a siphoning action. The siphoning was initially so great that no fluid left through the desilter underflow ports. This started the procedure of providing a vacuum breaker on the desilter discharge manifold.[2] The new shakers were much lighter and did not have lids like the first ones had (Fig. 20.A.8).

After that field test, the next field test was at Tilden, Texas using potassium chloride (KCl) drilling fluid as it was being developed. The two 4′ diameter shakers had lids to prevent splatter (Fig. 20.A.9). The mud cleaner not only removed drilled solids that would have been detrimental to drilling performance but also recovered a significant quantity of the very expensive liquid phase of the drilling fluid. Pioneer Centrifuge Company 4″ hydrocyclones were again mounted above the shakers. The underflow

Fig. 20.A.7 Mud cleaner field test #2.

---

[2] The lesson learned later was to avoid any threads at the upper end of the vacuum breaker. Units were operating on two platform rigs with threads at the upper end of the 1″ pipe. After about a month, they were examined. One had a water hose connected to it and the other one had a needle valve installed.

Fig. 20.A.8 Katy field test.

could be split between the two shakers or all of the underflow fed to just one of the shakers.

The two shakers were mounted on a large platform that was too heavy for the mud tanks but provided ample room for researchers to access all components of the system. Special legs had to be mounted under the platform to keep the platform from crushing the mud tanks.

The new arrangement provided ample work room for measurements and access to all components (Fig. 20.A.10).

After the Tilden test was completed, Sweco decided that they would organize a solids control company for the oil patch. Some of their board members had lost money on "new ideas" for the petroleum industry and were very skeptical about committing company funds to forming a new branch of their company. They were very successful in the chemical processing industry with their unique shaker (Fig. 20.A.11).

The first two commercial applications were at Exxon's Pecan Island field in Louisiana. The rigs did not have enough room on the mud tanks to mount the mud cleaners. Two Sweco mud cleaners were mounted outside of the mud room on a special platform. The underflow from the desilter bank does not have many viscosifiers in it. Solids cannot be transported with this fluid. To return the screen throughput back to the drilling fluid system, the overflow from the desilters was plumbed into a trough which caught the screen throughput. This provided sufficient velocity and viscosity to move the retained solids back to the drilling fluid system. This is the same process recommended when a main shaker is converted into a mud cleaner. The drilling fluid passing through the mud cleaner screen has no carrying capacity and additional flow must be used to return it to the mud tanks (Figs. 20.A.12 and 20.A.13).

Fig. 20.A.9 Mud cleaner arrangement #3. *(Picture taken from rig floor.)*

Fig. 20.A.10 Third field test of mud cleaner.

Just after the Pecan Island test, Brandt entered the market with hydrocyclones above their circular motion shale shaker. They used the Pioneer Centrifuge Company desilters (Fig. 20.A.14).

Sweco modified their shakers to provide a higher barrier around the shaker and removed the top (Fig. 20.A.15).

Sweco emphasized that their shaker screens needed to remain horizontal for the spinning motion of the desilter underflow to be properly screened. While drilling on Lake Maracaibo in Venezuela, the shakers were suspended from the overhead beams (Fig. 20.A.16). This allowed the barges to have a considerable heaving motion without affecting the flow of solids across the screen.

Fig. 20.A.11 View of tilden mud tanks from the crown block.

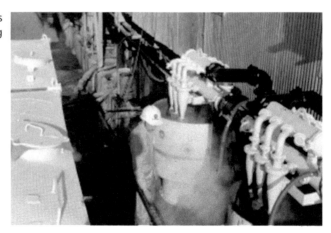

Fig. 20.A.12 Les Hansen, Sweco's mud cleaner manager, standing next to the first commercial unit.

The Sweco field hands complained about the heavy all metal hydrocyclones which were being used. Sweco modified them with light weight plastic cones which are now more of the standard for most companies (Fig. 20.A.17).

The only modification needed for the mud cleaner for an oil-based drilling fluid was a curtain to prevent oil splatter around the unit (Fig. 20.A.18).

When the linear motion shaker became popular on most large drilling rigs, the use of mud cleaners diminished to almost zero. Then, little by little, the mud cleaners were found to be a great insurance factor for processing all of the rig flow. This should have been obvious (but it wasn't) when considering the processing of unweighted drilling fluid. Plugged cones were (and are) common from solids that are too large to pass through

Fig. 20.A.13 First commercial unit on a barge drilling rig in Louisiana.

Fig. 20.A.14 Jimmy miller, Exxon researcher, examines cuttings coming from a brandt unit.

Fig. 20.A.15 A commercial installation of a new design mud cleaner.

Fig. 20.A.16 Installation in Lake Maricaibo.

Fig. 20.A.17 Sweco introduced their new light-weight hydrocyclones at OTC.

Fig. 20.A.18 Mud cleaner processing NADF for the first time.

Fig. 20.A.19 Mud cleaner using long body cones.

the lower apex of 4″ hydrocyclones. These solids are much larger than even solids which would be rejected by an API 20 (850 μm) screen.

Several companies are marketing mud cleaners with long body hydrocyclones mounted above shaker screens. These cones were popular in California and are now used more extensively (Fig. 20.A.19).

CHAPTER 21

Centrifuges

Introduction

Roy Bobo of Phillips introduced the centrifuge to the oil well drilling industry in the early 50s. In the initial applications, and in most of those during the following 50+ years, centrifuges were used almost exclusively to control the viscosity of weighted drilling fluids by separating viscosity-building colloidal and ultrafine particles. This is achieved by using the centrifuge to split the incoming mud into two streams: the overflow (centrate, effluent, light slurry) containing most of the liquid and the finest solids, and the underflow (cake, heavy slurry) consisting of the coarser solids and the liquid wetting them. The overflow (light) slurry is discarded (Fig. 21.1). This is *traditional centrifuging*; the removal of the finest solids together with the liquid phase of the drilling fluid. The underflow (heavy) slurry contains most of the barite and some drilled solids. The purpose of the centrifuge is to eliminate drilled solids.

The outer housing of the centrifuge rotates about 1600–3000+ RPM. Diluted drilling fluid enters along the shaft and centripetal acceleration move the fluid and solids to the outer wall. The larger solids (greater than about 5–7 μm) are discharged through one port and the smaller solids with most of the liquid are discharged through the other port. The liquid is pressed against the inner wall of the rotating housing. Solids are subjected to a large centrifugal force which accelerates the settling through the slurry. There is no shear within the slurry. The low-shear-rate viscosity of current drilling fluids is quite high (to help transport cuttings) and will impede the settling rate of solids in the drilling fluid. A dilution fluid is blended with the feed slurry to assist the settling of the particles in the fluid. An auger slowly rotates within the housing to move the cuttings to the solids

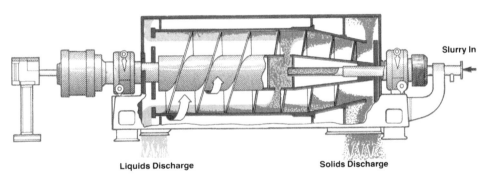

Fig. 21.1 Decanting centrifuge.

discharge port. The liquid (containing the very small particles) overflows through ports at the end of the housing.

The effects of the removal of excess fines include an increase in mud quality resulting in improved control of downhole filtration, and reduced torque and drag; both resulting from improved wall cake quality. Other important effects are a reduction in plastic viscosity, and a reduction in dilution requirements.

Centrifuging in the traditional manner permits the recovery of much of the barite because the underflow, consisting of the coarser solids, is returned to the mud, while the liquid and finer solids are discarded. This is a drilled solids removal piece of equipment and the discard stream is the important feature of the centrifuge.

Traditionally, centrifuges have been used primarily with weighted muds to reduce the concentration of fine particles, thereby reducing the need for dilution to control viscosity. There are a number of persistent myths about centrifuging drilling fluids that contribute to the general confusion about what it can be expected to accomplish. Among the worst, is the idea that centrifuges can be used to separate barite from low gravity solids.

Drilling fluids weighted with barite always contain both barite and drilled solids in the size range from less than 1 μm (μ) to 20 μm and larger. According to Stokes' Law, sedimentation and—therefore—centrifugal separation, is based on particle mass. Mass, in turn, is based on density and volume. Assuming that the average specific gravity of barite and low gravity solids particles are 4.20 and 2.60 respectively, the mass of a barite particle is equal to a low gravity particle approximately 50% larger. What this means is, in a given centrifuge, processing a given fluid, if most of the barite particles finer than, for example, 6μ remain in the overflow, most of the low gravity solids particles finer than 9 μ will also remain in the overflow. The larger particles–both barite and low gravity solids-will, of course, be found in the underflow. The separation is based primarily on the mass of the particles. A centrifuge does not *separate barite from low gravity solids, a centrifuge separates particles with larger mass from particles with smaller mass.*

Oilfield centrifuges make a separation of solids in the range of 5–10 μm and have very limited capacity; hence, they process only a small fraction of the drilling fluid. Therefore, centrifuges must be run all of the time to make a significant impact on the cleanliness of the drilling fluid. This also helps maintain constant drilling fluid properties. Typically the cut point is 6 to 10 μm for weighted muds and 5 to 7 μm for unweighted drilling fluids. The lighter (and smaller) particles are discarded in the "light" or "overflow" slurry. The heavier (and larger) particles are retained in the "heavy" or "underflow" slurry and returned to the active system. A centrifuge separates particles according to their mass. Frequently, Stokes' law is used to describe settling of particles. The derivation of the Stokes' Law relationship is presented in the Appendix to this chapter.

The economic justification for the use of centrifuges with weighted muds should be based on the removal of drilled solids NOT the cost of the barite recovered. A piece of pipe will recover 100% of the barite. What flows in one end will flow out the other—this is 100% recovery.

Traditional centrifuging has, unfortunately, become known as the "barite recovery" process. This is misleading and a pure advertising gimmick that confuses the novice. This terminology cannot be justified. The centrifuge rejects particles with small mass in the overflow (light) slurry. These particles are both the drilled solids and the small barite. A centrifuge separates by the mass on a particle independent of the species or value of the particle. Gold, silver, platinum, drilled solids, and barite are treated alike. The centrifuge returns much of the barite back to the drilling fluid because the particles have a larger mass and settle quickly in the centripetal acceleration field of a centrifuge. This is a cheaper process than discarding large quantities of drilling fluid as a means of decreasing the colloidal content of a drilling fluid.

In a weighted water-based drilling fluid, a centrifuge usually returns the heavy, or underflow, slurry to the active system and discards the light, or overflow, slurry. The purpose of a centrifuge is to remove very small solids. The concentration of drilled solids in the discard stream is a primary goal. The underflow (heavy) slurry is difficult to retort—just as is the mud cleaner screen discard. Both of these slurries have about the same solids concentration—around 60% solids and 40% liquid. If the slurry is weighed with a mud balance, an estimate can be made about the concentration of drilled solids (or low gravity solids) in the returning stream, (Fig. 21.2 and Fig. 21.3).

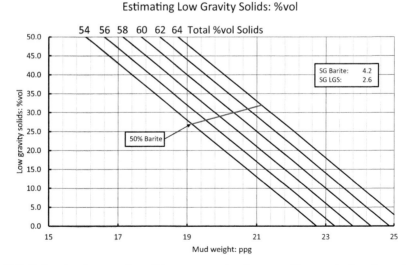

Fig. 21.2 Estimating low gravity solids concentration in a centrifuge underflow (heavy) slurry.

Viscosity can be effectively controlled by discarding a relatively small amount of colloidal size solids. Standard centrifuge applications take advantage of their ability to make a very fine cut on the order of 2–6 μm.

When treating weighted water-base drilling fluids, centrifuges are used intermittently to process a small portion circulated from the well bore to reduce the amount of colloidal and improve the flow properties of the mud. In order to remove these solids, the liquid fraction from the decanter (or the lighter slurry fraction from the perforated cylinder centrifuge) is discarded. The sand-size and silt-size semi-dry solids fraction from the decanter (or the heavier slurry fraction from the perforated cylinder centrifuge) is returned to the active system.

The centrifuge is installed downstream from all other solids control equipment. Ideally, suction for a centrifuge mud feed would be taken from the same pit or compartment which receives the returns from a mud cleaner.

The centrifuge underflow (solids) should be discharged to a well-stirred spot in the pit for thorough mixing with whole mud before the solids have a chance to settle out in the bottom of the pit. This is especially important with a decanter centrifuge, which discharges damp solids.

The underflow discharge should not be too close to the rig pump suction. The overflow (liquid/colloidal solids) gravity feed down a constantly sloping chute or pipe to waste. Sufficient working space should be provided for routine maintenance and operating adjustments to the centrifuge.

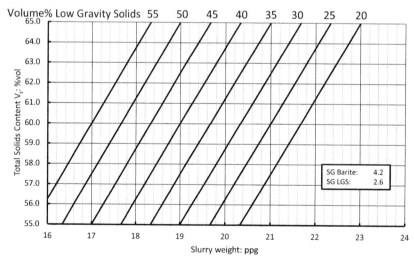

Fig. 21.3 Volume percent of low gravity solids for various mud weights and total solids concentrations.

The centrifuge should be used continuously during drilling. This relates to the standard purpose of the centrifuge to control plastic viscosity by removal of colloidal size particles. Centrifuges should be run to maintain a constant plastic viscosity. The flow rate into a centrifuge should be adjusted to accomplish this.

The maximum and minimum plastic viscosity limits should be established as part of the overall drilling fluid program. Viscosity will normally creep up when centrifuges are shut down due to the size degradation of low gravity solids. Processing too high of a flow rate through the centrifuge may remove too many of the necessary low-shear-rate viscosifiers and filtration control additives. Low-shear rate additives and various chemicals must be added back to the mud system. The amount of replacement may be calculated exactly from mass balance equations. "Under-centrifuging" simply will not achieve the desired reduction in viscosity. Both over-centrifuging and under-centrifuging should be avoided as the economics of operation quickly disappear under these circumstances.

Other popular applications of centrifuges are to reduce overall solids content, reduce average particle size and to reduce overall waste volume by "dewatering" the discharge from other solids control equipment. Environmental considerations may require special handling of the solid and/or liquid base. As disposal costs increase, waste volumes become a major component of overall well cost. Additional solids removal equipment-such as a decanting centrifuge-is often cost effective due to minimized waste volume. (In addition, better control of drilling fluid properties is also accomplished, further reducing well cost.)

Centrifuging NADF

Centrifuging non-aqueous drilling fluid (NADF) has become confusing. Much of this confusion arises from a similar concept to the barite recovery misunderstanding. The liquid phase of NADF is expensive and, frequently, considered too precious to discard.

To recover the non-aqueous fluid (NAF), a technique called "double-centrifuging" is often recommended, as shown in Fig. 21.4.

In double centrifuging, a slower speed centrifuge processes the active weighted drilling fluid, and the heavy, or underflow, fraction is returned to the active system, as is normally accomplished. The light, or overflow, slurry is sent to a second, higher speed, centrifuge to recover the liquid phase. The light, or overflow, fraction from the second centrifuge is returned to the active system, while the heavy, or underflow, fraction is discarded. Unfortunately, the light fraction contains the majority of the smallest particles, including colloids, with the result that the concentration of these ultra-fine particles in the active system climbs continuously. The total amount of drilled solids discarded by the second centrifuge is relatively small, as shown in Fig. 21.5.

The area between the two curves in Fig. 21.5 indicates the solids which would be discarded. The solids larger than the ones indicated by the separation curve for centrifuge #1 (the red curve) would be returned to the active system. The solids smaller than the

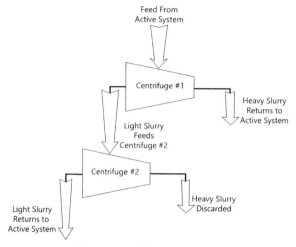

Fig. 21.4 Double centrifuging.

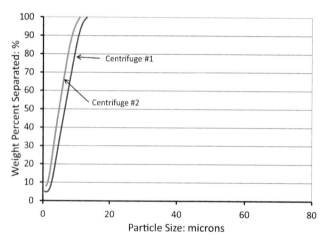

Fig. 21.5 Separation curves for double centrifuging.

ones indicated by the separation curve for centrifuge #2 (the green curve) would be returned to the active system. When the curves shown in Fig. 21.5 are plotted on semi-log paper, as in Fig. 21.6, the area between the curves is enhanced and double centrifuging appears to be more effective.

However, this distorts the real picture of the amount of drilled solids that is removed, which is very small. The result is that double centrifuging for the purpose of recovering NAF does actually recover much of the liquid, but it also recovers the detrimental small colloidal solids that centrifuges are designed to eliminate from the drilling fluid system.

The two-stage centrifugation process is based on two erroneous assumptions: first, that the centrifuge is capable of separating barite from low gravity solids; and, second, that it is

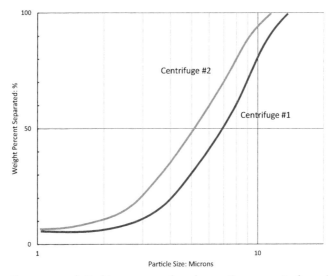

Fig. 21.6 Separation curves plotted to erroneously enhance the concept of good solids removal.

capable of producing an effectively solids-free liquid for return to the drilling fluid at the second stage. These assumptions are incorrect and ignore the physics of sedimentation.

Centrifugation is accelerated sedimentation. Rotation of the centrifuge bowl at high speed subjects the contents to increased gravitational forces and forces the sedimentation that would occur in a settling tank in hours, days, weeks, or months to take place in seconds. Stokes' Law defines sedimentation. It states that the sedimentation rate is directly proportional to the difference in mass between the settling particle and the surrounding liquid, and inversely proportional to the viscosity of the liquid, and provides a mathematical means of evaluating the effects of the individual variables.

In a non-aqueous drilling fluid, using the overflow (light) slurry for other applications to include gravel pack, cementing fluid, and packer fluid can be beneficial instead of returning the very detrimental particles to the active system.

A potential solution to the build-up of ultra-fine particles in the drilling fluid is to re-purpose the centrifuge overflow, or light, slurry, rather than return it to the active system. If the well is to be gravel-packed, the light slurry will contain very few large particles—if any. Changing the base fluid for gravel packing frequently has a large impact on the well bore stability. By using the same fluid for gravel packing that was used to drill the well, the shock to the well bore is minimized. While processing the NADF through the first centrifuge, store the overflow, or light, slurry in storage tanks for use during the completion mode.

Even if gravel packing fluid is not needed, the centrifuge overflow, or light, slurry can still be used for cementing and completions. Cement needs to displace the drilling fluid from the annulus to obtain the proper barriers to prevent flow behind casing. The drilling fluid normally used for drilling is not a good "cementing" fluid. Drilling fluid needs to have a high viscosity in the low-shear-rate flow pattern so that it can transport cuttings. This high viscosity, however, makes it difficult to remove in narrow annuli. Cementing recommendations normally result in treating the drilling fluid to decrease the yield point or the low-shear-rate viscosity. Such a fluid could be created from the centrifuge overflow, or light, slurry and be used as a packer fluid during completions operations.

Operating tips

Centrifuges are relatively easy to operate, but they require special skills for repair and maintenance. Rig maintenance of centrifuges is limited to routine lubrication of the unit.

Although operating procedures will vary in detail from model to model, a few universal principles apply to virtually all centrifuges:
- If the solids underflow is to return to the system, locate the centrifuge so the underflow falls into a well-stirred spot.
- If the solids underflow is to be discarded, locate the machine so the underflow can be removed periodically.
- Direct the liquid overflow to a well-stirred spot.
- Do not locate the machine solids or liquid returns too close to the rig pump suction. Allow time and space for adequate mixing.
- Liquid effluent lines should have a constant downward slope.

Summary

If the following were more generally understood, the use of centrifuges would be greatly improved:
- Centrifugal separation is based on the mass of the particles, not on their specific gravity.
- Centrifuges cannot separate barite from low gravity solids.
- The objective of centrifuging weighted drilling fluids is the removal of colloidal and near-colloidal particles, not the removal of low gravity solids.
- The most serious problems associated with the use of weighted drilling fluids are those arising from excessive concentrations of colloidal and near-colloidal particles; not necessarily from excessive concentrations of drilled solids. At higher mud weights, the problem particles are predominantly barite.
- Colloids are particles that are so fine that they will not settle through liquid. In pure water, particles finer than 2 μ are classified as colloids. In more viscous fluids, colloidal behavior is exhibited by much larger particles.
- Particles that will not settle (colloids) cannot be separated by centrifuging.

- Centrifuging, even at higher than normal rotational speeds, cannot separate the colloidal solids from the liquid. The return of centrifuge overflow to a mud system always involves the return of colloids; and, therefore, is always potentially damaging.

Rotary mud separator

Amoco Research introduced a concept to separate very small particles from the drilling fluid instead of using a decanting centrifuge. The new concept was called a rotary mud separator. Drilling fluid was introduced into the annulus of a cylinder and a rotating cylinder with holes in it. The fluid would move along the annulus to the discharge end of the centrifuge. The shaft rotating the perforated cylinder was hollow and had ports all along the length of the shaft. Fluid was withdrawn from the annulus and the hollow shaft. As large particles tried to pass through the rotating perforated cylinder as they moved along the annulus, the centrifugal force would reject their entry into the rotating cylinder. Only the very small particles would pass through the holes in the rotating cylinder. This separates the fluid into two streams: one containing very small light particles (the overflow, or light slurry) and the other fluid removed from the annulus containing the larger heavier particles (the underflow, or heavy slurry) (Fig. 21.7).

The big advantage of this system was the fact that both streams had the ability to be pumped. The centrifuge could be mounted on a trailer, located next to the mud tanks, suction and discard lines could be plumbed into the mud tanks system.

If a drilling operation unexpectedly found the colloidal content of the drilling fluid increasing, a crane was not needed to hoist a centrifuge to the top of the mud tank (Figs. 21.8 and 21.9).

The heavy, or underflow, slurry from a decanting centrifuge contains about 60% - volume solids and does not have the ability to be pumped. The centrifuge must be

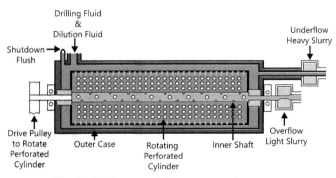

Fig. 21.7 Diagram of the rotary mud separator.

Fig. 21.8 Trailer mounted rotary mud separator.

Fig. 21.9 The heavy (or underflow) from the rotary mud separator.

located on top of the mud tanks in a location where agitators can blend this heavy slurry into the drilling fluid. The rotary mud separator discharge was easier to blend with the active drilling fluid system but had a totally different appearance. The liquid discharge of the heavy slurry seemed to indicate that the cut point of this device was not as low as a decanter. Several tests at different rig locations surprisingly indicated that the cut point was only slightly higher that the decanting centrifuge (Fig. 21.10).

The cut point curve was surprising and the D50 point seemed to low to be believable. The test was repeated on the active system and verified. However, to confirm the low D50 cut point, two additional units were located on other drilling rigs and tested. The additional tests verified the original data. The discharge looks so different from the decanter centrifuge that acceptance in field operation is not very encouraging.

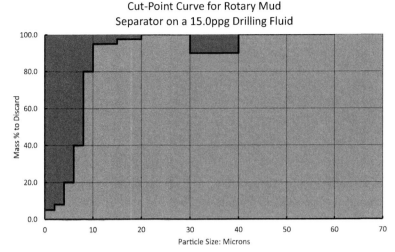

Fig. 21.10 Cut point curve for a 15 ppg water-based drilling fluid measured at the drilling rig.

Appendix 21.A
Stokes' law settling

Stokes' law describes the settling of spheres in a Newtonian fluid. A spherical particle placed in a Newtonian fluid will sink if the buoyant force does not match or exceed the gravitational force on the sphere. The net downward force on a sphere is the difference between the settling force and the buoyant force. From Newton's second law of motion, force applied to a mass causes the mass to accelerate:

$$F = ma = (\text{volume})(\text{density})a$$

$$\text{Net downward force} = \left[\frac{4}{3}\pi r^3 \rho_s - \frac{4}{3}\pi r^3 \rho_l\right]a = \left[\frac{4}{3}\pi r^3 (\rho_s - \rho_l)\right]_a$$

where m is volume × density.
r is the radius of the particle,
ρ_s is the density (g/cm^3),
ρ_l is the liquid density (g/cm^3), and
a is the acceleration of the particle.

When a spherical particle starts settling through a column of liquid, the drag force (or retarding force) can be calculated from the equation:

$$\text{Retarding force} = 6\pi\mu r v$$

where μ is the viscosity of the liquid, and v is the velocity of the particle.

When the particle acceleration causes the particle to reach a velocity which makes the retarding force equal to the net downward force, the particle will cease accelerating and have a constant velocity. This is called the terminal velocity (V_T):

$$6\pi\mu r V_T = \frac{4}{3}\pi r^3 (\rho_P - \rho_L)$$

This is Stokes' law which describes the terminal velocity of a sphere falling in a Newtonian fluid.

$$V_T = \frac{2}{9}\frac{r^2(\rho_P - \rho_L)}{\mu}$$

As discussed in the Sand Trap section, the settling rate in most NADF is very low because of the need to prevent settling in well bores which causes barite sag and/or while circulating cuttings from the hole. A centrifuge applies a centripetal acceleration to the solids which causes them to settle faster than they would in a quiescent pool. When a rock is whirled around on the end of a string, the motion is a circle because of the force applied to the rock by the string. This force is the tension in the string and makes the rock move in a circle. This force creates an angular acceleration (ω) which can be calculated from the equation:

$$\alpha = r\omega^2$$

where r is the radius of rotation, ft; and.
ω is the angular velocity, radians/min.

For a centrifuge bowl with a 2 foot diameter and spinning at 1600 rev/min, the angular momentum would be:

$$\alpha = (1\text{ ft})\left[\left(1600\frac{\text{rev}}{\text{min}}\right)\left(2\pi\frac{\text{radians}}{\text{rev}}\right)\left(\frac{\text{min}}{60\text{ s}}\right)\right]^2 = 28{,}073\frac{ft}{s^2}$$

This angular acceleration is frequently compared with the acceleration of gravity (g) and reported in g's:

$$\alpha = 28{,}073\frac{ft}{s^2}\left[\frac{g}{32.19\frac{ft}{s^2}}\right] = 873g$$

In this case, a particle would experience an acceleration of 873 times the acceleration it would experience in free fall toward the earth. This causes the particles to settle through the thin liquid phase in a centrifuge more quickly. This liquid layer is pressed against the inside wall of the spinning centrifuge bowl and experiences very little shear parallel to the surfaces. The inner and outer layers are moving with the same speed, which means this is a low-shear-rate environment. Usually, a dilution fluid is introduced into the feed stream to decrease the low-shear-rate viscosity. This helps separate the lower weight (or mass) particles from the larger weight (or mass) particles.

The particles in the drilling fluid are not spherical (or shouldn't be if the bore hole is being cleaned). The cross-sectional area which governs the retarding force depends upon the orientation of the cuttings and the shape of the cuttings. Both the mass of the particle and this cross-sectional area determine the exact settling rate within the thin layer of fluid in the centrifuge.

The average densities of barite and LGS are 4.2 and 2.6 g/cm^3, respectively. A 10 μm spherical barite particle has the same mass as an 11.7 μm spherical LGS particle:

Mass of barite spherical particle = Mass of LGS spherical particle

$$\frac{4}{3}\pi(10 \times 10^{-4} \text{ cm})^3\left(4.2 \frac{\text{g}}{\text{cm}^3}\right) = \frac{4}{3}\pi R^3\left(2.6 \frac{\text{g}}{\text{cm}^3}\right)$$

which yields: $R = 5.86 \times 10^{-4}$ cm, or a diameter of 11.7×10^{-4} cm.

If barite particles finer than 10 μm remain in the overflow, then LGS particles finer than 11.7 μm will also remain in the overflow; all of these will be discarded. The centrifuge can be used to separate larger barite particles from smaller LGS particles, but if the LGS and barite particles are of similar size, the separation efficiency will be very poor.

Problems

Problem 21.1

A well just south of Boondocks, Texas is using an 18 ppg water-based drilling fluid while drilling at 16,900 ft. The solids control equipment has been poorly maintained; solids control was primarily by dilution. The cuttings were large and the hole was clean. Before the weight-up, the drilling rate was somewhat lower than expected but the bits were lasting longer than anticipated. The 15 ppg drilling fluid in the hole prior to weight-up had the following properties: PV of 38cp, YP of 14 lb/100 sqft, gels 8/15 and 31 %vol solids. Before weight-up, 1,000 bbl of this drilling fluid was moved to a storage tank. Even so, the well has been running on budget but the company representative wants to set a record and also save money on the mud bill. He recommends a decanter centrifuge be used to recover the barite from the stored 15 ppg drilling fluid and used for maintenance of the 18 ppg drilling fluid. The Neverstuck Drilling Company has been proud of its name and record and recommended against using that barite source. The tool pusher agrees that the centrifuge separates the barite from the colloidal drilled solids but his experience tells him that stuck pipe follows such action. He can't explain why the "decanter-centrifuge" barite causes stuck pipe but agrees that it may be the fault of his centrifuge. The company representative called a rental outfit for a good barite recovery centrifuge. After the centrifuge was running for a while, the mud engineer reported that the barite slurry density in the underflow was 20.3 ppg and the discharge was very dry for a decanter centrifuge. The company representative rejoiced and assured the toolpusher that everything was under control (Fig. 21.P.1).

Is it?

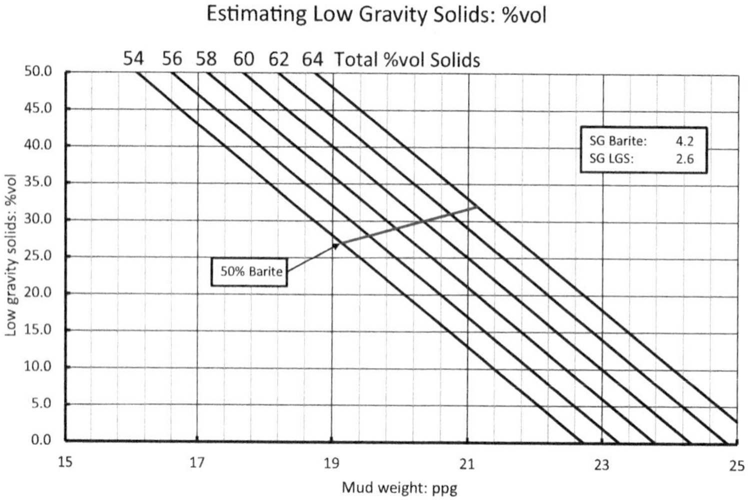

Fig. 21.P.1 Estimating low gravity solids concentration in a centrifuge underflow (heavy) slurry.

Problem 21.2

An 1800 RPM centrifuge is used to process a variety of fresh-water drilling fluids around a rig. The suction and discharge can be moved to different places.

Part 1) When a 12.5 ppg drilling fluid in the active system is centrifuged, the underflow weighs 22.0 ppg and the overflow weighs 10.1 ppg. Assume that the underflow solids content is 60% volume and the overflow solids content is 12% volume.

1. Is this centrifuge performing a useful function?

Part 2) A 15.0 ppg fresh-water base drilling fluid from another well is brought to the drilling rig. The rig is drilling with an 11.0 ppg drilling fluid that has 11% volume total solids. The mud weight needs to be increased to 12 ppg. The leftover drilling fluid from storage seems like an excellent source of barite. The centrifuge feed is switched to the stored drilling fluid. The underflow weighs 20 ppg and has a solids content of 62% volume.

1. Is this a good idea? (Compare the drilled solids in the drilling fluid with the drilled solids content of the underflow of the centrifuge.)
2. What results do you expect to see from saving so much money by using this "free" barite to increase the mud weight?

Part 3) Suppose the discharge from a mud cleaner weighed 20 ppg with the 62% - volume solids concentration.

1. Is the mud cleaner performing a useful function? (Calculate low gravity solids and barite in the discard.)

CHAPTER 22

Solutions to chapter problems

Preface quiz answers

1. Define viscosity:
   Viscosity is the ratio of shear stress to shear rate. If the shear stress is in dynes/sq.cm and the shear rate is in reciprocal seconds, the viscosity will be expressed in poise.
2. With Viscometer readings R600 = 60 & R300 = 45; calculate PV and YP.

$$R\,600 = 60$$
$$R\,300 = 45$$
$$PV = 15$$
$$YP = 30$$

3. Calculate the viscosity at each reading:

$$\text{Viscosity} = \frac{60}{600} \times 300 = 30 \text{ cp}$$

$$= \frac{45}{300} \times 300 = 45 \text{ cp}$$

4. Why would you want Plastic Viscosity to have a high value?
   PV should be as low as possible—this represents the viscosity of the fluid passing through the nozzles.
5. Plastic viscosity depends upon four things. They are:
   Size, shape, number of solids and the liquid phase viscosity
6. A centrifugal pump is connected directly to a joint of 7 inch casing standing next to the derrick. The casing is open at the top. When the pump takes suction from a tank filled with water, the water stands to a height of 20 ft above the liquid level in the tank. The water is drained from the system and the centrifugal pump suction is connected to a tank filled with 16.6 ppg drilling fluid (twice as heavy as water). When the pump is turned on, how high will the drilling fluid go in the casing? (circle the correct answer)
   b. Same height as the water
7. In problem 6, would the 16.6 ppg drilling fluid go any higher if the centrifugal pump motor horsepower was doubled? **No**
8. In problem 6, if another identical pump was connected to bottom of the casing (install a tee), would the fluid in the casing go almost twice as high? **No**

9. What is NPSH (Net Positive Suction Head)?
    Net positive suction head. The head at the suction of the pump above absolute zero.
10. What is a flounder point?
    The point at which the drilling fluid ceases to remove all of the cuttings from the bottom of the hole before the next row of cutters regrind them.
11. What drilling fluid parameters control the flounder point?
    Plastic viscosity, hydraulic impact or hydraulic power.
12. What are the Bingham Plastic model and the Power Law model for drilling fluid?
    Both are two parameter simple rheological models which approximate the shear stress/shear rate relationship of the fluid. The Bingham Plastic model helps separate the effect of solids from the effect of electrochemical behavior (or contamination).
13. How do you select the flow rate to use while drilling a well?
    All kinds of ways are used here.
14. How should you select a flow rate to use while drilling a well?
    Maximize the hydraulic impact or the hydraulic power of the fluid at the drill bit.
15. What is "energy" (definition)?
    Force times distance—also called "work".
16. What is energy per unit volume?
    Pressure or stress (units of pounds per square inch).
17. How do you make the drilling fluid have a low viscosity when it strikes the bottom of the hole to remove the cuttings and a high viscosity in the annulus to bring cuttings to the surface?
    Non-Newtonian flow causes a shear thinning fluid to have a high viscosity at low shear rates and a low viscosity at higher shear rates. Look at the answer to question #2 above.
18. A surface casing is set at 3000 ft in a 9.0 ppg drilling fluid. What is the head at the casing seat?
    3000 ft.
19. A barite recovery centrifuge will separate barite from low gravity solids. **F**
    Desanders, desilters and centrifuges exert a centrifugal force on the solids and separate by mass not density. Particles with the same mass are found together.
20. A 3000 RPM centrifuge exerts a high-shear rate on the drilling fluid. **F**
    Although the fluid is spinning with 3000 RPM, there is no shear in the fluid. The inside surface is not moving relative to the outside surface.
21. Mud guns can be used to stir all mud tanks if the fluid is clean drilling fluid in the suction compartment. **F**
    Any clean fluid returned to the removal section is blended with the "dirty" fluid and will require more removal capacity. No mud guns should be in removal section.
22. The shale shaker back tank (possum belly) should be dumped into the sand trap before each trip to prevent fluid from drying on the shaker screens. **F**

Do not dump shaker back tank into the active system. The solids do not settle and will plug desilters and desanders when pumping resumes.

23. The desilter overflow should be returned to a compartment up stream from its suction so that the desilter can "look at the mud twice". **F**

    The "clean" desilter overflow blends with the "dirty" drilling fluid reducing the process efficiency by almost half.

24. Fluid loss gives a good indicator for the quality of filter cake. **F**

    In a Newtonian fluid, the filter cake thickness is a function of fluid loss. In a Non-Newtonian fluid, it is not. As solids disintegrate, the surface area wetted by the liquid phase increases greatly. A low fluid loss drilling fluid can still have a very thick cake.

25. An API 200 (75 µm) screen has a cut-point of 75 µm. **F**

    A cut point is a performance indicator. Cut points of an API 200 can range from 75 µm for dry solid, to a very low number. While screening NADF, if the screen becomes water-wet, the openings are extremely small. This results in the large quantities of fluid leaving the system.

26. Drill strings will not experience differentially stuck pipe in an oil-base drilling fluid. **F**

    Oil-base drilling fluid can deposit very thick filter cakes and still have a low fluid loss. This is common in wells where the NADF has been used in one or two other wells.

27. Centrifuges and desilters separate drilled solids in the same size ranges in a weighted drilling fluid. **F**

    In a weighted drilling fluid, centrifuges separate drilled solids smaller than the barite (and the very fine barite); mud cleaners remove solids larger than the barite.

28. Since drilled solids have much less effect on Yield Point in an oil-based drilling fluid than they do in a water-based drilling fluid, good solids control is not necessary in an oil-based drilling fluid system. **F**

    NADF does not tolerate drilled solids. The visual change of yield point observed in water-based muds does not exist within NADF. This gives a false sense of security. Thick filter cakes can create stuck pipe. But, probably the most expensive effect is the failure of cement to develop a permanent barrier in the annulus. Have mud engineers run a HTHP cell with 500 psi differentiated pressure at room temperature and examine the filter cake in a well-used dirty NADF.

29. A centrifugal pump produces a constant pressure. **F**

    Centrifuge produces a constant head. The pressure depends on the mud weigh. $P = 0.052 \, (MW) \, (head)$

30. The flow rate from a centrifugal pump should be controlled with a valve on its suction. **F**

    Restricting the suction of a centrifugal pump will create cavitation and quickly destroy the pump.

31. Closing the valve on the discharge pipe of a centrifugal pump for a few minutes will damage the pump. **F**

Closing the centrifugal pump discharge for a short time will not harm the pump. It will use less horsepower than when it is pumping fluid. Closing the valve for a long period will cause the fluid to become very hot and may destroy the seals. A pressure gauge mounted below the valve makes a good diagnostic tool. Close the valve briefly and read the pressure. Check the pump curves to see if the head produced is correct.

32. The head at a surface casing seat at 3000 ft in 9.0 ppg drilling fluid is 1400 psi. **F**
    The head is 3000 ft. The pressure is 1400 psi.

Chapter 1 Rheology solutions

Problem 1.1 solution

FLUID #1:
R600 = 20
R300 = 10
PV = 10 cp
YP = 0 lb/100 sqft

FLUID #2:
R600 = 20
R300 = 15
PV = 5 cp
YP = 10 lb/100 sqft

FLUID #3:
R600 = 40
R300 = 35
PV = 5 cp
YP = 30 lb/100 sqft

FLUID #4:
R600 = 40
R300 = 15
PV = 25 cp
YP = -10 lb/100 sqft

Comments about these fluids:
Fluid #1 is a Newtonian fluid, the YP is zero.
Fluid #2 has a low PV and a larger YP so it will provide an excellent drilling fluid— good chip removal from beneath the drill bit and good cuttings transport up an annulus.

Fluid #3 would also provide a similar benefit; but will look very "thick"—like a milk shake.

Fluid #4 shows a negative yield point. A Bingham Plastic Fluid with this property would be difficult to keep in a container. It would start moving before a shear stress was applied. This type of fluid does exist—but it is not a Bingham Plastic Fluid. The fluid would be called a shear-thickening fluid. High concentrations of methyl-methacrylate or corn starch in water will produce a fluid that becomes more viscous as the shear rate increases. Drilling fluid should do the opposite; they should become less viscous as the shear rate increases. When these values are observed in a drilling fluid, it indicates that settling has occurred during the measurements. The fluid measured at the 600 RPM speed is not the same fluid measured at the 300 RPM speed because solids have settled. This is frequently observed in thin weighted drilling fluids. Examination of the cup after the measurement will discover a large quantity of barite settled on the bottom.

A negative yield point usually indicates the drilling fluid has changed while making the measurement. Specifically if the 600 RPM reading is made and the outer cylinder slows to 300 RPM, some solids may settle.

On a broader perspective, a negative yield point would have a shear rate/shear stress curve as shown in Fig. S1.1A.

The literal interpretation of this curve would indicate that the fluid would start moving before a shear stress is applied. Extreme extrapolation would indicate that thinking about moving the fluid would cause the fluid to jump out of the cup.

Even mud engineers know this is preposterous. A negative yield stress can occur in a real fluid because the fluid cannot be represented by the Bingham Plastic model. Drilling Fluid is shear-thinning—meaning the viscosity decreases. Methyl methacrylate and

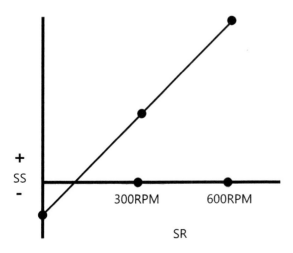

Fig. S1.1A

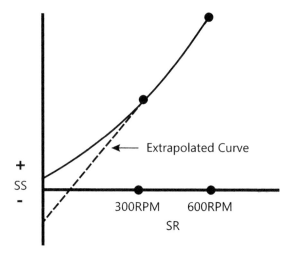

Fig. S1.1B

cornstarch solutions are shear thickening. Their shear stress/shear rate curve, shown in Fig. S1.1B would extrapolate to a negative eyelid stress.

The Bingham Plastic model provides a great method of monitoring drilling fluids. The plastic viscosity indicates the liquid phase viscosity and the size, shape, and number of particles. Basically, it keeps track of the solids. The yield point indicates the electro-chemical effects from containments or polymer interactions. It is the simplest rheological model. It does not actually follow the shear stress/shear rate behavior of most drilling fluids. For this reason, it should not be used for pressure loss calculations. However, it does allow proper treatment of drilling fluid.

Problem 1.2 solution

Speed, rpm	Reading	Visc cp
600	60	30
300	41	41
200	32	48
100	24	72
6	12	600
3	11	1100

Note that the calculated viscosities at the very low speeds (3 and 6 RPM) do not match the viscosities of the other readings because of the inability to read accurately in this low shear rate range.

$R600 = 60$

$R300 = 41$

PV = 19 cp
YP = 22 lb/100 sqft

The viscosity at the 600 RPM reading (or 1022 reciprocal seconds) is 30 cp; and the PV is 19 cp. (Again, PV is the viscosity the fluid has at an infinite shear rate; so it should be lower with a shear-thinning fluid).

Chapter 2 Mud tank arrangements solutions

Problem 2.1 solution

From Fig. 2.5; read an increase in mud weight needed = 0.8 ppg.
Mud weight of slug should be a minimum of 14.8 ppg.
Or calculate:

$$MW_{slug} = \frac{(MW_{orig})(100 \text{ ft} + H_{slug})}{H_{slug}}$$

$$MW_{slug} = \frac{(14.0)(100 \text{ ft} + 1690 \text{ ft})}{1690 \text{ ft}} = 14.82 \text{ ppg}$$

Problem 2.2 solution

From Fig. 2.7; read an increase of 1.8 ppg.
Or Mud weight should be 17.8 ppg.
Or calculate:

$$MW_{slug} = \frac{(MW_{orig})(100 \text{ ft} + H_{slug})}{H_{slug}}$$

$$MW_{slug} = \frac{(16.0)(100 \text{ ft} + 868 \text{ ft})}{868 \text{ ft}} = 17.84 \text{ ppg}$$

Problem 2.3 solution

Calculate the length of air column above the slugged pipe from the equation:

$$X = \frac{(MW_{slug})(H_{slug}) - (MW_{orig})(H_{slug})}{MW_{orig}}$$

Height of the slug from Table 2.2 for $4\frac{1}{2}''$ drill pipe is 2110 ft.

$$X = \frac{(15 \text{ ppg})(2110 \text{ ft}) - (14.0 \text{ ppg})(2110 \text{ ft})}{14.0 \text{ ppg}} = 151 \text{ ft}$$

The drill pipe will be dry because the fluid level will be 151 ft below the flow line. The driller could have saved some barite if the required mud weight had been calculated:

$$MW_{slug} = \frac{(MW_{orig})(100 \text{ ft} + H_{slug})}{H_{slug}}$$

$$MW_{slug} = \frac{(14.0)(100 \text{ ft} + 2110 \text{ ft})}{2110 \text{ ft}} = 14.7 \text{ ppg}$$

Problem 2.4 solution

Calculate the length of air column above the slugged pipe from the equation:

$$X = \frac{(MW_{slug})(H_{slug}) - (MW_{orig})(H_{slug})}{MW_{orig}}$$

Height of the slug from Table 2.2 for 6 5/8″ drill pipe is 868 ft.

$$X = \frac{(15 \text{ ppg})(868 \text{ ft}) - (14.0 \text{ ppg})(868 \text{ ft})}{14.0 \text{ ppg}} = 62 \text{ ft}$$

The drill pipe will not be dry because the fluid level will be only 62 ft below the flow line.

What should the mud weight have been for an effective slug?

$$MW_{slug} = \frac{(MW_{orig})(100 \text{ ft} + H_{slug})}{H_{slug}}$$

$$MW_{slug} = \frac{(14.0)(100 \text{ ft} + 868 \text{ ft})}{868 \text{ ft}} = 15.6 \text{ ppg}$$

Chapter 5 Drilled solids calculations solutions

Problem 5.1 solution

$$(\rho_B - \rho_{LGS}) V_{LGS} = 100\rho_f + (\rho_B + \rho_f) V_S - 12MW$$

For barite: $\rho_B = 4.2$, $\rho_f = 1.0$

Additional note:

To check the constants in the equations, calculate the mud weight in each equation for a fluid that has no solids. The low gravity solids would be zero and the total solids would be zero.

LGS specific gravity	Equation
2.3	$(4.2 - 2.3) V_{LGS} = 100 + 3.2 V_{LGS} - 12 \, MW$ $1.9 \, V_{LGS} = 100 + 3.2 V_{LGS} - 12 \, MW$ $V_{LGS} = 52.6 + 1.68 \, V_S - 6.32 \, MW$
2.4	$(4.2 - 2.4) V_{LGS} = 100 + 3.2 V_{LGS} - 12 \, MW$ $1.8 \, V_{LGS} = 100 + 3.2 V_{LGS} - 12 \, MW$ $V_{LGS} = 55.6 + 1.78 \, V_S - 6.67 \, MW$
2.5	$(4.2 - 2.5) V_{LGS} = 100 + 3.2 V_{LGS} - 12 \, MW$ $1.7 \, V_{LGS} = 100 + 3.2 V_{LGS} - 12 \, MW$ $V_{LGS} = 58.8 + 1.88 \, V_S - 7.06 \, MW$
2.6	$(4.2 - 2.6) V_{LGS} = 100 + 3.2 V_{LGS} - 12 \, MW$ $1.6 \, V_{LGS} = 100 + 3.2 V_{LGS} - 12 \, MW$ $V_{LGS} = 62.5 + 2.0 \, V_S - 7.50 \, MW$
2.7	$(4.2 - 2.7) V_{LGS} = 100 + 3.2 V_{LGS} - 12 \, MW$ $1.5 \, V_{LGS} = 100 + 3.2 V_{LGS} - 12 \, MW$ $V_{LGS} = 66.7 + 2.13 \, V_S - 8.00 \, MW$
2.8	$(4.2 - 2.8) V_{LGS} = 100 + 3.2 V_{LGS} - 12 \, MW$ $1.4 \, V_{LGS} = 100 + 3.2 V_{LGS} - 12 \, MW$ $V_{LGS} = 71.4 + 2.29 \, V_S - 8.57 \, MW$
2.9	$(4.2 - 2.9) V_{LGS} = 100 + 3.2 V_{LGS} - 12 \, MW$ $1.3 \, V_{LGS} = 100 + 3.2 V_{LGS} - 12 \, MW$ $V_{LGS} = 76.9 + 2.46 \, V_S - 9.23 \, MW$

For LGS specific gravity of 2.6:
$0 = 62.5 + 0 - 7.5 \, MW$.
Or $7.5 \, MW = 62.5$.
$MW = 8.33$ ppg which is the mud weight of water.

Problem 5.2 solution

2.3:
$$V_{LGS} = 52.6 + 1.68 \, V_s - 6.32 \, MW$$
$$52.6 + 1.68 \, (20) - 6.32 \, (13.0)$$
$$52.6 + 33.6 - 82.16$$
$$V_{LGS} = 4\% \text{ Volume}$$

2.6:
$$V_{LGS} = 62.5 + 2.0\ V_s - 7.5\ MW$$
$$62.5 + 2(20) - 7.5(13.0)$$
$$62.5 + 102.5 - 97.5$$
$$V_{LGS} = 5\%\ \text{Volume}$$

2.9:
$$V_{LGS} = 76.9 + 2.46\ V_s - 9.23\ MW$$
$$76.9 + 2.46(20) - 9.23(13.0)$$
$$76.9 + 126.1 - 119.99$$
$$V_{LGS} = 6.1\%\ \text{Volume}$$

Problem 5.3 solution

$$\rho_B = 4.1$$
$$\rho_B - \rho_f = 3.1$$
$$V_{LGS} = \frac{100}{\rho_B - \rho_{LGS}} + \frac{\rho_B - \rho_f}{\rho_B - \rho_{LGS}}\ V_S - \frac{12}{\rho_B - \rho_{LGS}}$$

For $\rho_{LGS} = 2.3$:
$$\rho_B - \rho_{LGS} = 4.1 - 2.3 = 1.8$$
$$V_{LGS} = \frac{100}{1.8} + \frac{3.1}{1.8}\ V_S - \frac{12}{1.8}\ MW$$
$$= 55.56 + 1.72\ V_S - 6.67\ MW$$
$$= 55.56 + 1.72(20) - 6.67(13.0)$$
$$= 55.56 + 34.44 - 86.71$$
$$V_{LGS} = 3.29\%\ \text{Volume}$$

For $\rho_{LGS} = 2.6$:

$$\rho_B - \rho_{LGS} = 4.1 - 2.6 = 1.5$$

$$V_{LGS} = \frac{100}{1.5} + \frac{3.1}{1.5} V_S - \frac{12}{1.5} MW$$

$$= 66.67 + 2.07(20) - 8(13)$$

$$= 66.67 + 41.4 - 104$$

$$V_{LGS} = 4.1\% \text{ Volume}$$

For $\rho_{LGS} = 2.9$:

$$\rho_B - \rho_{LGS} = 4.1 - 2.9 = 1.2$$

$$V_{LGS} = \frac{100}{1.2} + \frac{3.1}{1.2} V_S - \frac{12}{1.2} MW$$

$$= 83.3 + 2.58(20) - 10(13)$$

$$= 83.3 + 51.6 - 130$$

$$V_{LGS} = 4.9\% \text{ Volume}$$

Chapter 6 Cuttings transport solutions

Problem 6.1 solution

Part 1

The lowest annular velocity for this well is 60 ft/min. For the drilling fluid properties listed below, determine which wells will have a problem with hole cleaning.

Fluid #	MW, ppg	PV, cp	YP, lb/100 sqft	K, eff cp	CCI
1	9.0	10	10	340	0.46
2	9.0	5	5	140	0.19
3	9.0	5	10	580	0.78
4	15.0	30	5	60	0.14
5	15.0	30	20	370	0.83
6	10.0	10	15	740	1.11

$$CCI = K \, (MW) \, (AV)/400,000$$

With the annular velocity so low (as it might be in a large diameter hole or in a riser), none of these fluids can transport solids from the hole as they should be, except for the last fluid in the list.

Part 2

The yield point of a drilling fluid can be changed without a significant change in the PV. What should the YP be to make certain that all drilling fluids are cleaning the borehole?

Fluid #	MW, ppg	K needed, eff cp	YP, lb/100 sqft
1	9.0	740	14
2	9.0	740	11
3	9.0	740	11
4	15.0	450	23
5	15.0	450	23
6	10.0	670	14

Sample calculation:

$$K = 400,000/(\text{MW, ppg})(\text{AV, ft/min})$$

From the K-value graph, read the YP needed.

Problem 6.2 solution

This was a real problem on a well. The cubic pieces of shale coming off the end of the 200 mesh screen were something unexpected. The drilling fluid should have been transporting large cuttings and none were leaving the shaker. The actual solution to the problem became obvious when the rig made a connection. Large cuttings were revealed on the screen as the pool of liquid disappeared through the screen when the mud pumps were shut off. They could not climb the steep slope of the screen until they were pounded into smaller pieces. The screen was changed to a 100 mesh and lowered to a reasonable slope. Cuttings were quickly removed and the plastic viscosity started decreasing. After a couple of circulations, the screen was changed to a finer screen and more solids were removed. Eventually, the shaker was able to handle the entire flow on a 200 mesh screen.

One comment: The word "MESH" is no longer used but these screens were labeled 200 Mesh.

Problem 6.3 solution

First, calculate the CCI:
  MW = 11 ppg
  PV = 20 cp
  YP = 10 lb/100 sq.ft
  AV = ?

Annular velocity in a 16" casing with 5" drill pipe for a flow rate of 800 gpm. Fig. 6B.4 shows an annular velocity of 100 ft/min. The K-value is about 160 eff.cp.

$$\text{CCI} = (11 \text{ ppg})(100 \text{ ft/min})(160 \text{ eff.cp.})/[400,000] \text{ or } 0.44$$

Cuttings are not being transported to the surface until they become small and well-rounded. The high drag, observed when the mud pumps are stopped, is created by the cuttings in the annulus falling back on top of the stabilizers and drill bit. Shale does not creep into a well bore in a few minutes. This case has been observed in many places around the world and is used to validate the misconception that the well bore needs a higher mud weight to hold the shale back. Increasing the mud weight usually increases the CCI so the well is cleaned—but, with roller cone bits, the penalty is that the drilling rate decreases greatly.

Problem 6.4 solution

The annular velocity could be calculated:

$$\text{AV} = 24.51 \frac{Q}{(D_2)^2 - (D_1)^2}$$

$$\text{AV} = 24.51 \frac{600}{(19^2) - (5^2)} = 43.6 \text{ ft/min}$$

or from Chart 6B.4;

$$\text{AV} = 40 \text{ ft/min}$$

The "K" value can be calculated:

$$K = (511)^{1-n}(\text{PV} + \text{YP}) = 65$$

$$n = 3.322 \log \frac{2 \text{ PV} + \text{YP}}{\text{PV} + \text{YP}}$$

$$= 3.322 \log \frac{35}{20} = 0.81$$

$$K = (511)^{0.19}(\text{PV} + \text{YP}) = 65$$

or, use Fig. 6C.1; PV 15 cp, YP = 5 lb/100 ft^2, $K = 65$.

Cuttings are not being transported. They are grinding into much smaller debris and the plastic viscosity will increase. More significantly, the filter cake will become much thicker.

Calculate the value of K needed to clean the hole:

$$K = \frac{400,000}{(\text{AV})(\text{MW})} = \frac{400,000}{(40 \text{ ft/min})(15 \text{ cp})} = 667 \text{ eff cp}$$

From Fig. 6C.1: PV = 15 cp, a YP of 19.3 lb/100 ft^2 will provide a CCI of one and clean the hole.

Increase YP to 20 lb/100 ft²:
To confirm calculate K for PV 15 cp and YP 20 lb/100 ft²

$$n = 3.322 \log \frac{2\,PV + YP}{PV + YP}$$

$$= 3.322 \log \frac{50}{35} = 0.515$$

$$K = (511)^{1-n}(PV + YP)$$

$$= (511)^{0.485}(35) = 720 \text{ eff cp}$$

Increasing YP from 5 to 20 lb/100 ft² will bring cuttings to the surface.

Problem 6.5 solution

Gel/Lignosulfonate Drilling Fluid.
   MW = 10.5 ppg
   VS = 14% volume
   AV = 100 ft/min
   MBT = 27.5 lb/bbl
   PH = 10.5
Calculate viscosity at RPM:

$$\text{Viscosity} = \frac{R_{600}}{600} 300 = \frac{33}{600} 300 = 16.5 \text{ cp}$$

Viscometer readings		
RPM	Dial reading	Viscosity, cp
600	33	16.5
300	20	20
200	18	27
100	13	39
6	2	100
3	1	100

Calculate PV and YP:

$$PV = 13 \text{ cp}; \quad YP = 7 \text{ lb}/100 \text{ ft}^2$$

Calculate CCI:

$$K = 230 \text{ eff cp}$$

$$CCI = [(10.5 \text{ ppg})(100 \text{ ft/min})(230 \text{ eff cp})/400,000)] = 0.6$$

Describe cuttings:
- Cuttings will have no sharp edges.
- Calculate change that needs to be made:
- The yield point should be increased to bring the CCI up to a value of one.
- The K value needed to do this is:

$$K = \frac{400{,}000}{(10.5 \text{ ppg})\left(\dfrac{100 \text{ ft}}{\text{min}}\right)} = 380 \text{ eff cp}$$

For a PV of 13, YP needs to be 12 lb/100 ft^2. This will be a very small change in YP that will permit the cuttings to be delivered to the surface without grinding into small pieces.

Problem 6.6 solution

Drilling a 17½" hole at 6,000 ft below 20" casing using 5" drill pipe circulating a 9.0 ppg drilling fluid at 1000 gpm.

1. Calculate the carrying capacity (CCI) for the following drilling fluid properties:

PV, cp	YP, lb/100 sqft.	K, eff cp	CCI
15	10	190	0.3
15	15	400	0.6
15	20	730	1.1
10	5	100	0.2
10	10	290	0.5
10	15	740	1.2

AV for 5" drill pipe in 20" casing = 70 ft/min

$$\text{CCI} = \frac{(190 \text{ eff cp})(9.0 \text{ ppg})\left(70 \dfrac{\text{ft}}{\text{min}}\right)}{400{,}000} = 0.3$$

2. What "K" value is needed?
In all cases: the K value needed is

$$K = \frac{400{,}000}{(9 \text{ ppg})\left(70 \dfrac{\text{ft}}{\text{min}}\right)} = 635 \text{ eff cp}$$

Problem 6.7 solution

AV = 122 ft/min. MW = 10.0 ppg.

PV, cp	YP, lb/100 sqft	K, eff cp	CCI
20	20	520	1.6
20	15	300	0.9
20	10	150	0.5
15	20	720	2.2
15	15	400	1.2
15	10	180	0.6
10	15	700	2.1
10	10	300	0.9
10	5	100	0.3

What "K" value is needed?

$$\left[\frac{400,000}{(10.0)(122)}\right] = 328 \text{ eff cp}$$

Problem 6.8 solution

From Fig. 6C.1:

$$K = 140 \text{ eff cp}$$

$$CCI = \frac{(10 \text{ ppg})(100 \text{ ft/min})(140 \text{ eff cp})}{400,000} = 0.35$$

Cuttings will be well rounded and PV should continue to increase.

$$K_{needed} = \frac{400,000}{(10 \text{ ppg})(100 \text{ ft/min})} = 400 \text{ eff cp}$$

for PV = 25 cp, YP should be 17–18 lb/100 ft^2.

Problem 6.9 solution

From Fig. 6C.1:

$$K = 470 \text{ eff cp}$$

$$CCI = \frac{(10 \text{ ppg})(100 \text{ ft/min})(470 \text{ eff cp})}{400,000} = 1.2$$

Cuttings will have sharp edges.
If not, what changes should be made? No changes need to be made.

Problem 6.10 solution
From Fig. 6C.1:

$$K = 155 \text{ eff cp}$$

$$\text{CCI} = (15 \text{ ppg})(60 \text{ ft/min})(155 \text{ eff cp})/400,000 = 0.35$$

Cuttings are NOT being transported properly.

$$K_{needed} = 400,000/(15 \text{ ppg})(60 \text{ ft/min}) = 450 \text{ eff cp}$$

for PV = 20 cp, YP should be 18 lb/100 ft^2.

Chapter 7 Dilution solutions
Problem 7.1 solution
1. The volume of solids drilled can be calculated from the diameter of the hole and the length drilled.

$$\text{Volume drilled} = \{[(9.875^2)/1029]\text{bbl/ft}\}(44 \text{ ft/h})(24 \text{ h}) = 100 \text{ bbl}$$

With 60% SRE, 60 bbl drilled solids were discarded.

40 bbl drilled solids remained in the system

2. With 60 bbl drilled solids discarded as 35% volume of the total discard:

$$60 \text{ bbl} = 0.35 \text{ (volume of discard)}$$

$$\text{Volume of discard} = 171.4 \text{ bbl}$$

3. To keep the pit levels constant while drilling, 171.4 bbl of clean drilling fluid must be added.
4. The clean drilling fluid will have 2% volume bentonite (LGS) and must weigh 15.0 ppg.

$$V_{lg} = 62.5 + 2 V_S - 7.5 \text{ MW}$$

$$2 = 62.5 + 2 V_S - 7.5(15.0)$$

$$V_S = 26\%$$

$$LGS = 2\%$$

Consequently, $V_b = 24\%$ volume

This quantity of barite would be needed in the 171.4 bbl of clean drilling fluid which must be added. Volume of barite needed = 0.24 (171.4 bbl) = 41.1 bbl.

Barite weighs 1470 lb/bbl. This would be a barrel of solid barite—or 100%. The weight of barite needed would be (41.1 bbl) (1470 lb/bbl) or 60,417 lb. Barite comes in 100 lb sacks; 604 sacks of barite would be needed for the clean drilling fluid to keep the pit levels constant.

5. With 60% volume of the drilled solids discarded and 100 bbl reporting to the surface, 40% volume of the drilled solids, or 40 bbl, remained in the drilling fluid. Initially, the 2000 bbl of drilling fluid had 5% volume drilled solids, or 100 bbl. After drilling, the total volume was 2100 bbl. The fluid added while drilling had no drilled solids, so the remaining fluid (2000 bbl−171.4 bbl) would still have the 5% volume drilled solids or 0−(1828.6 bbl) = 91.43 bbl. The 40 bbl of drilled solids remaining in the system would mean that 131.43 bbl of drilled solids are now in the 2100 bbl system. The drilled solids concentration would be 131.43 bbl/2100 bbl or 6.26%.

6. Some drilling fluid must be removed from the system so that clean drilling fluid can dilute the remaining drilled solids back to the target 5% volume. The total drilled solids volume needs to be 0.05 (2100 bbl) or 105 bbl. Currently, the 2100 bbl system contains 131.43 bbl of drilled solids. Removal of 26.43 bbls of drilled solids is required to return the system to the target value of 5% volume drilled solids.

$$26.43 \text{ bbl} = 6.26\% \text{ (volume to be discarded)}$$

$$\text{Volume to be discarded} = 422.2 \text{ bbl}$$

$$\text{Volume of clean drilling fluid required} = 422.2 \text{ bbl}$$

7. The barite needed for the clean drilling fluid can be calculated from the volume fraction of barite in fluid containing 2% volume bentonite.

$$V_{lg} = 62.5 + 2 V_s - 7.5 \text{ MW}$$

$$2 = 62.5 + 2 V_s - 7.5 \text{ (15.0 ppg)}$$

V_s = 26% volume; therefore, barite must be 24% volume.

A total of 422.2 bbl of clean drilling fluid was added and it contained [0.24 (422.2 bbl)] of barite, or 101.3 bbl. Since barite weighs 1470 lb/bbl, then 148,950 lbs (1489 sacks) of barite have been added to the clean dilution fluid to decrease the drilled solids from 6.26% volume to 5% volume. To drill this interval, 2100 sacks (611 sacks + 1489 sacks) of barite were used.

The volume percent of solids in the drilling fluid containing 5% volume drilled solids can be calculated:

$$V_{lg} = 62.5 + 2 V_s - 7.5 \text{ MW}$$

where $V_{lg} = 7\%$ volume.

In this case, the volume percent solids measured by the mud engineer should be 28.5% volume.

When the interval finished drilling, before jetting the 422.2 bbl of drilling fluid, the low gravity solids had increased to 6.26% volume. The volume percent solids can again be calculated from the equation above, where $V_{lg} = 8.26\%$. In this case the mud engineer would report total solids to be 29.1% volume.

8. With a 50 cc retort, the liquid volumes for the two cases would not be very different. In the case of the 29.1% volume solids, (or 70.9% volume water), 35.45 cc of water would be captured in the measuring cylinder. In the case of the 28.5% volume solids (or 71.5% volume water), 35.75 cc of water would be captured in the measuring cylinder. Notice a change of only 0.3 cc still resulted in the use of 422 sacks of barite. We probably need a more accurate method of determining solids in a drilling fluid, especially when a 10 cc retort is used for the measurement.

Problem 7.2 solution

A practical rig-site solution.

Assume the Equipment Solids Removal Efficiency (SRE) is not known and needs to be determined. Using the same hole interval and using the information about additives to the drilling fluid, calculate the SRE.

Situation:

Drilling a $9^7/_8''$ hole at 44 ft/h for 24 h with a 15 ppg drilling fluid which has a target of 5% volume low gravity solids. The total volume of drilling fluid (active surface and hole) is 2000 bbl. Solids are discarded in a flow stream containing 35% volume drilled solids. Assume density of barite is 4.2 g/cc and of low gravity solids is 2.6 g/cc.

After drilling this interval, the mud engineer reports that 604 sacks of barite was required to keep the clean mud additions at 15.0 ppg.

1. How much clean drilling fluid volume would require this quantity of barite?

This clean drilling fluid would contain, 2% volume bentonite, an unknown percent barite, and water and weigh 15.0 ppg.

$$V_{lg} = 62.5 + 2\,V_s - 7.5\,\text{MW}$$

$$2 = 62.5 + 2\,V_s - 7.5\,(15.0\,\text{ppg})$$

V = 26% volume solids, V = 24% volume barite.

Since 604 sacks of barite were added and Barite weighs 1470 lb/bbl, the volume of barite added would be:

(604 sacks) (100 lb/sack) (bbl/1470 lb) or 41.1 bbl

The barite is 24% volume of the fluid added. Stating this in an equation:

$$(0.24)(\text{volume of fluid added}) = 41.1 \text{ bbl}$$

$$\text{Volume fluid added} = 171.2 \text{ bbl}$$

2. What volume of drilled solids was discarded while drilling this interval?

    This means that 171.2 bbl of the system was discarded by the solids removal equipment. If the concentration of drilled solids in the discard was 35%, then the drilled solids volume discarded would be 0.35 (171.2 bbl) or 59.9 bbl.

3. What is the solids removal efficiency?

    The hole volume drilled is 100 bbl. If 59.9 bbl of drilled solids are discarded, the removal efficiency would be 59.9%.

4. The mud engineer reports that the drilling fluid in the tanks after drilling this interval of hole has a solids concentration of 29.1% volume. Find the volume percent of drilled solids in this fluid.

    With the equation:

$$V_{lg} = 62.5 + 2\,V_s - 7.5 \text{ MW}$$

Solve for V_{lg}:

$$V_{lg} = 62.5 + 2\,(29.1) - 7.5(15.0 \text{ ppg}) = 8.2\% \text{ volume}$$

Bentonite contributes 2% volume, Drilled solids must be 6.2% volume.

5. How much fluid needs to be discarded to make room for clean drilling fluid to reduce the drilled solids back to the target value of 5% volume?

    The total system volume is now 2100 bbl. Currently there are 6.2% drilled solids in that volume or (0.062) (2100 bbl) = 130.2 bbl. The system needs to have only (0.05) (2100 bbl) or 105 bbl. So 25.2 bbl of drilled solids must be discarded by discarding a fluid which contains 6.2% volume drilled solids:

$$25.2 \text{ bbl} = (0.062) \text{ (volume to be discarded)}$$

$$\text{Volume to be discarded} = 406 \text{ bbl}$$

6. The mud engineer reported that 374 sacks of barite was used after the pit levels were decreased by 406 bbl to allow for the clean dilution drilling fluid to be added. Was this correct? How much barite must be used in the clean drilling fluid after discarding the 406 bbl?

    The clean drilling fluid will contain 2% volume bentonite and x volume of barite. Again using the equation:

$$V_{lg} = 62.5 + 2V_s - 7.5 \text{ MW}$$

$$2 = 62.5 + 2 \text{ VS} - 7.5\,(15 \text{ ppg})$$

$V_s = 26\%$ volume with 2% bentonite, the barite is 24% volume.
Barite needed = 0.24 (volume of clean drilling fluid added).
Barite needed = 0.24 (406 bbl) = 25.4 bbl.
Barite weighs 1470 lb/bbl.
Barite needed = 24.5 bbl (1470 lb/bbl) = 374 sks.

Problem 7.3 solution

1. The volume of discard was 1840 bbl and contained 35% volume drilled solids.
   The volume of solids is (0.35) (1840 bbl) or 644 bbl.
   The volume of hole drilled:

$$\text{Volume hole drilled} = \left[\frac{(14 \text{ in})^2}{1029}\right] 5000 \text{ ft} = 952 \text{ bbl}$$

$$\text{Volume of solids drilled} = (952 \text{ bbl})(0.90) = 856.8 \text{ bbl}$$

Removal efficiency is the ratio of solids remaining divided by solids presented to the equipment.

$$\text{Removal efficiency} = \frac{644 \text{ bbl}}{857 \text{ bbl}} = 0.75 \text{ or } 75\%$$

2. Solids remaining in fluid is the difference between the volume of solids drilled and the volume of solids discarded:

$$\text{Solids remaining} = 857 \text{ bbl} - 644 \text{ bbl} = 213 \text{ bbl}$$

To find the fraction of low gravity solids in the 12.0 ppg drilling fluid with 18% volume total solids:

$$V_{LG} = 62.5 + 2.0 \, V_s - 7.5 \text{ MW}$$

$$V_{LG} = 62.5 + 2.0 \, (18) - 7.5 \, (12.0 \text{ ppg})$$

$$V_{LG} = 8.5\% \text{ volume}$$

The drilling fluid has 2% volume bentonite as indicated by the MBT of 18 lb/bbl. Bentonite weighs 9.1 lb/bbl.

Drilled solids would be 8.5% volume − 2% volume or 6.5% volume.
The volume percent of barite would be 18% volume − 8.5% volume or 9.5% volume.
Barite has a density of 1470 lb/bbl.
The liquid required to dilute the remaining drilled solids (212 bbl) to 6.5% volume in the system can be calculated from the equation:

$$0.065(\text{liquid to dilute}) = 212 \text{ bbl}$$

$$\text{liquid to dilute} = 3262 \text{ bbl}$$

The diluted fluid would contain:
9.5% (3262 bbl) barite or 309.9 bbl
2% (3262 bbl) bentonite or 65.2 bbl
82% (3262 bbl) water or 2675 bbl
Clean fluid 3050 bbl

The remaining 212 bbl of drilled solid would bring the total up to 3262 bbl or rebuilt drilling fluid.

3. Liquid hauled from location (discards) = 1840 bbl
   The clean fluid needed to dilute the remaining drilled solids is 3262 bbl.
   The excess fluid which must be discarded is 3262 bbl−1840 bbl or 1422 bbl.
4. With 85% removal efficiency and 857 bbl of solids reaching the surface, the solids discarded would be 0.85 (857 bbl) or 686 bbl. The solids remaining would be 0.15 (857 bbl) or 129 bbl.
   The liquid required to dilute 128 bbl to 6.5% volume solids would be:

   $$0.065(\text{liquid to dilute}) = 129 \text{ bbl}$$

   $$(\text{liquid to dilute}) = 1978 \text{ bbl}$$

This is almost the same as the quantity of fluid discarded from the system. The volume discarded is 1840 bbl which is less than the volume needed by only 138 bbl. This system is almost at the optimum removal efficiency which will eliminate the excess dilution fluid needed to keep the drilled solids at the target value.

Problem 7.4 solution

With the 75% removal efficiency and 856 bbl of drilled solids reporting to the surface, 646 bbl will be discarded and 212 bbl will be returned to the pits, (as before). The returned drilled solids must be reduced to 4% volume in the pits, so the dilution volume will be (212 bbl/0.04) or 5300 bbl. Assuming the same concentration of drilled solids in the equipment discard of 35% volume, as before, the equipment discard (and the decrease in pit volume) will be 646 bbl/0.35, or 1840 bbl. The excess drilling fluid generated is 5300 bbl− 1840 bbl or 3460 bbl.

Now the cost: The cost of new drilling fluid added to the system is 5300 bbl × $90/bbl or $477,000.00. The cost of disposal for this interval is (1840 bbl + 3460 bbl) × $40/bbl, or $212,000. Total cost for this 6000 ft interval is $115,000 or $167/ft of hole.

Chapter 10 Centrifugal pumps solutions
Problem 10.1 solution

Forty 4″ cones are connected to a 8 × G × 14 Magnum Pump. Impellers is rotating at 1750 RPM. The feed manifold is Located 25 ft above the liquid level in the mud tanks. Assume 10 ft of head is lost in the plumbing.

What impeller size should be used?

Forty 4" cones will process 2000 gpm with 75 ft of head applied. The total head loss will be 75 ft + 25 ft + 10 ft or 110 ft. Plotting this point on the pump curve, shows that a 12" impeller would be required (the red dot on the pump curve below) (Fig. S10.1).

What Horsepower pump would be required to pump a 16.6 ppg drilling fluid?

An 80 horsepower motor would be required to pump water. The horsepower required to pump a 16.6 ppg drilling fluid would be (16.6 ppg/8.34 ppg) or 2 times that horsepower. A 160 horsepower motor would be required. This may indicate that the desilter bank should be split and two pumps used with smaller horsepower instead of the 40 cone bank.

What NPSH would be required?

NPSH from the graph is 18 ft.

Problem 10.2 solution

Solution A

What impeller should be used?

Twenty cones will process 1000 gpm at 75 ft head. The head loss to lift the fluid from the mud tanks to the desilter cones is 10 ft. This head loss must be added to the head loss in the piping. Use Tables 10.2 and 10.3.

Calculate head loss through the 8" piping:

(1.56 ft/100 ft) for 8" pipe with 1000 gpm Friction loss in pipe fittings in terms of equivalent feet of straight pipe

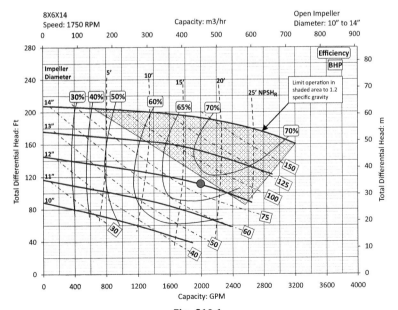

Fig. S10.1

29.9 ft for butterfly valve; 2 times 20 or 40 ft for elbows.
Equivalent length of pipe: 10 ft pipe + 29.9 ft + 40 ft = 79.9 ft.
Head loss for 8″ suction line = 79.9 ft pipe (1.56 ft/100 ft pipe) = 1.24 ft.
Calculate head loss through the 6″ piping:
(6.17 ft/100 ft) for 6″ pipe with 1000 gpm Friction loss in pipe fittings in terms of equivalent feet of straight pipe
22.7 ft for butterfly valve; 4 times 15.2 or 60.8 for elbows.
Equivalent length of pipe: 30 ft pipe + 22.7 ft + 60.8 ft = 113.5 ft.
Head loss for 6″ discharge line = 113.5 ft pipe (6.17 ft/100 ft pipe) = 7.00 ft.
Piping head loss for 1000 gpm = 1.24 ft + 7.00 ft = 8.24 ft.
Total head loss for pump = 75 ft for cones + 10 ft lift + 8.24 ft for plumbing = 93.24 ft.
A 11″ impeller would provide the required 1000 gpm at a 95 ft head.

Solution B

With this impeller, can two 1″ diameter mud guns be used to stir the suction tank if no agitator is available? If not, what impeller is needed? Assume 15 ft of 3″ diameter pipe is used to supply fluid to the mud guns.

Head loss for the 3″ mud gun:

Assume 200 gpm would be sufficient to agitate the suction compartment of the desilters. The 3″ line would have at least four elbows and a butterfly valve.

Calculate head loss through the 3″ piping:
(8.9 ft/100 ft) for 3″ pipe with 200 gpm Friction loss in pipe fittings in terms of equivalent feet of straight pipe
11.5 ft for butterfly valve; 4 times 7.67 ft or 30.68 ft for elbows.
Equivalent length of pipe: 15 ft pipe + 11.5 ft + 30.68 ft = 57.18 ft.
Head loss for 3″ suction line = 57.18 ft pipe (8.9 ft/100 ft pipe) = 5.09 ft.
If the lift to the mud gun line is minimal (line should be just above the drilling fluid in the tanks), the head available at the mud guns would be 95 ft−5.09 ft or 89.9 ft.
Flow rate through mud guns would be:

$$Q = 19.6 \, (\text{diameter})^2 \, (\sqrt{\text{head}}$$

$$Q = 19.6 \, (1")^2 \, (\sqrt{89.9} = 185.8 \text{ gpm}$$

One mud gun nozzle of 1″ in diameter would blend the tank. From the pump curve, this flow rate is easily achieved with the impeller selected for the desilter bank (Fig. S10.2).

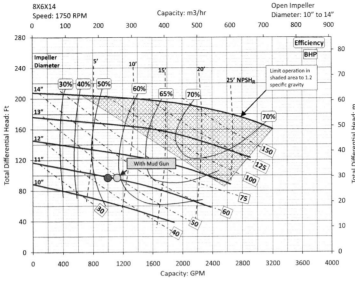

Fig. S10.2

Problem 10.3 solution

Twenty 4″ cones are connected to a 6 × 5 × 14 Magnum Pump. Impeller is rotating at 1750 RPM. The feed manifold is 30 ft above the liquid level in the mud tanks. Assume 15 ft of friction in the plumbing.

What impeller size should be used?

Twenty cones should process 1000 gpm with 75 ft of head applied.

Total head needed at pump: 75 ft + 30 ft + 15 ft or 120 ft of head (Fig. S10.3).

An 11 inch impeller will produce the flow rate at 120 ft of head.

What Horsepower pump would be required to pump a 16.6 ppg drilling fluid?

From the pump curve, about 43 BHP will be required for water and about twice that for the 16.6 ppg drilling fluid.

What NPSH would be required?

The NPSH is between 10′ and 15′ of head—Use about 12 ft for an estimate.

Problem 10.4 solution

Ten 4″ hydroclycones are connected to a 5 × 4 × 14 centrifugal pump. The impeller is rotating at 1750 RPM. The feed manifold is 20 ft above the liquid level in the mud tanks. Assume 10 feet of head loss in the pipe. What impeller should be used? What Horsepower pump would be needed to pump 16.6 ppg drilling fluid?

To supply 75 ft of head at the desilter manifold, the pump will need to produce 75 ft + 20 ft + 30 ft or 125 ft of head for the 500 gpm needed (Fig. S10.4).

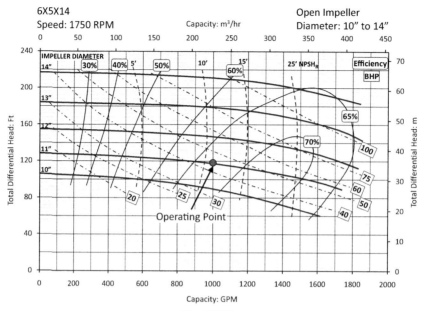

Fig. S10.3

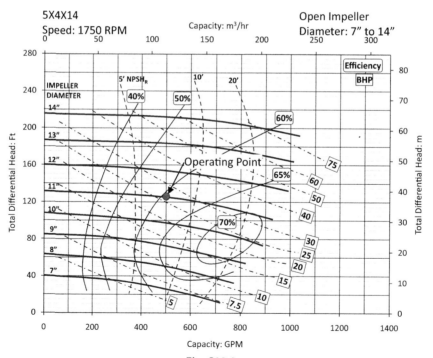

Fig. S10.4

An 11" impeller will provide the 500 gpm at 125 ft of head. With water, the horsepower required would be about 25 HP, for the 16.6 ppg drilling fluid, the horsepower required is 25 HP (16.6 ppg/8.34 ppg) or about 50 HP.

Chapter 11 Fraction of drilling fluid processed solutions

Problem 11.1 solution

Determine fraction of fluid processed by desander (Fig. S11.1A).

Analysis

The contract probably specified the desander capacity should be 100 gpm higher than the flow from the well.

$$\text{The Fraction of fluid processed} = \text{Fluid processed/Fluid entering tank}$$
$$= ((500 \text{ gpm})/(400 \text{ gpm} = 500 \text{ gpm})) = 0.56 \text{ or } 56\%$$

The dirty drilling fluid blends with the clean drilling fluid and only 56% has the drilled solids removed.

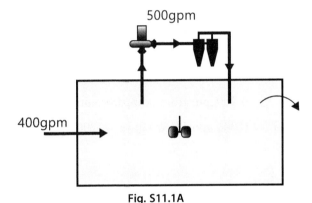

Fig. S11.1A

Solution

If the discharge line from the desander overflow is a hose, it can be moved to discharge into the next downstream compartment and the process efficiency will increase to 100%.

The plumbing needs to be changed to separate the clean drilling fluid in compartment B from the dirty drilling fluid in compartment A (Fig. S11.1B).

$$\text{Fraction processed} = \text{Fluid flow cleaned/fluid flow entering suction}$$
$$= 500 \text{ gpm}/400 \text{ gpm} + 100 \text{ gpm} = 1.0 \text{ or } 100\%$$

Fig. S11.1B

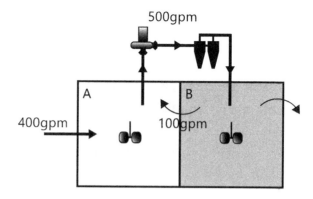

Problem 11.2 solution

Find fraction of fluid processed (Fig. S11.2).

First, find the flow rate between compartments A and (B)

600 gpm arrives from the well. 800 gpm leaves to flow through the desanders. The tanks will balance if 200 gpm flow back from compartment B to compartment A.

The fraction process is a ratio of the fluid processed divided by the flow entering the suction compartment.

$$\text{Fluid processed is 800 gpm}$$
$$\text{Fluid entering compartment A} = 600 \text{ gpm from well}$$
$$+200 \text{gpm from compartment B}$$
$$\text{Fraction processed} = ((800 \text{ gpm})/(600 \text{ gpm} = 200 \text{ gpm})) = 1.0 \text{ or } 100\%$$

Problem 11.3 solution

When Moore moved the desilter overflow upstream from the suction tank for the desilters, the volume entering the suction tank was 450 gpm + 500 gpm or 950 gpm. The fluid processed is 500 gpm (Fig. S11.3A).

Fig. S11.2

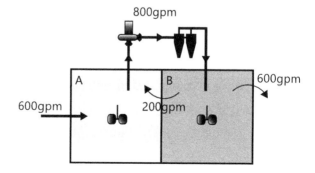

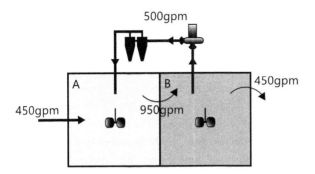

Fig. S11.3A

Analysis
The flow rate into tank A is 450 gpm from the well plus the 500 gpm from the desilters—or 950 gpm. This is also the flow rate into tank B, which is the suction tank for the desilters.

The removal efficiency is the ratio of the volume treated divided by the volume entering the suction tank. In this case the removal efficiency decreased to 500 gpm/950 gpm or 53%.

Solution
Move the desilter suction to tank A and the overflow to tank (B) Now 50 gpm will be flowing from tank B to tank (A) Now, entering tank A will be 450 gpm from the well and the 50 gpm backflow—500 gpm. The removal efficiency would be 500 gpm/(450 gpm + 50 gpm) or 100% (Fig. S11.3B).

Problem 11.4 solution
Analysis
All gauges indicated 39 psig for the 10.0 ppg drilling fluid. The rig had one 12″ desander and one bank of eight 4″ desilters. The rig was circulating 400 gpm downhole.

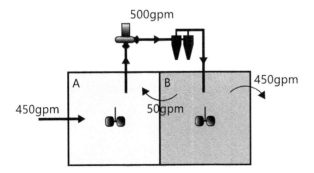

Fig. S11.3B

These hydrocyclones needed to have a 75 ft head at the inlet manifold.

P = 0.052 (mud weight) (head)

39 psi = 0.052 (10 ppg) (head)

Head = 75 ft

Consequently, the hydrocyclones should be working correctly.

He then examined the suction tanks for both the desander and the desilters. They were not well agitated, so he had two mud guns, made from 3″ × 1″ swedges, installed in both tanks. The drilling fluid was supplied to the mud guns from the suction compartment. Only clean drilling fluid agitated the tanks and sufficient fluid was pumped through the mud guns to keep the tanks well-agitated.

All tanks in the removal section should be well agitated to provide a homogeneous feed to the hydrocyclones. However, using "clean drilling fluid" does not mean that it will not dilute the "dirty drilling fluid" from the well. The additional flow volume needs to be processed.

The next step is to determine the flow rate into each of tanks. This is needed to determine the removal efficiency (Fig. S11.4A).

To determine the fraction of drilling fluid treated, the quantity of drilling fluid treated with the solids removal equipment is divided by the quantity of drilling fluid entering the suction compartment. The first task will be to determine the quantity of mud entering the suction tanks of the desander and the desilters.

The flow rate through the mud guns can be approximated from the equation:

$$Q = 19.6 \left(\sqrt{\text{Head, in ft.}}\right)(\text{Nozzle diameter, inches})^2$$

with 81 ft of head, each mud gun should pass 176 gpm into the comportment, so the total mud gun flow into each suction compartment should be 352 gpm. One 12′ desander will treat 500 gpm of fluid when 75 ft of head is applied. Each 4′ desilter cone will treat 50 gpm of fluid when 75 ft of head is applied.

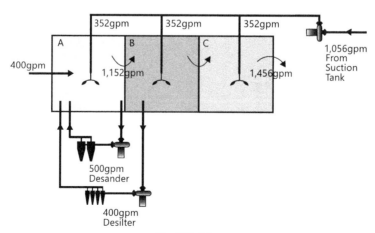

Fig. S11.4A

The flow rate into tank A is 400 gpm from the well + 352 gpm from the mud guns +500 gpm from the desanders + 400 gpm from the desilters or 1652 gpm. However, the net flow into tank A is 1152 gpm.

The flow rate into tank B is 1152 gpm from tank A plus 352 gpm from the mud guns or 1504 gpm.

The flow rate into tank C is the excess flow into tank B (1504 gpm minus the 400 gpm desilter flow or 1104 gpm) plus the 352 gpm mud gun flow rate which is 1456 gpm.

Calculate fraction processed by desander: Volume entering desander suction tank A:

352 gpm mud guns + 400 gpm from shaker + 400 gpm desilter + 500 gpm desander

$$= 1652 \text{ gpm}$$

$$\text{Fluid processed} = 500 \text{ gpm}$$

$$\text{Fraction processed} = 500 \text{ gpm}/1652 \text{ gpm or } 30\%$$

Calculate fraction processed by desilter: Volume entering desilter suction tank B

$$352 \text{ gpm mud guns} + 1652 \text{ gpm from tank A} = 2004 \text{ gpm}$$

$$\text{Fluid processed} = 500 \text{ gpm}$$

$$\text{Fraction processed} = 500 \text{ gpm}/2004 \text{ gpm or } 25\%$$

Obviously 100% of the fluid from the well bore is not being desanded and desilted.

Solution

The system needs to be replumbed so the desander processes its fluid into the next downstream tank and the desilter takes suction from that tank and processes into the next tank down stream (Fig. S11.4B).

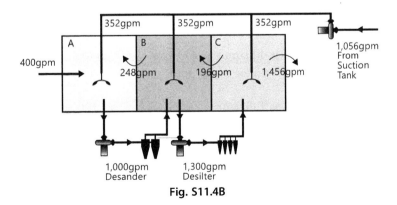

Fig. S11.4B

Assuming that mud guns will still be used for the required agitaton in the tanks, the number of hydrocyclones must be increased to handle all of the flow entering the suction compartments of the desander and desilter. The flow through the removal equipment must exit or discharge down stream. The desander must process more than the 400 gpm + 352 gpm (or 752 gpm) from tank A into tank (B) A bank of desanders processing 1000 gpm will provide a back flow of 248 gpm. This will grantee that all the fluid entering tank B has been desanded. The suction compartment for the desilters (tank B) will have a net flow entering the compartment of 752 gpm plus the mud gun flow of 352 gpm or 1104 gpm. If additional desilters are added to process 1300 gpm from tank B into tank C, a back flow of 196 gpm will guarantee that all the fluid entering tank C has been desilted.

Problem 11.5 solution

Note: To provide a positive feed head, centrifugal pumps are usually positioned at ground level NOT on top of the mud tanks. The drawings may give the appearance that the pumps are mounted above the tanks (Fig. S11.5A).

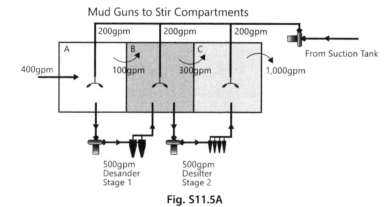

Fig. S11.5A

Analysis

Desander:
  Flow entering stage 1 pit is 400 gpm + 200 gpm = 600 gpm.
  Fluid processed through desander is 500 gpm.

$$\text{Desander Process Efficiency} = \frac{500 \text{ gpm}}{600 \text{ gpm}} \times 100 = 83\%$$

Desilter:
Flow entering stage 2 pit is 100 gpm + 500 gpm + 200 gpm = 800 gpm.
Fluid processed through desilter is 500 gpm.

$$\text{Desilter Process Efficiency} = \frac{500 \text{ gpm}}{800 \text{ gpm}} \times 100 = 60\%$$

Solution

The "mud gun" fluid can be processed if additional equipment is added. If the desander process rate is increased to 700 gpm, all of the fluid entering compartment #1 would be processed; this would be the 400 gpm rig flow rate, the 200 gpm mud gun flow rate and the back flow of 100 gpm. For the desilter processing, the fluid flow rate entering the compartment would be the 700 gpm from the desander, the 200 gpm from the mud guns, and a back flow from the downstream compartment of 100 gpm. From this value of 1000 gpm, 100 gpm would be subtracted because of the underflow leaving the second compartment and going to the desander suction compartment. The process efficiency of 100% then requires a desilter capacity of 900 gpm. For four inch balanced hydrocyclones, this would mean adding 8 additional cones to the 10 cones that would be required to provide the 500 gpm capacity (Fig. S11.5B).

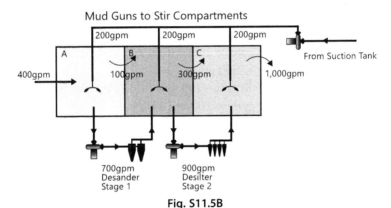

Fig. S11.5B

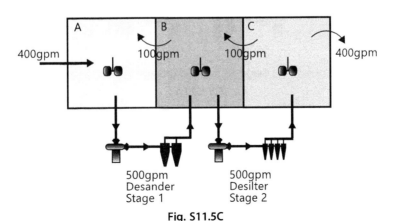

Fig. S11.5C

An alternate solution would be to add mechanical agitators to blend the tanks. Agitators used instead of mud guns (Fig. S11.5C).

Stage 1: Processed by desander = 500 gpm
Entering compartment A = 400 gpm + 100 gpm = 500 gpm
% Processed = 500 gpm/500 gpm × 100 = 100%
Stage 2: Processed by desilter = 500 gpm
Entering compartment B = 500 gpm−100 gpm + 100 gpm = 500 gpm
% Processed = 500 gpm/500 gpm × 100 = 100%

Solution—plumbing is correct. Note that the desander and desilter were sized to process 125% of the circulating rate (500 gpm/400 gpm) and thus satisfied the requirement, however the processing efficiency is less than 100%.

Problem 11.6 solution
Analysis
Agitators used instead of mud guns (Fig. S11.6).

Stage 1: Processed by desander = 500 gpm
Entering compartment A = 400 gpm + 100 gpm = 500 gpm.
% Processed = 500 gpm/500 gpm × 100 = 100%
Stage 2: Processed by desilter = 500 gpm.
Entering compartment B = 500 gpm−100 gpm + 100 gpm = 500 gpm.
% Processed = 500 gpm/500 gpm × 100 = 100%

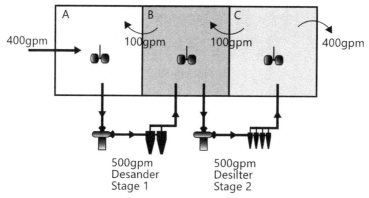

Fig. S11.6

Solution

Plumbing is correct.

Problem 11.7 solution
Analysis
First step to determine the fraction of drilling fluid processed is to determine the flow rate into and out of each tank (Fig. S11.7A).

Flow rate into tank 1 is 450 gpm from the well. The flow rate through the degasser is 500 gpm which causes a back flow from tank 2 to tank 1 of 50 gpm.

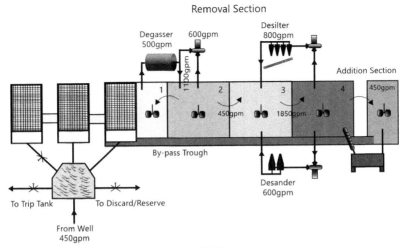

Fig. S11.7A

The flow rate into tank 2 is 500 gpm from the degasser. The power fluid for removing the degassed drilling fluid from the degasser is 600 gpm into and out of tank 2. (This is considered an "internal flow" and does not contribute to unprocessed drilling fluid entering or leaving the tank. For example, if an agitator is moving 1000 gpm around in a compartment, this flow is an "internal flow".)

The flow rate into tank 3 from tank 2 is 450 gpm (500 gpm−50 gpm) plus the 600 gpm from the desander plus the 800 gpm from the desilters or entering tank 4: 1850 gpm.

Solution

The fraction of fluid processed is the ratio of the fluid processing flow rate divided by the net flow rate into the suction compartment:

Fraction of fluid desanded = 600 gpm/1850 gpm = 32%

Fraction of fluid desilted = 800 gpm/1850 gpm = 43%

Neither the desander nor the desilter is processing all of the drilling fluid coming from the well. The system needs to be replumbed.

Note that the desander and desilter were sized to process greater than 125% of the circulating rate (600 gpm/450 gpm for the desander and 800 gpm/450 gpm for the desliter) and thus satisfied the requirement, however the processing efficiency is less than 100%.

The degasser, the desander, and the desilter should take suction from one tank and process the fluid into a tank down stream (Fig. S11.7B). If the equipment is processing more fluid than is entering the suction tank, the excess fluid will "back flow" into the suction compartment. The fluid from tank 2 should over flow into tank 1. The fluid

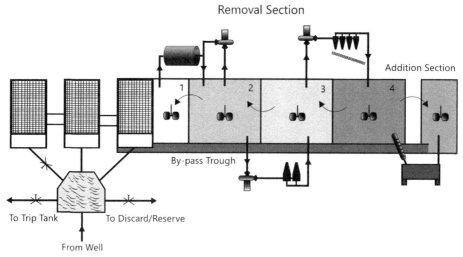

Fig. S11.7B

from tank 3 should have an underflow into tank 2 and the fluid from tank 4 should underflow into tank 3. The fluid from the removal section (Tanks 1,2,3, and 4) should overflow into the addition section down stream The advantage of doing this is to provide a constant submergence depth for the centrifugal pumps, as well as keep a constant volume in the removal section. This will allow a more sensitive detection of a kick or lost circulation because all the pit level changes will occur in the addition section.

Problem 11.8 solution

Note: To provide a positive feed head, centrifugal pumps are usually positioned at ground level NOT on top of the mud tanks. The drawings may give the appearance that the pumps are mounted above the tanks (Fig. S11.8A).

Analysis
Mud gun flow rate:

$$Q = 19.6 \left(\sqrt{\text{head}}\right)(\text{diameter}^2) = 170 \text{ gpm}$$

with 75 ft of head on the nozzles, the mud guns in each compartment will supply approximately 340 gpm of clean drilling fluid. (This is an approximation because the head loss in the feed lines is ignored. However, the actual flow rate will not be much different and these mud guns should not be there anyway.) (Fig. S11.8B).

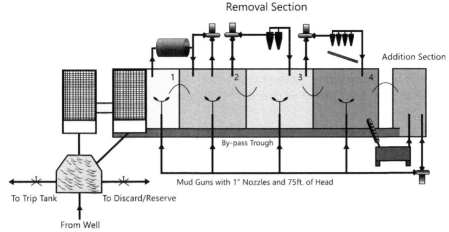

Fig. S11.8A

First step: determine flow rates into and out of tanks

Flow into tank 1: From well 500 gpm + mud guns 340 gpm = 840 gpm.
Flow from tank 1: 700 gpm degasser.
Overflow from tank 1 to tank 2: 840 gpm − 700 gpm = 140 gpm.
Flow into tank 2: 140 gpm overflow from tank 1 + 700 gpm thru degasser + 340 gpm mud gun = 1180 gpm.

(The 800 gpm "power" fluid for the degasser flows into and out of tank 2. Consider this an internal flow or zero contribution.)

Flow from tank 2: 600 gpm thru desander.

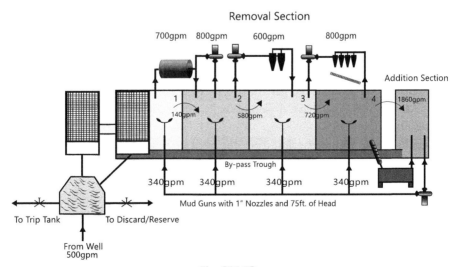

Fig. S11.8B

Net flow between tank 2 and tank 3: 1180 gpm in − 600 gpm out = 580 gpm (Underflow from tank 2 to tank 3).

Flow into tank 3: 500 gpm underflow from tank 2 + 340 gpm mud gun + 600 gpm desander = 1520 gpm desander = 1520 gpm.

Disilter processed flow from tank 3 = 800 gpm.
Underflow from tank 2 to tank 3 = 1520 gpm.

Second step: calculate the fraction of fluid processed

Degasser:
Fraction of fluid processed,

$$= (\text{Fluid processed})/(\text{Fluid into suction tank})$$

$$= \frac{700 \text{ gpm}}{500 \text{ gpm} + 340 \text{ gpm}} = 0.51 \text{ or } 51\%$$

Desander:
Fraction of fluid processed,

$$= \frac{600 \text{ gpm}}{140 \text{ gpm} + 700 \text{ gpm} + 340 \text{ gpm}} = 0.51 \text{ or } 51\%$$

Desilter:
Fraction of fluid processed,

$$= \frac{800 \text{ gpm}}{580 \text{ gpm} + 600 \text{ gpm} + 340 \text{ gpm}} = 0.53 \text{ or } 53\%$$

Conclusion

1. The degasser fails to remove gas from 49% of the flow from the well. Centrifugal pumps downstream may have difficulty maintaining the proper head for the desander and desilter.
2. Almost half of the fluid coming from the well does not pass through the hydrocyclones. The drilled solids concentration will start increasing and can only be decreased with dilution.
3. Even though the capacity of the degasser, desander and desilter exceeds the flow rate coming from the well (which may satisfy the conditions stipulated in a contract), the removal system needs to be changed.
4. Obviously many changes should be made and there are several solutions available to achieve 100% processing.

Problem 11.8 solutions A–D

Four different solutions are described below. The first one relates to the system which should have been installed on the drilling rig when it was built. Frequently, however, the systems in the field are plumbed as indicated in the problem presentation. If the rig is being used for only one or two wells, less invasive (and cheaper) solutions can work. Three alternate "temporary" solutions can be installed on the rig to allow the processing of all of the fluid.

The first solution requires a complete an extensive change to make the removal section perform correctly. This is the system which should have been installed on the drilling rig when it was built.

Problem 11.8 solution A

Two shale shakers are shown in the diagram. If API 140 screens, or finer, can be used, the desander will not be needed. A desander was invented to decrease the solids load on desilters when screens around API 80 and coarser were used on drilling rigs. Removal of the desander will decrease the number of compartments and decrease the volume in the removal section (Fig. S16.8C).

Mud guns should not be used in the removal section unless the additional volumes are accounted for with removal equipment or each centrifugal pump stirs its own suction. All active system compartments must be agitated. Individual agitators should be mounted in each compartment.

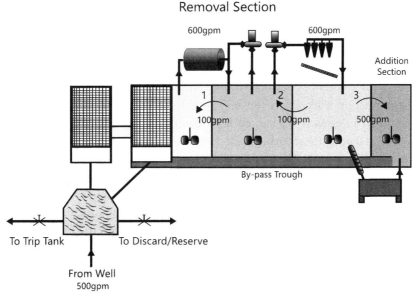

Fig. S11.8C

First step: calculate the removal efficiency

Flow into tank 1: From well 500 gpm.
Flow from tank 1: 600 gpm degasser.
Overflow from tank 2 to tank 1: 600 gpm − 500 gpm = 100 gpm.
Flow into tank 2: 600 gpm thru degasser − 100 backflow to tank 1 + 100 gpm from tank 3 = 600 gpm.

Second step: calculate the fraction of fluid processed

Degasser: Fraction of fluid processed,

$$= (\text{Fluid processed})/(\text{Fluid into suction tank})$$

$$= \frac{600 \text{ gpm}}{500 \text{ gpm} + 100 \text{ gpm}} = 1.0 \text{ or } 100\%$$

Desilter: Fraction of fluid processed,

$$= \frac{600 \text{ gpm}}{500 \text{ gpm} + 100 \text{ gpm}} = 1.0 \text{ or } 100\%$$

Not only is the desander not necessary but the number of desilter cones is reduced and the system still process 100% of the fluid reaching the surface.

Problem 11.8 solution B
Another solution to this problem would be to supply additional removal equipment to process the additional flow caused by the mud guns. The assumption here is that the main shale shaker can process the drilling fluid through an API 140 or finer screen. This eliminates the need for a desander—as in the preceding solution.

The new tank system needs only three compartments. The fluid flow rate entering the degasser suction tank (1) is the 500 gpm from the well and 340 gpm from the mud guns. If two degassers are use, each processing 500 gpm, the flow into the second tank will be 160 gpm more than is entering the degasser suction compartment. This will cause a 160 gpm back flow from tank 2 to tank 1.

The flow rate entering the desilter suction tank (3) will be 1000 gpm from the degassers plus 340 gpm from the mud guns minus the 160 gpm flowing from tank 2 into tank 1 or 1180 gpm. If enough desilters are added to process 1400 gpm, the desilters will process 220 gpm excess quantity of drilling fluid (1400 gpm − 1180 gpm). The excess 220 gpm will back flow from tank 3 to tank 2.

The flow into tank 3 is 1400 gpm from the desilters minus 220 gpm backflow plus 340 gpm from the mud guns or 1520 gpm (Fig. S11.8D).

With this arrangement, calculate the removal efficiency.

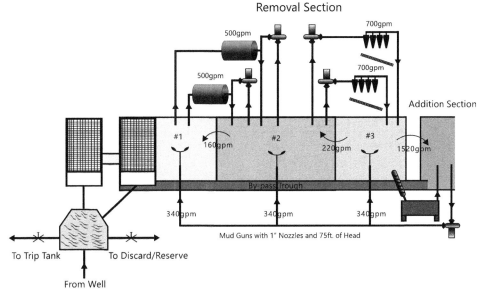

Fig. S11.8D

Flow into tank 1 (the degasser suction) From well 500 gpm plus 340 gpm from mud guns plus 160 gpm from tank 2 or 1000 gpm.
Degasser Fraction of fluid processed = (Fluid processed)/(Fluid into suction tank)

1000 gpm/1000 gpm = 1.0 or 100%

Flow into tank 2 (the desilter suction).

From mud guns 340 gpm, plus 1000 gpm from degasser plus 220 gpm from tank 3 minus the back flow of 160 gpm to tank 1 or 1400 gpm.

The desilters are processing 1400 gpm from tank 2 to tank 3.

The fraction of fluid processed is 1400 gpm/1400 gpm = 1.0 or 100%.

Problem 11.8 solution C

The removal section should not have mud guns adding excess fluid to be processed. However, conditions on some removal systems may not allow the installation of agitators because shale shakers and other removal equipment are mounted on the tanks with no room beneath them.

One possibility is to have each centrifugal pump stir its own suction compartment with a mud gun. If 6 × 8 × 14 1750 RPM centrifugal pumps are being used, their capacity will be sufficient to continue to supply fluid to the equipment as well as the mud guns.

In this case the degasser can be processing 600 gpm of fluid (100 gpm more than is entering tank 1). The head on the degasser discharge and the desilter manifold should

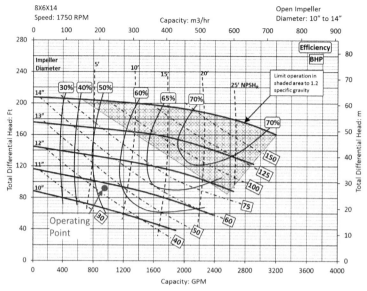

Fig. S11.8E

be around 75 ft. The pump should be producing around 90–100 ft of head to provide the 75 ft of head at the manifold. The flow from the centrifugal pump needs to be 600gpm plus the mud gun flow rate of 340 gpm or 940 gpm (Fig. S11.8E).

The pump curves for this pump indicate a relatively constant head is available for flow rates up to about 1000 gpm. A $10^3/_4''$ impeller in the pump would supply sufficient flow rate and head for the degasser. This solution is obviously a temporary expedient to correct processing efficiency while drilling. Welders can install lines from the centrifugal pumps to the mud guns while the rig is running casing or tripping pipe.

For the solution shown (Fig. S11.8F), calculate fraction of fluid processed:

Degasser Fraction of fluid processed,

$$= \text{(Fluid processed)}/\text{(Fluid into suction tank)}$$

$$= \frac{700 \text{ gpm}}{500 \text{ gpm} + 200 \text{ gpm} \quad \text{overflow from tank 2}}$$

$$= \frac{700 \text{ gpm}}{700 \text{ gpm}}$$

$$= 1.0 \text{ or } 100\%$$

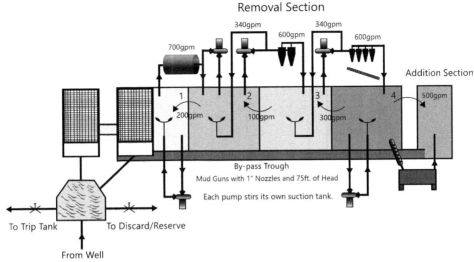

Fig. S11.8F

Desander Fraction of fluid processed,

$$= \frac{600 \text{ gpm}}{700 \text{ gpm from degasser} - 200 \text{ gpm to tank } 1 + 100 \text{ gpm underflow from tank } 3}$$

$$= \frac{600 \text{ gpm}}{600 \text{ gpm}}$$

$$= 1.0 \text{ or } 100\%$$

Desilter Fraction of fluid processed,

$$= \frac{800 \text{ gpm}}{600 \text{ gpm from desander} - 100 \text{ gpm underflow to tank } 2 + 300 \text{ gpm underflow from tank } 3}$$

$$= \frac{800 \text{ gpm}}{800 \text{ gpm}}$$

$$= 1.0 \text{ or } 100\%$$

This solution assumes that the main shakers are not able to process the drilling fluid through API140 or finer screens.

Problem 11.8 solution D

In the previous solutions, a new pump would be required to agitate the degasser suction tank #1. An alternative solution could be to make tank #1 small enough so it does not require additional agitation or is "self-agitating" (Fig. S11.8G).

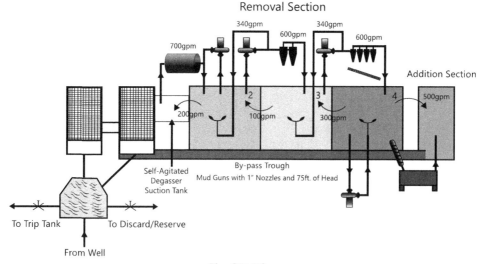

Fig. S11.8G

For example, if tank #1 is 8 ft deep, 4 ft wide and 4 ft long (Fig. S11.8H), it could be a self agitated suction tank for the degasser. If the drilling fluid is 7 ft deep, the volume of liquid will be 112 ft^3 or 20 bbl or 840 gal. In this problem, 500 gpm will enter tank #1 from the shakers. The degasser is processing 700 gpm which will cause an overflow from tank #2 to tank #1 of 200 gpm. The entry of 700 gpm into a volume of 840 gallons

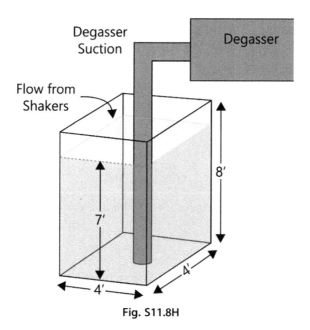

Fig. S11.8H

means that the resident time of the fluid will be 840 gal/700 gal/mm or 1.2 min. This technique has been used in a fluid test where the entire removal section was skid mounted and consisted of 30″ diameter casing standing vertically. Centrifugal pumps were mounted under each casing; consequently anything that settled was pumped to the next casing compartment. No mud guns or agitators were used. The drilled solids were maintained at a very low level with this system.

Problem 11.9 solution

Fluid entering degasser pit:
  From the shaker 400 gpm; 200 gpm overflow from desilter suction pit (Fig. S11.9A)
  (The degasser power pump is circulation unprocessed fluid—1000 gpm in and out)
  Fluid entering desilter pit (Fig. S11.9B):

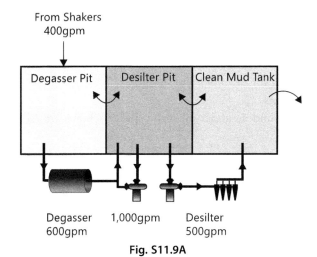

Fig. S11.9A

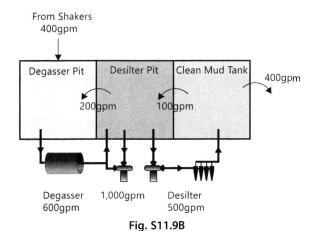

Fig. S11.9B

From the degasser = 600 gpm; From underflow from clean mud tank 100 gpm.
Fluid leaving desilter suction pit = 200 gpm (back flow)

$$\text{Fraction processed} = \frac{600 \text{ gpm} - 200 \text{ gpm} + 100 \text{ gpm}}{500 \text{ gpm}} = 1.0$$

All fluid entering the clean mud tank has been processed through the desilters.
If the shaker screens are API 100 or coarser, a desander is needed.
If the shaker screens are API 140 or finer, a desander is not deeded.
The tanks are not agitated, consequently the process efficiency might not be 100% and the cones may plug from slugs of large volumes of solids. Even though the process efficiency is 100%, the tanks must be well agitated to keep slugs of solids from plugging the desilter cones.

Possible replumbing
- Added sand trap (Fig. S11.9C).
- Proper sequencing of degasser, desander, and desilter.
- Proper fluid routing for desander and desilter.
- Stirred compartments.

Eliminate sand trap and desander pit if the shaker screens are API 140 or finer.

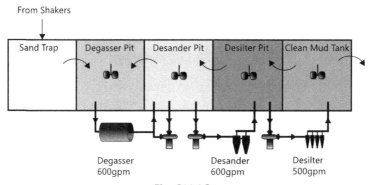

Fig. S11.9C

Problem 11.10 solution
Analysis
Flow from the well into shale shaker is 1000 gpm. Flow into the desander is 1200 gpm.
  Flow rate entering suction tank (C) of desander = 1000 gpm + 1200 gpm = 2200 gpm.
  Flow rate processed by desander = 1200 gpm.
  Process Efficiency = 1200 gpm/2200 gpm = 0.55 or 55% (Fig. S11.10A).
  Flow rate entering suction tank (B) of desander = 1000 gpm + 200 gpm = 1200 gpm.

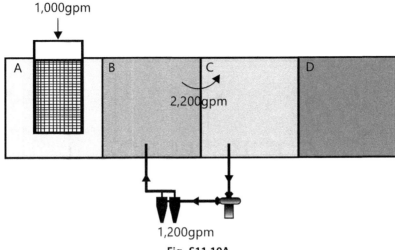

Fig. S11.10A

Flow rate processed by desander = 1200 gpm.
Process Efficiency = 1200 gpm/1200 gpm = 1.0 or 100% (Fig. S11.10B).

Problem 11.11 solution
Example rig #1
What problems exist with this mud tank arrangement? (Fig. S11.11A).
- Desander and desilter rigged up backwards, mud bypasses equipment
- Degasser downstream of desander and desilter, pumping gas cut mud
- Unstirred compartments

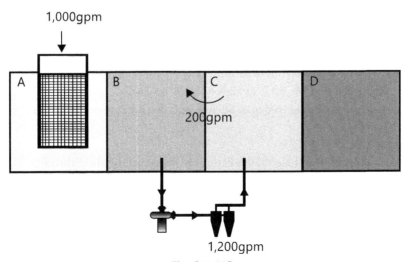

Fig. S11.10B

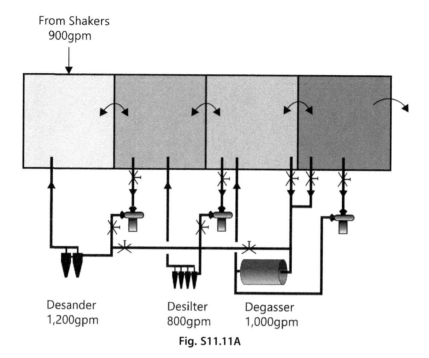

Fig. S11.11A

Calculate the process efficiency (Fig. S11.11B):

$$\text{Desander}: \frac{1200 \text{ gpm}}{2100 \text{ gpm} + 800 \text{ gpm}} = 0.41 \text{ or } 41\%$$

$$\text{Desilter}: \frac{800 \text{ gpm}}{2900 \text{ gpm} + 1000 \text{ gpm}} = 0.21 \text{ or } 21\%$$

The centrifugal pumps supplying the hydrocyclones will probably cease pumping fluid at the proper head because of the gas cut drilling fluid. The drilled solids should start increasing rapidly. The plastic viscosity will increase. The filter cakes will become thick and "sticky". Need to have fisherman on stand-by for stuck pipe.

What changes that could be made to improve solids removal?

Needs major replumbing to degas, desand, and desilt in proper order. Needs agitators in tank.

Up-grade (Fig. S11.11C):

Added sand trap if the shale shaker cannot process drilling fluid through an API 140, or finer screen. Proper sequencing of degasser, desander, and desilter. Proper fluid routing for desander and desilter. Stirred compartments.

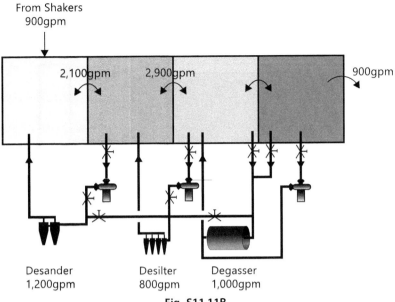

Fig. S11.11B

Problem 11.12 solution

Rig #2.

Analysis

Compartments are not agitated. Only the sand trap should have no agitation (Fig. S11.12A).

The pump supplying fluid to the degasser and desilters should not have a suction line in the sand trap.

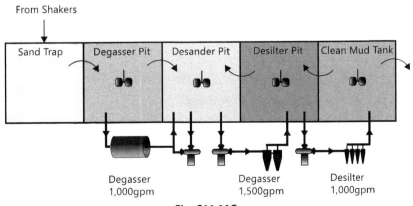

Fig. S11.11C

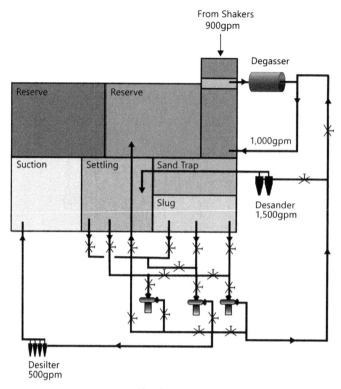

Fig. S11.12A

The degasser and desander are in parallel flow. Each should have its own pump.

The centrifugal pumps supplying fluid to the degasser, desanders, and desilters should be plumbed so that a pump can only supply fluid to one of them. Multiple options are not desirable and usually create a very low solids processing efficiency.

Solids removal should not be necessary for the slug tank. Only processed drilling fluid should be used in the slug tanks.

Solution
What problems exit with this mud tank arrangement?
- No back flow for degasser
- No simultaneous degassing and desanding
- Desilting newly treated mud
- Insufficient desilter capacity
- Complex plumbing
- Unstirred compartment

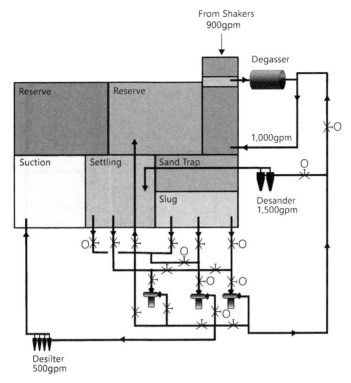

Fig. S11.12B

For the best processing with this system, which valves should be open? (Fig. S11.12B).

Find fraction of fluid processed
Degasser flow has no effect on desander process efficiency (same flow in and out of tank).

Desander
1500 gpm/(1500 gpm + 900 gpm + 1000 gpm − 1000 gpm) = 63%

Desilter
500 gpm/900 gpm = 56% (Fig. S11.12C).

Modifications needed to correct problems with tank arrangement:
- All compartments should be agitated.
- The multifold options of suction and discharges are eliminated: one pump—one function.
- Sequential treatment of fluid through degasser, then desander, then desilters.
- Slug tank is not involved with solids removal.
- All compartments will have underflow equalizers except for the degasser suction.

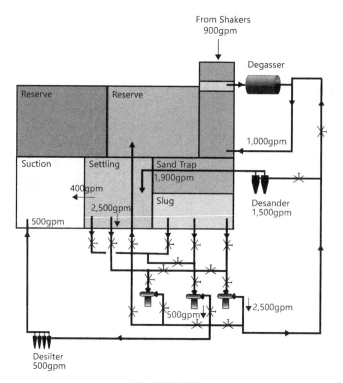

Fig. S11.12C

Add additional walls and agitators in each compartment. Flow into A is 900 gpm. Degasser removes 1000 gpm. Overflow B—A 100 gpm. 900 gpm entering B (desander suction).

Desander processing 1500 gpm—flow from C to B = 600 gpm.

Fraction processed = 1500 gpm/900 gpm + 600 gpm = 1.00 or 100%.

Desilters must process over 900 gpm not 500 gpm.

From C Fraction processed = 1000 gpm/(1500 gpm + 100 gpm − 600 gpm) = 1.0 or 100% (Fig. S11.12D).

Problem 11.13 solution
Example rig #3
Valves that should be open (Fig. S11.13A):

Find fraction of fluid processed by desander and desilter

The desander is processing 1,000 gpm with the suction from tank #1 and discharge into tank #2. Entering tank #1 is 900 gpm from the well and 100 gpm underflow from tank #2.

Fraction processed = 1000 gpm/900 gpm + 100 gpm = 1.0 or 100%.

Solutions to chapter problems 541

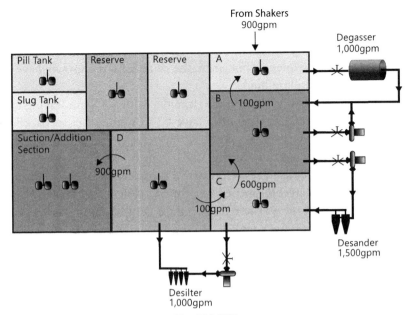

Fig. S11.12D

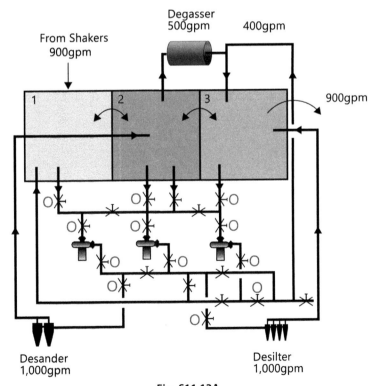

Fig. S11.13A

However, the centrifugal pump will probably cease pumping fluid because of the gas in the fluid coming from the well.

The desilter is processing 1,000 gpm with the suction in tank #2 and discharge into tank #3. This is a correct sequence, but again, the centrifugal pumps will have great difficulty pumping gas-cut drilling fluid.

Entering tank #2 is 1,000 gpm from the desander with 100 gpm leaving tank #2 to equalize tank #1. 500 gpm back flow from tank #3 from the degasser backflow. Flow entering Tank #2 is 900 gpm + 500 gpm.

Fraction of fluid Processed = 1000 gpm/1400 gpm = 71%.

Even if the processing efficiencies were correct the system needs to be greatly simplified. If the correct valves are not opened or properly closed, the efficiency will suffer. However, the hydrocyclones will not process fluid continuously. The degasser needs to be moved to take suction from tank 1.

What is wrong?
- Insufficient equipment capacity.
- Improper placement of degasser.
- Not enough process compartments.
- Unstirred compartments.
- Plumbing too complex.
- The probability of having the correct valves open or closed is very small.

Changes that need to be made to mud tank

The degasser needs to process more fluid than is entering tank 1. The hydrocyclones process more fluid than is entering their suction tanks. There is a backflow of at least 100 gpm from tank 3 to tank 2 and from tank 4 to tank 3 (Figs. S11.13B).

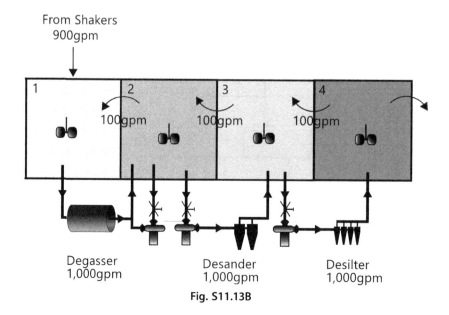

Fig. S11.13B

Problem 11.14 solution
Analysis
Flow through each mud gun (Figs. 11.14A and 11.14B):

$$Q = 19.6\left(64 \text{ ft}^{0.5}\right)\left(1''^2\right) = 159 \text{ gpm}$$

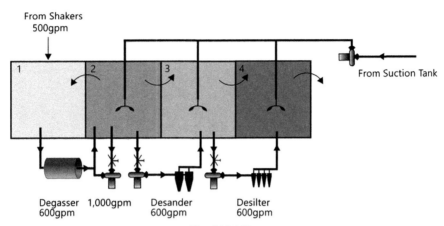

Fig. S11.14A

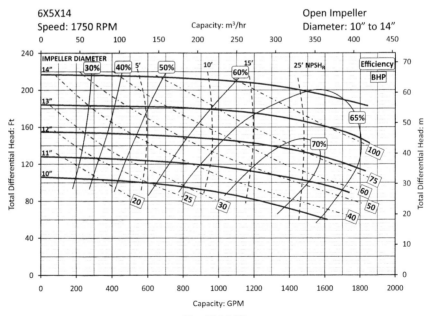

Fig. S11.14B

Backflow from #2 to #1

For Compartment #2 (desander suction):

318 gpm from mud guns entering tank.

600 gpm from degasser.

100 gpm leaving tank from #2 to #1.

1000 gpm power fluid from degasser into and out of compartment #2 without processing is an "internal flow" as would be the flow around agitator.

Net flow into tank #2 is 918 gpm.

Fraction processed by Desander:

Total entering #2 = 918 − 100 gpm = 818 gpm.

= 600 gpm/818 gpm = 0.73 or 73%

For Compartment #3 (desilter suction):

Fluid entering #3 from tank #2.

318 gpm that is not processed through the desander.

600 gpm from desander.

318 gpm from mud guns.

Total: 318 gpm + 600 gpm + 318 gpm = 1236 gpm.

Fraction processed by Desilter: = 600/1236 = 0.49 or 49% (Fig. S11.14C).

What changes need to be made?

Mud guns should only be used in the removal section unless each centrifugal pump stirs its own suction tank. Best solution would be to buy agitators for compartments. On many rigs, particularly jack-up rigs and platform rigs, the equipment is sitting on top of these tanks. Most centrifugal pumps will have sufficient additional capacity to handle the additional flow rate through two $1/2''$ nozzles. A two inch flow line could be attached to each centrifugal pump and supply fluid to the mud guns.

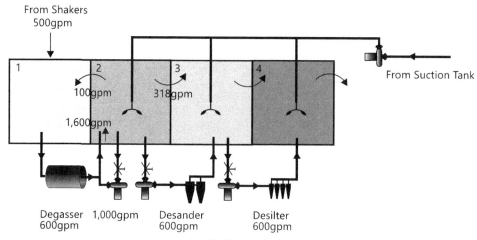

Fig. S11.14C

Problem 11.15 solution
Analysis
Overflow
    Tank B to A (Fig. S11.15A)
    Entering tank A = 500 gpm
    Leaving tank A = 600 gpm
    B to A = 100 gpm
    Flow entering tank B = 600 gpm − 100 gpm + 700 gpm − 700 gpm = 500 gpm
    Flow leaving tank B = 800 gpm + 600 gpm = 1400 gpm
    Underflow entering tank B = 500 gpm + 1400 gpm
    Fraction of fluid processed through desilter = 800 gpm/1900 gpm = 42%
    Fraction of fluid processed through desander = 600 gpm/1900 gpm = 32%

Process sequentially
Move desilters suction from B to C and desilters discharge from C to D (Fig. S11.15B).
    Fraction of fluid processed through desanders = 600 gpm/500 gpm + 100 gpm = 100%
    Fraction of fluid processed through desilters = 800 gpm/600 gpm − 100 gpm + 300 gpm = 100%.

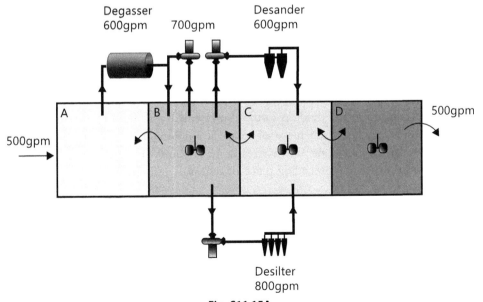

Fig. S11.15A

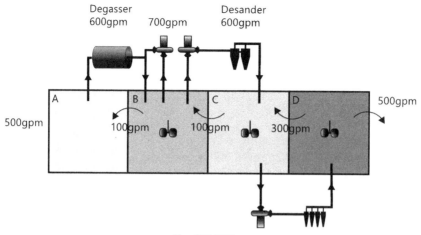

Fig. S11.15B

Problem 11.16 solution

Analysis
Desilter
Fraction of Mud Cleaned (Fig. S11.16A):
   Vol treated with Desilter/Vol into tank B = 700 gpm/(2200 gpm + 700 gpm) = 0.24 or 24%

Desander
Fraction of Mud Cleaned:
   Vol treated with Desander/Vol into tank B = 800 gpm/(2200 gpm + 700 gpm) = 0.28 or 28%

Solution
The degasser, desander, and desilters cannot be supplied with fluid by one pump. Two more pumps are needed. Each piece of removal equipment should process from one compartment to the next one down stream. Each should be processing fluid which was processed through the preceding piece of removal equipment.

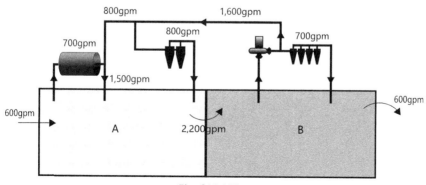

Fig. S11.16A

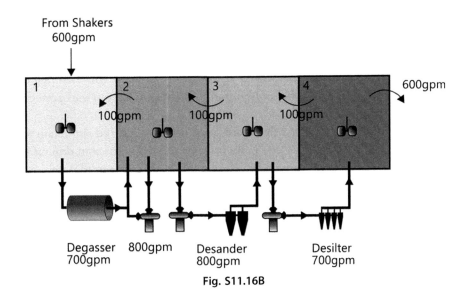

Fig. S11.16B

Things actually got better when the desander was turned off. In this case the desilter was processing 700 gpm/600 gpm + 1400 gpm = 0.35 or 35%. However, usually the desilters would plug up and no drilled solids were removed. This mud tank system needs to have 3 compartments in the removal section. The degasser, desander and desilter need their own pump. An addition/suction section must be added to have sufficient processed fluid to keep the drill pipe filled with a homogeneous fluid (Fig. S11.16B).

Chapter 12 Equipment solids removal efficiency solutions

Problem 12.1 solution

This is a fairly common example.

First, the hole volume, or volume of solids reaching the surface, will be calculated (1143 bbl).

$$\text{Hole Volume} = \frac{(14.0 \text{ inches})^2}{1029} \times 6000 \text{ ft} = 1143 \text{ bbl}$$

This assumes that the rock drilled has no porosity. The value of solids could be reduced by 20% to a value of 914 bbl, but the accuracy of the diameter measurement is insufficient to justify such minutia.

Second, the concentration of drilled solids in the drilling fluid must be determined. For this calculation, the density of the barite will be assumed to be 4.2 g/cc and the density of the low-gravity solids will be assumed 2.6 g/cc.

Volume fraction of low gravity solids = 62.5 + 2.0 [Volume fraction of solids] − 7.5 [Mud weight, ppg]

This calculation indicates that the low-gravity solids concentration is 8.5% volume. Since 2% volume is bentonite, the drilled solids concentration in this 12.0 ppg drilling fluid is 6.5% volume. The total solids concentration is 18% volume, so the barite concentration must be 9.5% volume.

The solids removal equipment will discard some drilled solids and also retain some in the system. The solids remaining in the tanks must be diluted with clean drilling fluid to produce 6.5% drilled solids concentration. The liquid added to the mud system to dilute the returned solids will be called dilution volume.

Drilled solids returned = (Drilled solids concentration)(Dilution volume)

This equation says that the drilled solids in the new drilling fluid added to the tanks will be diluted to the drilled solids concentration (6.5% vol).

The total fluid added to the mud system (which is the dilution volume) will consist of two components, the clean drilling fluid and the drilled solids returning to the system.

Dilution volume = Volume of clean drilling fluid + Drilled solids returned

or

Volume of clean drilling fluid added = Dilution volume − Drilled solids returned

$$\text{Volume of clean drilling fluid added} = \left[\frac{\text{Drilled solids returned}}{\text{Drilled solids concentration}}\right] - \text{Drilled solids returned}$$

$$\text{Volume of clean drilling fluid added} = [\text{Drilled solids returned}]\left[\frac{1}{\text{Drilled solids conc.}} - 1\right]$$

Third, Calculate the volume of drilled solids returned to the system by the solids control equipment. The Volume of Clean Drilling Fluid added while drilling this 6000 ft of hole was 7163 bbl. The drilled solids concentration in the drilling fluid was maintained at 6.5% volume.

From the equation above:

7163 bbl = Drilled solids returned [(1/0.065) − 1] or.

Drilled solids returned to the system = 497 bbl.

Fourth, calculate the removal efficiency of the surface equipment on this rig.

Since 1143 bbl arrived at the surface while drilling this interval, 646 bbl were discarded. The removal efficiency would be the ratio of the solids discarded and the solids arriving times 100, or 56.5%.

Costs!!

Without commenting on the hole problems associated with drilled solids, the cost of this inefficient removal system will perhaps be surprising. For this example, the drilling fluid will be assumed to cost $90/bbl and the disposal costs will be assumed to be $40/bbl.

How much drilling fluid was discarded during this 6000 ft interval? This would be the sum of the volume discarded with the solids removal equipment and the excess volume needed for dilution of the drilled solids.

This system discarded 646 bbl of drilled solids. These drilled solids were not dry; they carried with them some of the drilling fluid. For API 200 shaker screens and other equipment, a concentration of 35% volume drilled solids in the discard is a reasonable number. This means that the volume of discard would be (646 bbl/0.35) or 1846 bbl.

The pit levels would have decreased by 1846 bbl while drilling this interval. During this part of the well, 7163 bbl of clean fluid was added to the 497 bbl of drilled solids returned to the pits by the solids control equipment. The total drilling fluid volume built would be 7660 bbl. Since only 1846 bbl of fluid was removed from the pits, an excess volume of (7660−1846) or 5814 bbl was built.

The new drilling fluid would cost ($90/bbl × 7660 bbl) or $689,000. The discarded volume would be the sum of the discarded volume from the solids control equipment and the excess volume generated to dilute the returned drilled solids, or 1846 bbl + 5814 bbl, or 7660 bbl. This cost would be $40/bbl × 7660, or $306,000. Total drilling fluid cost for operating this system for 6000 ft of drilling would be $995,000 or $166/ft of hole.

Problem 12.2 solution

With the 56.5% removal efficiency and 1143 bbl of drilled solids reporting to the surface, 646 bbl will be discarded and 497 bbl will be returned to the pits, (as before). The returned drilled solids must be reduced to 4% volume in the pits, so the dilution volume will be (497 bbl/0.04) or 12,430 bbl. Assuming the same concentration of drilled solids in the equipment discard of 35% volume, as before, the equipment discard (and the decrease in pit volume) will be 646 bbl/0.35, or 1850 bbl. The excess drilling fluid generated is 12,430 bbl − 1850 bbl or 10,580 bbl.

Now the cost: The cost of new drilling fluid added to the system is 12,430bbl × $90/bbl or $1,118,700. The cost of disposal for this interval is (1850 bbl + 10,580 bbl) × $40/bbl, or $497,200. Total cost for this 6000 ft interval is $1,615,900 or $269/ft of hole.

Decreasing the drilled solids concentration by from 6.5% volume to 4% volume with this inefficient removal system increases the cost by $620,900.

Problem 12.3 solution

With 80% removal efficiency and 1143 bbl of drilled solids reporting to the surface, 914bbl would be discarded and 229 bbl returned to the pits.

The drilling fluid needed to dilute the 229 bbl to 4% volume would require adding (229 bbl/0.04) or 5725 bbl of new drilling fluid. This 5725 bbl would consist of 229 bbl of drilled solids and 5496 bbl of clean drilling fluid. The discard volume would be, assuming the 35% volume concentration of drilled solids in the discard, 914 bbl/0.35 or 2611 bbl. The excess volume of drilling fluid is (5725 bbl − 2611 bbl) or 3114 bbl.

The drilling fluid cost for 80% removal efficiency and 4% volume drilled solids would be:

(5725 bbl × $90/bbl) + [(3114bbl + 2611bbl) × $40] or $744,000. Increasing the removal efficiency from 56.5% to 80%, decreases the cost of the new fluid and disposal from $1,615,900 to $744,000 for this 6000 ft of hole. The $871,900 difference could justify significant changes in the drilling fluid system.

Problem 12.4 solution

With 80% removal efficiency and 1143 bbl of drilled solids reporting to the surface, 914 bbl would be discarded and 229 bbl returned to the pits.

The drilling fluid needed to dilute the 229 bbl to 6% volume would require adding (229 bbl/0.06) or 3817 bbl of new drilling fluid. This 3817 bbl would consist of 229 bbl of drilled solids and 3588 bbl of clean drilling fluid. The discard volume would be, assuming the 35% volume concentration of drilled solids in the discard, 914 bbl/0.35 or 2611 bbl. The excess volume of drilling fluid is (3817 bbl-2611 bbl) or 1206 bbl.

The drilling fluid cost for 80% removal efficiency and 6% volume drilled solids would be:

(3817 bbl × $90/bbl) + [(1206 bbl + 2611 bbl) × $40] or $496,000. With an 80% removal efficiency, increasing the drilled solids target concentration from 4% volume to 6% volume, decreases the cost of the new fluid and disposal from $744,000 to $496,000 (or $248,000) for this 6000 ft of hole.

15 Control drilled solids solutions

Problem 15.1 solution

This problem has multiple issues associated with it. Each issue will be discussed individually.

Issue #1

Losthole Drilling Company is drilling a turnkey wildcat through an old gas field and expects to see several hundred feet of drawn-down sands. The mud spec sheet calls for a very low fluid loss to prevent stuck pipe. The drilling fluid now has a fluid loss of 8 cc/30 min the "OUT" mud properties were: MW 10.6 ppg, Temp 155 F, FV 46, PV 14, YP 11, Gels 3/12, API 8.0, Solids 10, Ca 120, Cl 2000, pH 10.0, MBT 20.

The drilling foreman found several holes in the linear motion shaker screen. He noticed that during the past five days, the solids content had increased but the fluid loss was decreasing. The mud properties are now: MW 10.6 ppg, Temp 160 F, FV 45, PV 20, YP 15, Gels 8/25, API 5.5, Solids 12, Ca 120, Cl 2500, pH 10.0, MBT 25.

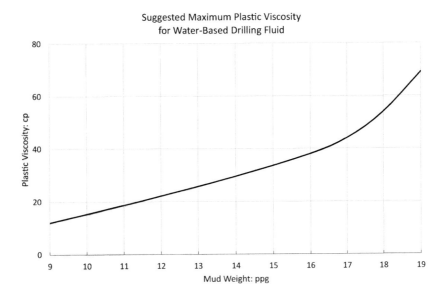

Fig. S15.1 Recommended maximum PV for a water-based drilling fluid.

Analysis of issue #1

Increase in solids content which decreases the fluid loss is perhaps a surprising effect of drilled solids. In a Newtonian fluid, the filter cake thickness can be related to the fluid loss. This is not true in a non-Newtonian fluid. Guidelines for plastic viscosity for a water-based drilling fluid were presented in Chapter 1, Fig. 1.19. Plastic viscosity depends upon four things: liquid phase viscosity, size, shape, and number of solids. Allowing the solids to increase raised the PV from 14 to 20 lb/100 sq.ft. The mud engineer did not report the filter cake thickness. The increase of solids concentration from 10% volume to 12% volume seems like a small increase but the impact is very large.

Issue #2

The mud program called for rather expensive filtration products since the bottom hole temperature was now over 300° F. Discussion with a mud engineer convinced him that he could achieve the 3–5 cc/30 min, if he would let the drilled solids increase only a little more. The rationale was that the filter cake needed a range of solids sizes to insure the maximum packing density. By allowing the drilled solids to provide this wide range of solid, all of the holes in the filter cake would be filled with solids. This maximum packing would decrease the fluid loss.

Analysis of issue #2

The maximum packing density may decrease fluid loss but the deposition of additional solids from a non-Newtonian fluid also decreases fluid loss. The filtration concepts normally discussed in engineering classes relate to Newtonian fluids and not to drilling fluid.

Issue #3
Two days later, the fluid-loss was down to 3 cc/min. The mud properties were: MW 10.6 ppg, Temp 160 F, FV 55, PV 25, YP 20, Gels 10/35, API 4.5, Solids 15, Ca 140, Cl 3000, pH 10.0, MBT 28.

Analysis of issue #3
The plastic viscosity has now increased to 25 cp, the percent solids to 15% volume, and the MBT from the original 20— 28 while the fluid loss has now decreased to a very low value. Again, the mud engineer did not report the filter cake thickness.

Issue #4
One more day of drilling remained before the well reached the first depleted sand with low pore pressure. He figured that he saved his company several tens of thousands of dollars in chemical costs. He also observed that he did not have to add much barite during the past 4 days. With all of the savings generated by this concept, he was expecting a big raise and promotion. He petitioned the office to allow him to implement this technique throughout the company. It could easily save a lot of barite and chemicals and therefore, a lot of money. His turnkey bids would always be lower than the competition. When he called his boss, what do you think he was told? Did the drilling foreman solve the problem cheaply? Can this technique be used to significantly decrease costs? What should be done, if anything, before that sand is reached?

Analysis of issue #4
This situation actually happened in the field—except the description of the problem was modified to protect the guilty. The drilling foreman did not call his boss and ask for advice or claim to have saved much money—and it wasn't a turnkey job. The pipe stuck as should have been expected. The fishing costs and NPT totally obliterated any savings in the cost of barite or not using the proper fluid loss additives. The mud engineer should have measured the HTHP fluid loss and given the foreman the THICK filter cake from the test.

Before the drawn-down sands were reached, the drilling fluid should have had a massive clean-up. The plastic viscosity (which depends upon the quantity of drilled solids in the drilling fluid) should have been decreased back to values below the recommended maximum value. The increase in MBT indicates that a considerable amount of reactive clay was being retained in the drilling fluid and was probably not dispersing.

Problem 15.2 solution
Solids removal efficiency
Drilled solids = 0.03 (new drilling fluid built).
   New drilling fluid volume built = drilled solids + clean drilling fluid.
   New drilling fluid built = 0.03 (new drilling fluid built) + clean drilling fluid.

0.97 (new drilling fluid built) = clean drilling fluid.
New drilling fluid built = 1000 bbl/0.97 = 1031 bbl.
Returning to the first equation:
Drilled solids remaining = 0.03 (new drilling fluid built).
Drilled solids remaining = 0.03 (1031 bbl) = 30.9 bbl.

Since 100 bbl of drilled solids reported to the surface and 30.9 bbl remained in the system, the system eliminated 69.1 bbl of drilled solids.

The solids removal efficiency would be the ratio of the discarded solids to the recently drilled solids presented to the equipment; or 69.1 bbl/100 bbl which is 69.1%.

How much excess drilling fluid was discarded to keep the pit levels constant?

Since 200 bbl of solids and drilling fluid were discarded, the pit levels would decrease by only 200 bbl. The 1000 bbl clean drilling fluid added to the system would be 800 bbl more than the system needed. This 800 bbl would need to be stored in reserve pits or otherwise eliminated from the active system.

The drilled solids concentration in the discard would be the ratio of 69.1 bbl/200 bbl or 34.5%.

What should be the solids removal efficiency to eliminate generating the excess drilling fluid (i.e. the lowest cost removal efficiency)?

To eliminate the excess fluid generation, the volume of clean drilling fluid added to the system should be equal to the volume of discarded drilling fluid and solids.

Drilled solids= (target drilled solids concentration) (volume new mud built).

(A) Volume of clean drilling fluid can be calculated from the volume of new mud built.
Volume of new mud built = drilled solids + clean drilling fluid.
Volume of clean drilling fluid = volume of new mud built − drilled solids remaining.
Volume of clean drilling fluid = (Drilled solids/target solids) − drilled solids.
Volume of clean drilling fluid = [(1-drilled solids conc./drilled solids conc)](drilled solids).
In this case, Volume of clean drilling fluid = {(1−0.03)/0.03}(drilled solids remaining).
Drilled solids remaining = (1−SRE) (rock volume drilled).
So, Volume of clean drilling fluid = (361) (1−SRE) (rock volume drilled)

(B) Volume of drilling fluid and solids discarded = (SRE) (rock drilled)/(concentration of drilled solids in discard).
In this case, Volume of discard = (SRE) (rock drilled)/0.345.
Equating the volumes in (A) and (B).
Volume clean drilling fluid = Volume of discard.
(361) (1−SRE) (rock volume drilled) = (SRE) (rock drilled)/0.345.

124.5−124.5SRE = SRE.
SRE = 124.5/125.5 = 99.2%

This is an extremely difficult efficiency to obtain. The cost required to keep the drilled concentration at such a low value will be great. The requirement for a 3% volume drilled solids concentration should be examined.

Chapter 21 Centrifuge solutions
Problem 21.1 solution

The 15 ppg drilling fluid probably had an excessive concentration of low-gravity (or drilled) solids. The plastic viscosity (PV) of 38 cp for a 15 ppg drilling fluid exceeds the recommended range (Fig. 1.19). PV depends upon the liquid phase viscosity (1 cp for water) and the size, shape and number of solids. The stored drilling fluid has too many low-gravity solids. The very dry discharge from the centrifuge indicates that the total solids concentration is around 60%−62%. The low gravity solids will be 31%−35% in the recovered slurry. The barite concentration will be between 25% vol and 31% vol. This means that about one-half of every barrel of solids recovered will be drilled solids.

Note that accurate calculations could be made to determine the exact ratio but the decision making process does not need that. The estimated value is sufficient to see this is not a good procedure. No one would buy a sack of barite that contained 50% of the volume in drilled solids.

Problem 21.2 solution
Part 1

When a 12.5 ppg drilling fluid in the active system is centrifuged, the underflow weighs 22.0 ppg and the overflow weighs 10.1 ppg. Assume that the underflow solids content is 60% volume and the overflow solids content is 12% volume. Is this centrifuge performing a useful function?

A centrifuge is used to discard colloidal solids from a drilling fluid. The discard stream is the light slurry. Calculate the low −gravity solids in the discard,

$$(\rho_B - \rho_{lg}) V_{lg} = 100\rho_f + (\rho_B - \rho_{lg}) V_s - 12 \text{ MW} \tag{22.1}$$

where:
ρ_B = Density of High Gravity Solids (Barite is 4.2 g/cc)
ρ_{lg} = Density of Low Gravity Solids,
V_{lg} = Volume Percent Low Gravity Solids,
ρ_f = Density of Filtrate,
V_S = Volume Percent Undissolved Solids, and

MW = Mud Weight, ppg.

For a fresh water drilling fluid, assuming the density of the drilled solids is 2.6 gm/cc, the density of the barite is 4.2 gm/cc, and the water density is 1.0 gm/cc, the equation becomes:

$$V_{lg} = 62.5 + 2\,V_s - 7.5(MW) \qquad (22.2)$$

$$V_{lg} = 62.5 + 2(12) - 7.5(10.1)$$

$$V_{lg} = 10.75\% \text{ volume}$$

This concentration of 10.75% volume low-gravity solids is mostly colloidal and very detrimental to drilling. Only 1.25% volume of the discard is barite. However, colloidal barite is not a desirable additive even though it costs money to add to the drilling fluid. In this situation, the centrifuge is doing a great job and should be continued. (Notice we did not calculate how much barite was being "saved". This is a secondary point.)

Part 2

A 15.0 ppg fresh-water base drilling fluid from another well is brought to your drilling rig. You are drilling with an 11.0 ppg drilling fluid that has 11% volume total solids. The mud weight needs to be increased to 12 ppg. The leftover drilling fluid from storage seems like an excellent source of barite. The centrifuge feed is switched to the stored drilling fluid. The underflow weighs 20 ppg and has a solids content of 62% volume. Is this a good idea? (Compare the drilled solids in the drilling fluid with the drilled solids content of the underflow of the centrifuge.) What results do you expect to see from saving so much money by using this "free" barite to increase the mud weight?

$$V_{lg} = 62.5 + 2\,V_s - 7.5(MW) \qquad (22.3)$$

For the drilling fluid, the low-gravity solids content would be:

$$V_{lg} = 62.5 + 2(11) - 7.5(11.0)$$

$$V_{lg} = 2\% \text{ volume}$$

For the centrifuge underflow, the low-gravity content of the slurry added to the drilling fluid is:

$$V_{lg} = 62.5 + 2(62) - 7.5(20.0)$$

$$V_{lg} = 36.5\% \text{ volume}$$

Since the slurry has 62% volume solids, and 36.5% volume is low-gravity, only 25.5% volume is barite. This will destroy a very clean drilling fluid. No one would buy a sack of barite if more than half of the sack volume contained low-gravity (drilled) solids. Why put this into the drilling fluid even if it is free? Actually, this will probably result in slower drilling rates, possible stuck pipe, possible lost circulation and problems that will cost much more than the small amount saved by not purchasing a few sacks of barite. Also, note that we calculated what was added to the drilling fluid here and not in the first part of this problem. In the first part of this problem the fluid processed came from the mud pits. Only material was removed. In that case, additional drilled solids could not be added to the drilling fluid. In the second part of this problem, solids could be added from an outside source. AND in the case above, it would definitely destroy a good drilling fluid.

Part 3

Suppose the discharge from a mud cleaner weighed 20 ppg with the 62% volume solids concentration. Is the mud cleaner performing a useful function? (Calculate low gravity solids and barite in the discard.)

In part 3, the calculation would be the same as in part 2. The discard contains more drilled solids by volume than barite. The decision is clear. The mud cleaner is discarding a healthy amount of drilled solids and should continue. If the drilling fluid contains a large quantity of drilled solids, even an equal quantity of barite and drilled solids in the discard would not be objectionable.

Appendices

Appendix A
pH changes with temperature

The negative logarithm of the hydrogen ion concentration is called pH. The neutral point on a pH scale is the point at which there are as many hydrogen atoms as hydroxyl molecules. Normally at room temperature this occurs at a pH of 7. As the temperature increases, however, the neutral point also decreases. As shown in Table A.1, fluid #2 could have a pH of 8.61 at 176 °F, but the pH at room temperature (70 °F) would be 10.0

This effect frequently comes as a surprise to many mud engineers. It is particularly important when drilling with aluminum drill pipe. The depth capability of a rig can be extended by using aluminum drill pipe instead of steel. When drilling rigs become scarce, this is a common ploy to drill some deep holes with available drilling rigs. Generally, the mud engineer is warned about raising the pH of the drilling fluid above 9.0 or 9.5. Caustic is as detrimental to aluminum as acid is to steel. With so much pressure and serious admonitions about taking care of the pH, a diligent mud engineer may take the pH meter to the flow line to obtain an "accurate" value. The temperature adjustment on the pH meter does not account for the change in neutral point of the hydrogen/hydroxyl ions. A string of aluminum drill pipe can be lost quickly because of this effect.

Table A.1 Effect of temperature on pH.

Temperature		pH					
°C	°F	Fluid 1	Fluid 2	Fluid 3	Fluid 4	Fluid 5	Fluid 6
24	70	9.5	10.0	10.50	11.00	11.50	12.00
30	86	9.33	9.83	10.33	10.83	11.33	11.83
35	95	9.18	9.68	10.18	10.68	11.18	11.68
40	104	9.03	9.53	10.03	10.53	11.03	11.53
45	113	8.90	9.40	9.90	10.40	10.90	11.40
50	122	8.76	9.26	9.76	10.26	10.76	11.26
55	131	8.64	9.14	9.64	10.14	10.64	11.14
60	140	8.52	9.02	9.52	10.02	10.52	11.07
65	149	8.40	8.90	9.40	9.90	10.40	10.90
70	158	8.30	8.80	9.30	9.80	10.30	10.80
75	167	8.20	8.70	9.20	9.70	10.20	10.70
80	176	8.11	8.61	9.11	9.61	10.11	10.61
85	185	8.02	8.52	9.02	9.52	10.02	10.52
90	194	7.95	8.45	8.95	9.45	9.95	10.45

Appendix B
Drilling fluid suggestions

1. Compare the filter cakes from filtration tests at room temperature with a 100 psi and a 900 psi pressure differentials. Use a HTHP cell for the high pressure test. Filter cake thickness is depends upon the type of solids in the fluid.
2. A drilling fluid with very low concentrations of drilled solids (1–2% volume) decreases the chance of stuck pipe and lost circulation.
3. The API fluid loss does not correlate with the cake thickness. Fluid loss can decrease with an increase in drilled solids.
4. In water, the pH is a function of temperature. At 75 °F, the concentration of hydroxyl ions and hydrogen ions are equal at a pH of 7.
5. Entrained gas or air in a drilling fluid can decrease the rig pump efficiency. One case where the pump efficiency was measured, only 6% volume air reduced the pump efficiency to 85%. This frequently explains problems with low stand pipe pressure and hole cleaning.
6. Measure the true mud weight with a pressurized mud balance. If one is not available, add some defoamer to a mud cup full of drilling fluid. Pour through a funnel viscometer two or three times. Measure the mud weight with a rig mud balance. The mud weight will be close to the pressurized mud balance reading.
7. Barite plugs are used to stop underground blowouts. Settled barite looks like an impermeable mass. It is not. Settled barite still has porosity and permeability. Many test have shown that a settled barite plug will eventually fail. Non-settling barite plugs with good filtration control should be used to stop underground blowouts. Another method of sealing an underground blowout is to cause barite to settle at the bottom of the well. This method has been used many times; however, there are draw backs to this technique. As the barite settles, the bed of solids has over 35-%volume porosity. As the barite starts settling and the fluid is no longer supporting the solids, the hydrostatic pressure at the bottom of the hole decreases. This means that the pressure in the well bore may drop considerably below the formation fluid pressure. Although the barite has a low permeability, gas can flow through this bed. This technique can buy time, some cases, but with the pressure decreasing as the barite settles, flow will eventually follow.
8. Water-based drilling fluids should have a calcium concentration of 100–300 ppm if bentonite is used in the fluid. Values lower than 50 ppm usually mean that there is no free calcium in the system.
9. Bentonite should be prehydrated for a minimum of 12 h (preferably 24 h) before being added to a drilling fluid. The prehydration tank should contain only water and bentonite. Do not add caustic or lignosulfonate to a prehydration tank.

10. Lignosulfonate deflocculates bentonite and does not disperse the clay. It actually prevents clay from dispersing.
11. Any drilling fluid added to an active system should be filtered through the shale shaker screens. Fluid from trip tanks or fluid from reserve pits should be added through the shale shakers.
12. In drilling fluid reports, "sand" designates a particle size not a material containing quartz.
13. If the sand content is larger than a trace in the suction tank, either the solids removal equipment is not functioning correctly or barite has just been added to the system. After one complete circulation after adding fresh barite, the sand content should be a trace or lower.
14. H_2S measure with the Garrett Gas Train can be removed from a drilling fluid by adding one pound per barrel of zinc carbonate for every 500 ppm sulfide or using "ironite sponge".
15. Densities of NADF should be measured at the same temperature daily and that temperature reported. Every 10 °F change in temperature will result in changes of as much as 0.07 ppg.
16. The density of water is also a function of temperature, Table B.1.
17. Annular velocity does not cause hole erosion (see Appendix D)

Table B.1 Density of water changes with temperature.

Density of water			
Degrees, °C	Density, kg/m³	Degrees, °F	Density, ppg
10	999.6996	50.00	8.342896
12	999.4974	53.60	8.341208
14	999.2444	57.20	8.339097
16	999.9430	60.80	8.344927
18	998.5956	64.40	8.333683
20	998.2041	68.00	8.330415
22	997.7705	71.60	8.326797
24	997.2965	75.20	8.322841
26	996.7837	78.80	8.318562
28	996.2335	82.40	8.313970
30	995.6473	86.00	8.309078
32	995.0262	89.60	8.303895
34	994.3715	93.20	8.298431
36	993.6842	96.80	8.292695
40	992.2158	104.00	8.280441
44	990.6280	111.20	8.267190
46	989.7914	114.80	8.260208
48	988.9273	118.40	8.252997
49	988.4851	120.20	8.249306
50	988.0363	122.00	8.245561

Solids control suggestions and thoughts

1. Hydrocyclones frequently have plugged orifices because of large solids. This indicates that some solids have by-passed the shaker screens.
2. Plugged hydrocyclones will greatly decrease the drilled solids removal efficiency.
3. Before a trip for a new bit, the back-tank (or possum belly) of the shale shakers should NOT be dumped into the active system or the sand trap. The large solids do not settle and will plug desilter underflows.
4. Even though the main shakers are processing fluid through API 170 or API 200 screens, a mud cleaner will still remove a significant quantity of drilled solids. It makes a good "insurance" package.
5. Almost all hydrocyclone manufacturers recommend 75 ft of head for proper operation in an unweighted drilling fluid. Some hydrocyclones have been modified from the original design. Check with the manufacturer to determine the head required and the resulting flow rate.
6. The hydrocyclones on a mud cleaner, processing a weighted drilling fluid can operate properly with a lower head than 75 ft. The underflow will contain more liquid but the liquid will be filtered through the screen.
7. The head can be calculated from the measured pressure at the manifold from the equation:

$$\text{Pressure, psi} = 0.052(\text{mud weight, ppg})(\text{head, ft})$$

$$\text{Head(ft))} = \text{Pressure (psi)}/[0.052(\text{Mud weight (ppg)})]$$

8. If the underflow from hydrocyclones is too dry and the screen cannot properly separate the solids, a small stream of drilling fluid can be sprayed on the screen to assist the separation. Frequently, a small hose and valve can be mounted on the hydrocyclone manifold to provide the spray. The liquid phase of the drilling fluid (water or NAF) is not recommended because it causes too much dilution.
9. There is no commercial solids removal equipment available to use in the active system which will separate barite from drilled solids.
10. A centrifuge is used to "remove" drilled solids to control PV and filter cake quality.
11. In a weighted drilling fluid, centrifuges can remove solids smaller than about 5 μm. These solids could be drilled solids, barite, filtration control additives, or rheology modifiers.
12. In a weighted drilling fluid, filtration additives and the rheology modifiers should be replaced to keep the drilling fluid properties required according to the program.
13. In an unweighted drilling fluid, centrifuges can remove most of the solids larger than about 10 μm.

14. A centrifuge does NOT RECOVER BARITE. It is used to control colloidal particles In a weighted drilling fluid. These particles increase plastic viscosity and change filter cake quality.
15. Both the light (overflow) slurry and the heavy (underflow) slurry from a centrifuge will contain drilled solids and barite.

Appendix C
Solids control

Linear motion and balanced elliptical motion shakers place screens on an incline. The liquid pool provides a head which assists fine screens in separating more solids. However, if the slope of the screen is too high, solids can be degraded in the pool before they can bounce their way out of the liquid. The screen can be tilted to a large angle during bottoms-up to keep from losing thick drilling fluid off the end of the shaker. However, after bottoms-up, the shaker screens should be lowered to a more reasonable elevation.

Solids problem without an increase in solids content

Plot the plastic viscosity daily or more frequently, like every four to eight hours. If solids are grinding into colloidal sizes, the PV will gradually increase. The total percent of solids may not increase appreciably but if they break apart, the total number increases significantly. PV depends upon the liquid phase viscosity, the size, shape, and NUMBER of particles. These solids could be ground in the annulus on the way out of the hole. These solids could have by-passed the solids removal equipment and recirculated down hole. These solids could increase because of an undetected hole in the shaker screen or drilling fluid is by-passing the shaker screen. If the back tank (possum belly) of the shale shaker is dumped into the active system before tripping pipe, the valve may not be completely closed. These solids could increase because the desilters cones are plugged. If the cones are plugged, the solids causing the plug are usually larger than the openings in the shaker screen. Drilling fluid would be by-passing the shakers.

Appendix D
Hole erosion

Frequently, in technical meetings, a comment is made stating with great certainty that high annular velocities erode well bores. As proof, a well bore diameter was not as large after the annular flow rate was decreased. Tests conducted in a field with IADC bits 537 indicate that other factors than annular velocity erodes the well bore. When the annular flow rate is decreased, the nozzle velocity decreases, the shear rate of the fluid through the nozzle decreases, the hydraulic impact force decreases, and the hydraulic power through the bit decreases.

When the flow rate is decreased by 10%, the nozzle flow rate decreases by 10%. The hydraulic impact force decreases by 20% because hydraulic impact force is calculated by multiplying the density times the nozzle velocity times the flow rate. The hydraulic power is decreased by 30% because it is calculated by multiplying the force times the velocity. The hydraulic force will always change by twice the change in flow rate. The hydraulic power will always change by three times the change in flow rate. The guideline suggested in the SPE paper #30497[1] to decrease hydraulic hole erosion was to keep the nozzle shear rate below 100,000 s^{-1}, or the nozzle impact force below 2000 N, or the hydraulic power below 200 kW. In the field, drilling tests reported in the SPE paper, five wells were drilled with the same rig, the same bits (IADC 537), the same mud weight program, annular velocity, and the same crew. Two of the wells were drilled with oil mud and three with water-based drilling fluid. The annular velocity in all wells was maintained at 90 m/min or slightly over 180 ft/min. Only the flow rate through the nozzles was varied. Caliper logs indicated that there was hole erosion in all wells.

In very soft sediments like unconsolidated sands, the nozzle velocity may erode the wellbore. However, for consolidated formations, the hydraulic impact or power seems to be the cause of wellbore erosion.

Appendix E
Significant figures

In Chapter 3, a comment was made about the accuracy of the calculation and the effect of the measurements on the iteration process of calculating flow rates. One problem seems to be pervasive through the drilling industry: significant figure accuracy. Usually, most calculations are made with the assumption that all of the numbers are significant and then a "safety" factor is added to the answer.

Multiply two times three and of course the answer is six. Each number has one significant figure and the answer has one significant figure.

$$\begin{array}{r} 3 \\ \times\ 2 \\ \hline 6 \end{array}$$

---

[1] Chemerinski B, Robinson L. Hydraulic wellbore erosion while drilling. SPE Paper # 30497. In: Presented at 1995 SPE annual technical conference and exhibition, Dallas, Texas, 22—25 October, 1995.

Suppose the first number is measured more accurately and is really 3.1212. Now a calculator will show the answer to be different.

$$
\begin{array}{r}
3.1212 \\
\times 2 \\
\hline
6.2424
\end{array}
$$

This number has five significant figures. Since the second significant figure is not known for the number 2, place an "X" there instead.

$$
\begin{array}{r}
3.1212 \\
\times 2X \\
\hline
XXXXX \\
6.2424 \\
\hline
6.YYYYX
\end{array}
$$

The value of "Y" is not known. Now the answer once again has only one significant figure, the "2424" numbers, which the calculator provides, are totally meaningless.

If the value of "2" is measured more accurately and is 2.1, a calculator will show the multiplication of 3.1212 times 2.1 to be 6.55452. Again, place an "X" at the end of each number to represent the unknown number and multiply "long-hand":

$$
\begin{array}{r}
3.1212X \\
\times 2.1X \\
\hline
XXXXXX \\
31212X \\
62424X \\
\hline
6.5YYYYYX
\end{array}
$$

The number of significant figures in the answer is two [6.5] and the last four numbers in the result (YYYYX) have no meaning.

There is an exception to the rule that the maximum number of significant figures in the answer is the minimum number in the lowest accuracy number in the multiplication. Multiply 6 times 7 and the answer is 42. Now, evaluate how many numbers are significant figures are in the answer.

```
        7.X
  x     6.X
  _____
       42.XX
```

If the first two numbers multiply to give a value larger than nine, there can be one more significant figure accuracy in the answer than the number of significant figures in the lowest number of one of the multipliers.

To check this concept, assume the value of the "6" is really 6 and the value of "7" is really 7.1212. The calculator will provide an answer of 42.4606052. Now, multiply the numbers using the "X" value to indicate the next unknown value:

In this case, there are two significant figures in the answer because the accurate correct answer is a "binomial". Again, suppose the seven was measured to four decimal places.

```
       7.1212
  x        6X
  _____
        XXXXX
      42.7272
  _____
      42.YYYYX
```

Again, the answer still has only two significant figures.

If the value of "6" is really 6.1, the number of significant figures in the answer should increase.

```
        7.1212X
    x    6.1X
    ─────────
         XXXXXX
        71212X
       427272X
    ─────────
      43.4YYYYYX
```

In this case, the answer is 43.4, not 43.43932, as would be indicated by a calculator.

Significant figure concept applied

Measurements are usually made with some limitation on accuracy. For example, consider the calculations used to calculate bottom hole pressure. The equation used in well control procedures is well accepted and universally used:

$$\text{Pressure, psi} = (0.052)(\text{mud weight, ppg})(\text{depth, ft})$$

A mud balance could read the mud weight as 12.4 ppg. This number has a three significant figure accuracy. An accurate mud weight measurement could indicate the mud weight is really 12.398 ppg. For the purposes of this discussion, the number could be written as 12.4X ppg. If this number is multiplied by a depth of 12,432 ft, a calculator will indicate an answer of 15,4156.8. If the numbers are multiplied "long hand", the answer is 154ZYY.YX.

```
         12432
    x    12.4X
    ─────────
         XXXXX
         49728
         24864
         12432
    ─────────
       152ZYY.YX
```

where $Z = 20 + X$

In other words the numbers 156.8 are meaningless, and the value should be reported as 154×10^3.

In the discussion of the concept, the mud weight was 12.4 ppg and the well depth was 12,432 ft. What would be the bottom hole pressure? The equation used in the calculation (from well control procedures) is:

$$P, \text{ psi} = 0.052 \text{ (mud weight, ppg)(depth, ft)}$$

$$P, \text{ psi} = 0.052(12.4 \text{ ppg})(12,432 \text{ ft}) = 8016.1536 \text{ psi}$$

This is impressive accuracy. Now, examine the values in terms of the significant figure accuracy just discussed. How accurate is the coefficient 0.052? The equation is derived from the basic concept that pressure is energy per unit volume. Energy or work, in this case would be the potential energy of the column of 12.4 ppg drilling fluid at the depth of 12,432 ft.

$$Pm = mgh/\text{volume}$$

mg (or mass times gravity) is weight (in lb)
weight/volume is density of the fluid (in lb/ft^3)
h would be the depth in ft. $P = [W/vol] h$

Pressure is usually reported in pounds per square inch or psi and the density in pounds per gallon.

$$P, \text{ psi} = [W, \text{ lb/vol, gal}](h, \text{ ft})$$

The units on the right side of the equation must be changed to calculate pressure in pounds per square inch. One gallon is 231 $in.^3$ and one foot is 12 in. These equivalent values are accurate to several significant figures.

$$P; \text{ psi} = W, \text{ lb/vol gal} \left[\text{gal}/231 \text{ in.}^3\right](h, \text{ ft})[12 \text{ in.}/\text{ft}]$$

$$= 12/231 \text{(mud weight, ppg)(depth, ft)}$$

$$P, \text{ psi} = 0.051948 \text{(MW, ppg)(depth, ft)}$$

From the previous calculation where depth is

12,432 ft and mud weight is 12.4 ppg

$$P, \text{ psi} = 0.05195 \text{ (MW, ppg)(depth, ft)}$$

The constant (0.05195) now has four significant figures. The mud weight still has only three significant figures; consequently there is no reason to include five significant figure accuracy in the constant.

$$
\begin{array}{r}
0.051948 \\
\times \quad 154\text{ZYY} \\
\hline
\text{XXXXX} \\
\text{XXXXX} \\
\text{XXXXX} \\
207792 \\
259740 \\
51948 \\
\hline
798\text{Z.YYYYYX}
\end{array}
$$

$$Z = 18 + X = 20$$

Multiplying with the calculator would create the answer 8005.6592 psi. The pressure calculation could indicate a pressure of 800x psi where the x indicates an unknown number. When the annular flow rate is decreased, the nozzle velocity decreases, the hydraulic impact force decreases and the hydraulic power through the bit decreases.

The equation normally used is:

$$P = 0.052(\text{MW})(\text{depth})$$

or

$$P = 0.052(154\text{ZYY})$$

Long hand multiplications will indicate:

$$
\begin{array}{r}
154\text{ZYY} \\
\times \ 0.052\text{X} \\
\hline
\text{XXXXXX} \\
308\text{XXX} \\
770\text{XXX} \\
\hline
90\text{X.YYYX}
\end{array}
$$

This would be 9.0×10^3 psi or only two significant figure accuracy, because the conversion unit of 0.52 was only two significant figures.

The pressure which would be calculated from the equation:

$$\text{Pressure, psi} = 0.052(12.4 \text{ ppg})(12,432 \text{ ft})$$

$$\text{Pressure, psi} = 8016.1536 \text{ psi}$$

A mud balance could read the mud weight as 12.4 ppg. This number has a three significant figure accuracy. An accurate mud weight measurement could indicate the mud weight is really 12.398 ppg. For the purposes of this discussion, the number could be written as 12.4X ppg. If this number is multiplied by a depth of 12,432 ft, a calculator will indicate an answer of 15,4156.8. If the numbers are multiplied "long hand", the answer is 152ZYY.YX.

```
        12432
    x   12.4X
    ─────────
        XXXXX
        49728
        24864
        12432
    ─────────
       152ZYY
```

where $Z = 20 + X$

In other words, the numbers 156.8 in the "calculator answer" are meaningless, and the value should be reported as 154×10^3.

The mud weight usually used in these calculations is the value measured at the surface at some temperature lower than the bottom hole temperature. A water-base drilling fluid (without gas/air) is incompressible but the density (or mud weight) decreases with an increase in temperature. A Non-Aqueous Drilling Fluid (NADF) is compressible, which means the density (or mud weight) will increase with pressure. NADF density also decreases as the temperature increases. With either fluid, the mud weight measured a the surface will probably not be the effective mud weight at the bottom of the well bore.

However, to calculate the mud weight required to kill the well, the shut-in drill pipe pressure will be measured through the fluid with uniformly changing density.

The drill pipe must be filled with a homogenous drilling fluid (which may have a change in density with depth). When a kick is detected, the BOP is closed. If the drill pipe pressure reads 500 psig, a kick my have been validated. If the casing pressure at the surface is zero, the drill pipe pressure may be high because of cuttings in the annulus. If the surface casing pressure is not zero, a kick has probably occurred. The 300 psig reading can be used to calculate the additional mud weight needed.

$$P = 0.05195 \, (MW)(depth)$$

In this example, assume the original mud weight was 12.0 ppg and the well was 12,000 ft. Deep. The additional mud weight needed would be:

$$300 \text{ psi} = 0.015195 \, (\Delta MW)(12,000 \text{ ft})$$

$$\Delta MW = 0.48 \text{ ppg}$$

Because of the uncertainties of calculating the actual pressure, usually an additional pressure of 200–300 psi is applied. In this case, the safe increase in mud weight would be:

$$300 \text{ psi} + 200 \text{ psi} = 0.015195(\Delta MW_{Safe})(12,000 \text{ ft})$$

$$\Delta MW_{Safe} = 0.82 \text{ ppg}$$

Final point: The calculation of the bottom hole pressure before the influx would normally be:

$$P = 0.05195(12,000 \text{ ft})(12.0 \text{ ppg})$$

$$= 7480.8 \text{ psi}$$

Considering the significant figure of the mud weight, the answer is only correct to three significant figures or:

$$= 7.48 \times 10^3 \text{ psi}$$

Since the effect of temperature is not included in the density of the fluid in the drill pipe, the bottom-hole pressure calculation would limit the accuracy of this number.

The increase in mud weight (0.48 ppg) was accurate to three significant figures. However, because of the effect of temperature with a waster-based drilling fluid and the effect of temperature and pressure with a NADF, the 200 psi safety factor becomes an intelligent choice.

Appendix F

Acceleration of gravity

You are standing on a beach at the equator. How fast are you moving? Obviously, relative to the beach, you are not moving. Relative to the sun, however, you are on a rotating sphere. The earth is about 8000 miles in diameter at the equator. The circumference would be about 25,100 miles. The earth rotates one turn every 24 h. Your speed would be 25,100 miles/24 h, or about 1000 miles/h. If you were a short distance from either the north or South Pole and the circumference was only 100 miles, your speed would be only about 4 miles/h. However, relative to the sun, your speed at the poles would be about 20,000 miles/h relative to the sun. The earth's orbit, about 90×10^6 miles from the sun, circles once every 8760 h.

The earth is not a sphere but has a greater circumference around the equator than the circumference on a line around the north and south poles. The fact that there is a significant difference in centripetal acceleration between the poles and the equator possibly accounts for this change from a spherical shape.

The acceleration of gravity is not constant over the surface of the earth. The acceleration of gravity is higher at the equator than near the north or South Pole, Table F.1. This has an impact on calculation of the gravitational force (or weight) of a body. For example, the moon's acceleration of gravity is one-sixth of the acceleration of gravity (g) on the earth. If you weigh 180 pounds on earth, you would weigh only 30 pounds on the moon. On the other hand if you have a drill string with a mass of 1000 kg at the equator, the weight, in SI units, would be calculated by the product of mass times the acceleration of gravity (mg):

Table F.1 Acceleration due to gravity at sea level

Latitude, degrees	ft/s²	cm/s²
0	32.08730	978.0327
5	32.08858	978.0719
10	32.09240	978.1884
15	32.09865	978.3786
20	32.10712	978.6370
25	32.11757	978.9556
30	32.12969	979.3249
35	32.14310	979.7337
40	32.15741	980.1698
45	32.17218	980.6199
50	32.18696	981.0704
55	32.20130	981.5074
60	32.21476	981.9178
65	32.22694	982.2890
70	32.23746	982.6096
75	32.24599	982.8698
80	32.25228	983.0616
85	32.25614	983.1791
90	32.25744	983.2186

1000 kg times 9.832186 m/s^2, or 9832 N. This same mass near the north or south pole would weigh 9780 N. Clearly, using SI units requires an adjustment in the value of the acceleration of gravity used to calculate weight.

As a matter of interest, the English system of units has a corresponding unit for weight — called the "poundal". The 1000 kg drill string would have a mass of 22,000 pounds. At the equator the weight would be calculated the same way as using the SI unit system:

At the equator, the weight would be 22,000 lb times 32.25744 ft/s^2, or 7097 × 10^2 poundals; at the poles, the weight would be 7059 × 10^2 poundals.

Appendix G
Final thoughts or "the musings of a curmudgeon"

At any gathering of people in the drilling industry, many tales are told during breaks. Some might be worth capturing (Fig. G.1).

New processes and equipment are difficult to introduce to the drillers working of rigs. George Stonewell Orsmby was involved with introducing hydrocyclones to a skeptical audience. He told about one of the first deployments of desilters on a drilling rig close to Houston, Texas, in the 1940s. The rig crew would not allow him to install the bank of desilters on their mud pits but made him install it on the berm of their reserve pit [In those days, a large pit was dug close to the mud tanks and all excess fluids were dumped into this pit.] He had to provide his own centrifugal pump and motor. By mid-afternoon, the desilter was discarding a large quantity of drilled solids. He remained with the unit until late that evening and it was performing superbly. He drove back home for the night and called the rig early the next morning to see how it was working. The tool pusher told him to come get that piece of ★#@#&# (an oilfield versatile noun) because it wasn't working

Fig. G.1 Leon Robinson:—explosive drilling capsule—circa 1968.

any more. When George arrived at the rig, he couldn't find his desilter bank. He asked the tool pusher if they had hauled it off. The tool pusher told him that it was right where he left it. The discharge line from the desilter bank had so many solids that it had plugged. The solids continued to build until the bank of cones buried themselves. He cleaned the solids from the discharge and the unit then continued to remove drilled solids.

In another illustration of introducing a new appliance to the oil patch, a gentleman by the name of "Mr. Martin" visited my office to discuss the latest in measurements currently being made on drilling rigs. He regaled a tale about how he and another fellow named "Decker" had introduced a device to the drilling rig that would measure how heavy the drill string was. He said the drillers resisted using the device because they didn't need it. The weight indicator was superfluous for them because they could "feel everything they needed to know" by the way the brake acted.

Clearly, it would be almost impossible to drill wells now without the weight indicator. Still the Martin-Decker weight indicator was not welcomed when it first was offered to the drilling rigs.

Not all new things introduced to the industry resulted in a quest to solve the problem which they solve. Mac McKinley was assigned a research project to help reservoir engineers map production zones. Usually, when drilling, a field geologist has to insert several faults to make their maps agree with the formations as they are drilled. Mac decided to try to use some micro-seismic techniques to see if faults could be located at some distance from well bores. He wanted to detonate a small explosive in a well bore and listen for the return echo. This would allow discoveries of faults in various directions away from the well bore and also indicate their distance from the well bore. While the explosive charge was being developed, he designed and built the listening device. In preparation for deployment, he found an abandoned well and gained permission to lower the new electronics into the well. When the sonde reached the bottom of the hole, he thought the electronics had completely failed at that pressure and temperature, because he had so much static. As he pulled the sonde from the hole, the static ceased. Puzzled about the electronics, he lowered the sonde back down the hole and the static began again. At one point in the well, he could start and stop the static. Looking at an old log of the well, that point was adjacent to a very permeable sand formation. There was flow behind the pipe. This, then, became the noise log used to validate that cement in the annulus does form a barrier.

Cement bond logs look at the interface between the cement and the pipe and the interface between the cement and the formation. Holes created by the migration of gas through the cement as it sets cannot be observed with cement bond logs. The noise log can detect the failure of the cement to form a barrier for flow in the annulus. This discovery was so important that he never did have the opportunity to return to his original assignment.

Bibliography

[1] Chopey NP. Handbook of chemical engineering calculations. United States: McGraw Hill; 1994.
[2] Zeidler U. Fluid and drilled particle dynamics related to drilling mud carrying capacity [Ph.D. dissertation]. Tulsa, Oklahoma: University of Tulsa; 1974.
[3] Peden JM, Luo Y. Settling velocity of variously shaped particles in drilling and fracturing fluids. SPEDE December 1987:337—43.
[4] Bizanti MS, Robinson S. Transport ratio can show mud-carrying capacity. Oil Gas J June 1988;39: 43—6.
[5] Bizanti MS, Robinson S. PC program speeds slip velocity calculations. Oil Gas J November 1988: 44—6.
[6] Iyoho AW, Horeth JM, Veenkant RL. A computer model for hole-cleaning analysis. JPT September 1988:1183—92.
[7] Chin WC. Exact cuttings transport correlations developed for high angle wells. Offshore Mag May 1990:67—70.
[8] Gavignet AA, Sobey IJ. Model aids cuttings transport prediction. JPT September 1988:916—21.
[9] Uner D, Ozgen C, Tosun I. Flow of a power-law fluid in an eccentric annulus. SPEDE September 1989:269—72.
[10] Chin WC. Advances in annular borehole modeling. Offshore Mag February 1990:31—7.
[11] Haciislamoglu M, Langlinais J. Discussion of flow of a power-law fluid in an eccentric annulus. SPEDE March 1990:95.
[12] Sifferman TR, Becker TE. Hole cleaning in full-scale inclined wellbores. SPEDE June 1992:115—20.
[13] Sifferman TR, Myers GM, Haden EL, Wahl HA. Drill-cutting transport in full-scale vertical annuli. JPT November 1974:1295—302.
[14] Bridges S (Derrick Equipment Company), Robinson L (Retired), Morrison R (Derrick Equipment Company). Incremental dilution calculations. In: 2008 AADE fluids technical conference and exhibition. AADE Houston, TX Chapter. Paper # AADE-08-DF-HO-36.
[15] Williams CC, Bruce GH. Carrying capacity of drilling muds. Trans AIME 1951;189:111—20.
[16] Taggert AF. Handbook of mineral dressing-ores and industrial minerals. New York City: John Wiley & Sons Inc.; 1953. p. 7—72.
[17] Manohar L, Hoberock LL. Solids-conveyance dynamics and shaker performance. SPEDE December 1988:385.

Index

'*Note*: Page numbers followed by "f" indicate figures and "t" indicate tables.'

A

AC. *See* Alternating Current (AC)
Accelerations, 385
Accuracy requiring for specific gravity of solids, 456–457
Active surface processing system design, 311–322
    flow line, 312
    removal section, 312–319, 312f
Addition section, 53f, 185. *See also* Mud guns
    mud hoppers, 49–53, 50f
        hopper "Air Separator", 53f
        recommendations, 53–54
    of mud processing system, 31f
    plumbing, 58f
AFE. *See* Approved for expenditure (AFE)
Agitation, 32
    compartments, 98
    examples, 97–98, 101–103
    impellers, 88–93
    mechanical agitators, 87
    mixing and blending drilling fluid, 87
    motors, 88
    mud guns, 103
    natural frequency determination, 96
    proprietary blades, 93–98
    sizing agitators, 98–99
    turnover rate, 99–103
Agitators, 98
    mechanical, 87
    shaft and blades, 97f
Air bladder screen retention system, 372
Airplane wings, 49–50, 50f
Alternating Current (AC), 397–398
American Petroleum Institute Recommended Practice 13C (API RP13C), 267, 298, 354, 361–363, 366
    confusion caused by "non-standard" screen, 363f
    manufacturer labeling screens as "conforming to API RP13C", 362
    test screen mounted in holder and Ro-Tap® with test screen and sieve stack, 364f
    tutorial, 361–362

Amplitude of motion, 376
Annular velocity, 145–147, 151, 152f–155f
API 20 shaker screens, 407, 409–410
API 80 screens, 410
API 140 (fine screen shakers), 314–316, 358
API 170 shaker screens, 359, 407
API 200 screens, 317
    shaker screens, 407, 410
    standard screen, 357–358
API RP13C. *See* American Petroleum Institute Recommended Practice 13C (API RP13C)
API Screen Designation for shaker screens, 330
API Screen Number, 357–358, 361–365
API water loss test (APIWL), 22
Approved for expenditure (AFE), 454
ASTM E-11 Wire Cloth Standard, 354, 364t
ASTM Screens, 300
Atmospheric degasser, 414–416, 416f
Atmospheric pressure, 195, 218–219
Axial flow impeller, 93f, 95

B

"Back-ground" gas, 413
Balanced elliptical motion, 410
    shakers, 313
Balanced-activity, 328–329
Barite, 18–19, 95, 439, 441–442, 447–448
    API specifications for, 18t, 448–449
    and low gravity solids, 134
    size distribution, 443f
    weighting agents, 22
Barium sulfate, 18
Bayou Sale Shaker, 463–464, 464f
Bentonite, 19, 133, 149, 322
    API specifications for, 18t
Bernoulli Principle, 32, 49, 51
BHA. *See* Bottom hole assembly (BHA)
Bingham Plastic Model, 9, 12–14, 16, 490
    numbers, 141–142
Bottom hole assembly (BHA), 332
Bottom hole pressure, 195
Buoyant force, 408

576 Index

By-pass troughs, 182–183
    fluid flow sizing charts, 182t
    after shale shakers, 182–183
Bypassing shale shaker, 411

C

Calendaring, 349, 349f
Canted blade impeller, 91–92
Capacity limits, 383
Carrying capacity index (CCI), 141–143, 150, 503t
Cascade arrangement, 400
Cascade systems, 398–400
Cased hole method, 29
Casing drilling, 328–329
Cation exchange concentration (CEC), 133
Cavitation, 221–224, 222f
CCI. See Carrying capacity index (CCI)
CEC. See Cation exchange concentration (CEC)
Cement placement, 335–336
Cementing, 338
Centrifugal pumps, 3, 32, 54, 187, 193–196, 197f–198f, 200f, 229, 229f, 426, 435, 489, 520, 544
    application for desilting drilling fluid, 214–217
    atmospheric pressure, 218–219
    cavitation, 221–224, 222f
    constant head device, 196–197
    curves, 200–206, 202f, 204f–205f
        application, 206–208
    description, 197–199
    efficiency curves, 205f
    flow into, 219–221
    head curves, 230–242
    head producing by, 206f
    NPSH, 218
    operating point, 208–213
    practical operating guidelines, 224–228
        head of fluid, 228f
        preventing air from entering suction of pump, 225f
        screening suction line, 227f
        straightening flow pattern before fluid enters pump, 226f
    pressure for different mud weights, 207t
    sizing impellers, 208
    solutions, 510–515
Centrifugation, 481

Centrifuges, 178, 187, 197, 312–313, 318, 407–408, 475–476
    cut points of, 302–305
    decanting, 475f
    estimating LGSs concentration, 477f
    maximum and minimum plastic viscosity limits, 479
    oilfield, 476
    operating tips, 482
    operational guidelines for, 318–319
    rig up of, 318
    rotary mud separator, 483–484, 483f
    solutions, 554–556
Centrifuging, 476
    double-centrifuging, 479, 480f
    NADF, 479–482
    traditional, 475, 477
Chaser fluid, 37
Chemicals, 49
    measurements, 24–25
        pH scale, 24t
    treatment, 269
Circular motion, 313
    shaker, 379, 379f
Circular style desilter package, 430f
Clean drilling fluid, 518
    addition system design, 322
Cleaning
    empirical correlation for, 139
    process efficiency, 248–250
Colloidal solids, 271, 407
Compartments, 98
Complex fluids, 16
Compression screen retention system, 372
Concentric cylinder viscometer, 9, 10f
    calculating viscosity for, 12t
    shear stresses, 9, 11
Conductance, 354–357
    equipment used to determining conductance, 356f
    one laboratory arrangement to measure screen conductance, 356f
    partial list of ASTM E-11 wire cloth standard test sieves, 364t
Cones. See Hydrocyclones
Contour blade impeller, 92f, 94f
Control drilled solids solutions, 550–554. See also Drilled solids
    discarding excess drilling fluid, 553

solids removal efficiency, 552–554
Controlling solids with dilution, 454–455
Conventional shale shakers, 387
Critical polymer concentration (CPC), 149
Cut points, 295–297
　of centrifuges, 302–305
　comment of particle size presentations, 305–310
　curve, 442
　curve for 15 ppg water-based drilling fluid, 484, 488f
　curves for solids control equipment, 295f
　of desilters, 303f
　of hydrocyclone, 300–302, 426t, 431–432
　of shale shakers, 298–300
　of solids control equipment, 441f
　solids removed by solids control equipment, 295t
Cuttings transport
　using correlation, 143–144
　diagnostics, 144–145
　empirical correlation, 140–143
　　for cleaning, 139
　historical perspective, 140
　hole cleaning, 139
　　for highly deviated wells, 147–150
　laboratory data for, 140f
　practical suggestions, 145–146
　solutions, 499–504
Cyclones. See Hydrocyclones

D

D100 cut point of screen, 362, 369
D40 cut point, 295–296
D50 cut point, 295–298, 431–432, 484
　of screen, 362
Darcy's law, 354–355
Deep ocean floor temperatures, 22
Degassers, 178, 315, 413, 530
　atmospheric, 416f
　effects of gas-cut drilling fluid, 417
　installation, 419–420
　operational guidelines for, 316
　procedure for operation, 420
　removing gas bubbles, 418
　rig up of, 315–316
　vacuum, 415f–416f
　vertical, 415f
Desanders, 178, 187, 316, 425, 531, 539–542

　of hydrocyclones, 432
　parallel processing of, 249f
　processing efficiency, 248–250
Desilters, 180–182, 246, 423, 424f, 439, 444, 531, 539–542
　circular style desilter package, 430f
　comparison of conventional desilter with long body desilter, 432f
　discharge upstream from hydrocyclone, 249f
　of hydrocyclones, 424f, 434–435
　overflow from, 248f
　parallel processing, 249f
　processing efficiency, 248–250
　transparent, 425f
Desilting drilling fluid, application for, 214–217
Dewatering unit, 389–390
DF. See Dilution factor (DF)
Dilution, 163–164, 276
　application, 168–170
　diluting as means for controlling drilled solids, 268–269
　principles, 164–168
　quantity calculation method, 292
　solutions, 505–510
　true costs of drilling well, 163–164
　volume, 165
Dilution factor (DF), 292
Dirty drilling fluid, 518
Discharge mass flow rates, 296
Dispersant, 322
Disposal costs, 336
Distribution chamber, 191
Ditches. See By-pass troughs
Double-centrifuging, 479, 480f
　separation curves, 480f
Downward force, 408
Drag force, 408
Drill pipe displacement, 40t
Drilled cuttings, 166–167
Drilled solids, 22, 105–106, 139, 147–149, 163, 325, 329
　calculations solutions, 496–499
　carrying capacity, 342–343
　cement placement, 335–336
　cementing, 338
　concentration, 169f–170f
　controlling, 331
　derivation of formula, 121
　diluting as means for controlling, 268–269

Drilled solids (*Continued*)
    discussion, 108−111
    disposal costs, 336
    drilling costs increasing in, 164f
    drilling performance, 339−341
    economics of solids control, 325−327
    example, 132
    extra note, 335
    filter cakes in NADF, 334−335, 334f−335f
    formation damage, 338
    invisible NPT or trouble costs, 337
    LGS
        alternate method, 131−136
        calculation, 107−108
        calculation volume fraction, 128−130
        procedure for determination, 128−130
    log interpretation, 337
    lost circulation, 336−337
    low concentration, 163
    in mud, 164
    NADF, 123−125
    PDC bits, 341
    potassium chloride drilling fluids, 128
    removal, 163
        60% removal, 278f, 281
        70% removal, 278f, 281
        80% removal, 277f, 280
        90% removal, 277f, 280
        100% removal, 279
        reasons for, 268
    retorts, 111−121, 111f
        size required for accurate solids calculation, 113−115
    sensitivity to measurements, 119−120
    sizes, 328−331
    solids calculations, 133−134
    solids fractions, barite and low gravity solids, 134
    specific gravity of, field measurement, 120−121
    stuck pipe, 331−334
    surge and swab pressure, 336
    tolerant, 331
    validation of equation, 121−123
    visible NPT or trouble costs, 331
    wear, 336
Drilled solids removal factor (DSRF), 292
Drillers, 163−164, 195
Drilling
    costs, 447
    NPT, 338
    performance, 339−341
    rigs, 204
Drilling fluid surface systems
    active surface processing system design, 311−322
    addition section, 319−320
        operational guidelines for, 320
        rig up of, 319−320
    centrifuges, 318
    clean drilling fluid addition system design, 322
    degassers, 315
    hydrocyclones, 316
    mud cleaners, 317
    mud guns, 321
    prehydration tank design, 322
    sand traps, 314
    shale shakers, 313
    suction section, 320−322
    surface drilling fluid system, 311
    trip tank system design, 321−322
Drilling fluids, 11, 16−17, 87, 139, 141, 145−146, 163−165, 165f−166f, 185, 188, 245, 252, 359, 425, 476, 483
    additives, 17−19, 87, 178
    agitators, 87
    costs, 166
    functions, 3−5
    measurements, 19
    measuring properties, 9−11
    mixing and blending, 87
    and moved drilled rock, 274f
    mud cleaner location in drilling fluid system, 447−448
    process efficiency, 246−252
        overflow from desilters, 248f
        tank arrangement, 247f
    processing system, 327−328, 401
    properties, 106
    quantity of clean drilling fluid correlation, 171f
    rheology review, 3
    standing in long tube, 220f
    unweighted water-base, 119
    viscometer readings for, 14f
    viscosity, 410
    in well bore and surface equipment, 273f
Dryer, 445
    shakers, 389−391
DSRF. *See* Drilled solids removal factor (DSRF)

E

ECD. *See* Equivalent circulating pressures (ECD)
Effective viscosity, 15, 150–151
Efficiency, 267
Electric stability, 25
Elliptical type motion, 377
Empirical correlation for cleaning, 139
Equation derivation, 285
Equipment solids removal efficiency (SRE), 507
   60% removal of drilled solids, 281
      removal efficiency, 278f
   70% removal of drilled solids, 281
      removal efficiency, 278f
   80% removal of drilled solids, 280
      removal efficiency, 277f
   90% removal of drilled solids, 280
      removal efficiency, 277f
   100% removal of drilled solids, 279
   API method, 292–293
      DSRF, 292
   chemical treatment, 269
   costs, 549
   diluting as means for controlling drilled solids, 268–269
   dilution quantity calculation method, 292
   drilled solids removal, reasons for, 268
   effect, 267–268
   equation derivation, 285
   estimation
      for unweighted drilling fluid from field data, 286–289
      for weighted drilling fluid, 289–291
   examples of effect, 276–284
   mechanical separation-basics, 270–271
   mechanical treatment, 269–270
   problems, 293–294
   relationship to clean drilling fluid needed, 272–273
   solids removal
      efficiency to dilute drilled solids, 284–286
      system performance effect, 271–272
   solutions, 547–549
Equivalent circulating pressures (ECD), 5
"Equivalent cp", 150
Excess lime, 130
Expendables, 113
Exxon Production Research, 20–21

F

"Fann" brand viscometer, 9, 10f
Feed pressure, 423
Filter cake
   compressibility, 23
   in NADF, 334–335, 334f–335f
Filtration rates, 22–23
Fine screen shakers, 381–383
   unweighted drilling fluid with, 188f
   weighted drilling fluids with, 179, 180f
Flat blade impeller, 88, 90f
Flocculants, 269
Flounder point, 490
Flow
   into centrifugal pump, 219–221
   distribution chamber, 179, 180f, 188, 190f
   distributor, 179
   line, 312
      diameters, 183t
   rate, 147
Fluids, 49
   loss, 491
      control, 17
   velocity, 147, 198
Force, 20
Forcing function, 386
Formation damage, 338
Founder point, 339
   drilled solids effect, 339
Fraction of drilling fluid processed solutions, 246–247, 515–522, 524–540, 545–547
Fresh-water drilling fluid, 107–108
Friction loss, 208
   for pipe fittings, 212t
   in pipe fittings, 512
   for water in steel pipe, 209t–211t
Funnel viscometer, 13, 16f
Funnel viscosity, 13–14

G

g/cc. *See* Specific gravity (SG)
"G" factor
   determination, 384–389
      G-force and angle determination, 388f

"G" factor (*Continued*)
  relationship to stroke and speed of rotation, 389
Gas bubbles removal, 418
Gas discharge lines offshore, 418
Gas-busters. *See* Mud/gas separators
Gas-cut drilling fluid effects
  in degassers, 417
  example, 417–418
Gel strength, 22
Generic systems
  for unweighted drilling fluid, 177–178, 177f
  for weighted drilling fluid, 178–179
"Go-no-go" gauge, 349
Gumbo, 177–178, 185–186
  busters, 372–373
  removal, 393
  remover, 393, 393f
  solids problems, 372–373

H

Head, 195
  curves, 203f, 204–205, 230–242
  loss, 201, 202f, 207, 207f, 213t
  of mercury, 219
  producing by centrifugal pump, 206f
Heavy coarse solids, 423–424
Heavy drilling fluids, 457–458
Hematite (Fe_2O_3), 18–19, 447–448
  API specifications for, 18t
  weighting agents, 22
HHP. *See* Hydraulic horsepower (HHP)
High temperature high pressure (HTHP), 334
High-performance linear motion shale shakers, 401
Hole cleaning, 139, 163
  for highly deviated wells, 147–150
Hooke's law, 148
Horizontal holes, 148–149
HTHP. *See* High temperature high pressure (HTHP)
Hudson-Boucher automatic shale separator, 375f, 377
Hydraulic horsepower (HHP), 204–205
Hydraulic(s), 339
  impact, 163
  optimization procedures, 33
Hydrocyclones, 312–313, 316, 407–408, 423, 424f, 425, 439
  air cylinder collapsing, 428f
  bank of, 428f
  circular style desilter package, 430f
  comparison of conventional desilter with long body desilter, 432f
  cut points, 300–302, 426t, 431–432
  desanders, 432
  desilters, 424f, 434–435
  flow rates through, 427t
  internal geometry, 427
  maintenance, 435–436
  operating efficiencies, 429
  operating tips, 436–437
  operating with spray discharge, 429
  operational guidelines for, 317
  rig up of, 316–317
  rope discharge, 429–431
  separation curve for 4" hydrocyclone, 442, 442f
  tanks and arrangements, 436
  ten commandments, 438
Hydrophobic surface, 328

I

Impellers, 88–93, 193, 195, 196f, 197, 199f
  axial flow, 93f, 95
  canted blade, 91–92
  contour blade, 92f, 94f
  flat blade, 88, 90f
  radial flow, 88, 90f
  vortex formation in fluid, 92f
Integral unit
  with multiple vibratory motions, 400
  with single vibratory motion, 400
Internal flow, 523
Invisible NPT, 325, 337

J

Jack-up rigs, incorrect plumbing on, 251–252
Jet pump discharges, 414

K

Katy field test, 467, 467f
"Kick", 38
  contingency method, 29–30
    capacity of internal upset drill pipe, 30t
Kinetic energy, 194

L

Labeling requirements, 366–368
  horizontal labeling format, 368f
  vertical labeling format and example, 367f

LGS. *See* Low gravity solids (LGS)
Lignosulfonate, 319
Lime, 130
Linda "K" rotary shale extractor, 374–376, 375f
Linear motion, 313, 377–378, 381, 382f, 410
　shakers, 389–390
　shale shakers, 188, 444
Linear motion dryer, 391
Liquid additives, 316
Liquid capacity limit, 383
Log interpretation, 337
Lost circulation, 336–337
Low gravity concentration (V_{lg}), 449
Low gravity solids (LGS), 107
　alternate method in unweighted, water-based drilling fluid, 131–136
　barite and, 134
　calculation, 107–108
　　volume fraction, 128–130
　concentration estimation, 477f
　equations for, 109t–110t
　fraction of, 132–133
　procedure for determination, 128
　volume percent, 110t
Low-shear-rate viscosities (LSRV), 17
Lower shear rate range, 12
LSRV. *See* Low-shear-rate viscosities (LSRV)

M

MARKET GRADE (MG), 402
Marsh funnel, 13, 20–21
Mass of sphere, 408
Mechanical agitators, 87
Mechanical separation-basics, 270–271
Mechanical treatment, 269–270
Median cut, 432
Median size particle. *See* Median cut
Mercury barometer, 217, 219f
Mesh, 346, 362–363
MG. *See* MARKET GRADE (MG)
Misconception, 106–107
Mixed metal hydroxides fluid (MMH fluid), 149
Mobil mud
　centrifuge mounted on trailer, 305f
　separator, 305f
　　cut point curve of, 306f
Motion types, 376–378, 377f
Motors, 88
　horizontal drive for radial flow, 88f
　horizontal motor mount, 89f
　vertical motor mount, 89f
Mud cleaners, 187–188, 317, 439–445, 468f, 471f–472f
　accuracy requiring for specific gravity of solids, 456–457
　barge drilling rig in Louisiana, 471f
　commercial, 440f
　conversion of main shaker to, 440f
　in drilling fluid system, 447–448
　economics, 454–456
　estimating ratio of low gravity solid volume and barite volume, 449–451
　field test, 466f
　heavy drilling fluids, 457–458
　history of development, 462–473
　installation, 459–461
　　in Lake Maricaibo, 472f
　light-weight hydrocyclones, 472f
　using long body cones, 473f
　non-oilfield usage, 461–462
　operating, 448–449
　operational guidelines for, 318, 458–459
　performance, 451–454
　relative size solids in drilling fluid, 441f
　rig up of, 317
　situations, 447
　Sweco's mud cleaner manager, 470f
　third field test, 469f
　tilden mud tanks from the crown block, 470f
　uses, 445–447
Mud guns, 49, 52, 54–56, 55f, 87, 103, 185, 321, 521. *See also* Addition section
　application
　　removal section, 82–83
　　suction section, 83
　centrifugal pump curve, 59f
　educator, 57f–58f
　finding head at each mud gun nozzle, 60–61
　finding mud gun nozzle sizes, 62
　flow rate equation derivation, 84–85
　fluid flow calculation, 70–82
　friction loss, 60f–61f
　head calculation, 62–68
　operational guidelines for, 321
　plumbing for swivel mounting, 56f
　rig up of, 321

Mud guns (*Continued*)
   rotating nozzles, 57f
   selecting nozzles and centrifugal pumps, 56—58
   situation, 58—82
Mud hoppers, 49—53, 50f
   hopper "Air Separator", 53f
   recommendations, 53—54
Mud tank
   changes need to be made, 542—544
   volume, 28
Mud tank arrangements for safety
   agitation, 32
   arrangement and equipment selection, 27—28
     surface drilling fluid processing plant, 27f
   cased hole method, 29
   drill pipe displacement, 40t
   going in hole, 39—40
   history, 39
   kick contingency method, 29—30
     capacity of internal upset drill pipe, 30t
   mud tank volume, 28
   mud weight increase guideline, 37
   plugged bit method, 28—29
   plumbing, 30—31
     addition and suction sections of mud processing system, 31f
     required to properly use trip tanks, 40—41
     scenarios, 41—43
   pulling out of hole, 38—39
   slug tanks, 32—37
     height of slugs for drill pipe diameters, 34t
     increase in mud weight needed for slug, 35f—36f
   solutions, 495—496
   suction section, 28—43
   trip tanks, 38
Mud weight (MW), 19, 84, 108, 448
   increase guideline, 37
   mud balance, 20, 21f
     force diagram for, 21f
     pressurized, 20—21, 21f
Mud/gas separators, 418, 420—421
   basic piping to and from, 421f
   mud outlet with over-the-top connection to possum belly, 422f
Multiple deck shale shakers, 398
MW. *See* Mud weight (MW)

N

NADFs. *See* Non-aqueous drilling fluids (NADFs)
NAF. *See* Non-aqueous fluid (NAF)
Natural frequency determination, 96
Near-vertical boreholes, 139
Net Positive Suction Head (NPSH), 206, 218, 221f, 490
Net Positive Suction Head Available ($NPSH_A$), 223—224
Net Positive Suction Head Required ($NPSH_R$), 223—224
Net positive suction pressure, 218
Newton's second law of motion, 386, 485
Newtonian fluid, 3, 15, 59—60, 485
   shear stress *vs.* shear rate, 7f
Newtonian liquid, 148
Newtonian Viscosity, 7
Non-aqueous drilling fluids (NADFs), 22, 37, 87, 123, 182—183, 298, 327—328, 330—331, 395—396, 411, 434, 459, 491
   calculation volume fraction of low gravity solids
     derivation of equation, 128
   centrifuging, 479—482
   filter cakes in, 334—335, 334f—335f
   LGS
     calculation in water-based drilling fluid, 107—108
     derivation of equation, 123
     procedure for determination, 128
Non-aqueous fluid (NAF), 17, 372—373, 479—480
Non-blanked area, 363
Non-flooded suction, 223—224, 223f
Non-Newtonian
   flow, 490
   fluids, 3
Non-oilfield usage of mud cleaners, 461—462
Non-pre-tensioned screens, 370
Non-productive time (NPT), 106, 113, 163—164, 167—168, 185, 325
   drilling, 338
   invisible, 325, 337
   visible, 325, 331
Non-shaking shale shaker, 389
NPSH. *See* Net Positive Suction Head (NPSH)
$NPSH_R$. *See* Required net positive suction head ($NPSH_R$)
NPT. *See* Non-productive time (NPT)

O

Oil, 16
  oil-base drilling fluid, 491
  oil-based fluids, 19
Oilfield
  centrifuges, 476
  cyclones, 425–426
Operating point, 208–213, 213f
Optimum solids removal efficiency, 284–285
Overslung Crowned deck screen retention system, 371–372

P

Panel screens, 347–349, 348f
PDC bits. *See* Polycrystalline Diamond Compact bits (PDC bits)
Permeability, 355
Pilot testing, 25
Pipe rotation, 147
PIT. *See* Pressure integrity test (PIT)
Plain square weave, 351
  wirecloth, 352f
Plastic viscosity (PV), 9, 12–13, 395, 489, 554
Platform rigs, incorrect plumbing on, 251–252
Plugged bit method, 28–29
Plugged cones, 437
Plugged desilter cones, 444
Plumbing, 30–31
  addition and suction sections of mud processing system, 31f
  poor plumbing, field example of, 250
  required to properly use trip tanks, 40–41
  scenarios, 41–43
Polycrystalline Diamond Compact bits (PDC bits), 341
  drill-off test for, 341f
Polymer drilling fluid, 8, 8f
"Poor-boy" degassers. *See* Mud/gas separators
Possum belly, 313, 404
Potassium chloride (KCl), 467
  drilling fluids, 128
  solution densities, 127t
Potential Energy, 194
Power, 204
  systems, 396–398
Power Law
  equation, 150
  Model, 14, 16, 17f, 342, 490
  Rheological Model, 141
PR13C, 363, 365
Pre-tensioned screen panels, 370
Prehydration tank design, 322
Pressure, 195, 207
  and flow, 54
  loss, 201
    through pipe, 51, 51f
Pressure integrity test (PIT), 335
Proprietary blades, 93–98
  installation, 95–96
  variable pitch blades, 93
PV. *See* Plastic viscosity (PV)

R

Radial flow impeller, 88, 90f
Removal efficiency, 509
Removal section, 53f, 82–83, 185, 312–319, 312f
  distribution chamber, 191
  with multiple fine screen shakers for weighted drilling fluids, 189f
  plumbing, 58f
  unweighted drilling fluid, 185–187
  weighted drilling fluid, 187–190
Required net positive suction head ($NPSH_R$), 91
Retorts, 111–121, 111f
  size required for accurate solids calculation, 113–115
Rheology, 3
  Bingham Plastic Model, 12–13
  chemical measurements, 24–25
    pH scale, 24t
  drilling fluid
    additives, 17–19
    functions, 3–5
    measurements, 19
    measuring properties, 9–11
  electric stability, 25
  filtration rates, 22–23
  funnel viscosity, 13–14
  gel strength, 22
  mud weight, 19
    mud balance, 20, 21f
  pilot testing, 25
  PV, 13

Rheology (*Continued*)
  review, 3
  rheological models, 14–17
    data compared with, 19f
  sand content, 23–24
  solids content, 23
  solutions, 494–495, 515–517
  special note, 19–22
  viscosity
    calculation, 11–12
    dilemma, 5–9
    measurement, 9
Rig shakers, 380
Rig site performance, 366
Ro-Tap®, 363–364
  stack of sieves for API RP 13C D100 determination, 353f
  with two ASTM standard screens, 352, 353f
Rope discharge, 429–431
Rotary mud separator, 483–484, 483f
  trailer mounted, 484f
"Roughneck-proof" flow meter, 300–302, 462
RP13E, 363, 365–366

S

Safety, 27–28
Salvaging barite, 459
Sand content, 23–24
Sand traps, 178, 314, 315f, 407, 411, 445–446
  bypassing shale shaker, 411
  operating guidelines for, 314–315
  rig up of, 314
  settling rates, 407–411
Sandwich screen, 347, 348f
Scalping shaker, 177–179, 186, 378–381
Screen
  conductance, 362
  mesh, 346
  tensioning mechanism, 372
  types, 370–372
  weaves
    plain square weave, 351
    rectangular weave, 352
Sedimentation, 481
Self-priming pumps, 223
Settling tank, 445–446

SG. *See* Specific gravity (SG)
Shakers, 179–182, 270, 439. *See also* Shale shakers
  capacity, 383
  conversion of main shaker to mud cleaner, 181f
  users guide, 401–404
    installation, 401–402
    maintenance, 403
    operating guidelines, 403–404
    operation, 402
Shale shakers, 179, 313, 345
  API RP13C document, 361–363
  API screen number determination, 363–365
  blinding and plugging of screens, 373–376
  cascade systems, 398–400
  comparison with RP13E, 365–366
  conductance, 354–357
  confusion caused by "non-standard" screen, 363f
  cut points of, 298–300
  design, 376
  dryer shakers, 389–391
  equivalent aperture opening size, 358
  fine screen shakers, 381–383
  "G" factor
    determination, 384–389
    relationship to stroke and speed of rotation, 389
  gumbo and water-wet solids problems, 372–373
  gumbo removal, 393
  history, 374–376
  label to use, 368
  labeling requirements, 366–368
  maintenance, 394–396
  motion types, 376–378, 377f
  operational guidelines for, 313–314
  panel screens, 347–349, 348f
  power systems, 396–398
  questions regarding RP13C, 369
  rig site performance, 366
  rig up of, 313
  scalping shakers, 378–381
  screen design, 349–351
    B60 screen, 350f
    double layer screen, 351f
    screening surfaces, 350–351
  screens, 345–347
    labeling, 357–358
    openings, 352–354
    types and tensioning systems, 370–372
    weaves, 351–352

screens fluid, 358—361
  Honey flows slowly through fine screen, 359f
shaker capacity, 383
shaker users guide, 401—404
shaker without shaking, 394
solids conveyance, 384
test screen mounted in holder and Ro-Tap® with test screen, 364f
three dimensional, 370f
triple deck shakers, 391—392
Shallow tanks, 311
Shear rate (SR), 6—7, 6f
  of Newtonian fluid, 7f
Shear stress (SS), 6—7, 6f, 148, 150
  on concentric cylinder viscometer, 9, 11
  of Newtonian fluid, 7f
Shear thinning fluid, 8, 8f, 11
Simple-to-use trip tank arrangement, 27—28
Sizing
  agitators, 98—99
  exercise, 102f
  impellers, 208
Slug effectiveness derivation, 44—45
Slug tanks, 32—37
  height of slugs for drill pipe diameters, 34t
  increase in mud weight needed for slug, 35f—36f
"Smear" effect, 328—329
Sodium chloride solution
  data for, 135t
  densities, 126t—127t
  weight and volume, 129t
Soft landing, 407
Solids, 426—427, 475—476. *See also* Drilled solids
  capacity limit, 383
  content, 23
  control equipment, 267
  control equipment arrangement, 163
  conveyance, 384
  plugging desilters, 188
  problem, 330—331, 342—343
Solids removal
  efficiency, 166—168, 267
    to dilute drilled solids, 284—286
  equipment, 296
  system performance effect, 271—272
Special note, 19—22

Specific gravity (SG), 98—99, 108
  of drilled solids, field measurement, 120—121
SR. *See* Shear rate (SR)
SRE. *See* Equipment solids removal efficiency (SRE)
SS. *See* Shear stress (SS)
Standpipes, 199
Stokes' law, 408, 410, 476, 481
  settling, 485—487
Stroke detector, 388
  readings and troubleshooting, 390t
  sticker, 388f
Stuck pipe, 331—334
Suction, 436
Suction section, 28—43, 83, 185, 320—322
  of mud processing system, 31f
  operating guidelines for, 320—321
  rig up of, 320
Super-desilter, 187
Surface drilling fluid systems, 177, 311
  alternate system, 179—182
  by-pass troughs after shale shakers, 182—183
  generic systems
    for unweighted drilling fluid, 177—178, 177f
    for weighted drilling fluid, 178—179
Surge and swab pressure, 336
Surplus shaker, 188—190
Synthetic polymers, 395—396

T

Tailing blades, 374—376, 375f
Tank depth, 311
Tensioning systems, 370—372
Terminal velocity, 486
Thompson shale separator, 376—377, 376f
Three dimensional shaker screen, 370f
Tilden test, 468
Time of rollover (TOR), 99—100, 102
Traditional centrifuging, 475, 477
Transparent desilter, 425f
Trip tanks, 38
  system design, 321—322
Triple deck shakers, 391—392, 392f
Trouble costs, 113, 331, 337
Troughs, 182—183
True costs of drilling well, 163—164
Turnover rate, 99—103
Two-stage centrifugation process, 480—481

U

Unbalanced elliptical shaker, 378–379, 378f
　motion shakers, 313
Under-centrifuging, 479
Unweighted drilling fluid
　with fine screen shakers, 188f
　generic surface drilling fluid processing system for, 186f
　generic systems for, 177–178, 177f
　processing plant with flow distributor, 181f
　removal section, 185–187
　water-base drilling fluid, 119

V

Vacuum chamber, 187
Vacuum degassers, 414, 415f–416f
Velocity head loss, 201–203, 203f
Vertical boreholes, 139
Vertical degasser, 414, 415f
Very high shear rate range, 12
Vibration
　amplitude, 386
　angle, 387
VIF. See Volume increase factor (VIF)
Viscoelastic behavior of drilling fluids, 148
Viscosifiers, 444–445
Viscosity, 141, 142f, 148, 150, 408, 478, 489
　calculation, 11–12
　　for concentric cylinder viscometer, 12t
　dilemma, 5–9
　measurement, 9
Visible NPT, 325, 331
Volume increase factor (VIF), 292

W

Waste disposal, 163–164
Waste volumes, 163
Water, 16
　water-based fluids, 19
　water-wet solids problems, 372–373
Water-based drilling fluids (WBDFs), 182–183, 327, 330–331
Water-wet solids problems, 372–373
WBDFs. See Water-based drilling fluids (WBDFs)
Wedge retention system, 371
Weight on bit, 163
Weighted drilling fluid
　with fine screen shakers, 180f
　generic systems for, 178–179
　generic tank arrangement for, 189f
　processing plant, 30–31, 30f
　removal section, 187–190
Weighting agents, 108t
Whole mud alkalinity, 130

X

XC-polymers, 149

Y

Yield point (YP), 9, 12